Fundamentals of the
FUNGI

3RD EDITION

Fundamentals of the
FUNGI

Elizabeth Moore-Landecker

Glassboro State College

Prentice Hall, Englewood Cliffs, New Jersey 07632

Library of Congress Cataloging-in-Publication Data

MOORE-LANDECKER, ELIZABETH.
    Fundamentals of the fungi / Elizabeth Moore-Landecker.—3rd ed.
        p.   cm.
    Includes bibliographical references.
    ISBN 0-13-332941-0
    1. Mycology.   2. Fungi.   I. Title.
QK603.M62   1990
589.2—dc20   89-23043
                                                    CIP

Editorial/production supervision
    and interior design: *Kathleen M. Lafferty*
Cover design: *Ben Santora*
Manufacturing buyer: *Paula Massenaro*

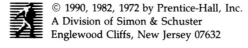 © 1990, 1982, 1972 by Prentice-Hall, Inc.
A Division of Simon & Schuster
Englewood Cliffs, New Jersey 07632

Printed in the United States of America

10 9 8 7 6 5 4 3 2

ISBN 0-13-332941-0

Prentice-Hall International (UK) Limited, *London*
Prentice-Hall of Australia Pty. Limited, *Sydney*
Prentice-Hall Canada Inc., *Toronto*
Prentice-Hall Hispanoamericana, S.A., *Mexico*
Prentice-Hall of India Private Limited, *New Delhi*
Prentice-Hall of Japan, Inc., *Tokyo*
Simon & Schuster Asia Pte. Ltd., *Singapore*
Editora Prentice-Hall do Brasil, Ltda., *Rio de Janeiro*

To Richard P. Korf,
who introduced me to the fungi

Contents

PART II PHYSIOLOGY AND REPRODUCTION

Preface

The science of mycology had modest beginnings in the eighteenth century, often as a Sunday hobby for physicians. Now fungi are the concern of the taxonomist, morphologist, geneticist, ecologist, plant pathologist, physician, biochemist, and commercial microbiologist. They are becoming increasingly important as organisms used as tools in basic research, such as that in molecular biology. This list reflects the ubiquity of the fungi, their usefulness as research organisms, and their involvement in many facets of our everyday lives.

Fundamentals of the Fungi provides a broad introduction to the field of mycology; it explores the morphology, taxonomy, evolution, physiology, ecology, and commercial utilization of the fungi. The level of this book is suitable for the person who has a basic understanding of biology and chemistry. It is hoped that this book will enrich the background of those who will not study mycology further and will serve to provide a firm foundation for those who choose to study mycology more intensively.

The third edition has been modified to place greater emphasis on the cell biology of the fungi and their evolution. This has resulted in an expanded treatment of cell structure and molecular biology in Chapter 1, the addition of information concerning nutrient uptake in Chapter 8, cellular events occurring in reproduction as well as an expanded section on rhythms in Chapter 10, a section on mycoviruses in Chapter 14, and the role of recombinant DNA research in the biotechnology of the fungi (Chapter 16). New sections on speciation and phylogeny of the fungi are incorporated into Chapter 7, which also has new material on cytoplasmic inheritance in the fungi.

Many people assisted in the preparation of the first and second editions. In addition, for the preparation of the third edition, Dr. Lekh Batra loaned me some of his personal library materials. Mr. William Garrabrant obtained numerous references on interlibrary loan and provided other helpful bibliographic services. The production editor, Ms. Kathleen M. Lafferty, cheerfully provided expert guidance during the processing of the manuscript. Mr. Dieter Ott provided untiring assistance during the collection of references in the library. Mrs. Beverly Wichess dedicated countless hours to the preparation of the manuscript, and many companies and individuals provided photographs or permission to use published materials, and are acknowledged in the text. I also wish to acknowledge the helpful suggestions of Dr. Sandra Anagnostakis, Dr. James B. Anderson, Dr. Harold Burdsall, and Dr. Robert Newland. All of these individuals and concerns are gratefully thanked.

Elizabeth Moore-Landecker

PART I MORPHOLOGY AND RELATIONSHIPS OF THE FUNGI

1

An Introduction to the Fungi

Laypersons are generally familiar with many fungi such as mushrooms in forests or fields, bracket fungi on the side of decaying logs or standing trees, puffballs which send forth clouds of spores when kicked by a playful child, and perhaps colorful cup fungi on logs or soil. Homemakers use yeast to leaven their bread, and brewers use this same organism to form alcohol in beer, wine, or whiskey. Perhaps farmers drown their sorrows with this alcoholic brew after their crops are destroyed by a devastating fungus disease such as corn smut or black stem rust of wheat. The city dweller who may not be familiar with fungi in their natural habitats may be waging a battle with ringworm or athlete's foot, fungus diseases of humans. Perhaps the same city dweller is involved in the manufacture of antibiotics from fungi, of fungus-processed cheeses, or of drugs which are chemically modified and perfected by chemical activities of fungi. Meanwhile, a tribal rite may be taking place in Mexico in which hallucinogenic mushrooms are consumed and vivid dreams are experienced. Many fungi rarely come to the attention of the layperson. These include microscopic inhabitants of almost every habitat, including those fungi which trap nematodes (roundworms). Some fungi eject their spores over a precise trajectory and with a precision force, much as a missile is precisely fired into space. They may become partners with algae, insects, or higher plants and enter into complex biological relationships with these other organisms.

GENERAL CHARACTERISTICS OF FUNGI

What are the fungi, and how do they differ from other organisms? Although the fungi are a diverse group of organisms, one feature they have in common is their mode of nutrition. Like animals, fungi are **heterotrophic** organisms that must consume preformed organic matter. They may live as *saprophytes*, which

1

digest and consume dead plants or animals, their parts, or their wastes. Alternatively, fungi may live as *parasites* and assimilate tissues of living plants and animals. In all cases, digestive enzymes are liberated from the fungal cell into the immediate environment, where food molecules are simplified and the nutrients pass into the fungal cell as a watery solution.

Other heterotrophic organisms include bacteria and animals. How does a fungus differ from these organisms? Bacteria lack a nucleus, while fungi possess a nucleus similar to that in higher plants and animals. Unlike fungi, animals ingest large masses of food which are then digested within the organism.

The fungi may be briefly characterized further. Most fungi produce thread-like cellular strands, the *hyphae*, that characterize the vegetative stage of their life cycle (Fig. 1–1). Like almost all fungal cells, each hypha is surrounded by a rigid cell wall. In nature, the hyphae are usually buried within a *substratum* that can provide food. The hyphae of a saprophytic fungus may occur within the soil, decaying plant debris, or animal wastes such as dung, while the hyphae of a parasitic fungus would occur in a living plant or animal. These hyphae produce digestive enzymes and take up nutrients, translocating them to the growing hyphal tips, where reproductive structures differentiate. Fungi reproduce by forming *spores*. A spore is a specialized cell that is capable of being separated from its parent thallus and disseminated, usually by wind or water. Terrestrial fungi produce spores having thick cell walls and food reserves such as oil droplets. Unlike the seeds produced by green plants, spores lack an embryo. When the spore germinates, it gives rise to a unicellular protoplast or, alternatively, to a *germ tube* that grows and differentiates into a new thallus (Fig. 1–1).

As a group, the fungi have been classified within different kingdoms: the plant kingdom in the older classification schemes, which were based on only two kingdoms (plants and animals), and later in the Kingdom Protista with other relatively simple organisms. Whittaker (1969) argued that the fungal mode of nutrition, which is based on extracellular digestion and absorption, is fundamentally different from that of autotrophic plants. Further, the eukaryotic structure of the fungal cell makes the fungi fundamentally different from the prokaryotic bacteria and blue-green algae, while their walled cells make them different from most protozoa. Whittaker devised a five-kingdom classification system in which the fungi are assigned to their own kingdom. This reflects his belief that the primitive protozoan groups gave rise to three evolutionary lines—plants, animals, and fungi—which are distinguished primarily by their mode of nutrition.

The slime molds have sometimes been included within the fungi, but these organisms have characteristics unlike those of true fungi. They have a naked

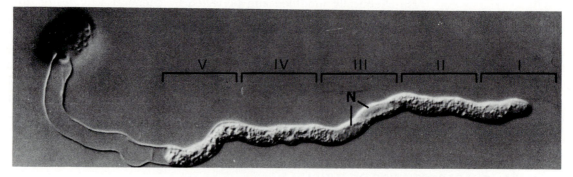

Figure 1–1 Germinating spore produces elongated germ tube which will develop into the hyphae. The regions designated I to V indicate zones that are significantly different from each other in terms of their major components. Regions I and II correspond to the hyphal tip (just forward of the nuclei). × 700. [From H. C. Hoch and R. C. Staples, 1983, Mycologia 75: 795–824.]

mass of protoplasm which has animal-like motility and creeps about, and they feed by engulfing and ingesting their food, as do protozoa. When slime molds reproduce, they form spores with thick walls that may be windblown, as do most of the fungi. Although Whittaker (1969) includes these organisms in the Kingdom Fungi, he states: "The slime molds cross the distinctions of the kingdoms in both nutrition and organization, and offer a free choice of treatment as aberrant fungi, eccentric protists, or very peculiar animals." Because these organisms are not clearly related to the vast majority of the fungi, they will not be treated further in this book.

We shall classify the fungi in the Kingdom Myceteae (Fungi) and recognize the subdivisions proposed by Ainsworth (1973). The major character important in delimiting the Divisions is the presence or absence of swimming spores, the *zoospores*. In the Division Mastigomycota, which forms zoospores, further separation into classes relies on zoospore morphology. In contrast, the Division Amastigomycota, which lacks zoospores, is subdivided into major taxa on the basis of sexual reproduction and life cycles. The classification that we shall follow is:

> Kingdom Myceteae
> (nutrition by extracellular digestion and absorption: eukaryotic, vegetative cells typically with a thick wall and usually elongated)
> > Division Mastigomycota (forming swimming zoospores; the following classes are based on the morphology of zoospores and are explained in Chapter 2)
> > > Class Chytridiomycetes
> > > Class Hyphochytridiomycetes
> > > Class Plasmodiophoromycetes
> > > Class Oomycetes
> > Division Amastigomycota (nonmotile, not forming zoospores)
> > > Subdivision Zygomycotina (forming zygospores as the sexual spore)
> > > Subdivision Ascomycotina (forming ascospores as the sexual spore)
> > > Subdivision Basidiomycotina (forming basidiospores as the sexual spore)
> > > Form-subdivision Deuteromycotina (lacking sexual spores)

Attributes of the fungi that help distinguish them from other organisms and make their biology unique include their cellular structure, organization at the cellular level, and their unique pattern of sexual reproduction. Exploration of these topics follows in the remainder of the chapter.

THE FUNGAL CELL

The fungi have some unique characteristics that can be determined from a study of their cells. Fungi typically have walls containing chitin, mitotic divisions differing in several respects from most organisms, and some other unique cytological and molecular features. Cytological features of the various types of fungal cells are generally similar.

Cell Surfaces

The majority of fungal cells are surrounded by a rigid cell wall. Cell walls of fungi are comparatively thin in vegetative cells, where the cell walls are usually only about 0.2 micrometers thick, but may be thicker in some specialized cells and spores (Fig. 1–2). A variety of polysaccharides usually represent about 80

- I — Mucilage
- II — KOH-soluble glucan
- III — Chitin fibrils in β-1,3-glucan matrix

Inside

- 380 nm
- 180 nm
- 20 nm

𝒳 chitin fibers

⦚ short fibers

• protein

°° coarse-grained matrix (containing melanin)

Figure 1–2 Diagrammatic representation of wall structure in *Agaricus brunnescens.* Top: mycelial wall; bottom: spore wall. The third layer is a thin mucous covering. In addition to the components illustrated, lipid occurs throughout the spore wall. Note that the thickness of the different layers is given and that the spore wall is about seven times as thick as the mycelial wall. [Top: from G. O. Michalenko, H. R. Hohl, and D. Rast, 1976, J. Gen. Microbiol. 92:251–262; bottom: from D. Rast and G. O. Hollenstein, 1977, Can. J. Botany 55:2251–2262, National Research Council of Canada.]

to 90% of the wall. Other important constituents include protein, which is often conjugated with the polysaccharides, and a small amount of lipid. The most important polysaccharides are chitin and cellulose, both polymers of glucose (Fig. 1-3). The majority of the fungi have chitin but not cellulose in their walls, whereas in a few fungi the reverse is true. Microfibrils of chitin or cellulose are intertwined and embedded in an amorphous matrix which cements them together (Fig. 1-4), providing a skeletal framework that gives walls their morphological form. The matrix contains protein and polysaccharides such as glucans or mannans (polymers of glucose and mannose, respectively). The wall typically consists of two or more layers (Fig. 1-2). The skeletal framework of chitin or cellulose microfibrils typically occurs in a single layer, while the remaining layers appear to be homogeneous and consist primarily of amorphous proteins and/or carbohydrates. In a few cases the glucan may be organized into a coarse reticulate pattern or may occur as a mosaic of short rodlets (Bartnicki-Garcia, 1968; Burnett, 1979). Wall layers are particularly pronounced in spores that produce

A portion of the cellulose molecule

A portion of a chitin molecule

Figure 1–3 Major components of the cell walls of fungi.

Figure 1–4 Microfibrils from the internal surface of a fungal cell wall. × 55,000. [From S. Bartnicki-Garcia and E. Reyes, 1968, Biochim. Biophys. Acta 165:32–42.]

a conspicuous inner layer, the *endospore*, which surrounds the protoplast and in turn is surrounded by the *epispore*, the outer layer. The entire spore may be surrounded by a *perispore*, an external layer(s) that can be removed without impairing the functions of the spore (Fig. 1–5).

Figure 1–5 Section of a spore of a terrestrial fungus. The endospore (En) appears as a broad light band next to the cytoplasm, and it is bordered by the epispore (E), which appears as a thinner and darker band. Both are surrounded by the perispore, which is a thick ribbed band. Organelles within the cytoplasm are the nuclei (N) and mitochondria (M). × 5600. [Courtesy R. J. Lowry and A. S. Sussman. From A. S. Sussman and H. O. Halvorson, 1966, Spores: their dormancy and germination, Harper & Row, New York.]

Some fungal cells representing a specific phase of the life cycle sometimes lack a cell wall. These include some fungi that occur parasitically within other living cells and zoospores. A thin granular layer, the *cell coat,* covers the zoospores of some fungi (Powell and Bracker, 1986).

Whether or not a cell wall or coat is present, the cell is surrounded by the *plasmalemma,* which forms the outermost boundary of the cytoplasm. It is a unit membrane appearing as a three-layered structure in electron micrographs and consists of lipids and proteins. Like the plasmalemma of other organisms, it consists predominantly of proteins and lipids (see also Chapter 9). The plasmalemma of zoospores of *Blastocladiella emersonii* lends itself particularly to detailed chemical analysis because a cell wall is absent and the zoospores lack an endoplasmic reticulum, which could contaminate the sample. The plasmalemma of these cells contains approximately equal proportions of protein and lipid by weight. The majority of the lipids are glycolipids (47%) and neutral lipids (42%), with phospholipids occurring in lesser quantities (11%) (Leonards and Haug, 1980).

The Cytoplasm

The cytoplasm contains a number of components that are composed of unit membranes. The most important of these are the endoplasmic reticulum, Golgi apparatus, and various types of vacuoles and vesicles (Figs. 1–5 to 1–7). Other constituents are the ribosomes, often associated with the endoplasmic reticulum, mitochondria, and the cytoskeleton.

CYTOPLASMIC MEMBRANES AND RIBOSOMES

A major structural feature is the *endoplasmic reticulum,* which consists of a pair of unit membranes separated by a space, the *lumen* (Figs. 1–6 and 1–7). The endoplasmic reticulum of the fungi, unlike that of the higher plants and animals, is usually comparatively sparse. It is a continuation of the nuclear envelope, but continuity of the endoplasmic reticulum with the plasmalemma is either rare or absent in the fungi. Most often the unit membranes of the endoplasmic reticu-

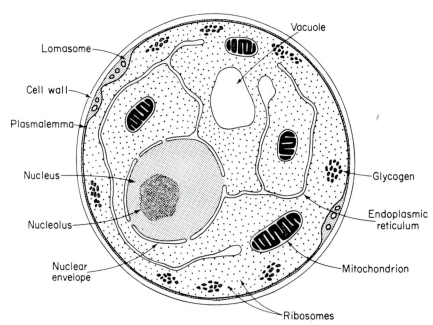

Figure 1–6 Diagram of a cross section of a generalized fungus cell.

Figure 1–7 Ultrastructural detail of a portion of a fungal cell. Mb, microbody; Mt, mitochondrion; G, Golgi apparatus; ER, endoplasmic reticulum; R, ribosomes. The cristae in the mitochondrion are tubular. × 63,700. [Courtesy I. B. Heath.]

lum are sheetlike, but sometimes they may be in the form of tubules. Frequently, the endoplasmic reticulum produces isolated vesicles that appear to form by pinching off of the membranous segments of the endoplasmic reticulum. A vesicle is a bubblelike organelle surrounded by a single unit membrane. They may contain materials that were previously within the lumen of the endoplasmic reticulum and may also transport these materials to other parts of the cell. Some of these vesicles may be involved in wall growth (see below).

The morphology of the endoplasmic reticulum may vary with the age and physiological condition of the cell as well as with the role of the cell. The endoplasmic reticulum is usually more abundant in young or metabolically active cells than in aged or inactive cells. In cells which may be secreting a material that is emptied outside the cell, the endoplasmic reticulum may be a branching system of smooth tubules or it may be sheetlike but dilated and swollen so that the paired membranes are far apart from each other. Dilated endoplasmic reticulum may also occur in cells in which wall deposition is occurring. Closely stacked endoplasmic reticulum does not occur commonly but may be present in some metabolically active cells, such as those involved in secretion or those that are forming wall material. The endoplasmic reticulum in such metabolically active cells is most often of the *rough* type; that is, its surface is covered by ribosomes. However, most endoplasmic reticulum is of the *smooth* type, lacking ribosomes.

Ribosomes, similar to those found in higher plants and animals, occur in the cytoplasm of fungal cells as well as on the surface of the endoplasmic reticulum (Figs. 1–6 and 1–7). They are the sites where polypeptide synthesis occurs, accounting for their abundance in metabolically active cells. The ribosomes have a sedimentation value of about 80 S, similar to that of other eukaryotes but differing from prokaryotes. Ribosomes consist of about equal parts of ribonucleic acid and protein, which are organized into two subunits, a 60 S and a 40 S subunit. Four distinct types of RNA occur, while it is reported that there are between 31 and 45 different kinds of protein in the 60 S subunit alone, and another 21 to 34 in the 40 S subunit (Russell and Wilkerson, 1980).

Golgi dictyosomes have been detected in some fungi, although they are not widely distributed among the fungi (Figs. 1–7 and 1–8). When present, the Golgi dictyosome is generally similar to that occurring in plants or animals. It consists of a stack of adjacent membranous *cisternae,* which appear to be pinching off *vesicles,* round bodies surrounded by a single unit membrane. Unlike the dictyosomes occurring in the higher plants and animals, only two to five cisternae occur in fungal dictyosomes. In some fungi lacking the stacked cisternae, there are sometimes single dilated cisternae that can easily be confused with the endoplasmic reticulum but always have smooth membranes. Cytochemical tests in some fungi demonstrate that these single cisternae are equivalent to those in the Golgi dictyosome and are presumably carrying out the same roles. The Golgi

(c)

Figure 1–8 (a,b) Events leading to wall formation or enlargement in the fungi. Vesicles, which may originate from a Golgi apparatus as diagrammed here, may fuse to form a new wall (sequence *a*), or may empty their contents between the plasmalemma and the existing wall by fusing with the plasmalemma (sequence *b*). (c) Electron micrograph of hyphal tip showing apical vesicles (A) concentrated at growing points, mitochondria (M), and Golgi body (G). × 7700. [(c) From H. C. Hoch and R. C. Staples, 1983, Mycologia 75:795–824, New York Botanical Garden.]

dictyosome carries out various roles in the fungi. In some fungi the Golgi apparatus produces vesicles that are involved in wall formation (e.g., Grove et al., 1970; Heath, 1976) (Fig. 1–8), secretion of extracellular materials such as the cell coat, and the cleavage of spores out of a multinucleate cytoplasm (Dorward and Powell, 1983; Powell and Bracker, 1986; Taylor and Fuller, 1981).

Wall growth is a dynamic process that is associated with the membranous cytoplasmic organelles. This process is conspicuous during many parts of the life of a fungus. The walls of vegetative cells such as hyphal tips grow by extension, new septa may form in hyphal cells or in sexual cells, new walls may be initiated when spores are formed endogenously within a cell, or walls may thicken. Both the endoplasmic reticulum and Golgi apparatus apparently are capable of producing vesicles that are involved in wall growth. Vesicles have been found in unusually high numbers in those parts of the cell where wall growth or deposition is occurring. In electron micrographs some of these vesicles appear to be emptying their contents between the plasmalemma and cell wall or to be merging (Fig. 1–8). Some vesicles contain enzymes that will soften the existing wall, making it pliable and capable of extension, while other vesicles contain wall precursors. Extremely small vesicles, the *chitosomes*, contain chitin synthetase, an enzyme that assists in the assembly of a microfibril of chitin from its precursors and is active where the wall is being constructed (Bartnicki-Garcia et al., 1979; Gozalbo et al., 1987).

The fungal cell may also contain microbodies (Fig. 1–7), a special class of small vesicles that are similar to those found in higher plants, where they originate from the endoplasmic reticulum (Frederick et al., 1975). The microbodies are surrounded by a single unit membrane, are usually round or oval in shape, and have somewhat dense contents with a granular matrix and possibly a crystalline core. The microbodies all contain catalase as well as a variety of other enzymes. Some contain enzymes which are involved in fatty acid oxidation and in the glyoxalate cycle (discussed further in Chapter 9) (Heath, 1976; Powell, 1976; Powell and Bracker, 1986).

Some vesicles contain the hydrolytic enzyme acid phosphatase as well as other hydrolytic enzymes, and because of their distinctive contents, they are known as *lysosomes*. Some lysosomes originate as Golgi vesicles (Aliaga and Ellzey, 1984). Lysosomes are often responsible for the controlled breakdown of cellular components. Examples include septum breakdown, which may allow nuclear migration to take place between cells (Wessels and Sietsma, 1979), breakdown of a host cell wall by a parasitic fungus (Armentrout and Wilson, 1969), and breakdown of some spore-bearing sacs (asci), which become an ooze and release the spores (Rosing, 1984; Wilson and Stiers, 1970).

Vacuoles are often apparent in fungal cells (Fig. 1–6) and, like the vesicles, are surrounded by a single unit membrane. There is sometimes no sharp delineation between the use of the terms *vesicle* and *vacuole* for a given structure. Usually, the term *vacuole* is reserved for structures that appear to be conspicuously large and have light contents. The vacuoles may originate from the enlargement and fusion of small vesicles. Vacuoles become increasingly numerous and enlarge in maturing cells. Fungal vacuoles do not, however, achieve the prominence of the large vacuoles in mature plants. The vacuolar contents are aqueous and may include amino acids, pigments, or hydrolytic enzymes. In the ''inky cap'' mushroom, in which the gills undergo autolysis and drip away as an ooze (Fig. 5–43), the hydrolytic enzyme chitinase has been found in the vacuoles and is apparently responsible for breakdown of the gills (Iten and Matile, 1970).

Membranous configurations are frequently observed in electron micrographs of fungal cells. These configurations include the *lomasomes* (Fig. 1–9), which consist of membranous tubules, vesicles, or parallel sheets in a matrix between the plasmalemma and the cell wall. The matrix may be part of the cell

Figure 1–9 Complex membranous configuration (mc) containing both tubules and vesicles is adjacent to the fungal cell wall (w). Wall material is apparently permeating the spaces between the vesicles so that they become embedded within the wall as a lomasome (l). × 112,700. [From I. B. Heath and A. D. Greenwood, 1970, Gen. Microbiol. 62:129–137.]

wall itself or of some other composition. The lomasomes occur singly or in groups. Usually, the plasmalemma becomes domelike over the lomasomes, but it may protrude inward to form a pouch. Sometimes the membranes within the lomasome may be continuous with that of the plasmalemma. Membranous configurations resembling the lomasomes are frequently observed in the cytoplasm, where they may be attached to the plasmalemma, isolated within the cytoplasmic matrix, or occur within vacuoles. The significance of these various membrane configurations is not known and could vary according to the species or physiological condition of the fungus involved as well as the nature of the chemical fixation treatment used in preparation for electron microscopy. Marchant and Moore (1973) maintain that at least some of these lomasomes and membranous configurations are normal cell constituents.

MITOCHONDRIA

Mitochondria occur in the cytoplasm of fungal cells (Figs. 1–6 and 1–7). They may be circular, oval, elongated, or even branched in the yeast cells, which contain only a single mitochondrion (Fig. 1–10). Each mitochondrion has a

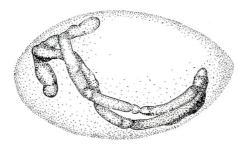

Figure 1–10 Unicellular yeastlike fungus, *Bullera alba,* with its single mitochondrion. The diagram was prepared from computer reconstructions of electron micrographs of serially sectioned cells. [From J. W. Taylor and K. Wells, 1979, Exptl. Mycol. 3:16–27.]

smooth outer membrane and an inner membrane that extends into cristae that penetrate into the matrix. The cristae are usually platelike in form, but tubular cristae also occur, particularly in the Division Mastigomycota (Fig. 1–7). The matrix contains ribosomes as well as several circular molecules of DNA, comprising 1 to 20% of the DNA occurring in the cell. The size of these DNA molecules varies greatly from one fungus to another, and values ranging from about 19 to 176 kilobase pairs have been reported (Scazzocchio, 1987).

The DNA codes for the ribosomal RNA and several transfer RNA molecules specific to the mitochondrion. Mitochondrial ribosomes are involved in the synthesis of some proteins that occur in the matrix and cristae. As is the case for other organisms, the tricarboxylic acid cycle occurs in the matrix, while electron transport and ATP production occur in the cristae. These processes are important, as they provide energy to the cell.

Mitochondrial Plasmids. In the common yeast, about 1 to 5% of the cellular DNA occurs in *plasmids,* circular DNA molecules of about 2 micrometers in circumference that can replicate autonomously. Plasmids have been detected in some filamentous fungi as well as in the yeasts, where they usually occur within the mitochondria in addition to the regularly occurring large DNA molecule. The mitochondrial plasmids are of two basic types (Nargang, 1985). One type has some genetic homologies with part of the larger mitochondrial DNA molecule from which they were derived. These small circular plasmids are associated with respiratory deficiencies (page 250). The second type has no apparent relationship to the large mitochondrial DNA molecule. These plasmids can be linear or circular and are normal components of healthy cells. Their function, if any, is unknown. Naturally occurring plasmids in fungi are of great interest to molecular biologists because they can potentially be used to transfer genes from one organism to another, a useful tool in biotechnology (page 523).

CYTOSKELETON AND CYTOPLASMIC MOVEMENT

The cytoskeleton of fungal cells consists primarily of *microtubules,* minute hollow tubes that are polymerized from a special protein, *tubulin,* and microfilaments of the contractile protein *actin.* Microtubules are dispersed throughout most of the cytoplasm and are oriented primarily parallel to the long axis of the hypha (Fig. 1–11). They may be evenly dispersed in the cytoplasm, or concentrated into bundles, particularly in the outermost part of the cytoplasm. Cross-linkages between adjacent microtubules form a network. Although their length may also vary, the microtubules are considerably shorter than those occurring in animals. Within the cytoplasm, the microtubules are often closely adjacent (but not in physical contact) to cytoplasmic organelles such as mitochondria, nuclei, and lysosomal vesicles. The distribution of the actin is much more difficult to determine, but might form a diffuse network by cross-linking. Actin is particularly abundant at the tips of growing hyphae. The cytoskeleton provides some mechanical strength to the cytoplasm, maintains the position of the organelles, and is responsible for cytoplasmic and organelle movement (McKerracher and Heath, 1987).

Movement of cytoplasm and organelles within fungal cells can readily be observed and three types distinguished (Heath and Heath, 1978; McKerracher and Heath, 1987). There is a generalized, steady cytoplasmic streaming together with all cytoplasmic constituents toward the hyphal tip, where elongation occurs. The rate of cytoplasmic streaming occurs at about the same rate as hyphal tip growth is taking place (from about 1 to 10 micrometers per minute). Although all organelles can be carried along in the moving cytoplasm, the nuclei and mitochondria are also capable of long, steady, independent movement, which typically occurs at about the same rate as cytoplasmic streaming, but can be faster or slower. Migration of the nuclei is highly ordered, and the nuclei

(a)

(b)

Figure 1–11 Microtubules in hyphal apices of *Uromyces phaseoli*. (a) Section through the hyphal apex showing the longitudinal orientation of the microtubules. Numerous apical vesicles (A) are associated with the growing point. × 21,800. (b) Enlarged view of the microtubules, × 71,500, with cross-linkage between microtubules shown in the inset, × 206,100. [From H. C. Hoch and R. C. Staples, 1983, Mycologia 75:795–824, The New York Botanical Garden.]

maintain a particular position relative to each other and to the growing hyphal tip. A third type of movement is the rapid (1 to 10 micrometers per second) and erratic behavior of small lipid droplets, vesicles, and other small unidentified particles. These particles move in a pattern characterized by starts, stops, and bidirectional changes.

Although the precise manner in which the cytoskeleton regulates movement in fungal cells is unknown, McKerracher and Heath (1987) present some speculations based on current evidence. They suggest that a type of amoeboid movement may be involved in cytoplasmic streaming. The amoeboid movement would be generated by cytoplasmic contraction involving actin and myosin. The cross-linkages holding the actin together in a structural network would be lost, caused by an influx of calcium. Then the actin would be free to form a network with myosin, and the actin–myosin complex becomes gelled and contracted. The actin is anchored to the plasmalemma and cell wall, and this contraction would literally pull the cytoplasm forward. The microtubules appear to be directly involved with organelle movement. They may form tracks along which the smaller organelles can move rapidly and bidirectionally, as well as regulate the positioning and movement of the larger organelles, such as mitochondria and nuclei. The slow, steady organelle movement may involve the association of the organelle with some microtubules associated with only part of the cytoskeleton. These microtubules presumably could slide to move organelles, or remain static to hold the organelles in their appropriate position.

Some structural features of the fungal nucleus are similar to those of other eukaryotes. The nucleus is surrounded by a nuclear envelope consisting of two unit membranes separated by the *perinuclear space*. A number of pores exist where the inner and outer membranes are joined. The nuclear envelope may be continuous with the endoplasmic reticulum. The nuclear envelope surrounds the nucleoplasm, which contains the nucleolus, both readily visible in electron micrographs.

The nuclei of fungi are unusually small and are often near the limit of resolution with the light microscope. The genomes of fungi are correspondingly small and were found to range between 1.7 and 9.3 × 10⁷ nucleotide pairs in a survey of a number of fungi (Ullrich and Raper, 1977). They have a DNA content intermediate between that of prokaryotes and the higher eukaryotic plants and animals (Fig. 1–12). The DNA content of the cultivated mushroom is typical for the low values occurring in the fungi. The size of the haploid genome of *Agaricus brunnescens* is 3.42 × 10⁷ nucleotide pairs (Arthur et al., 1982), approximately eight times as much DNA as that of the bacterium *Escherichia coli,* but only about 1% of the amount occurring in a haploid human cell.

Like the DNA of other eukaryotes, fungal DNA can consist of single-copy or multiple-copy (redundant) nucleotide sequences. The single-copy sequences code for mRNA, while most redundant sequences code for ribosomal RNA, tRNA, and chromosomal proteins, all needed in large quantities. Fungi have a much lower percentage of redundant DNA than is found in the higher plants and animals. Typically, about 10 to 20% of the DNA in fungi is redundant, representing about 100 copies of a gene, while as much as 80% of the DNA may be redundant in other eukaryotes (Dutta, 1974), representing from 100 to 1 million copies of a gene. Replication of DNA in the fungi is generally similar to that occurring in other organisms (Berry and Berry, 1976).

The DNA is associated with proteins and forms the *chromatin*, which is

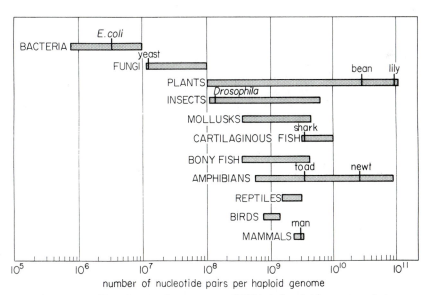

Figure 1–12 Variation in the amount of DNA in haploid genomes from a variety of organisms. Note the relatively small amount of DNA that the fungi have in comparison with other eukaryotes. The common yeast has the smallest known genome of any eukaryote, and is less than 0.5% that of a haploid cell in humans. [From B. Alberts, D. Bray, J. Lewis, M. Raff, K. Roberts, and J. D. Watson, 1983, Molecular biology of the cell, Garland Publishing, New York.]

sometimes visible as the chromosomes condense during nuclear division. The protein may be histones, basic proteins of the nature found in other eukaryotic organisms, or acidic proteins. The nature of the protein content and the quantity varies considerably among the fungi.

Chromosome number is extremely difficult to determine because of their small size and lack of condensation. Estimates of chromosome numbers in fungi range from about 2 to 18 in haploid cells. The common yeast *Saccharomyces cerevisiae* has 17 chromosomes, which are extremely small; the smallest is about the size of a T4 virus genome (Fangman and Zakian, 1981).

NUCLEAR DIVISION

The cytological features of nuclear division are unique in the fungi. In mitosis, the nuclear envelope does not disappear but instead constricts in a dumbbell-like fashion and eventually separates into daughter nuclei (Fig. 1–13). Similarly, the nucleolus is usually retained during mitosis and may be stretched out and divided between the daughter nuclei. In some fungi, the nucleolus either is cast out of the nucleus, disperses and disappears within the nucleus, or disperses within the nucleus but remains detectable during mitosis.

Either centrioles or *spindle pole bodies* may be associated with the nucleus during interphase. Centrioles contain distinct triplets of microtubules, while the spindle pole bodies lack microtubule triplets but rather are platelike, globular, or of some other form. During nuclear division, the centrioles or spindle pole bodies replicate and migrate to opposite poles of the nucleus, where they will lie next to or within the nuclear envelope. The microtubules that will comprise the spindle originate from either the centrioles or spindle pole bodies, and elongate by polymerization. Spindle formation may be initiated from the centrioles or spindle pole bodies either while they are migrating or after they have reached their final position at opposite poles of the nucleus. The spindle typically lies within the nuclear envelope (Figs. 1–13 and 1–14). Some of the microtubules of the spindle are attached to chromosomes. The chromosomes themselves are often not observed at all or appear only as patches of chromatin. In most fungi, the chromosomes are randomly distributed during metaphase and do not define the equatorial plate that is commonly observed in other eukaryotes.

Meiosis conforms more closely to the norm for other organisms than does mitosis. In meiosis both the nucleolus and nuclear envelope degenerate, usually near the end of prophase, although the nucleolus may degenerate later. As in mitosis, the chromosomes are usually irregularly distributed on the spindle at

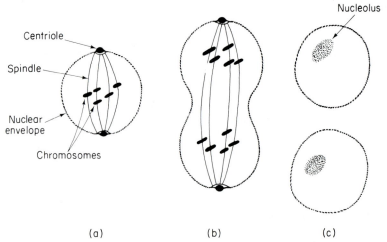

(a)   (b)   (c)

Figure 1–13 Some stages in fungal mitosis: (a) metaphase; (b) telophase; (c) interphase.

Figure 1–14 Early metaphase of mitosis in *Saprolegnia ferax*. The centrioles (at the large arrows) lie outside the nuclear envelope at the spindle poles. Some of the spindle fibers terminate at kinetochores and their chromatin (small arrows). [From I. B. Heath, 1978, *in:* Heath, I. B., Ed., Nuclear division in the fungi, Academic Press, New York, pp. 89–176.]

metaphase, but rarely they may be aligned on an equatorial plate. Sometimes the chromosomes are so small and poorly defined even in meiosis that studies of cytology may be extremely difficult. This has led to some confusion in understanding certain fungal life cycles if the time of meiosis cannot be precisely determined by conventional cytological techniques. Other methods, such as quantitative analysis of DNA, may be required to pinpoint the place or time of meiosis.

In addition to their role in nuclear division, centrioles are important in motile cells formed by members of the Division Mastigomycota. This topic is discussed further in Chapter 2.

CELLULAR ORGANIZATION

Some fungi such as the yeasts and chytrids consist of one, or at most a few cells throughout most of their life. Unlike these *unicellular* fungi, the vast majority of the fungi are *multicellular,* producing vegetative or reproductive structures composed of a number of cells. Hyphae are the most common multicellular structures. Normally, large numbers of hyphae develop as a result of repeated branching and growth. The massive quantities of hyphae produced by a fungus are collectively termed a *mycelium* (plural, *mycelia*). If one examines a fungal culture, the hyphae can be distinguished microscopically as individual cellular tubes, while the mycelium often appears as a mass resembling cotton.

Probably the majority of fungi not only produce mycelia, but also produce multicellular tissues which are derived by differentiation from hyphae. These tissues may consist of hyphae that have aggregated but retain their typical elongated appearance, or they may range to types in which the hyphae have formed rounded, more or less isodiametric cells. The latter tissue type resembles the parenchyma of higher plants and is often called *pseudoparenchyma* (Fig. 1–15). Formation of tissues typically results in the development of solid, often macroscopic fungal structures specialized for particular roles.

Hyphae are the fundamental structures from which all other multicellular fungal structures are derived. The organization and biology of the hyphae are of great importance.

Organization and Structure of Hyphae

Hyphae consist of a series of elongated tubular cells. Cross walls (*septa*) occur within hyphae, dividing them into separate cells or cytoplasmic compartments. The hyphae of some fungi are almost completely nonseptate when healthy and growing, but a septum can separate aging or damaged cells or reproductive structures from the remainder of the hypha. Most fungi, however, have hyphae that are regularly divided into individual cells by septa. These cells can be uninucleate, binucleate, or multinucleate, depending on the particular fungus and the life-cycle stage.

There are some interesting differences in septal structure among the fungi. Some members of the Division Mastigomycota (Chapter 2) form *pseudosepta,*

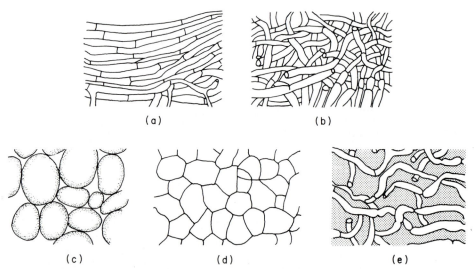

(a)   (b)

(c)   (d)   (e)

Figure 1–15 Representative fungal tissues. One type may merge into another, and modifications of the types shown here may occur. (a) Tissue consisting of hyphae that remain distinct and are more or less parallel to each other; (b) tissue consisting of intertwined hyphae; (c) tissue consisting of rounded cells with intercellular spaces; (d) tissue with cells that are angular, owing to mutual pressure and lacking intercellular spaces; (e) tissue consisting of hyphae interspersed through a watery gel (shaded).

septa that are perforated by so many pores that they are sievelike. In the majority of the fungi, the septum is perforated only by a single pore or a few pores, which are small but nevertheless allow nuclei and cytoplasm to pass from cell to cell. Two major septum types are the *simple septum*, formed by many members of the Ascomycotina, and the *complex septum*, formed by many members of the Basidiomycotina.

In the hyphae of many members of the Ascomycotina and Deuteromycotina, the septum is relatively simple and its wall tapers gradually and evenly toward the pore, forming a wedge in cross section. The pore is about 0.05 to 0.5 micrometer in diameter. About three or four *Woronin bodies* are adjacent to the pore and remain near it even when cytoplasmic streaming occurs through the septum. A Woronin body is a rounded organelle that is surrounded by a double unit membrane and has granular contents consisting largely of protein. While the hypha is healthy, the Woronin bodies remain in their usual position in the cytoplasm adjacent to a pore (Fig. 1–16), but when a hyphal cell ages or becomes damaged, the Woronin bodies move into the pore and become a plug. This plug effectively separates the cytoplasm of the aged or damaged cell from the cyto-

(a)

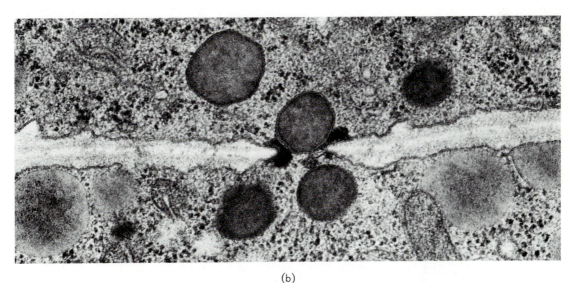

(b)

Figure 1–16 Simple septum formed in hypha of a member of the Ascomycotina: (a) diagrammatic interpretation of the simple septum together with a Woronin body near the septal pore; (b) Woronin bodies in the vicinity of the septal pore, × 55,000. [(a) Adapted from R. T. Moore and J. H. McAlear, 1962, Am. J. Botany 49:86–94; (b) courtesy T. M. Hammill.]

plasm of cells that are still healthy and can develop quickly, as demonstrated by Collinge and Markham (1985). They cut hyphae with a razor blade and determined that plugging of the septum could occur within 4 seconds, and plugging would occur not only in the cells directly damaged by the cutting but would also occur in the septa of the next few cells. Although the septum type appears to be comparatively uniform in the vegetative hyphae of the Ascomycotina, morphologically elaborate septa can be found in the multicellular reproductive bodies and vary in structure from one fungus to another (Kimbrough and Curry, 1986).

A highly structured and complex septum occurs in most members of the Basidiomycotina and some members of the Deuteromycotina (Chapters 5 and 6). Near the center of the septum there is an enlarged ring, which surrounds the septal pore and resembles a doughnut in its overall form. The most distinctive feature of the septum is the formation of a thickened electron-dense *septal pore cap* covering the septal swelling and septal pore. This cap is formed from the endoplasmic reticulum, lying close to and parallel to the cross walls on both sides (Fig. 1–17). The septal pore cap may be continuous or have either small or large pores. Structure of the septal pore cap sometimes differs slightly from one part of the fungus to another (Gull, 1978; Wells, 1978). Cytoplasm and small organelles can pass through the septum from cell to cell, but nuclear migration presumably can occur with some modification of the pores. In *Rhizoctonia solani*, the nuclei and other organelles constrict as they pass through the discontinuities in the septal pore cap and subsequently through the septal pore, which enlarges to accommodate their passage (Bracker and Butler, 1964). Alternatively, the elaborate septum may be degraded by hydrolytic enzymes to leave a simple septum (with no thickenings or caps), resulting in an enlargement of the central pores

(a)   (b)

Figure 1–17 (a) Electron micrograph of a complex septum from the hypha of a member of the Basidiomycotina, × 11,500; (b) diagrammatic interpretation. C, septal pore cap, CD, discontinuity in septal pore cap; ER, endoplasmic reticulum; S, septal swelling; SP, septum. [Courtesy C. E. Bracker and E. E. Butler, 1963, Mycologia 55:35–58.]

to the extent that nuclei can pass from one cell to another (Marchant and Wessels, 1974). Blockage of the septum may occur if the central ring expands to close the pore entirely or if some electron-dense material plugs the openings (Wells, 1978).

Biological Attributes of the Hyphae

A *coenocyte* is a special type of cell having several nuclei in a common cytoplasm. In a coenocyte, the numerous nuclei all exert genetic control over the cytoplasm that they share. The hyphae that are predominantly nonseptate are clearly coenocytic, but the hyphae having septa can also be considered to be coenocytes because there is cytoplasmic continuity across the septum. The septa themselves have one or several comparatively large pores which allow the cytoplasm to pass readily from one cell to another, and in addition, the septa are often enzymatically degraded. The hypha then has the genetic attributes of a coenocyte because nuclear control is exerted not only by the one or more nuclei within a particular cell, but also by the nuclei in neighboring cells. A second major characteristic is that the hypha permits relatively unimpeded flow of the cytoplasm and nuclei from one cell to another.

The vegetative mycelium has an important role in the life cycle of the fungus which depends on its ability to transport nutrients rapidly through the hyphae. Translocation of the nutrients toward the hyphal tip occurs in conjunction with cytoplasmic streaming (discussed further in Chapter 8). The major role of the mycelium is to support nutritionally the development of the reproductive structures, permitting either sexual or nonsexual reproduction to take place.

REPRODUCTION

Sexual reproduction involves the union of two haploid cells (N). In the fungi, sexual reproduction can involve undifferentiated hyphae, gametes, or gametangia, all determined by the particular fungal species. A gamete is a single cell, either motile or not, that functions as a sex cell by combining with another cell, contributing its nucleus and cytoplasm. Unlike the gametes, the gametangia are differentiated multinucleate cells that either produce gametes or function directly in mating with another cell. All of these structures have haploid nuclei. Mating can occur in any combination involving the hyphae, gametes, or gametangia (i.e., mating may involve the union of a gamete with another gamete, with a hypha, or with a gametangium).

Mating is actually a two-stage process. First, the protoplasm of the haploid cells mingles, and then later, the haploid nuclei join to form a diploid *zygote* nucleus. In most organisms, the two steps occur so quickly both in time and in the common fusion product that the entire sequence is known as *fertilization*. The mating can occur in a similar manner in many fungi with the direct production of the diploid cell. However, the majority of the fungi have life cycles that are characterized by a delay and the additional production of cells between the two steps. Mycologists term the stage at which the gametes or other sex cells fuse and the protoplast union occurs *plasmogamy*. The second stage, union of the nuclei to form the zygote, is called *karyogamy* and occurs much later. This now involves a new concept—what intervenes between the haploid and diploid stages?

The gametes or other sex cells are haploid. They are further characterized by having either a single nucleus or perhaps several nuclei but of the same genetic constitution. Therefore, these sex cells are said to be *monokaryotic*. When plasmogamy between two sex cells occurs, their cytoplasm mingles together and their nuclei exist in the common cytoplasm. Because these nuclei are genetically

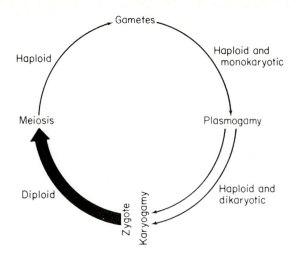

Figure 1–18 Succession of events and nuclear stages in a typical sexual reproductive cycle in a fungus. The relative lengths of these stages vary from fungus to fungus. Gametes and plasmogamy may be absent and a true dikaryon may not be produced if different nuclear types are not brought together in plasmogamy.

different, the cells having them are said to be *dikaryotic,* or $N + N$. The dikaryotic cells proliferate, producing new cells having both nuclear types (all $N + N$). Eventually, two haploid nuclei undergo karyogamy, forming the zygote ($2N$). Either that diploid zygote or its diploid progeny undergoes meiosis to restore the haploid state. Although sexual reproductive cycles typically progress through the haploid, dikaryotic, and diploid phases, the relative length of each varies with the species (Fig. 1–18). Life cycles involving a long dikaryotic phase occur in many members of the Ascomycotina and Basidiomycotina.

Some fungi do not produce either gametes or gametangia, and it is often not known whether or not plasmogamy between hyphae has actually taken place. In many of these fungi, karyogamy and meiosis do take place, just as in other fungi that produce conspicuous gametes or gametangia and in which plasmogamy is known to occur. Sexual reproduction is said to have taken place in such fungi, even in the apparent absence of sexual organs or plasmogamy. The use of the term *sexual reproduction* in conjunction with fungi implies only that karyogamy and meiosis occur.

Not all fungi reproduce sexually and some must rely entirely on *nonsexual reproduction.* Nonsexual reproduction may take place during the haploid, dikaryotic, or diploid phases of the sexual cycle or in organisms that apparently lack a sexual cycle. Unlike sexual reproduction, karyogamy does not take place at a precise stage in the life cycle (it may not occur at all), and meiosis never takes place in nonsexual reproduction.

Both sexual and nonsexual reproduction result in the production of spores. The spores may be borne upon or within hyphae similar to the vegetative hyphae or within isolated single cells. In many of the more complex fungi, multicellular reproductive bodies bear the spores—such a spore-bearing reproductive body is a *sporocarp.* You have probably seen colorful mushrooms while hiking through a forest. The visible mushroom is a sporocarp because it bears spores derived from a sexual reproductive cycle and is nourished by the mycelium, which occurs within the soil or plant debris and is invisible to the naked eye.

REFERENCES CITED

AINSWORTH, G. C. 1973. Introduction and keys to higher taxa. *In:* Ainsworth, G. C., F. K. Sparrow, and A. S. Sussman, Eds. The fungi: an advanced treatise. Academic Press, New York, Vol. 4A, pp. 1–7.

ALIAGA, G. R., and J. T. ELLZEY. 1984. Ultrastructural localization of acid phosphatase and alkaline phosphatase within oogonia of *Achyla recurva.* Mycologia 76:85–98.

ARMENTROUT, V. N., and C. L. WILSON. 1969. Haustorium–host interaction during myco-parasitism of *Mycotypha microspora* by *Piptocephalis viriginiana*. Phytopathology 59:897–905.

ARTHUR, R., F. HERR, N. STRAUS, J. ANDERSON, and P. HORGEN. 1982. Characterization of the genome of the cultivated mushroom, *Agaricus brunnescens*. Exptl. Mycol. 7:127–132.

BARTNICKI-GARCIA, S. 1968. Cell wall chemistry, morphogenesis, and taxonomy of fungi. Botan. Rev. 22:87–108.

——, J. RUIZ-HERRERA, and C. E. BRACKER. 1979. Chitosomes and chitin synthesis. *In:* Burnett, J. H., and A. P. J. Trinci, Eds. Fungal walls and hyphal growth. Cambridge University Press, Cambridge, pp. 149–168.

BERRY, D. R., and E. A. BERRY. 1976. Nucleic acid and protein synthesis in filamentous fungi. *In:* Smith, J. E., and D. R. Berry, Eds. The filamentous fungi: biosynthesis and metabolism. Edward Arnold, London, Vol. 3, pp. 238–291.

BRACKER, C. E., and E. E. BUTLER. 1964. Function of the septal pore apparatus in *Rhizoctonia solani*. J. Cell Biol. 21:152–157.

BURNETT, J. H. 1979. Aspects of the structure and growth of hyphal walls. *In:* Burnett, J. H., and A. P. J. Trinci, Eds. Fungal walls and hyphal growth. Cambridge University Press, Cambridge, pp. 1–25.

COLLINGE, A. J., and P. MARKHAM. 1985. Woronin bodies rapidly plug septal pores of severed *Penicillium chrysogenum* hyphae. Exptl. Mycol. 9:80–85.

DORWARD, D. W., and M. J. POWELL. 1983. Cytochemical detection of polysaccharides and the ultrastructure of the cell coat of zoospores of *Chytriomyces aureus* and *Chytriomyces hyalinus*. Mycologia 75:209–220.

DUTTA, S. K. 1974. Repeated DNA sequences in fungi. Nucleic Acids Res. 1:1411–1419.

FANGMAN, W. L., and V. A. ZAKIAN. 1981. Genome structure and replication. *In:* Strathern, J. N., E. W. Jones, and J. R. Broach, Eds. The molecular biology of the yeast *Saccharomyces:* life cycle and inheritance. Cold Spring Harbor Laboratory, Cold Spring Harbor, N.Y., pp. 27–58.

FREDERICK, S. E., F. J. GRUBER, and E. H. NEWCOMB. 1975. Plant microbodies. Protoplasma 84:1–29.

GOZALBO, D., F. DUBON, J. SCHWENCKE, and R. SENTANDREAU. 1987. Characterization of chitosomes in *Candida albicans* protoplasts. Exptl. Mycol. 11:331–338.

GROVE, S. N., C. E. BRACKER, and D. J. MORRÉ. 1970. An ultrastructural basis for hyphal tip growth in *Pythium ultimum*. Am. J. Botany 57:245–266.

GULL, K. 1978. Form and function of septa in filamentous fungi. *In:* Smith, J. E., and D. R. Berry, Eds. The filamentous fungi. Wiley, New York, Vol. 3, pp. 78–93.

HEATH, I. B. 1976. Ultrastructure of freshwater Phycomycetes. *In:* Jones, E. B. G., Ed. Recent advances in aquatic mycology. Wiley, New York, pp. 603–650.

——, and M. C. HEATH. 1978. Microtubules and organelle movements in the rust fungus *Uromyces phaseoli* var. *vignae*. Cytobiologie 16:393–411.

HOCH, H. C., and R. C. STAPLES. 1983. Ultrastructural organization of the non-differentiated uredospore germling of *Uromyces phaseoli* variety *typica*. Mycologia 75:795–824.

ITEN, W., and P. MATILE. 1970. Role of chitinase and other lysosomal enzymes of *Coprinus lagopus* in the autolysis of fruiting bodies. J. Gen. Microbiol. 61:301–309.

KIMBROUGH, J. W., and K. J. CURRY. 1986. Septal structures in apothecial tissues of the tribe Aleurieae in the Pyronemataceae (Pezizales, Ascomycetes). Mycologia 78:407–417.

LEONARDS, K. S., and A. HAUG. 1980. Isolation and characterization of the plasma membrane of *Blastocladiella emersonii* zoospores. Exptl. Mycol. 4:207–220.

MARCHANT, R., and R. T. MOORE. 1973. Lomasomes and plasmalemmasomes in fungi. Protoplasma 76:235–247.

——, and J. G. H. WESSELS. 1974. An ultrastructural study of septal dissolution in *Schizophyllum commune*. Arch. Mikrobiol. 96:175–182.

McKERRACHER, L. J., and I. B. HEATH. 1987. Cytoplasmic migration and intracellular organelle movements during tip growth of fungal hyphae. Exptl. Mycol. 11:79–100.

NARGANG, F. E. 1985. Fungal mitochondrial plasmids. Exptl. Mycol. 9:285–293.

POWELL, M. J. 1976. Ultrastructure and isolation of glyoxysomes (microbodies) in zoospores of the fungus *Entophylctis* sp. Protoplasma 89:1–27.

——, and C. E. BRACKER. 1986. Distribution of diaminobenzidine reaction products in zoospores of *Phytophthora palmivora*. Mycologia 78:892–900.

ROSING, W. C. 1984. Ultracytochemical localization of acid phosphatase within deliquescing asci of *Chaetomium brasiliense*. Mycologia 76:67–73.

RUSSELL, P. J., and W. M. WILKERSON. 1980. The structure and biosynthesis of fungal cytoplasmic ribosomes. Exptl. Mycol. 4:281–337.

SCAZZOCCHIO, C. 1987. The natural history of fungal mitochondrial genomes. *In:* Rayner, A. D. M., C. M. Brasier, and D. Moore, Eds. Evolutionary biology of the fungi. Cambridge University Press, Cambridge, pp. 53–73.

TAYLOR, J. W., and M. S. FULLER. 1981. The Golgi apparatus, zoosporogenesis, and development of the zoospore discharge apparatus of *Chytridium confervae*. Exptl. Mycol. 5:35–59.

ULLRICH, R. C., and J. R. RAPER. 1977. Evolution of genetic mechanisms in fungi. Taxon 26:169–179.

WELLS, K. 1978. The fine structure of septal pore apparatus in the lamellae of *Pholiota terrestris*. Can. J. Botany 56:2915–2924.

WESSELS, J. G. H., and J. H. SIETSMA. 1979. Wall structure and growth in *Schizophyllum commune*. *In:* Burnett, J. H., and A. P. J. Trinci, Eds. Fungal walls and hyphal growth. Cambridge University Press, Cambridge, pp. 27–48.

WHITTAKER, R. H. 1969. New concepts of kingdoms of organisms. Science 163:150–161.

WILSON, C. L., and D. L. STIERS. 1970. Fungal lysosomes or sphaerosomes. Phytopathology 60:216–227.

Additional Recommended References

BALLOU, C. E. 1982. Yeast cell wall and cell surface. *In:* Strathern, J. N., E. W. Jones, and J. R. Broach, Eds. The molecular biology of the yeast *Saccharomyces:* metabolism and gene expression. Cold Spring Harbor Laboratory, Cold Spring Harbor, N.Y., pp. 335–360.

BRACKER, C. E. 1967. Ultrastructure of fungi. Ann. Rev. Phytopathol. 5:343–374.

CHIONISKI, J. S., and J. T. MULLINS. 1977. Ultrastructural and enzymatic evidence for the presence of microbodies in the fungus *Achyla*. Am. J. Botany 64:593–599.

DARGENT R., J.-M. TOUZÉ-SOULET, J. RAMI, and C. MONTANT. 1982. Cytochemical characterization of Golgi apparatus in some filamentous fungi. Exptl. Mycol. 6:101–114.

FLEGLER, S. L., G. R. HOOPER, and W. G. FIELDS. 1976. Ultrastructural and cytochemical changes in the basidiomycete dolipore septum associated with fruiting. Can. J. Botany 54:2243–2253.

FULLER, M. S. 1976. Mitosis in fungi. Intern. Rev. Cytol. 45:113–153.

GROVE, S. N., and C. E. BRACKER. 1970. Protoplasmic organization of hyphal tips among fungi: vesicles and spitzenkörper. J. Bacteriol. 104:989–1009.

——, ——, and D. J. MORRÉ. 1968. Cytomembrane differentiation in the endoplasmic reticulum–Golgi apparatus–vesicle complex. Science 161:171–173.

HAWKER, L. E., M. A. GOODAY, and C. E. BRACKER. 1966. Plasmodesmata in fungal cell walls. Nature 212:635.

HEATH, I. B. 1978. Experimental studies of mitosis in the fungi. *In:* Heath, I. B., Ed. Nuclear division in the fungi. Academic Press, New York, pp. 89–176.

——. 1981. Nucleus associated organelles of fungi. Intern. Rev. Cytol. 69:191–221.

——, and A. D. GREENWOOD. 1970. The structure and function of lomasomes. J. Gen. Microbiol. 62:129–137.

McLAUGHLIN, D. J. 1974. Ultrastructural localization of carbohydrate in the hymenium and subhymenium of *Coprinus:* evidence for the function of the Golgi apparatus. Protoplasma 82:341–364.

MURRIN, F., W. NEWCOMB, and I. B. HEATH. 1987. Ultrastructure of microbodies in protoplasts and hyphae of *Entomophaga aulicae* (Zygomycetes). Mycologia 79:559–564.

NEHEMIAH, J. L. 1973. Localization of acid phosphatase activity in the basidia of *Coprinus micaceus*. J. Bacteriol. 115:443–446.

POWELL, M. J. 1974. Fine structure of plasmodesmata in a chytrid. Mycologia 66:606–614.

ROSE, A. H. 1976. Chemical nature of membrane components. *In:* Smith, J. E., and D. R. Berry, Eds. The filamentous fungi. Edward Arnold, London, Vol. 2, pp. 308–328.

ROSENBERGER, R. F. 1976. The cell wall. *In:* Smith, J. E., and D. R. Berry, Eds. The filamentous fungi. Edward Arnold, London, Vol. 2, pp. 328–344.

WERGIN, W. P. 1973. Development of Woronin bodies from microbodies in *Fusarium oxysporum* f. sp. *lycopersici*. Protoplasma 76:249–260.

2

Division Mastigomycota

The Division Mastigomycota contains those fungi that produce nonsexual motile *zoospores*. The zoospores help make it possible for these fungi to live in aquatic environments such as oceans or in freshwater bodies, including bogs, ponds, lakes, and streams. Many of these fungi occur in damp soil, where the zoospores can swim in accumulated water.

These fungi may frequently be collected from their natural habitats by "baiting" techniques. Many fungi will grow preferentially on substrata such as fruits, twigs, seeds, feathers, hair, cellophane, pollen grains, and insects. Water, soil, or debris may be collected in the field and placed in small dishes to which a little bait is added. The choice of bait depends on the type of fungus being sought, but half of a boiled hemp seed or a few dead fruit flies are often used. Usually, some fungi will colonize the bait within a few days, and then they can be studied further. Some can be directly isolated by the inoculation of agar with soil, water, or other debris. For further information on baiting and culturing, consult Sparrow (1960).

ZOOSPORE MOVEMENT AND STRUCTURE

The zoospores are propelled by one or two *flagella* (Fig. 2–1). A flagellum is a hairlike structure that undulates. The path travelled by zoospores is determined by movement of the flagella. The zoospores may travel in a nearly straight or in a circular path with an even gliding motion without rotations, in either a spiraled or more or less straight path accompanied by clockwise rotations, or in an irregular path accompanied by hopping and darting. Zoospores having a single flagellum swim by transmitting a succession of waves, more or less in a single plane, through that flagellum. When zoospores have two flagella, one flagellum extends anteriorly and the other extends posteriorly while swimming. In this case,

Figure 2–1 (a) Swimming zoospores showing amoeboid movement of zoospore body and flagellum; (b) diagram of the cross section of a flagellum showing the two central microtubules and nine pairs of microtubules near the periphery. [(a) From J. S. Karling, 1948, Am. J. Botany 35:503–510; (b) from I. R. Gibbons and A. V. Grimstone, 1960, J. Biophys. Biochem. Cytol. 7:697–716.]

(a)   (b)

each flagellum undulates in a more or less single plane that is different from that of the other flagellum. Undulation is often more prominent in the anterior flagellum, which has led to suggestions that the anterior flagellum actually propels the zoospore.

Structurally zoospores are highly specialized cells and have some components not found in vegetative cells. The membrane surrounding the flagellum is continuous with the plasmalemma of the cell. The flagellum may be smooth (*whiplash type*) or may bear filaments extending laterally from the main axis (*tinsel type*) (Fig. 2–2). Like flagella in higher plants and animals, the flagella of the Division Mastigomycota consist of a core of two central microtubules that are surrounded by nine pairs of microtubules (Fig. 2–1). These microtubules are contractile and are responsible for flagellar movement. Near the point of attachment of the flagellum to the zoospore, the two central microtubules disappear and the outer microtubules extend into the *kinetosome,* a centriolelike body with nine groups of microtubules arranged in triplets and lying near the nucleus (Fig. 2–3). The point of juncture between the flagellum and kinetosome may be marked by a platelike structure, the *terminal plate.* In some zoospores, the kinetosome is connected to the nucleus by a large, fibrous *rhizoplast* consisting of fused

Figure 2–2 Portions of a whiplash flagellum (top) and tinsel flagellum bottom). × 1250. [Courtesy A. P. Kole and K. Horstra, 1959, Konikl. Ned. Akad. Wetenschap. Proc. Ser. C 62:404–408.]

Figure 2–3 Top: base of a zoospore and its flagellum. The flagellum (*f*) is arising from the kinetosome (*k*). Note that the central microtubules are visible only within the flagellum and not within the kinetosome. A rootlet (*r*) is visible at the left and extends from the kinetosome to other organelles. Both a terminal plate (*tp*) and props (*p*) are present at the kinetosome base. A nonfunctional centriole (*c*) is present at the right. × 77,000. Bottom: cross-sectional view of the kinetosome and nonfunctional centriole (left), the flagellum in the vicinity of the props (center), and the flagellum that is exterior to the body of the zoospore (right). Note the transition from the spiral of nine triplets of microtubules in the kinetosome to the ring of nine doublets in the flagellum. [From D. J. S. Barr and V. E. Hadland-Hartmann, 1978, Can. J. Botany 57:887–900.]

microtubules or strands (Fig. 2–5). Structures similar to the rhizoplast, the *striated rootlets,* may be present and extend to other organelles such as the centriole, second kinetosome, or mitochondria. Individual microtubules may also be prominent as part of the rootlet system. Anchoring of the flagellum to cellular structures by rootlets is important, as otherwise the vigorous movement of the flagellum would tear it away from the cell. Uniflagellate zoospores swim more vigorously than biflagellate zoospores, and predictably the flagellum of a uniflagellate zoospore is anchored to a much larger mass of cellular organelles and by a more extensive rootlet system than are the flagella of biflagellate zoospores. This presumably allows the more vigorous swimming of the uniflagellate zoospores (Heath, 1976).

The structures mentioned in the preceding paragraph were related to flagellation. Other organelles are important in the zoospore (Figs. 2–4 and 2–5). Zoospores may contain a single giant mitochondrion enveloping the kinetosome and the smaller end of the nucleus or a number of smaller mitochondria scattered throughout the cytoplasm. Ribosomes may be scattered throughout the cytoplasm, attached to the endoplasmic reticulum, or present as a *nuclear cap*. A nuclear cap is a loose aggregation or membrane-enclosed aggregation of ribosomes surmounting the large end of the nucleus. Lipid, which provides energy to the zoospore, may be dispersed in the cytoplasm in several droplets or present as a single large droplet. If the species possesses Golgi bodies in the vegetative cells, one or two Golgi bodies are normally found in the zoospore. Uniflagellate zoospores may contain a *rumposome*, a highly organized disklike structure of which the outer surface is appressed to the cell membrane while the inner surface is connected to the nuclear cap by two or three lamellae extending through the cytoplasm. The rumposome does not occur in vegetative cells, and

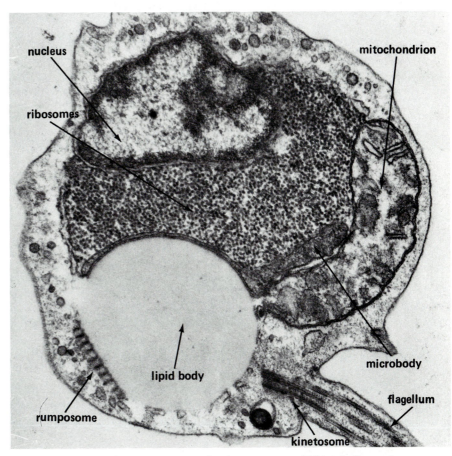

Figure 2–4 Electron micrograph of a zoospore of *Rhizophydium subangulosum* (Chytridiales). This zoospore contains a large single mitochondrion, ribosomes clustered into an area near the nucleus, a pleomorphic microbody, a single large lipid body, and a rumposome adjacent to the lipid body. The kinetosome and flagellum are at the lower right. ×32,400. [Courtesy D. J. S. Barr. From D. J. S. Barr and V. E. Hadland-Hartmann, 1978, Can. J. Botany 56:2380–2404.]

Figure 2–5 Electron micrograph of the zoospore of *Rhizophlyctis rosea* (Chytridiales). Compare this zoospore with that in Fig. 2–4 and the flagellar apparatus with that in Fig. 2–3. This zoospore has several small mitochondria (*m*), scattered ribosomes throughout the cytoplasm, several clustered lipid bodies (*l*); it lacks a rumposome and has a conspicuous rhizoplast (*rh*) extending from the nucleus (*n*) to the flagellum (*f*). × 22,000. [Courtesy D. J. S. Barr. From D. J. S. Barr and V. E. Hartmann, 1977, Can. J. Botany 55:1221–1235.]

its function in the zoospore is unknown. *Contractile* vacuoles that function in osmoregulation by expelling water have been observed in some biflagellate zoospores and may originate as vesicles produced by the Golgi bodies. Some species have a *side body*, a simple pleomorphic organelle which resembles microbodies and is bounded by a double membrane. *Gamma particles* have been observed in zoospores of *Blastocladiella emersonii* (Fig. 2–21), but not of other species. The gamma particle is a cup-shaped organelle consisting of an inner, electron-dense core surrounded by a single unit membrane. DNA, RNA, and the enzyme chitin synthetase have been found in the gamma particles. Chitin synthetase is involved in cell wall synthesis, and it has been suggested that gamma particles may play a role in the synthesis of a new wall when the zoospore encysts (Truesdell and Cantino, 1970).

ZOOSPOROGENESIS

Zoospores are formed within a saclike structure, the *zoosporangium* (Fig. 2–6). Zoosporogenesis has been studied in detail in a number of fungi (e.g., Armbruster, 1982-a,b; Hoch and Mitchell, 1972; Martin and Miller, 1986-a,b). At the beginning of zoosporogenesis, the zoosporangium contains cytoplasm and various

(a)

(b)

(c)

Figure 2–6 Zoosporangia and zoospores of *Saprolegnia:* (a) zoospores packing a mature zoosporangium; (b) zoospores escaping from the zoosporangium through an apical opening; (c) nearly emptied zoosporangium containing a few zoospores. [Courtesy Carolina Biological Supply Company.]

cytoplasmic organelles and is multinucleate. Cleavage of the multinucleate cytoplasm into smaller uninucleate units follows and may be accomplished in one of two ways.

The first method involves the formation of vesicles, which may arise from the Golgi apparatus or possibly from some other source (Fig. 2–7). The vesicles surround a uninucleate portion of the cytoplasm, and the membranes of the vesicles fuse with each other to form a double unit membrane. This membrane

Figure 2–7 Zoosporogenesis in *Olpidium brassicae*. In this species zoosporogenesis occurs by fusion of vesicles. (a) Zoosporangium before cleavage; (b) part of zoosporangium showing cleavage of zoospores; (c) flagellum in cytoplasm of zoosporangium; (d) part of mature zoosporangium. The flagella are wrapped around the zoospores. *A* = axoneme (shaft) of flagellum, *CV* = cleavage vacuole, *ET* = exit tube, *L* = lipid globule, *M* = mitochondrion, *N* = nucleus, *TW* = thallus wall, *V* = vacuole. [Courtesy J. H. M. Temmink and R. M. Campbell, 1968, Can. J. Botany 46:951–956.]

becomes the plasmalemma of the zoospore and cleaves the zoospore, which is then free within an essentially empty zoosporangium.

In the second method, a large central vacuole is formed in the zoosporangium. This vacuole increases in size both by the synthesis of additional membrane material and by absorption of water from the cytoplasm, and the enlarging vacuole extends in various directions. Meanwhile the plasmalemma is pulling inward, and extensions of the plasmalemma meet and fuse with the vacuolar membrane. These membranes delimit the uninucleate zoospores and become the plasmalemma of the zoospores.

Development of the flagellum is an important phase of zoosporogenesis and occurs concurrently with cleavage (Heath, 1976). Two centrioles lie in the vicinity of each nucleus in the uncleaved zoosporangium. Each centriole contains nine groups of triplet fibers, and a centriole may elongate to form a kinetosome. In turn, the fibers of the kinetosome elongate to form the fibers in the remainder of the flagellum. Once the kinetosome is formed in the uncleaved zoosporangium, it may lie at rest until cleavage is completed and the young zoospore is surrounded by a plasmalemma. The elongating flagellum is then covered by a membrane which is an extension of the plasmalemma. Alternatively, the kinetosome may not rest but rather extend into the flagellum while cleavage is taking place. If this happens, the flagellum lies within one of the vesicles surrounding the uninucleate mass of cytoplasm that will become the zoospore, and the membrane surrounding the flagellum is derived from some unknown source, possibly the vesicles. In biflagellate fungi, both centrioles form flagella. In uniflagellate fungi, the second centriole either does not undergo change or elongates into a kinetosome that does not undergo further development.

ZOOSPORE DISCHARGE AND GERMINATION

Zoospores may be released from the zoosporangium in a number of ways. In the simplest cases, the zoosporangial wall simply disintegrates. Usually, however, a localized nipplelike protrusion, the *papilla*, is formed on the surface of the zoosporangium or at the end of an elongated *exit tube*. The papilla may or may not have a more or less distinct plug within it. Enzymatic breakdown of the papilla causes it to open. The plug may also be disintegrated or pushed out. The zoospores then escape through the opening.

When a zoospore has completed its swarming stage, it is ready to settle down, germinate, and produce a vegetative thallus. In most cases, the flagellum is withdrawn into the zoospore and subsequently broken down and dispersed within the cell. The zoospore becomes converted into a nonmotile walled cell, the cyst. The wall material is initially contained in vesicles, which may arise from the Golgi bodies. The vesicles release the wall material that forms the outermost wall of the cyst. The inner wall of the cyst is subsequently deposited from the cytoplasm. After encystment occurs, the organelles undergo a transition to the form typical for vegetative structures. The rootlets are lost within a few minutes, the pear-shaped nucleus becomes spherical, the ribosomes in the nuclear cap become dispersed, the Golgi body moves away from its typical position adjacent to the nucleus or vacuole, and any specialized structures such as the rumposome break down. If a single large mitochondrion is present initially, it fragments into smaller mitochondria. In most cases, when the cyst germinates, the inner wall of the cyst extends and ruptures through the outer wall to form an extension that will become the germ tube. The germ tube will give rise to the vegetative thallus.

CLASSIFICATION

The Division Mastigomycota may be divided into four classes on the basis of the number, position, and type of flagella (Sparrow, 1973-a, 1976). The uniflagellate fungi are separated into the Class Chytridiomycetes (posterior flagellum) and the Class Hyphochytridiomycetes (anterior flagellum). The members of the two remaining classes are biflagellate. In the Class Plasmodiophoromycetes both flagella are of the whiplash type, while a tinsel and a whiplash flagellum occur in the Class Oomycetes. Further division into orders is based primarily on thallus structure and type of sexual reproduction. The taxa follow:

Division Mastigomycota
Class Chytridiomycetes
Order Chytridiales
Order Blastocladiales
Order Harpochytriales
Order Monoblepharidales
Class Hyphochytridiomycetes
Order Hyphochytriales
Class Plasmodiophoromycetes
Order Plasmodiophorales
Class Oomycetes
Order Saprolegniales
Order Lagenidiales
Order Peronosporales

Class Chytridiomycetes

Fungi with posteriorly uniflagellate zoospores are grouped within a single class, the Chytridiomycetes. The single flagellum is of the whiplash type. These are small, comparatively simple fungi. Principal differences among members of this group are to be found in complexity of thallus structure and the type of sexual reproduction.

ORDER CHYTRIDIALES

Members of the Chytridiales, commonly known as ''chytrids,'' are small, relatively simple fungi that produce zoospores with a single posterior flagellum and usually contain a conspicuous refractive lipid globule. The chytrids are also characterized by formation of a simple coenocytic thallus either entirely or partially within the substratum upon which they are feeding.

Chytrids occur principally in fresh water, although they also occur in marine waters, moist terrestrial environments such as bogs, and even dry, hot soils. Fossil chytrids have been recovered from 50-million-year-old shale (Bradley, 1967). Chytrids apparently live most often as saprophytes, some may live either as saprophytes or parasites, and some seem to live only as parasites. They occur on or in algae, other chytrids or aquatic fungi, microscopic animals such as rotifers or mosquito larvae, eggs of microscopic animals, cast insect exoskeletons, roots of higher plants, and plant debris. One can recover chytrids by collecting and examining the above materials or by baiting water samples with bits of hair, snakeskin, cellophane, filter paper, or boiled pollen or plant stems. Chytrids which were parasitic within their algal hosts were often mistakenly thought by early biologists to represent part of the algal life cycle. If environmental conditions are favorable in lakes, parasitic chytrids may multiply so quickly that natural populations of their host are infected in epidemic proportions, which results in the decline of the populations. Roots of terrestrial plants may be parasitized by chytrids. *Olpidium brassicae* may damage grass roots, and *Synchytrium endobioticum* causes the black wart disease of potato tubers.

Thallus. The morphology of the chytrids is extremely varied, and many distinctive features may occur in various combinations (Figs. 2–8 to 2–11). In some species, the thallus may be elongated or ellipsoidal and without *rhizoids*, while in others it may have rhizoids. Rhizoids are determinant, tapering, thread-like extensions of the thallus. They may arise from a broad globose or turbinate part of the thallus, the *apophysis*. Also, the rhizoids may be unbranched, slightly branched, or extensively branched. They absorb nutrients from the food source. The thallus may occur entirely within the cell of the living or dead substratum in which case it is termed *endobiotic*. In other species, the thallus is partly within

Figure 2–8 Chytrid, *Chytridium lagenaria*, growing as a saprophyte on a dead algal cell. × 2000. [Courtesy D. J. S. Barr.]

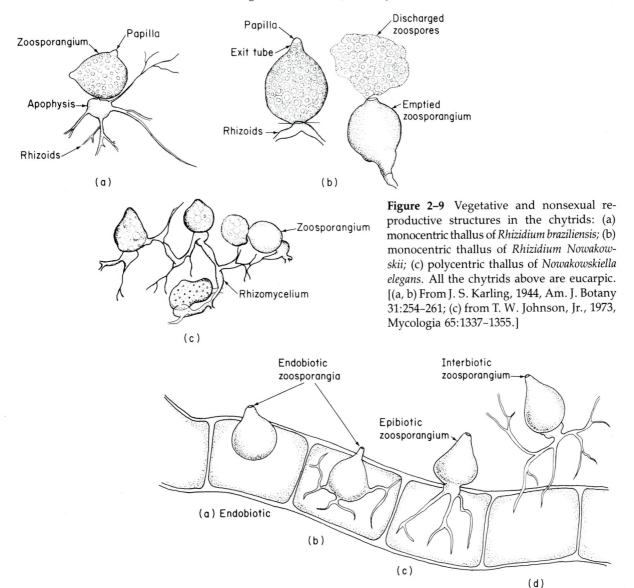

Figure 2–9 Vegetative and nonsexual reproductive structures in the chytrids: (a) monocentric thallus of *Rhizidium braziliensis*; (b) monocentric thallus of *Rhizidium Nowakowskii*; (c) polycentric thallus of *Nowakowskiella elegans*. All the chytrids above are eucarpic. [(a, b) From J. S. Karling, 1944, Am. J. Botany 31:254–261; (c) from T. W. Johnson, Jr., 1973, Mycologia 65:1337–1355.]

Figure 2–10 Thallus types in the chytrids. All are monocentric types [a polycentric type is illustrated in Fig. 2–9(c)]. (a) Holocarpic thallus showing complete conversion into zoosporangium; (b–d) eucarpic thalli showing zoosporangia with rhizoids.

Figure 2–11 Developmental types in the chytrids: (a) zoosporangium of *Olpidium allomycetos* in algal host cell; (b) from left to right, successive developmental stages of *Entophlyctis confervae-glomeratae* f. *marina*; (c) successive stages in development from left to right of *Chytridium lagenaria* on its algal host; (d) development of *Rhizidium braziliensis*. Successive stages in the germination of the zoospore are at the left. A zoosporangium with two exit tubes is free in the water and is attached to rhizoids that have penetrated the substratum wall (indicated by a horizontal line) in the center. Right, a resting spore has germinated by producing a zoosporangium. [(a) From J. S. Karling, 1948, Am. J. Botany 35:503–510; (b) from Y. Kobayashi and M. Ôkubo, 1954, Bull. Natl. Sci. Museum (Tokyo) 1:62–71; (c) from J. S. Karling, 1936, Am. J. Botany 23:619–627; (d) from J. S. Karling, 1944, Am. J. Botany 31:254–261.]

and partly outside the substratum, and that portion outside the substratum is *epibiotic*. In still other species, only a few rhizoids actually penetrate the substratum and the reproductive structures are not in contact with the substratum (in this case the thallus is *interbiotic*). The vegetative thallus is coenocytic and usually appears lustrous because it contains numerous vacuoles and refractive globules (probably lipid) (Sparrow, 1960). When reproduction is to take place, the thallus of some endobiotic chytrids will become entirely converted into a single reproductive structure (*holocarpic* type). A holocarpic thallus lacks rhizoids. Other chytrids have a thallus that only partially converts into reproductive structures (*eucarpic* type); these are the endobiotic species or species with epibiotic reproductive structures and rhizoids. In at least some chytrids, it is known that the portion of the thallus that will become a reproductive structure enlarges, is delimited from the vegetative part of the thallus by a septum, and continues to differentiate. Most of the chytrids form only a single reproductive structure on a thallus (are *monocentric*), while some produce several at different locations on a thallus (are *polycentric*). A polycentric thallus typically has a more extensively branched system of rhizoids than does a monocentric thallus (Figs. 2–9 and 2–10).

The chytrids may be divided into four basic groups on the basis of their development (Sparrow, 1960). Each developmental type is designated by the name of a genus in which it is typical, although it may occur in members of other genera.

1. *Olpidium* type. A zoospore withdraws its flagellum, encysts on the substratum surface, and a penetration tube enters the cell. The cytoplasm of the zoospore passes through the penetration tube and is deposited within the substratum cell. A wall forms around the cytoplasm, which has become a spherical or ellipsoidal thallus without rhizoids. At maturity, the entire thallus converts into a single reproductive structure (endobiotic and holocarpic) [Fig. 2–11(a)].

2. *Entophylctis* type. The zoospore encysts on the substratum surface, producing a penetration tube that develops into a branching rhizoidal system within the substratum. Usually, the zoospore and penetration tube become emptied of cytoplasm and disintegrate. Limited portions of the thallus lying just beneath the substratum wall expand to form the rudiments of the reproductive structures, which then continue to expand and differentiate. Either one or several reproductive structures form on a single thallus, and the rhizoidal system remains distinctly vegetative throughout. The cytoplasm from the rhizoids travels back and drains into the reproductive structures, which are then separated by septa from the empty rhizoids. These chytrids are therefore endobiotic, eucarpic, and either mono- or polycentric [Fig. 2–11(b)].

3. *Chytridium* type. An encysted zoospore on the substratum surface produces a penetration tube that enters the substratum. The penetration tube develops into rhizoids or remains as an unbranched peg and meanwhile utilizes some of the cytoplasm from the zoospore. After the rhizoids are established, a segment of the thallus immediately below the cyst may become enlarged as the apophysis (the apophysis is formed by some species only). Cytoplasmic contents from the rhizoids and apophysis then flow back into the cyst, which enlarges as the cytoplasm from the vegetative system is concentrated within it. The cyst is then separated from the emptied vegetative thallus by a septum and subsequently differentiates into a reproductive structure. This chytrid type is eucarpic and the reproductive structures are epibiotic. [Fig. 2–11(c)].

4. *Rhizidium* type. Chytrids of this type are unique because the encysted

zoospore does not necessarily rest directly on the substratum but germinates in the water and produces one or more rhizoids. These rhizoids eventually penetrate the substratum. The rhizoids become established, develop further, and convey materials back to the cyst. The cyst enlarges and becomes the reproductive structure. This type is eucarpic and interbiotic, as reproductive structures are not directly situated on the substratum surface [Fig. 2–11(d)].

Nonsexual Reproduction. The reproductive structures mentioned above are almost always zoosporangia, which are nonsexual. Zoosporangia occur in a variety of shapes, such as globose, ellipsoidal, tubular, and pyriform. Crowding of zoosporangia within the substratum may alter zoosporangial shape (Roane, 1973). In nature, the size of zoosporangia of a given species may vary, possibly because of competition for available nutrients, space available, or nature of the substratum (Sparrow, 1960). In his study of *Endochytrium operculatum*, Karling (1937) noted that size and shape of the zoosporangia, opercula, and resting spores vary on different host species. Studies of chytrids in culture under carefully controlled conditions show that zoosporangial size can be modified by pH, temperature, vitamins, and nutrients provided (Miller, 1976).

Zoospores are cleaved from the cytoplasm within the zoosporangium. The size of the zoosporangium can influence the number of zoospores formed within it. In *Endochytrium operculatum*, the number of zoospores in a single zoosporangium ranges from fewer than 20 in the smallest to an estimate of thousands in the larger (Karling, 1937). Once the zoospores have matured, they may be discharged immediately or may be retained within the zoosporangium for a relatively long period of time. Sometimes an abrupt change in surrounding conditions, such as temperature fluctuation or transfer to clean water with a higher level of oxygen, will stimulate zoospore discharge (Sparrow, 1960).

In a few chytrids (e.g., some *Rhizophydium* species), the entire zoosporangial wall may deliquesce or split to release the zoospores. In most chytrids, the zoospores are released through one or more *discharge papillae*, each a lens- or nipple-shaped protrusion. The discharge papillae may occur directly on the surface of the zoosporangium or at the end of an elongated exit tube. A hyaline, refractive, apparently gelatinous mass occurs inside the apical wall of the papilla. This gelatinous mass remains distinct from the remainder of the cytoplasm

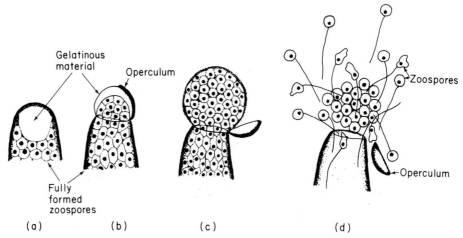

Figure 2–12 Exooperculation: (a) fully developed discharge papilla; (b) dehiscence of the discharge papilla; (c) a slightly later stage showing the envelope of gelatinous substance surrounding the passive zoospores; (d) zoospores swimming away, with detached operculum nearby. [From I. J. Dogma, Jr., 1973, Nova Hedwigia 24:393–411.]

Chap. 2 / Division Mastigomycota

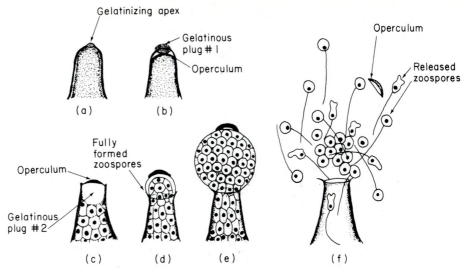

Figure 2–13 Endooperculation: (a) papilla with gelatinizing apex; (b) first gelatinous plug in the pore; and the operculum having formed within the papilla; (c) after the disappearance of the first gelatinous plug, and appearance of the second gelatinous plug; (d–f) successive stages in zoospore release. [From I. J. Dogma, Jr., 1973, Nova Hedwigia 24:393–411.]

where the zoospores are formed (Figs. 2–12 to 2–14). Eventually an opening is formed in the wall of the discharge papilla. The mechanism for forming the opening varies, and three major types are recognized. In the *operculate* type, a lidlike flap of wall material (the *operculum*) is formed and is thrown aside to leave an opening through which the zoospores escape. In turn, the operculate type may be subdivided into the *exooperculate* and *endooperculate* types, depending on whether the operculum is formed on the surface of the papilla or within the exit tube. The *inoperculate* zoosporangia lack an operculum and open by a simple pore. Dogma (1973) has described development of discharge papillae and has distinguished four types. These include exooperculation, two types of endooperculation, and inoperculation. We consider three of these further.

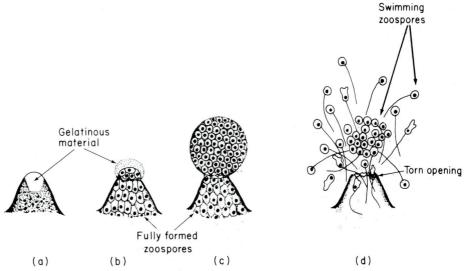

Figure 2–14 Inoperculation: (a) discharge papilla; (b–d) successive stages in zoospore discharge. A torn opening remains behind in (d). [From I. J. Dogma, Jr., 1973, Nova Hedwigia 24:393–411.]

Classification

Exooperculation. In this type of mechanism (Fig. 2–12), the wall at the tip of the discharge papilla develops a line of weakness which delimits a circular cap, the operculum. This type of operculum has been called an *exooperculum,* and has been considered by some mycologists to be the "true" operculum. When the operculum dehisces, the gelatinous mass is extruded outward, forming an envelope around the zoospores which are passively released [Fig. 2–12(b,c)]. The zoospores rest quietly at the mouth of the papilla for a brief period, and then begin to dart away. The lidlike operculum may simply be bent back, remaining attached to one edge of the papilla as though it were on hinges, or may be pushed off by the emerging mass of zoospores.

Endooperculation. In endooperculation (Fig. 2–13), the operculum is formed within the papilla and is sometimes called an *endooperculum.* The endooperculum may be formed either near the tip of the papilla or near its base. In the type in which the operculum is to be formed within the exit tube near the tip, the apex of the papilla inflates slightly and the wall dissolves [Fig.2–13(a)]. The resulting pore is then plugged with a gelatinous mass. Beneath the plug, the cytoplasm becomes modified to form the operculum, which is continuous with the wall of the papilla [Fig. 2–13(b)]. A second gelatinous plug is formed inside the operculum [Fig. 2–13(c)]. Dehiscence of the operculum occurs, and the gelatinous plug is exuded and surrounds the zoospores, which are passively released. The operculum is carried away by the discharged mass of zoospores [Fig. 2–13(f)].

Inoperculation. In this mechanism (Fig. 2–14), the wall at the tip of the discharge papilla becomes thin, and eventually the tip ruptures to form a pore. The hyaline, gelatinous mass is expelled [Fig. 2–14(b)], expands, and surrounds the zoospores, which are passively ejected [Fig. 2–14(c)]. After a brief period of inactivity, the zoospores swim away [Fig. 2–14(d)]. After the zoospores have left the zoosporangium, the pore has jagged, irregular edges.

Once the zoospores are released, they swim with a hopping or jerking motion, sometimes abruptly pausing and then becoming amoeboid in shape. Each zoospore is capable of producing a thallus if it lands on a suitable substratum.

Sexual Reproduction. Sexual reproduction has been found in comparatively few chytrids and is probably absent in many chytrid species. When it occurs, plasmogamy has been reported to involve fusion of gametes, gametangia, or rhizoids of thalli. Morphological variations are numerous, and sexual structures in a given chytrid species may be unique for that species as far as is known. Fusion of motile isogametes is illustrated in the life cycle of *Synchytrium brownii* in Fig. 2–17, and gametangial fusion occurring in *Polyphagus euglenae* is illustrated in Fig. 2–18. The gametangia in *Polyphagus euglenae* are not necessarily typical for chytrids. For example, encysted zoospores may function as gametangia in species of *Rhizophlyctis* and allied genera (Dogma, 1974); furthermore in *Dangeardia mammillata* the male gametangium strongly resembles an immature zoosporangium while the female gametangium resembles a mature zoosporangium (Canter, 1946). Fusion of rhizoids as the sexual mechanism is exemplified by *Chytriomyces hyalinus* (Fig. 2–15). In *C. hyalinus,* the rhizoids of two thalli approach each other and anastomose. At the point of fusion, a zygote is initiated and the contents of the contributing thalli flow through the rhizoids into the developing zygote. The zygote enlarges rapidly and at first has two nuclei. The two nuclei move to the central position in the cell and undergo karyogamy (Moore and Miller, 1973; Miller, 1977). Some chytrid species may be heterothal-

(a)

(b)

(c)

Figure 2–15 Sexuality in the chytrid *Chytriomyces hyalinus:* (a) germinated zoospores, × 2750; (b) fusion between the rhizoids of two thalli; a zygote that will also function as a resting spore is being initiated at the point of anastomosis (arrow), × 3000; (c) the developing resting body is enlarged (arrow) while the two contributing thalli have become empty, × 3600. [From C. E. Miller, 1977, Bull. Soc. Botan. France 124:281–289.]

lic, as gametes of two genetically different strains are required for fusion to take place (Mullins, 1961).

The Resting Spore. After the zygote is formed, it is converted into a *resting spore*. Resting spores may also be formed from zoosporangia under unfavorable conditions. The resting spore is characterized by its guttulate contents and thick wall, which may be smooth or ornamented with spines, blunt knobs, lobes, or undulations. It may be colorless or may range from light to dark shades of yellow or brown. Upon germination, the resting spore cracks open or forms a pore and its contents emerge. In some chytrids, a thin-walled zoosporangium is formed which protrudes through the opening in the resting spore wall (Fig. 2–16). Subsequently the protoplasm is emptied from the resting spore into the zoosporangium in which the zoospores are formed. In a few chytrids the zoospores are formed within the resting spore (which is therefore functioning as a zoosporangium) and are then released. Each zoospore may give rise to a new thallus.

Figure 2–16 Successive stages in germination of the smooth resting spore of *Nowakowskiella macrospora*, showing emergence of a zoosporangium (center) followed by cleavage of zoospores. [From J. S. Karling, 1954, Am. J. Botany 32:29–35.]

Representative Chytrids. We will discuss the life cycles of two chytrids, *Synchytrium brownii* and *Polyphagus euglenae*. Both these chytrids have a well-established sexual cycle.

Synchytrium. There are about 150 species in this genus. These holocarpic chytrids are parasites of algae, mosses, ferns, and flowering plants (Karling, 1964). These fungi live within the plant cells, typically causing gall formation on the diseased parts. Galls are enlarged structures consisting primarily of host cells that have been stimulated to enlarge or divide abnormally or both. The black wart disease of potato is an example of these diseases and is caused by *Synchytrium endobioticum*. In this disease pulpy irregular protuberances are formed on the underground parts of the plant. These galls blacken with age owing to the maturation of the resting spores of the parasites that they contain.

We will examine the life cycle of *Synchytrium brownii*, which occurs on the leaves of *Oenothera laciniata*. Although this particular *Synchytrium* is of little or no economic importance, its life cycle has been carefully studied and is well known (Lingappa, 1958-a,b,c). Lingappa (1958-a) believes that *S. brownii* may be the same organism that Kusano (1930) described as *S. fulgens* and in which he discovered a sexual cycle.

Synchytrium brownii. *S. brownii* has nonsexual and sexual cycles that may occur concurrently, perhaps even in the same host cell.

The nonsexual cycle begins when a zoospore settles down on the surface of the plant and becomes firmly attached (Fig. 2–17). The zoospores then produce a

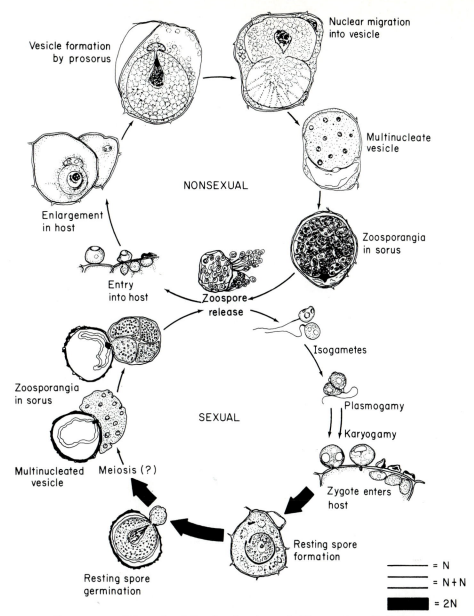

Figure 2–17 Life cycle of *Synchytrium brownii*. [Figures from B. T. Lin-gappa, 1958, Am. J. Botany 45:116–123, 613–620.]

narrow protoplasmic extension, which enters the epidermal cell, and the proto-plasm from the zoospore flows into the plant, leaving an emptied membrane on the outside. The thallus, now within a host cell, develops into a *prosorus*. During this development the thallus becomes located near the base of the host cell, enlarges, and undergoes cytological changes including vacuole formation. De-velopment can be completed in as little as 6 days at 30°C but requires twice as long at 20°C. The mature prosori are usually spherical or subspherical in shape and do not completely fill the host cell. The prosorus will then give rise to a *sorus*, a cluster of zoosporangia. In this process, the cytoplasm of the prosorus flows slowly out through a papilla into a vesicle. Vesicle formation requires about 6 hours. The single nucleus from the prosorus elongates and passes through the papilla into the vesicle, which will become the sorus. Successive nuclear divisions take place until eventually more than 100 nuclei are formed. After these nuclear divisions have started, a plug forms between the emptied prosorus and the developing sorus, perhaps to prevent the protoplasm from flowing back into the prosorus. Cleavage of the protoplasm into approximately 40 to 60 multinucleate segments takes place, beginning at the periphery of the sorus and progressing inward. Each segment acquires a hyaline wall and matures into a zoosporangium. Maturation of the zoosporangium requires about 8 hours. These zoosporangia may be forcibly discharged from the galls and may be scattered on the leaf surface in nature. There is some question about whether these exposed zoosporangia germinate. However, zoosporangia under appropriate conditions will cleave internally, and zoospores are released about an hour after zoosporogenesis begins. The zoospores are released and are then able to reinfect the host and produce another nonsexual generation involving the formation of prosori, sori, and zoosporangia. A complete developmental cycle takes about 7 days at 30°C, but requires a longer time at lower temperatures.

The sexual cycle is initiated when some of the zoospores become gametes. The gametes are morphologically similar but differ slightly in behavior. One gamete becomes sedentary and absorbs its flagellum; then a second, more active flagellate gamete comes into contact with it. Plasmogamy occurs about 1 to 3 hours after pairing, and karyogamy occurs within another half hour. The zygote has the flagellum of the active gamete for a while but is nonmotile. The zygote penetrates the plant and forms an intercellular thallus in the same manner as described above for infection initiated by a zoospore. The thallus enlarges and develops into a resting spore. Maturation of the resting spore requires about 12 days at 27°C and involves the development of a thick, orange outer wall and oily contents. The contents of the host cell disintegrate and may appear as residue on the resting spore. The resting spores must go through a dormancy period before germinating. Upon germination, the resting spores function as the prosori. Each forms an exit tube that penetrates through the host wall and produces a sorus in essentially the manner described above. The single diploid nucleus in the resting spore migrates into the developing sorus. This nucleus undergoes a division, presumably meiotic, and several mitotic nuclear divisions follow. Cleavage into multinucleate zoosporangia, zoosporogenesis, and release of zoospores follow.

Zoospores derived from the nonsexual or sexual cycle are indistinguishable from each other. Presumably any of these zoospores can function as gametes or can initiate a nonsexual cycle in *S. brownii*.

Polyphagus euglenae. *Polyphagus euglenae* parasitizes cells of species of *Euglena* that have encysted and are nonmotile. Like motile cells of *Euglena*, the zoospores of *P. euglenae* are positively phototactic, swimming toward a lighted area. The phototactic response of *P. euglenae* is advantageous, as the zoospores move into regions where the host cells have also accumulated.

This species follows the *Rhizidium* type of development. The zoospore germinates in the water and forms a profuse, interbiotic thallus. Only a few of the rhizoidal tips penetrate an individual *Euglena* cell, and a single thallus may be in contact with as many as 50 *Euglena* cells. After the *Euglena* cell is infected, the chlorophyll changes in color from green to yellow and eventually the protoplasm disintegrates. The encysted zoospore functions as a *prosporangium* from which the zoosporangium develops. The prosporangia lie free in the water and are usually fusiform in shape. The zoosporangium forms at the side of the prosporangium, is usually elongated and tubular in form, and is inoperculate. From a single zoospore to hundreds are cleaved out.

Sexual reproduction is initiated when the ends of the thalli become differentiated into elongated gametangia (Fig. 2–18). Antheridia and the female gametangia are formed on different thalli. The gametangia may be tubular or irregular in form, or alternatively the antheridium may be spherical or club-shaped, while the female gametangium is saclike. Cytoplasm from the thallus accumu-

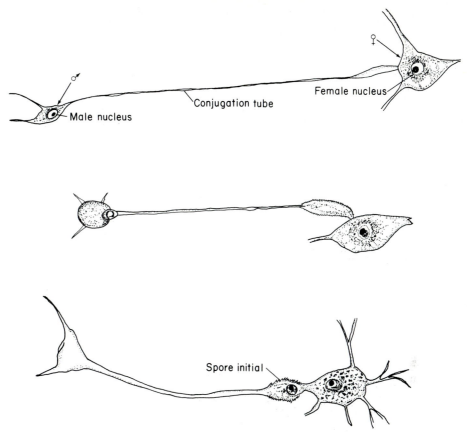

Figure 2–18 Progressive sexual stages in *Polyphagus euglenae*. Top: the antheridium is on the left and the female gametangium is on the right; they are separated by a very long conjugation tube, the tip of which is just beginning to swell to form the resting spore; middle: the male nucleus has begun its migration from the antheridium into the conjugation tube; bottom: the male nucleus has reached the resting spore initial, and the female nucleus is ready to begin its migration. [From H. Wager, 1913, Ann. Botany (London) 27:173–202.]

lates in the gametangium. An elongated conjugation tube is formed from the antheridium, which then makes contact with the female gametangium. Each gametangium contains a single nucleus that functions as the gamete. The nucleus from the antheridium and a second larger nucleus from the female gametangium pass into the conjugation tube. These nuclei do not fuse but lie near each other in the distal portion of the conjugation tube, which will undergo transformation to form a resting spore (Fig. 2–19). The male nucleus enlarges until it is the same size as the female nucleus and the nuclei move apart to opposite sides of the cell, which now becomes covered with a thin inner wall and a thick spiny outer wall and becomes the resting spore. When the resting spore germinates, the thick outer wall cracks and the inner wall emerges through the crack and becomes a zoosporangium. The two nuclei move out into the zoosporangium and undergo karyogamy to form a diploid nucleus, the zygote. Meiosis is presumed to take place immediately. The resulting nuclei, now haploid, multiply to form a multinucleate zoosporangium. Zoospores are formed within the zoosporangium and released. It has been suggested that the zoospores ultimately formed in the zoosporangium will produce male and female thalli in equal numbers.

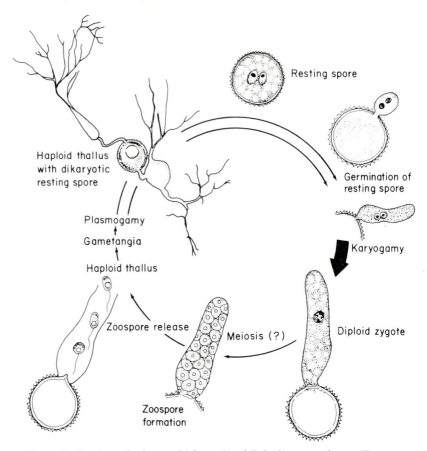

Figure 2–19 Sexual phase of life cycle of *Polyphagus euglenae*. [Figure of thallus from A. F. Bartsch, 1945, Mycologia 37:553–570; remaining figures adapted from H. Wager, 1913, Ann. Botany (London) 27:173–202.]

ORDER BLASTOCLADIALES

Most members of this order are saprophytes in the soil or in aquatic environments and have a relatively well-developed thallus (Fig. 2–20). Thallus morphology ranges from naked, endobiotic thalli without rhizoids in *Coelomomyces*, a parasite of arthropods, to mycelial, branching thalli in *Allomyces*. Most of the thalli have well-developed tapering and branching rhizoids, much like those formed by the chytrids. The rhizoids not only anchor the thallus but also gather food from the substratum.

Those characteristics unifying the members of this group are (1) production of zoospores with a single posterior whiplash flagellum and a prominent nuclear cap (Fig. 2–21), and (2) formation of resting spores in either nonsexual or sexual cycles. Resting spores are ovoid, spherical, or pyriform. In *Allomyces neo-moniliformis*, Skucas (1967) demonstrated that the thick outer wall layer contains melanin and lipid in addition to chitin and other components that normally occur in the hyphal walls. Unlike the other portions of the thallus, the resting spores can withstand unfavorable conditions such as freezing or drying. When the resting spore germinates, the outer wall cracks open and the zoospores escape.

Zoospores are also formed in colorless, thin-walled zoosporangia. In several species of this order, but apparently not in other orders, a refractive, pulley-shaped mass of pectic material plugs the discharge papilla in the thin-walled zoosporangia. When the zoospores are about to be discharged, first the lower portion and then the upper portion of the plug are digested (probably by enzy-

(a)

(b)   (c)

Figure 2–20 Representative members of the Blastocladiales: (a) rhizoids and resting sporangia of *Catenaria allomycis* in host, *Allomyces*; (b) thallus of *Blastocladiella cystogena*; (c) thallus of *Blastocladia pringsheimii* with resting sporangia. [(a) From J. N. Couch, 1945, Mycologia 37:163–193; (b) from J. N. Couch and A. J. Whiffen, 1942, Am. J. Botany 29:582–591; (c) from R. Thaxter, 1896, Botan. Gaz. 21:45–52.]

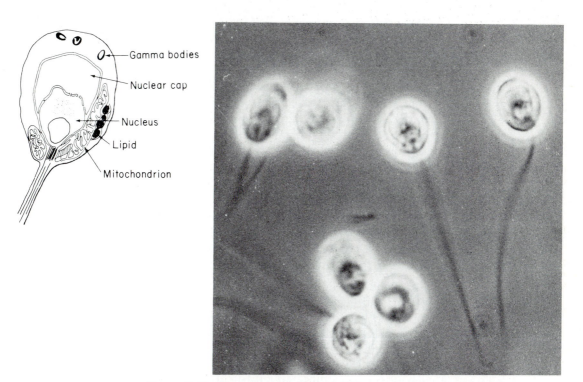

Gamma bodies

Nuclear cap

Nucleus

Lipid

Mitochondrion

Figure 2–21 Zoospores of *Blastocladiella emersonii*, a member of the Blastocladiales. Note the gamma bodies and large nuclear cap (the nuclear caps are visible in the photograph). [Diagram adapted from E. C. Cantino, J. S. Lovett, L. V. Leak, and J. Lythgoe, 1963, J. Gen. Microbiol. 31:393–404; photograph courtesy L. C. Truesdell and E. C. Cantino.]

matic action). The upper portion is digested from the inside, while the remaining outer shell expands to form a vesicle. The vesicle bursts and leaves an opening through which the zoospores can escape. This entire process takes about 30 seconds (Skucas, 1966). Typically, the zoospores are spherical when they escape from the zoosporangium but become ovoid as they swim. Sometimes the zoospores produce pseudopods and move in an amoeboid fashion. The nuclear cap is visible in the swimming zoospores as a turbinate or pyramidal structure which gleams faintly. A side body is also present in some zoospores and is likewise visible as a faintly gleaming structure.

Sexual reproduction is not known in all members of the Blastocladiales. When nonsexual reproduction takes place, it occurs through the union of motile isogametes or anisogametes. Alternation of morphologically similar gametophytic (haploid) and sporophytic (diploid) thalli occur in some life cycles (e.g., in species of *Coelomomyces* and *Allomyces*).

Coelomomyces. Species of *Coelomomyces*, representing the family Coelomomycetaceae, are obligate parasites within the body cavity of arthropods. There they form a coenocytic mycelium that is either irregularly or dichotomously branched and lacks a cell wall. The thallus is holocarpic and also lacks septa that separate rhizoids. At least some species that have been studied in detail are unique, as they are the only fungi known that require two arthropod host species to complete their life cycle. The gametophytic thallus occurs in a copepod, while the sporophytic thallus occurs in a mosquito (Whisler, 1979).

Infection of the hosts in the life cycle was first established in *Coelomomyces psorophorae* (Whisler, 1979). Let us begin with a haploid zoospore which is swimming about in the water. The zoospore invades the copepod and establishes the naked, gametophytic thallus. At first, the copepods swim actively and do not seem to be affected by the fungal invasion. They then slow down abruptly, fall to the bottom of the water, and die. Within a few minutes, the entire body cavity is swarming with the motile, uniflagellate isogametes. Union of isogametes takes place. The zygote, now biflagellate, swims away and its protoplasm must invade a mosquito larva to complete the life cycle. The zygote comes into contact with a mosquito larva and becomes attached to the cuticle with a gluelike substance which is released from cytoplasmic vesicles. The zygote then encysts, forms an appressorium from which a narrow tube extends, and penetrates the cuticle. The protoplasm passes from the cyst into an epidermal cell, and eventually the fungus completes its development in the hemocoel. The sporophytic thallus is morphologically similar to the naked, coenocytic gametophyte. Diploid resting spores are eventually formed. Upon germination, the resting spore undergoes meiosis and cracks along a preformed groove; the inner wall and contents protrude, and the four haploid zoospores formed as a result of meiosis are released. The zoospores are capable of infecting a copepod but are not capable of infecting another mosquito larva.

Blastocladiella. Members of this genus typically are soil inhabitants in warm climates such as that of Texas or Mexico but have been found in Great Britain and even in Iceland. The morphology of members of this genus was described by Couch and Whiffen (1942). The relatively simple thallus consists of a basal cell anchored by branched rhizoids at one end and supporting a single reproductive cell from the other end. The sporophytic thallus (presumably diploid) is observed most frequently. The sporophytic thallus produces a terminal zoosporangium which has a thin wall and is colorless or orange. The zoosporangium discharges its zoospores through one or more pores produced upon deliquescence of the papilla. Zoospores formed from these thin-walled zoosporangia are only capable of forming additional sporophytes upon germination. Resting spores with thick, dark, sculptured walls may be formed by the sporophyte. Upon germination, the resting spore releases zoospores. The behavior of the zoospores varies depending upon the species. In *B. emersonii,* the zoospores simply form additional sporophytic plants. In *B. cystogenes,* a zoospore encysts and upon germination produces four haploid isogametes. Fusion occurs only between isogametes from different cysts. These isogametes fuse in pairs after the release, and the resulting biflagellate zygote germinates to form a sporophytic thallus. In *B. variabilis,* the presumably haploid zoospores settle down and each germinates into a gametophytic thallus. The male thallus bears a single orange gametangium, while the female thallus bears a single colorless gametangium. The gametes are released and fuse in pairs, and the resulting zygote germinates to form a sporophyte.

Because *Blastocladiella emersonii* has been the subject of extensive physiological studies, we have some insight into the manner in which metabolic changes are correlated with morphological differentiation in *B. emersonii.* This topic is discussed further in Chapter 9.

Allomyces. The genus *Allomyces* is of special interest to us because some members of the subgenus *Euallomyces* are among the few fungi known to have alternative generations of sporophytic and gametophytic thalli (Fig. 2–22) and have been the subjects of extensive genetic and cytological studies. The diploid and haploid thalli are morphologically similar, consisting of a stout, trunklike portion anchored to the substratum by rhizoids and bearing slender branches. Branching of the thallus is more or less dichotomous. Pseudosepta with large triangular openings occur at intervals throughout the thallus, which appears to be jointed because it is constricted at each pseudoseptum. Reproductive organs are formed at the tips of the hyphal branches and are separated from the remainder of the thallus by true septa. These reproductive structures may be zoosporangia, gametangia, or resting spores.

The gametophytic thallus bears orange, globular male gametangia (*antheridia*) and colorless, somewhat cylindrical or ovoid female gametangia at the tips of the branches. The antheridia may be immediately above the female gametangia as in *A. macrogynus,* while in *A. arbuscula* the position is reversed. Both the male and female gametes are uniflagellate single cells. The male gametes produced in the antheridium are orange and are smaller and more active than the colorless female gametes produced in the female gametangium. After their release from the gametangia, the gametes pair with the flagellated ends together, and their cytoplasm mingles. The united gametes swim about and then settle down, their nuclei fuse, and the resulting zygote develops into a sporophytic thallus. The first germ tube formed from the zygote develops into rhizoids, while the second germ tube develops into the remainder of the thallus. The sporophytic thallus is similar to the gametophytic thallus except that it produces dark-walled resting spores and colorless zoosporangia. Zoospores are produced in the zoosporangia, and each zoospore may develop into a new sporophytic thallus, thus allowing many generations of nonsexual development to occur. Meiosis takes place in the resting spores when they germinate. The walls, consisting of thin outer and inner layers and a thick middle layer, crack open, a prominent papilla is formed from the inner wall as it pushes through the thick wall, and four haploid zoospores are released. A haploid zoospore may then settle down and give rise to a gametophytic thallus. The female gamete, the diploid zoospore, and the haploid zoospore are morphologically indistinguishable, but differ in their origin and subsequent behavior and development. Gamma particles have been observed in the haploid zoospore but not the diploid zoospore at the ultrastructural level (Olson, 1973).

ORDER HARPOCHYTRIALES

This order has been recently established to include only a few genera, *Harpochytrium* and *Oedogoniomyces* (Emerson and Whisler, 1968). The thallus is typi-

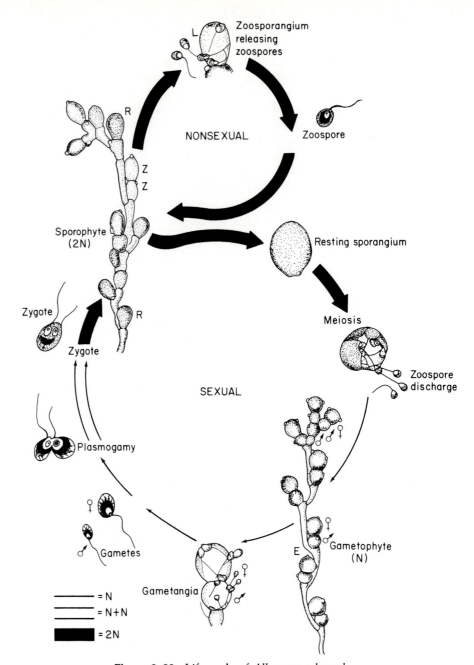

Figure 2–22 Life cycle of *Allomyces arbuscula.*

cally an unbranched single, coenocytic filament, which is usually curved in an arc or spiral. The thallus is attached to the substratum by an acellular, excreted holdfast, which is usually globular, discoid, or cup-shaped but may have other shapes [Fig. 2–23(b)]. Cellular holdfasts or rhizoids, found in some other fungi in this class, are lacking. Members of *Oedogoniomyces* are epiphytic on shells of living snails and on dead or living plant material such as fruits or twigs. *Harpochytrium* species occur as epiphytes on living algae. The substratum is not penetrated by the holdfast, and these fungi apparently obtain their nutrients in the form of a solution from the surrounding water. These fungi are found in fresh water and have been recovered from ponds that undergo periodic drying.

In members of the Order Harpochytriales, parts of the vegetative filamentous thallus function as zoosporangia and the cytoplasm cleaves into spores. In *Harpochytrium,* zoospores are formed only in the upper half of the thallus, which

Figure 2–23 *Oedogoniomyces lymnaeae:* (a) chain of gemmae breaking apart; (b) basal parts of the thalli, showing an especially elaborate hold-fast and germinated zoospores with discoid holdfasts; (c) apical zoosporangium with zoospores; the next zoosporangium is delimited but without zoospores; (d) zoosporangia formed later; the top zoosporangium has been ruptured to release zoospores, and the line of cleavage is forming in the second one at the arrow; (e) the same zoosporangium dehisces to release zoospores; (f) an especially short zoosporangium in the process of dehiscence. [From Y. Kobayashi and M. Ôkubo, 1954, Bull. Natl. Sci. Museum (Tokyo) 1:59–66.]

is now functioning as a zoosporangium. The zoospores are released when an operculum is formed by a circular rupture of the zoosporangial wall just beneath the apex. In contrast, the young vegetative thallus of *Oedogoniomyces* cleaves into separate cells, each functioning as a zoosporangium. An elongated apical cell is first delimited and forms zoospores, and meanwhile a second cell is being delimited by septum formation beneath it [Fig. 2–23(c)]. Zoosporangium delimitation continues in this basipetal direction until approximately one-half the thallus has been converted into zoosporangia. The apical zoosporangium of *Oedogoniomyces* opens by an operculum, as does *Harpochytrium.* All zoosporangia beneath the apex open by a unique mechanism [Fig. 2–23(e)]. A transverse cleavage line is formed around the wall near the upper part of the zoosporangium. This cleavage line dehisces and thereby releases a tubular section of the thallus consisting of segments of walls of two adjacent zoosporangia and the septum that originally divided them. The zoospores escape from the broken zoosporangium, and the zoosporangium immediately beneath it is ready to form zoospores.

Once released, the zoospores swim with a jerky motion. The zoospore swarms, encysts, and germinates. Upon germination, the encysted zoospore produces a single germ tube, which usually remains unbranched and develops into the single vegetative filament. Sexual reproduction is unknown.

No resistant bodies have been detected for *Harpochytrium.* One species of *Oedogoniomyces, O. lymnaeae,* however, produces chains of gemmae which are shorter and thicker-walled than the zoosporangia and are cylindrical cells which appear cuboidal from the side (Fig. 2–23). At maturity, they separate, fall into the mud, and overwinter. A gemma can germinate when conditions are favorable by releasing zoospores through a slit in its wall (Kobayashi and Ôkubo, 1954).

ORDER MONOBLEPHARIDALES

The Order Monoblepharidales is comparatively small, containing only three genera, *Gonapodya, Monoblepharis,* and *Monoblepharella* (Sparrow, 1960,

1973-a). *Monoblepharis* is most frequently recovered from dead, submerged twigs collected from cool, freshwater habitats and incubated in a cool environment. Species of *Gonapodya* are also recovered from twigs but are found most frequently on submerged fruits such as those of the apple or rose. *Monoblepharella* can be recovered from tropical soils by baiting water cultures with plant material. None of these fungi are of economic importance but they have fascinated mycologists because of their unique mode of sexual reproduction. Sexual reproduction is *oogamous*, involving the fertilization of a nonflagellated egg by flagellated, motile male gametes. We will examine *Monoblepharis* as an example of this order as it perhaps is the best-known genus.

Monoblepharis. These fungi form a eucarpic thallus with distinct hyphae. Zoospore germination is bipolar, with one germ tube giving rise to the rhizoids, which function as holdfasts, and the second germ tube giving rise to the remainder of the thallus. The thallus may be branched or unbranched, depending upon the species, and is nonseptate. The thallus typically has a foamy appearance because there are numerous vacuoles and refractive granules regularly distributed through the cytoplasm, producing a characteristic lustrous quality.

Usually nonsexual and sexual reproductive structures are not found on the same thallus, as their formation is controlled by temperature. They are found together only if the thallus has grown within a narrow temperature range allowing formation of both types.

Zoosporangia are delimited by septa from the terminal portions of the thallus. The zoosporangia are usually elongated, cylindrical, and slightly wider than the hyphae from which they were formed. A circular pore is formed by deliquescence at the tip of the zoosporangium, allowing the posteriorly uniflagellate zoospores to escape. The zoospores creep through this pore in an amoeboid fashion, remain temporarily caught in the vicinity of the pore by their tangled flagella, jerk away, and then swim away with a gliding motion. The zoospores have refractive granules that move to the anterior end of the zoospore as it begins to swim. Once a zoosporangium has been emptied of its zoospores, a new zoosporangium may form within the walls of the old one (this process is called *internal proliferation*).

Sexual reproduction involves the formation of antheridia and oogonia. The arrangement of the antheridia and oogonia varies among the species. In some species, including *Monoblepharis insignis* (Fig. 2–24), the antheridium is formed when a terminal portion of the hypha is cut off by a septum. Following this, a section of the hypha beneath the antheridium enlarges to form an intercalary oogonium. At maturity, the small antheridium appears as if it is inserted on the oogonium. In some other species, the oogonia are first formed in a terminal position and this is followed by formation of the antheridium from a hyphal protrusion below. In still other species, antheridia may be in the terminal position on some branches while oogonia are terminal on other branches. Additional gametangia may be produced beneath those already formed so that eventually a chainlike succession of gametangia results. Four to eight sperm are formed within each antheridium and escape through a pore, which opens by deliquescence. The sperm are morphologically similar to the zoospores except that they are somewhat smaller and usually move in an amoeboid fashion.

As the uninucleate oogonium matures, scattered small oil droplets coalesce into large droplets, which move to the center of the oogonium. Each oogonium produces a single, uninucleate oosphere. The oosphere moves towards the apex of the oogonium, and the nucleus becomes located especially close to the apex. When the oosphere is ready for fertilization, the oogonium dilates and opens at its tip, and a mucilaginous material is present that is more or less continuous with the cytoplasm of the oosphere. When the sperm reaches the oogonium, it

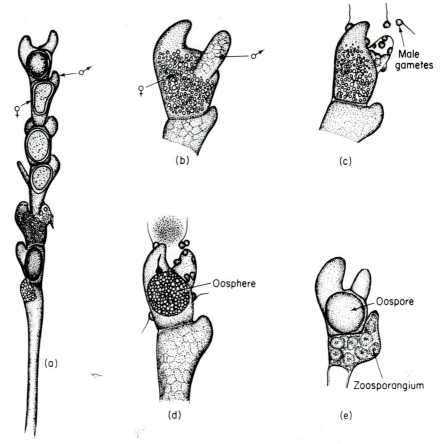

Figure 2–24 *Monoblepharis insignis:* (a) hypha bearing terminal oogonia with antheridia; (b) immature oogonium and antheridium; (c) antheridium has matured and some male gametes have escaped; oogonium is yet immature; (d) male gametes entering oogonium to unite with the differentiated oosphere; (e) oospore within the oogonium. Immediately beneath the oogonium is a zoosporangium in which zoospores are differentiating. [From R. Thaxter, 1895, Botan. Gaz. 20:433–440.]

rests momentarily and is then engulfed by the mucilaginous material; subsequently, both the mucilaginous material and sperm appear to be absorbed by the oosphere. After plasmogamy has been completed, the oosphere may remain within the oogonium in some species, while it is extruded to the outside in others. If the oosphere is extruded, it remains attached to the oogonium by a narrow hyaline collar. A thick wall with regularly spaced protuberances eventually develops around it, converting the oosphere into an *oospore*. The male and female nuclei remain side by side until the thickening of the oosphere wall reaches an advanced stage, and then karyogamy occurs. The mature oospore is uninucleate, and the single nucleus is presumably diploid. Oospores germinate by forming a germ tube (this process has rarely been observed). The timing of meiosis has not been conclusively established.

Class Hyphochytridiomycetes

ORDER HYPHOCHYTRIALES

All fungi with a single tinsel flagellum, which is inserted anteriorly, belong to the Hyphochytridiomycetes. This class contains a single order, the Hyphochytriales. In all major respects other than flagellation, the small number of fungi in this group parallel the members of the Chytridiales in their structure and life cycle. Although nonsexual reproduction is essentially like that in the chytrids, an important difference among members of this group is that in some the zoospores are completely differentiated in the zoosporangium, while in others the undifferentiated protoplasm is extruded through the orifice of the zoosporangium and the differentiation of the zoospores is completed outside the zoosporangium. Sexual reproduction is generally unknown in this group, and resting spores have been found only in representatives of certain genera.

Hyphochytrium catenoides. One of the best known fungi in this group is *Hyphochytrium catenoides*. Karling (1939) has described the development of this fungus, which occurs as a weak parasite and saprophyte in root hairs and parenchyma cells of corn. Within the corn cells, the thallus consists of hyphalike tubular segments connecting intercalary enlargements and oval or spherical zoosporangia. In agar cultures, the intercalary enlargements later mature into zoosporangia. The connecting tubular segments vary in length from approximately 10 to 138 micrometers but are relatively constant in width. The thallus type, therefore, is usually polycentric and eucarpic.

When the zoosporangia mature, they form one to four extremely long exit tubes that may be branched, straight, curved, coiled, or irregular. The tip of the exit tube eventually deliquesces and the protoplasm streams out, taking about 15 to 32 seconds to emerge. The protoplasm is naked and is apparently undifferentiated at first. About 120 seconds after the cytoplasm emerges, cleavage furrows that will eventually delimit zoospores are first evident. The partially delimited cytoplasmic segments begin to move individually and glide over each other. Meanwhile, the flagella are formed and become active, making the entire mass of protoplasm vibrate. Eventually the fully formed zoospores separate from each other and swim away; this occurs about 8 minutes after the entire process started. Although zoospores are most often formed outside the zoosporangium, in unusual cases they may be partially or entirely delimited within the zoosporangium.

The zoospores are oval and somewhat flattened, have numerous conspicuous vacuoles and granules, and have a single anterior flagellum. The flagellum moves rapidly, with the shaft of the flagellum extending more or less straight in front and the tip lashing back and forth. The period of activity usually lasts 20 to 80 minutes.

The zoospores round up and germinate by producing a germ tube that penetrates the host cell and then swells to form an extension [Fig. 2–25(b)], which will enlarge even further [Fig. 2–25(c)] and eventually form hypha-like extensions that will develop into the remainder of the thallus [Fig. 2–25(d)]. The original swelling will eventually become a zoosporangium. As the thallus develops, intercalary enlargements that will become zoosporangia are formed and the cytoplasm from the hyphae accumulates within them, leaving most of the hyphal segment devoid of cytoplasm [Fig. 2–25(d)].

Class Plasmodiophoromycetes

ORDER PLASMODIOPHORALES

Fungi with two unequal whiplash flagella are classified in the Plasmodiophoromycetes. Approximately 35 species are known and are contained within a single family, the Plasmodiophoraceae (Karling, 1968; Waterhouse, 1973-a). These fungi are obligate parasites and most often occur within cells of higher plants, but some also attack algae or aquatic fungi. Usually, these parasites cause marked enlargement of infected parts of the host. This may result simply from abnormal enlargement of the host cells, from stimulated division of the host cells, or from a combination of these events. Two serious diseases caused by

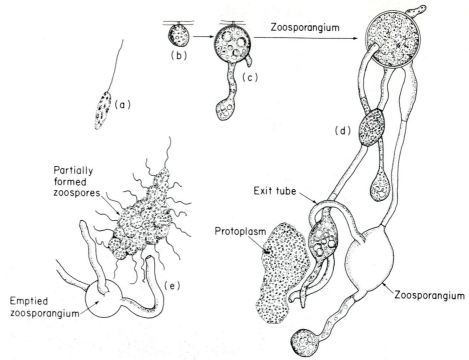

Figure 2–25 *Hyphochytrium catenoides;* (a) zoospore; (b–e) successive developmental stages in the hair of corn; (b) initial invasion of the hair has occurred; (b–c) approximately 5 hours later, the original enlargement of the thallus has formed a hyphalike extension, which is forming an additional enlargement; (d) approximately 32 hours after the original infection, the first enlargement has expanded into a zoosporangium with a short exit tube, and its contents have undergone cleavage internally into zoospores; below, an oval zoosporangium has discharged its undifferentated protoplasm through a large, curving exit tube; (e) zoospores are becoming delimited in a mass of exuded protoplasm, and their flagella are beating. [From J. S. Karling, 1939, Am. J. Botany 26:512–519.]

members of this group are powdery scab of potato (caused by *Spongospora subterranea*), in which scabby lesions form on the tubers and galls form on the roots and stems, and clubroot of crucifers (caused by *Plasmodiophora brassicae*), in which roots become shortened and distorted from the formation of tumorlike galls.

The Plasmodiophorales produce a naked *plasmodium* within the host cells (Fig. 2–26), a characteristic that further distinguishes these fungi. A plasmodium is a multinucleate mass of protoplasm that lies embedded within the host protoplasm, from which it is separated only by its plasmalemma. Once initiated, the plasmodium increases in size by growth that is associated with nuclear divisions. These nuclear divisions have a unique crosslike appearance, leading to the term *cruciform division*. In cruciform divisions, the nucleolus elongates and divides into two parts. While the nucleolus is stretched out into a dumbbell-like shape, it intersects the chromatin lying on the equatorial plane, giving the appearance of a cross. The nuclear envelope persists throughout metaphase and anaphase (Dylewski and Miller, 1983). Similar division figures have been noted in protozoa but not in other fungi. Following the last cruciform division in the plasmodium is the *akaryote* phase, during which the nuclei cannot be stained and therefore seem to disappear temporarily.

Plasmodiophora brassicae. *Plasmodiophora brassicae* is perhaps the most economically important member of this group. It causes the clubroot disease of

Figure 2–26 Amoeboid cells of *Plasmodiophora brassicae* in enlarged cells of cabbage root. [Courtesy Plant Pathology Department, Cornell University.]

cabbage and its relatives and occurs worldwide. Entire crops may be lost in badly infested fields, which may be virtually useless for cultivation of crucifers for several years because of the longevity of the resting spores (which may remain viable for up to 8 years). The roots of diseased plants have galls that may be as large as a human fist. Water transport in these plants is affected, and often the first indication of disturbance is that leaves of the infected plants wilt on hot and dry days. As the disease progresses, the leaves turn yellow and the entire plant appears stunted.

The life cycle of *P. brassica* is typical for members of this group (Fig. 2–27).

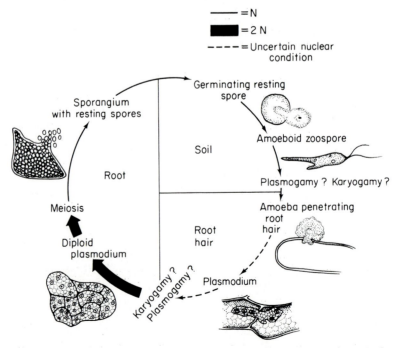

Figure 2–27 Life cycle of *Plasmodiophora brassicae*. [Figures from J. S. Karling, 1942, The Plasmodiophorales.]

It is initiated when a thick-walled haploid resting spore germinates to form a single zoospore. The zoospores usually become intermittently amoeboid and move about in the water or a water film until they come into contact with a host plant. A zoospore comes to rest on the root hair, rounds up and encysts, and then quickly penetrates the root hair. The penetration process is rather intricate—a flattened, adhesive ending of a tubular extension of the zoospore is applied to the surface of the root hair, and then the wall of the root hair is pierced by a specialized bullet-shaped organelle that is ejected through this tubular extension. Subsequently, the cytoplasm from the zoospore passes into the host cell (Aist and Williams, 1971). Once within the root hair, a plasmodium with more than 100 nuclei develops. The plasmodium contains numerous large lipid droplets, mitochondria, Golgi dictyosomes, endoplasmic reticulum, and ribosomes (Williams and McNabola, 1967). Eventually, the plasmodium cleaves into the uninucleate units, the zoosporangia. The nucleus in each zoosporangium divides mitotically, and at maturity about 4 to 10 zoospores are formed. The zoospores are discharged to the outside or remain within the root. Some investigators (e.g., Tommerup and Ingram, 1971) believe that these zoospores function as gametes, undergoing plasmogamy either outside the plant or within the plant. If plasmogamy occurs outside the plant, the zygote would reenter the plant. The exact manner in which plasmogamy occurs and its timing has been a controversial point for the Plasmodiophorales generally, with some investigators suggesting instead that these additional plasmodia originate from zoospores and are therefore haploid. According to this view, karyogamy occurs between nuclei within a haploid plasmodium (Karling, 1968). Nevertheless, the plasmodia are diploid at maturity. Garber and Aist (1979) determined that as growth of the plasmodium ceases, it enters the akaryote stage, during which prophase I of meiosis occurs. The remaining steps of meiosis occur as the resting spores are cleaved out of the plasmodium by coalescence of vacuoles and deposition of the cell wall. The resting spores are released into the soil when the plant decays.

A number of cellular changes occur in the infected host cells (Williams, 1966; Yukawa et al., 1966; Karling, 1968). The invaded host cells divide normally, although they eventually lose their ability to divide but continue to enlarge. The nucleus of the host cell enlarges as DNA synthesis is stimulated by the infection. Starch accumulates temporarily in infected cells. Cell and nuclear enlargement and starch accumulation all reach approximately 10 times normal levels. The nucleoli increase in number.

Class Oomycetes

The Class Oomycetes is a large, heterogenous group of fungi. Morphologically, the thalli range from simple holocarpic, endobiotic forms to eucarpic, freely branching mycelial forms. Habitats are equally diverse. The Oomycetes occur as parasites within or on simple animals such as rotifers, simple plants such as algae, fungi, or higher plants or animals. Many of these fungi live as saprophytes. They occur predominantly in aquatic environments or moist soil, but some also occur on the aerial parts of plants.

The Oomycetes are characterized by (1) *oogamy*, in which nonmotile eggs are formed, and (2) biflagellate zoospores. One flagellum is of the tinsel type and is directed forward while swimming; the second flagellum is of the whiplash type and is directed backward. Two morphological types of zoospores occur, depending on the particular life cycle. These are zoospores with flagella inserted anteriorly, which are typically pip- or pear-shaped when they swim, and zoospores with the flagella inserted laterally (Fig. 2–28). Usually, the laterally flagellate zoospores are shaped like a bean (reniform) or grape seed and have a long groove along the side from which the flagella emerge.

An individual member of the Oomycetes may lack sexual reproduction but

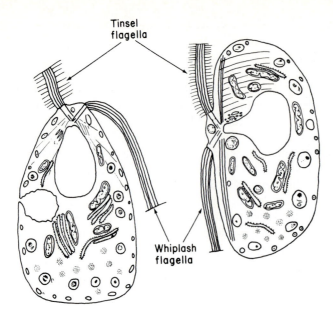

Figure 2–28 Oomycete zoospores. The apically flagellate *primary* type is at the left and the laterally flagellate *secondary* type is at the right. [From S. A. Holloway and I. B. Heath, 1977, Can. J. Botany 55:1328–1339.]

Tinsel
flagella

Whiplash
flagella

be classified in this group because its biflagellate zoospores resemble those of other Oomycetes. Alternatively, some members may lack motile zoospores but have oogamous sexual reproduction, as do others in this class. Unlike the majority of the fungi, several Oomycetes have walls that contain cellulose rather than chitin. Both chitin and cellulose have been found in walls of some of the members (Aronson and Lin, 1978).

Sexual Reproduction.   The female gametangium is the *oogonium,* typically a spherical or pear-shaped organ. The antheridium may occur in various forms, depending on the species, but is most often an elongated organ formed at the tip of hyphal branches. The antheridium may be on the same thallus producing the oogonium or on a different thallus.

One or more *oospheres* in the oogonium will function as female gametes (Fig. 2–29). The oosphere consists of a haploid nucleus surrounded by a small amount of cytoplasm. Depending on the species, the oosphere may be either a distinct membrane-bounded cell that lies free within the oogonium or a naked gamete that lies within differentiated surrounding protoplasm, the *periplasm.* The oospheres may develop without fertilization into thick-walled *oospores.* Alternatively, the oospheres may be fertilized by nuclei from the antheridium and then mature into oospores. During maturation, a reorganization of the protoplasmic contents occurs. This especially involves lipid drops and reserve globules, which are of poorly known composition but apparently contain reserve food. The oospore may fill the oogonium entirely, or there may be a space left between the oospore and the oogonial wall. If periplasm is present, it disappears as the oospore matures. The oospore then functions as a resting spore (Dick, 1969).

Although the above general characteristics apply to the entire Class Oomycetes, we know comparatively little about sexuality in many members of this group. This is especially true of the reduced chytridlike forms. Studies of sexuality have concentrated on the larger, filamentous fungi in the families Leptomitaceae and Saprolegniaceae in the Order Saprolegniales and the Order Peronosporales. Further discussion of sexuality will be based on these better-known fungi.

Until the early 1960s, mycologists generally believed that meiosis occurred in the oospore before germination to form a haploid thallus. In 1962, Sansome

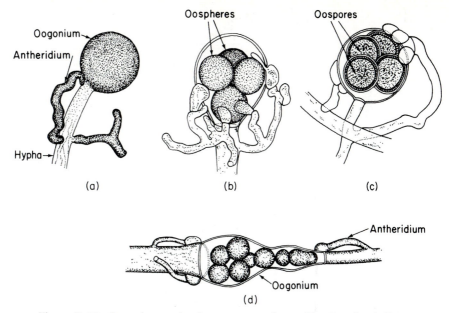

Figure 2–29 Sexual reproductive structures formed by *Saprolegnia litoralis:* (a) young terminal oogonium, × 310; (b) terminal oogonium with oospheres; the antheridia have not yet discharged their contents, × 310; (c) oogonium with mature oospores, × 310; (d) intercalary oogonium, × 170. [From W. C. Coker, 1923, The Saprolegniaceae, University of North Carolina Press, Chapel Hill, N.C.]

and Harris published a paper reporting that they had induced polyploids in three members of this group and had observed multivalent chromosomal configurations typical of meiosis in the gametangia. Therefore, the thallus would be diploid, contrary to the former concept. As in other fungi, the chromosomes are extremely small in the Oomycetes, which generally makes cytological analysis difficult and contributed to the early misinterpretations. Following the publication by Sansome and Harris (1962), other investigators have reexamined numerous members of the two major Oomycete orders, the Saprolegniales and Peronosporales. Cytological investigations showed that meiosis occurred in the gametangia of all those fungi which were critically reexamined (Dick and Tin, 1973). Light-microscopic studies were supported by ultrastructural observations of meiosis in gametangia (e.g., Ellzey, 1974; Howard and Moore, 1970). This interpretation has been supported by microspectrophotometric analysis of changing DNA levels during the life cycle. In microspectrophotometric studies of two members of the Saprolegniales, the DNA content of nuclei just before division in the gametangia was shown to be four times that of the nuclei resulting from division and twice that of nuclei in the vegetative hyphae. These data are consistent with the interpretation that gametangial division is meiotic (Bryant and Howard, 1969; Howard and Bryant, 1971). Genetic data have been obtained which can best be interpreted as involving segregation of characters in a Mendelian fashion at meiosis during gametogenesis, with the resulting production of a diploid thallus (Dick and Tin, 1973). We may conclude that a predominantly diploid life cycle (Fig. 2–30), including a diploid thallus, therefore occurs widely among the Oomycetes, at least within the Saprolegniales and Peronosporales.

Numerous investigators have studied the morphology of the sexual organs and their functions in the Saprolegniaceae (Order Saprolegniales) and Peronosporales (Figs. 2–29 and 2–31). Light-microscopic observations have been

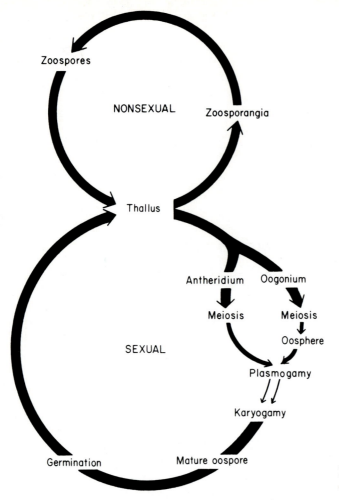

Zoospores

NONSEXUAL   Zoosporangia

Thallus

Antheridium   Oogonium

Meiosis   Meiosis

Oosphere

SEXUAL

Plasmogamy

Karyogamy

Germination   Mature oospore

Figure 2–30 Generalized Oomycete life cycle.

supplemented by recent electron-microscopic observations, and it has been established that these organisms generally share a common pattern of reproduction (Dick, 1969; Dick and Tin, 1973), which is described below.

The oogonium and antheridium originate as multinucleate swellings on a diploid thallus. Each gametangium is eventually separated from the thallus by a septum. The gametangia are packed with nuclei and cellular organelles, including the endoplasmic reticulum, mitochondria, Golgi bodies, and various types of vesicles. These vesicles include some which contain lipid or dense bodies. Meiosis occurs more or less simultaneously in the gametangia. Some of the excess nuclei may abort before, during, or after meiosis.

The antheridium and oogonium come into contact with each other, forming a more or less distinct adhesion zone at the point of contact (Hemmes and Ribeiro, 1977), perhaps at an early stage of development. Microbodies have been observed in the cytoplasm of both gametangia near the point of contact (Howard and Moore, 1970) or in the antheridium but not the oogonium (Hemmes and Ribeiro, 1977). Plasmogamy is initiated when a pore is formed between the two organs or when the antheridium penetrates the oogonial wall with one or more elongate *fertilization tubes*. Howard and Moore (1970) suggested that the microbodies may contain enzyme(s) that are responsible for development of a pore through which the fertilization tube passes. A fertilization tube reaches each egg, and a male nucleus is deposited in the vicinity of the egg to accomplish plasmogamy. The antheridium may be either partially or completely emptied.

The oogonium is meanwhile undergoing internal changes which lead to the formation of one or more eggs. There are two oogonial types. In one type,

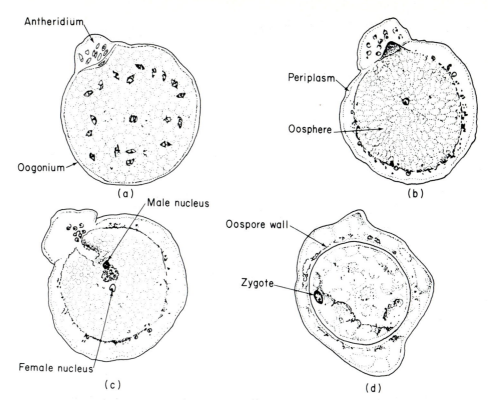

Antheridium

Oogonium

(a)

Periplasm

Oosphere

(b)

Male nucleus

Female nucleus

(c)

Oospore wall

Zygote

(d)

Figure 2–31 Fertilization in an Oomycete; (a) final nuclear division in both the antheridium and oogonium; within the oogonium, the nuclei have started to migrate toward the periphery; (b) the periplasm is delimited, and nuclear disintegration is beginning to occur; (c) the male nucleus is passing into the oosphere; (d) oospore wall is beginning to form, and karyogamy has presumably occurred. [Figures from K. M. Safeulla, M. J. Thirumalachar, and C. G. Shaw, 1963, Mycologia 55:819–823.]

the mature oogonium will have the eggs differentiated as discrete, membrane-bounded oospheres floating within an oogonium which appears to be empty. The entire oosphere will form an oospore. In the second type of oogonium, a naked egg is surrounded by *periplasm*, which will not be incorporated into the oospore. Let us examine ultrastructural observations which exemplify each type.

1. *Saprolegnia* species form numerous free oospheres in oogonia. Within the young oogonium, a large vacuole is formed when numerous vesicles containing dense bodies enlarge and their membranes fuse. The vacuole expands further by taking up water from the surrounding cytoplasm, which simultaneously decreases the volume of the cytoplasm. The vacuole eventually forms furrows that extend through the peripheral cytoplasm and join with the plasmalemma. The oospheres are cleaved out by this action, and the tonoplast of the vacuole becomes the plasmalemma of the oospheres. After plasmogamy, a wall is laid down around each oosphere to form the oospore (Gay et al., 1971; Howard and Moore, 1970).

2. *Pythium* species have oogonia in which a single naked egg lies within the periplasm. During the early stages, the oogonium has continuous cytoplasm, with the organelles more or less evenly distributed. The large central vacuole is absent. The organelles, including the nuclei, move to the peripheral cytoplasm in the oogonium. One of these nuclei will function as the female gamete. Plasmogamy occurs, and the per-

iplasm becomes delimited at approximately the time of plasmogamy; therefore an oosphere is not discernible. The periplasm contains vacuoles and a great deal of endoplasmic reticulum. The oospore wall begins to form at the inner face of the periplasm as soon as plasmogamy occurs, delimiting the oospore from the periplasm. While the oospore wall is forming, the vacuole in the periplasm increases and all other organelles disintegrate. The zone between the oospore wall and oogonium eventually becomes empty, and the disintegrated periplasm may form a thin covering over the oospore and inside the oogonium (Haskins et al., 1976; McKeen, 1975).

After plasmogamy in species of *Pythium* and *Saprolegnia,* the male nucleus lies near the female nucleus, and the developing oospore is binucleate. The timing of karyogamy varies, depending on the particular strain or species. The male and female nuclei may undergo karyogamy soon after plasmogamy or it may be delayed until the oospore has developed its thickened wall or, rarely, until the oospore begins to germinate.

A number of changes occur inside the oosphere as it matures into an oospore. There are several types of vesicles present in the oosphere. Some of these contain electron-dense material (a lipid-phosphoglucan complex), while others are microbodies (presumably glyoxysomes) that are associated with lipid (Beakes, 1980-b). The vesicles containing the electron-dense material coalesce to form a reserve globule. There is partial utilization of the neutral lipid and reorganization, so that at maturity the reserve dense globules and a variety of vesicles lie near the periphery of the oospore. Meanwhile many cytoplasmic organelles are lost.

The oospore undergoes a resting period and then germinates, usually by forming a germ tube terminated by a small functional zoosporangium. In his ultrastructural studies of oospore germination in *Saprolegnia ferax,* Beakes (1980-a) noted that the inner oospore wall becomes thinner while a new wall is secreted around the protoplast. The germ tube forms from the oospore. Germination may occur while the oospore is still in the oogonium, and then the elongating germ tube ruptures the oogonial wall. Meanwhile a large central vacuole has been forming in the oospore as a result of cytoplasmic breakdown (Beakes, 1980-a). The dense reserve globule disintegrates and the reserve glucans and lipids are utilized. Small granules from the disintegrating reserve globule are released into the central vacuole and peripheral vesicles, many of which become associated with mitochondria. Nuclear division occurs and the nuclei migrate into the elongating germ tube. Some of the protoplasmic constituents required for the early growth of the germ tube are probably derived by autolysis of the oospore contents by lysosomes (Beakes, 1980-b).

ORDER SAPROLEGNIALES

The Saprolegniales may be divided into six families (Sparrow, 1976). These fungi share at least some, but not necessarily all, of the following characteristics: (1) they have both apically and laterally biflagellate zoospores or equivalent stages; (2) they form one to several oospores that lie free in the oogonium and lack periplasm; and (3) the oospores usually have one or several reserve globules in the peripheral position (Sparrow, 1976).

The most conspicuous fungi are in the families Saprolegniaceae and Leptomitaceae. These fungi are often called *water molds* and are commonly found as saprophytes on plant and animal debris in water and soil. Although they typically inhabit fresh, unpolluted water, some members of the Leptomitaceae thrive in waters high in organic wastes, such as those polluted by industrial wastes. Although the majority are saprophytes, some species are parasitic on aquatic animals or on the roots of higher plants. Members of *Saprolegnia* are sometimes parasites on fish, causing superficial lesions and possibly damage to the viscera.

The thallus of members of the Saprolegniaceae is a well-developed mycelium consisting of a highly branched mass of hyphae of indeterminate growth. The hyphae lack constrictions and are typically nonseptate except where reproductive structures are delimited. Frequently the hyphae are visible macroscopically as a pustule or dense filamentous mass on the substratum after sufficient growth has occurred (Fig. 2–32). The thallus of the Leptomitaceae is generally similar to that of members of the Saprolegniaceae except that pseudosepta are formed at more or less regular intervals. The pseudosepta are partially plugged by a refractive material, *cellulin*. Because these hyphae are constricted at the sites where pseudosepta occur, the hyphae have a characteristic jointed and segmented appearance (Fig. 2–33). These hyphae are coenocytic with continuous cytoplasm.

Members of the remaining families are simple, chytridlike fungi that may lack sexual reproduction. The best known among these simple fungi are those in the family Ectrogellaceae. Species in this family are endobiotic within algae (especially diatoms) or other fungi. The thallus is unicellular and unbranched or sparingly branched. It may develop to the extent that it pushes the valves of a diatom host apart. At maturity, the holocarpic thallus develops into a zoosporangium. Sexuality in this family is poorly known but may occur (Fig. 2–34).

Sexual Reproduction. There is a great deal of morphological variation among the gametangia and oospores in members of the Saprolegniales. This

Figure 2–32 Colony of *Achlya* growing in water culture. [Courtesy Carolina Biological Supply Company.]

(a)   (b)

Figure 2–33 Gametangia of *Sapromyces androgynus*, a member of the Leptomitaceae: (a) plasmogamy between an antheridium (arrow) and an oogonium; note the constrictions at the base of the antheridium and oogonium; (b) after the completion of plasmogamy, an oospore formed within the oogonium. [Courtesy D. Gotelli and K. Mitchell, *in:* Fuller, M. S., Ed., Lower fungi in the laboratory.]

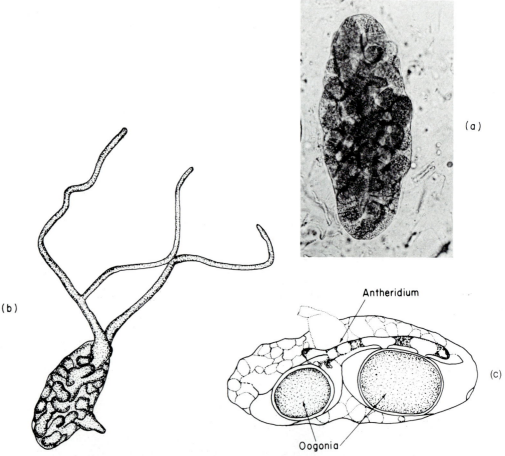

Figure 2–34 *Aphanomycopsis sexualis,* a member of the Ectrogellaceae: (a) holocarpic thallus that converted into a zoosporangium which almost entirely fills a midge egg, × 235; (b) branched discharge tubes arising from the zoosporangium, × 160; (c) antheridium in contact with two oogonia; each oogonium contains an oospore, × 360. [(a) Courtesy W. W. Martin; (b, c) from W. W. Martin, 1975, Mycologia 67:923–933.]

variation is of taxonomic importance and is used in delimiting genera and species (e.g., Dick, 1973-a,b; Johnson, 1956; Scott, 1961). Plasmogamy, karyogamy, and a general life cycle are described on pages 56–60.

The antheridium may simply be a cell beneath the oogonium [Fig. 2–35(a)]. Usually the antheridial apparatus consists of a branched or unbranched filament terminated by an antheridial cell. An antheridium may occur on the same thallus producing its oogonium and may arise from the oogonium itself, the stalk supporting the oogonium [Fig. 2–35(c)], or a vegetative hypha supporting the oogonium [Fig. 2–35(b)]. In some species, the antheridium arises from a different hypha than that supporting the oogonium [Fig. 2–35(d)].

In its simplest form, the oogonium may simply be similar to a hyphal segment. The oogonium may occur in an intercalary position (within the hypha) [Fig. 2–35(e)], on a short lateral branch [Fig. 2–35(b,d)], or terminating the main hypha [Fig. 2–35(f,g)]. The shape of the oogonium may vary from elongate to angular [Fig. 2–35(g)]. Usually, the oogonia are pyriform [Fig. 2–35(b,f)] or spherical [Fig. 2–35(a,c)]. Oogonial walls may be smooth [Fig. 2–35(a–f)] or orna-

Figure 2–35 Morphological variations of sexual structures in the Saprolegniales: (a–d) variations in antheridial position, oogonia are all lateral; (e–f) variations in oogonial position, intercalary and terminal; (a–g) variations in shape of oogonium; (g) highly branched and swollen antheridia; (a, c–e) oosphere stage; (b, f, g) oospore stage; (f) centric oospores; (g) eccentric oospores. [Figures (a–f) from R. L. Seymour, 1970, Nova Hedwigia 19:1–124; figure (g) from T. W. Johnson, Jr., 1974, Am. J. Botany 61:244–252.]

mented [Fig. 2–35(d)]. Although some species produce a single oosphere in an oogonium [Fig. 2–35(d)], most produce more than one, with a range of 2 to 100 or more in a single oogonium [Fig. 2–35(g)].

Oospores have their reserve materials located in various more or less constant locations and occur in two principal arrangements. These are the *centric* oospore, in which one or two layers of small droplets surround the central ooplasm [Fig. 2–35(b,f)], and the *eccentric* oospore, which has one or a few large drops on one side of the ooplasm only [Fig. 2–35(g)].

In the genera *Dictyuchus* and *Achlya*, some thalli will have only antheridia, while others will have only oogonia. Therefore, one thallus is "male" and the second is "female." Because fertilization must occur between two thalli, they are *heterothallic*. Most species of the Saprolegniales are *homothallic*, with fertilization occurring between gametangia on the same thallus. Homothallism and heterothallism are discussed further in Chapter 7. Differentiation of the male and female thalli in heterothallic species of *Achlya* is under complex hormonal control, discussed in Chapter 10.

Nonsexual Reproduction. The zoosporangia are usually elongate or spherical and occur most often in a terminal position on the thallus. The zoosporangium may proliferate internally; that is, when a zoosporangium has been emptied after the discharge of its zoospores, another zoosporangium may form within it.

There is a great deal of variation in the type and behavior of the zoospores. Species may have *monomorphic* zoospores, that is, form only one type of zoospores in their life cycle. Other species may have *dimorphic* (literally meaning "two-forms") *zoospores* and form anteriorly biflagellate zoospores followed later by laterally biflagellate zoospores. Because of their order of appearance, the anteriorly biflagellate zoospores are generally termed *primary* zoospores while the laterally biflagellate zoospores are termed *secondary* zoospores (Fig. 2–28).

We first examine representatives of *Saprolegnia*, which have dimorphic zoospores. The primary zoospore squeezes through the papilla and then swims slowly for about 5 to 10 minutes (Fig. 2–6). The zoospores then withdraw their flagella, round up, and form a cyst. The cyst rests for up to 4 hours or more and then germinates. The secondary zoospore squeezes out through a pore at the mouth of an exit papilla and emerges as a cell with a much larger volume than the cyst. It remains attached to the emptied cyst for about 15 to 30 minutes and then swarms for another 30 minutes. The secondary zoospore darts and glides quickly; it is a more active swimmer and swarms longer than the primary zoospore. It then slows down and jerks, and a beadlike protrusion is formed on the flagella, which then fall off the zoospore within 1 to 2 minutes. Encystment occurs after the flagella are lost (Crump and Branton, 1966). Unlike the primary zoospore, which encysts only once, the secondary zoospore may encyst repeatedly. When a cyst formed by a secondary zoospore germinates, it may release another secondary zoospore or may form a germ tube. At the ultrastructural level, it may be seen that there are a number of changes in the arrangement and morphology of the cellular organelles as the zoospore goes through its cycle involving the alternation of primary zoospore, cyst, secondary zoospore, cyst, and germination (Holloway and Heath, 1977). These changes include elongation of mitochondria in the motile stage and their shortening in the cysts. The cisternae of the rough endoplasmic reticulum are stacked in the primary zoospore but randomly arranged in both types of cysts and in the secondary zoospore. The organelles have a precise arrangement in both the primary and secondary zoospores but appear to be randomly arranged in the cysts.

Unlike species of *Saprolegnia*, which have pronounced swarming of both primary and secondary types of zoospores, other members of the Saprolegniales may have suppression of either one or the other of the stages. Members of the genus *Pythiopsis* have monomorphic zoospores and form only the primary type. These swarm, encyst, and germinate to form a new thallus. In species of *Achlya* and some species of *Ectrogella*, which are dimorphic, the primary zoospores typically emerge from the zoosporangium and immediately encyst at the mouth of the zoosporangium without swarming [Fig. 2–36(a)]. These cysts then germinate to release secondary zoospores that swarm. In species of *Dictyuchus*, there is further suppression and the primary zoospore stage is represented by cysts within the zoosporangium. These cysts remain in the zoosporangium and release secondary zoospores through a papilla that penetrates the zoosporangial wall and in which a pore is formed. The empty cysts, still contained within the zoosporangium, give this cell a netlike pattern [Fig. 2–36(b)]. Motile stages are regularly absent in members of one genus, *Geolegnia*, or absent under unfavorable environmental conditions in other fungi (e.g., *Achlya*). The zoospores are represented by cysts in the zoosporangium, and these cysts are usually released when the wall disintegrates [Fig. 2–36(c)]. Sometimes the cysts will begin to germinate while in the zoosporangium [Fig. 2–36(d)]. The cysts germinate by forming a germ tube.

(a)   (b)   (c)   (d)

Figure 2–36 Nonsexual reproductive structures in the Saprolegniales: (a) *Achlya*-type with encysted primary zoospores at mouth of the zoosporangium; (b) cysts and emptied cyst cases in zoosporangia of *Dictyuchus*; (c) row of cysts remaining behind after disintegration of zoosporangial wall in *Brevilegnia*; (d) cysts germinating before their release from the zoosporangium in *Achlya*. [From W. C. Coker, 1927, J. Elisha Mitchell Sci. Soc. 42:207–226.]

Gemmae may be produced by some members of the Saprolegniales. Gemmae are hyphal segments which have been delimited by septa, are often irregular in form, and have dense cytoplasm. These structures may be separated from the parent mycelium and when conditions are favorable, may produce a new thallus. Gemmae germinate by forming a germ tube or by functioning as a zoosporangium or even as an oogonium.

ORDER LAGENIDIALES

The Order Lagenidiales is a group of simple, holocarpic, and endobiotic fungi which may or may not be closely related (Sparrow, 1960, 1973-c). They are facultative parasites and occur chiefly within algae and other aquatic fungi, although some are parasitic within various forms of animal life such as nematodes, daphne, mosquito larvae, and other insects. These fungi occur principally in freshwater habitats, although there are some marine forms.

The multinucleate thallus may be a single, more or less amoeboid mass as in *Olpidiopsis*, or it may be a tubular or filamentous structure as in *Lagenidium* or *Haliphthoros* (Fig. 2–37). It may be either unbranched or sparingly branched and is nonseptate at first but may later become septate. In unicellular thalli such as those formed by species of *Olpidiopsis* (Fig. 2–38), each thallus is converted into a zoosporangium or gametangium. In the filamentous or tubular thalli, each cell will convert into a zoosporangium or gametangium at maturity.

In some species, the contents of the zoosporangium cleave into biflagellate zoospores before their discharge from the zoosporangium. The zoospores exit through the usually single papilla which leads from the zoosporangium to the exterior. In contrast, others discharge the undifferentiated zoosporangial protoplasm to the exterior, where it continues its differentiation into zoospores. In some species, the outer portion of the undifferentiated protoplasm forms an evanescent, delicate bubble or saclike structure, the *vesicle*. The remainder of the

Figure 2–37 *Lagenidium rabenhorstii*, a member of the Lagenidiales: (a) vegetative thallus in cell of algal host; (b) an isolated zoosporangium has formed a vesicle and zoospores; (c) early stage in union of the gametangia; (d) a thallus has become septate and divided into zoosporangia, an antheridium, and oogonium. An oospore is visible within the oogonium. [From W. Zopf, 1884, Nova Acta Acad. Leopoldina 47:143–236, plates 12–21.]

Figure 2–38 Member of the Lagenidiales, *Olpidiopsis* within a hypha of its host. Note the zoosporangia that are discharging zoospores (left) and the small antheridium that is in contact with the spiny oogonium (right). × 447. [From W. C. Coker, 1923, The Saprolegniaceae, University of North Carolina Press, Chapel Hill, N.C.]

protoplasm within it then continues to differentiate into zoospores, which are released when the vesicle bursts. In other members of this family, a vesicle is not formed around the extruded protoplasm.

The zoospores are somewhat variable in shape, but typically they are of the secondary type and are reniform, with the two flagella inserted in the concave elongate indentation. The zoospore is monomorphic, but it may undergo repeated encystment and emergence.

Sexual reproduction is only known to occur in some members of this group. Typically, a portion of the thallus may either function as or differentiate into gametangia. Martin and Miller (1986-b) obtained ultrastructural evidence that meiosis occurs in the gametangia of *Olpidiopsis varians*, confirming that the thallus of these minute organisms is diploid, but otherwise there are few records of meiosis. The gametangia may be similar in size, or more often, the antheridium may be smaller than the oogonium (Fig. 2–37). Plasmogamy occurs when the contents of the antheridium flow into the oogonium through either a fertilization tube or pore. A thick, often spiny wall forms around the zygote, now an oospore. The oospore usually lies free within the gametangium. In a few species some cytoplasm surrounds the oospore after it matures (this remaining cytoplasm is the periplasm). In some instances, the oospore germinates as a zoosporangium. Parthenogenetic development of the thick-walled oospore is common.

ORDER PERONOSPORALES

In the Order Peronosporales, there are some fungi that inhabit water or damp soil and live as saprophytes or weak parasites (Waterhouse, 1973-b). Many of these fungi, however, have abandoned the aquatic habitat and live on the dry, aerial parts of plants as parasites. Plant parasites in this order include species of *Pythium* that cause damping-off diseases of seedlings, *Peronospora* and other genera that cause downy mildew diseases of various plants, *Albugo*, which causes the white rust disease, and *Phytophthora*. *Phytophthora infestans* causes late blight of potatoes, a disease responsible for the potato famine in Ireland, and which thus changed the course of history in that country (see Chapter 13).

Characteristics delimiting this group of fungi from others are: (1) Only monomorphic zoospores of the secondary type may be formed; (2) periplasm surrounds the egg in the oogonium; and (3) the oospore bears a centrally located reserve globule (Sparrow, 1976). Sexual reproduction usually involves the formation of a single oospore within an oogonium. Morphological and cytological details of sexual reproduction are essentially similar to those detailed on pages 56–60 for the Saprolegniales.

The coenocytic hyphae are more slender than those in the Saprolegniales and are freely branching. Septa may be formed to delimit reproductive organs and in older mycelia. In parasitic species, hyphae may occur principally between the cells (intercellularly) and derive their nourishment through peglike hyphal branches which they send into the host cell. These specialized food-absorbing organs are called *haustoria*. In some members, an intracellular hypha which penetrates the cells may also be produced.

Many aspects of nonsexual reproduction are unique for this order and some species may show an adaptation to the drier terrestrial environment and a requirement for spore dissemination by air rather than water. Most of the species form zoosporangia, which range from an undifferentiated hyphal filament to spherical, oval, or pyriform cells. The zoosporangia are often borne on a specialized hypha, which is morphologically different from the vegetative hyphae and is called a *sporangiophore*. The zoosporangia usually germinate when they form zoospores. Depending on the species, the zoosporangia may (1) release their zoospores directly into the environment, (2) release their mature zoospores into a vesicle, or (3) release apparently undifferentiated cytoplasm into the vesicle, where it then differentiates into zoospores (Fig. 2–39). Alternatively, the zoosporangium may not form zoospores at all but germinate by forming a germ tube. The latter type of germination may be either environmentally or genetically controlled. If the zoosporangium germinates by forming a germ tube, it is functioning as does a *conidium*, a wind-disseminated nonsexual spore. Conidia are common in all groups of the higher fungi, which are discussed in the follow-

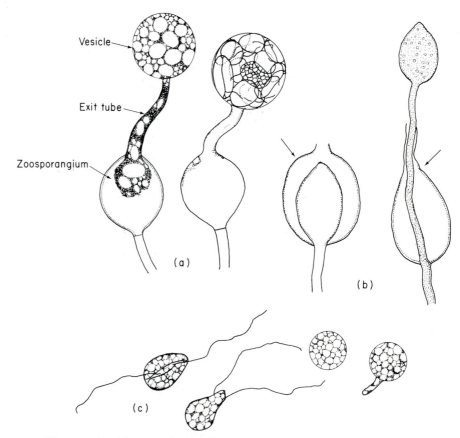

Figure 2–39 Nonsexual reproduction in *Pythium:* (a) zoosporangia with vesicles; a vesicle with undifferentiated protoplasm is on the left, and a later stage with delimited zoospores is on the right; (b) two types of proliferation of zoosporangia; the original, older zoosporangium is at the arrow; (c) from left to right, swarming, cyst, and germination of zoospores. [From J. T. Middleton, 1943, Mem. Torrey Botan. Club 20(1):1–171.]

ing chapters. Differences in nonsexual reproduction are of great importance in delimiting the genera in the Peronosporales. Next we examine some representative genera, particularly noting nonsexual reproduction.

Pythium. *Pythium* is a large genus. Its species are distributed worldwide and include saprophytes in water or soil and parasites on algae, other fungi, or higher plants (Middleton, 1943). The coenocytic thallus consists of an irregularly branching mycelium. In higher plants, the mycelium is most often found growing within the cells (intracellularly) of roots or stems. This invasion results in the rotting of the plant part. One well-known disease of this type is damping off, which attacks the roots of emerging seedlings and may be caused by *Pythium debaryanum* (Fig. 2–40). The zoosporangia are intercalary or terminal in position. In shape, they are usually filamentous and only slightly wider than the mycelium or spherical. The papilla may be sessile or may be borne upon a discharge tube of varying length. Upon germination, a membrane-bounded vesicle is produced at the end of the discharge tube [Fig. 2–39(a)]. The undifferentiated proto-

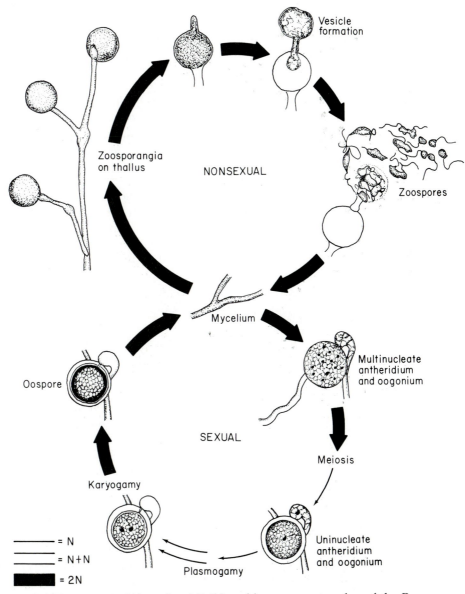

Figure 2–40 Life cycle of *Pythium debaryanum*, a member of the Peronosporales.

plasm from the zoosporangium passes into it. Within about 15 to 20 minutes, the protoplasm differentiates into zoospores. The zoospores move about and are finally released when the vesicle ruptures. The zoospores are of the secondary, laterally biflagellate type and usually encyst only once, but they may undergo repeated swarming and encystment. If the cyst is to release a zoospore, it does so by forming a small vesicle at the mouth of an exit tube, and subsequently a single zoospore emerges from the vesicle. Proliferation of the emptied zoosporangium may occur when a new zoosporangium forms inside the old one. Sometimes the supporting hypha will grow through the emptied zoosporangium, producing another zoosporangium at its tip [Fig. 2–39(b)].

Phytophthora. The genus *Phytophthora* includes some species that are capable of living saprophytically in the soil until a host plant is available, while others are known to occur only on their host plants. Many species are able to attack the aerial parts of the plants, causing the blights or rots of affected parts and quickly killing infected tissues. *Phytophthora infestans* causes late blight of potatoes, resulting in the rotting of the tubers. The mycelium of these plant parasitic fungi may grow either within the cells (intracellularly) or between the cells (intercellularly). Haustoria are produced by intercellular mycelium. The sporangiophores are usually branched and may be formed on the surface of the soil or, in the case of fungi on plants, may emerge from the host through the epidermis or stomates. The zoosporangia are oval, ellipsoidal, or somewhat lemon-shaped. For those species occurring within the soil, the zoosporangia usually lack a papilla, undergo proliferation, and are not shed. In contrast, the zoosporangia of those species that occur above ground on plants have a broad, conspicuous papilla at the apex, do not undergo proliferation, and are deciduous. The zoosporangia which have been shed may then be wind-disseminated. The zoosporangia are then capable of germinating in one of two ways: (1) indirectly by forming zoospores that are released without a vesicle being formed, or (2) directly by forming a germ tube. Once released, the zoospores are capable of repeated swarming and encystment stages.

The mode of germination of the zoosporangia (either direct or indirect) may be determined by age or environmental conditions. In *P. infestans*, indirect germination by zoospore formation occurs when temperatures are low (optimum at 12°C) and direct germination occurs at higher temperatures (optimum at 25°C). This differential germination contributed greatly to the Irish potato famine, as there were a number of cool summers in succession, which resulted in more propagative units being formed.

Ultrastructural studies have been made of zoosporangium germination in *Phytophthora parasitica*. When the zoosporangium cleaves into zoospores in indirect germination stimulated by environmental conditions, cleavage of the zoospores by vesicles occurs. These vesicles are apparently derived from the Golgi apparatus. Cleavage of the zoospores occurs in the manner which is typical for this group (Hohl and Hamomoto, 1967). If the zoosporangium is to undergo indirect germination, formation of the flagella and the initial steps in fusion of vesicles and cleavage occur as they would for zoosporogenesis. However, this is usually followed by the abnormal disappearance of the flagella, and cleavage is rarely completed. Sometimes the zoospores may be cleaved out but not released. An additional wall is formed around the inside of the zoosporangial wall. The Golgi-derived vesicles that would ordinarily form the plasmalemma of the zoospores in indirect germination appear to be involved in the transport of precursor materials for the new wall. A germ tube is formed which penetrates through either the zoosporangial wall or the fibrous plug in the discharge tube. Lomasomes, possibly derived from the Golgi apparatus, appear to be active during the penetration of the wall by the germ tube (Hemmes and Hohl, 1969).

Albugo. Members of the genus *Albugo* are parasites of plants, causing the white rust disease. This disease is especially common on the plant called shepherd's purse (*Capsella*). The mycelium is intercellular and produces haustoria. The sporangiophores and zoosporangia* are produced in a palisadelike mass (*sorus*), which is formed between the mesophyll and epidermis of the host plant and subsequently ruptures the epidermis (Fig. 2–41). The exposed pustule appears white. The sporangiophores are upright, club-shaped, and unbranched. Zoosporangia are formed at the tip of the sporangiophores in chains, with the youngest zoosporangia at the base of the chain. In this process, a zoosporangial initial buds out from the tip of the sporangiophore, and after it has achieved a certain size, it is delimited from the sporangiophore by a septum and is now a separate zoosporangium. This zoosporangium is then pushed upward when another one is delimited in the same fashion immediately under it. The zoosporangia are separated by a *disjunctor*, consisting of a double septum with an unusually thick middle wall (Khan, 1977). When the disjunctors rupture, the zoosporangia are shed. When the zoosporangia fall into water, perhaps forming an aqueous film on a plant, they germinate by cleaving into zoospores that then escape from the zoosporangium. These zoospores encyst and reinfect a host plant when they form a germ tube. The sexual stages are found within the host tissue.

Peronospora and Other Downy Mildew Fungi. Members of *Peronospora* are among the fungi that cause downy mildew diseases of vascular plants (Waterhouse, 1973-b). These fungi are specialized parasites, apparently occurring only on vascular plants. The mycelium is intercellular and haustoria are pro-

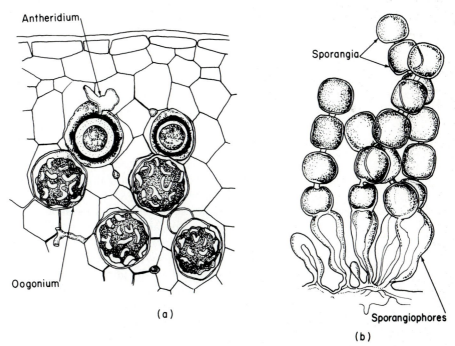

Antheridium

Oogonium

(a)

Sporangia

Sporangiophores

(b)

Figure 2–41 White rust fungus, *Albugo candidus:* (a) sexual reproductive structures within the host tissue; an antheridium, oospores, and oogonia are visible; (b) chains of sporangia on surface of the host. [From A. N. Berlese, 1898, Icones Fungorum, Phycomycetes, published by author.]

*These zoosporangia are called *conidia* by some mycologists. Refer to Chapter 6 for a discussion of this term.

Figure 2–42 Three members of the Peronosporales (downy mildew fungi): (a) *Basidiophora entospora,* which has simple, unbranched stout sporangiophores emerging through the epidermis of its host; an oogonium lies buried within the tissues (arrow); (b) apical portion of a slender, dichotomously branching sporangiophore of *Peronospora potentillae;* (c) tips of sporangiophores of *Bremia.* [From A. N. Berlese, 1898, Icones Fungorum, Phycomycetes, published by author.]

duced. Sporangiophores protrude through stomates of the host, giving the appearance of a ''down'' that varies from white to gray in color. The sporangiophores branch in a way that is characteristic for the genus (Fig. 2–42). For example, the branches of *Peronospora* species are produced dichotomously and are gently curving. Zoosporangia are formed at the tips of the sporangiophores and are always deciduous and wind-disseminated. In most species, the zoosporangia cleave into zoospores when they germinate. However, in *Peronospora* and *Bremia* the zoosporangia typically germinate by producing a germ tube. In members of these two genera, the zoosporangium almost always germinates in the same manner as a conidium (a term used by many mycologists to describe these structures) and its role as a zoosporangium is essentially absent. The downy mildews have therefore bridged the gap between the aquatic and terrestrial environments as they do not occur in the aquatic habitat, and the motile zoospore stage is all but absent in species of two of the genera. The sexual stages (gametangia and oospores) are found within the host tissues (Fig. 2–42).

REFERENCES CITED

AIST, J. R., and P. H. WILLIAMS. 1971. The cytology and kinetics of cabbage root hair penetration by *Plasmodiophora brassicae.* Can. J. Botany 49:2023–2043.

ARMBRUSTER, B. L. 1982-a. Sporangiogenesis in three genera of the Saprolegniaceae. I. Mycologia 74:433–459.

———. 1982-b. Sporangiogenesis in three genera of the Saprolegniaceae. II. Mycologia 74:975–999.

ARONSON, J. M., and C. C. LIN. 1978. Hyphal wall chemistry of *Leptomitus lacteus.* Mycologia 70:363–369.

BEAKES, G. W. 1980-a. Electron microscopic study of oospore maturation and germination in an emasculate isolate of *Saprolegnia ferax.* 1. Gross changes. Can. J. Botany 58:182–194.

———. 1980-b. Electron microscope study of oospore maturation and germination in an emasculate isolate of *Saprolegnia ferax.* 3. Changes in organelle status and associations. Can. J. Botany 58:209–227.

BRADLEY, W. H. 1967. Two aquatic fungi (Chytridiales) of Eocene age from the Green River formation of Wyoming. Am. J. Botany 54:577–582.

BRYANT, T. R., and K. L. HOWARD. 1969. Meiosis in the Oomycetes. I. A microspectrophotometric analysis of nuclear deoxyribonucleic acid in *Saprolegnia terrestris.* Am. J. Botany 56:1075–1083.

CANTER, H. M. 1946. Studies on British chytrids. I. *Dangeardia mammillata* Schroder. Trans. Brit. Mycol. Soc. 29:128–134.

COKER, W. C. 1923. The Saproegniaceae with notes on other water molds. University of North Carolina Press, Chapel Hill, N.C., 201 pp.

COUCH, J. N., and A. J. WHIFFEN. 1942. Observations on the genus *Blastocladiella.* Am. J. Botany 29:582–591.

CRUMP, E., and D. BRANTON. 1966. Behavior of primary and secondary zoospores of *Saprolegnia* sp. Can. J. Botany 44: 1393–1400.

DICK, M. W. 1969. Morphology and taxonomy of the Oomycetes, with special reference to Saprolegniaceae, Leptomitaceae and Pythiaceae. 1. Sexual reproduction. New Phytologist 68:751–755.

———. 1973-a. Saprolegniales. *In:* Ainsworth, G. C., F. K. Sparrow, and A. S. Sussman, Eds. The fungi: an advanced treatise. Academic Press, New York, Vol. 4B, pp. 113–144.

———. 1973-b. Leptomitales. *In:* Ainsworth, G. C., F. K. Sparrow, and A. S. Sussman, Eds. The fungi: an advanced treatise. Academic Press, New York, Vol. 4B, pp. 145–158.

———, and W. TIN. 1973. The development of cytological theory in the Oomycetes. Biol. Rev. Cambridge Phil. Soc. 48:133–158.

DOGMA, I. J., JR. 1973. Developmental and taxonomic studies on Rhizophlyctoid fungi, Chytridiales. I. Dehiscence mechanisms and generic concepts. Nova Hedwigia 24:393–411.

———, 1974. Developmental and taxonomic studies on Rhizophlyctoid fungi, Chytridiales. IV. *Karlingia granulata, Karlingia spinosa,* and *Karlingiomyces dubius.* Nova Hedwigia 25:91–105.

DYLEWSKI, D. P., and C. E. MILLER. 1983. Cruciform nuclear division in *Woronia pythii* (Plasmodiophoromycetes). Am. J. Botany 70:1325–1339.

ELLZEY, J. T. 1974. Ultrastructural observations of meiosis within antheridia of *Achlya ambisexualis.* Mycologia 66:32–47.

EMERSON, R., and H. C. WHISLER, 1968. Cultural studies of *Oedogoniomyces* and *Harpochytrium,* and a proposal to place them in a new order of aquatic phycomycetes. Arch. Mikrobiol. 61:195–211.

GARBER, R. C., and J. R. AIST. 1979. The ultrastructure of meiosis in *Plasmodiophora brassicae* (Plasmodiophorales). Can. J. Botany 57:2509–2518.

GAY, J. L., A. D. GREENWOOD, and I. B. HEATH. 1971. The formation and behaviour of vacuoles (vesicles) during oosphere development and zoospore germination in *Saprolegnia.* J. Gen. Microbiol. 65:233–241.

HASKINS, R. H., J. A. BRUSHABER, J. J. CHILD, and L. B. HOLTBY. 1976. The ultrastructure of sexual reproduction in *Pythium acanthicum.* Can. J. Botany 54:2193–2203.

HEATH, I. B. 1976. Ultrastructure of freshwater Phycomycetes. *In:* Jones, E. B. G., Ed. Recent advances in aquatic mycology. Wiley, New York, pp. 603–650.

HEMMES, D. E., and H. R. HOHL. 1969. Ultrastructural changes in directly germinating sporangia of *Phytophthora parasitica*. Am. J. Botany 56:300–313.

———, and O. K. RIBEIRO. 1977. Electron microscopy of early gametangial interaction in *Phytophthora megasperma* var. *sojae*. Can. J. Botany 55:436–447.

HOCH, H. C., and J. E. MITCHELL. 1972. The ultrastructure of *Aphanomyces euteiches* during asexual spore formation. Phytopathology 62:149–160.

HOHL, H. R., and S. T. HAMOMOTO. 1967. Ultrastructural changes during zoospore formation in *Phytophthora parasitica*. Am. J. Botany 54:1131–1139.

HOLLOWAY, S. A., and I. B. HEATH. 1977. An ultrastructural analysis of the changes in organelle arrangement and structure between the various spore types of *Saprolegnia*. Can. J. Botany 55:1328–1339.

HOWARD, K. I., and T. R. BRYANT. 1971. Meiosis in the Oomycetes. II. A microspectrophotometric analysis of DNA in *Apodachyla brachynema*. Mycologia 63:58–68.

———, and R. T. MOORE. 1970. Ultrastructure of oogenesis in *Saprolegnia terrestris*. Botan. Gaz. 131:311–336.

JOHNSON, T. W., JR. 1956. The genus *Achlya*: morphology and taxonomy. University of Michigan Press, Ann Arbor, Mich., 180 pp.

KARLING, J. S. 1937. The structure, development, identity, and relationship of *Endochytrium*. Am. J. Botany 24:352–364.

———. 1939. A new fungus with anteriorly uniciliate zoospores: *Hyphochytrium catenoides*. Am. J. Botany 26:512–519.

———. 1964. *Synchytrium*. Academic Press, New York, 470 pp.

———. 1968. The Plasmodiophorales, 2nd ed. Hafner Press, New York, 256 pp.

KHAN, S. R. 1977. Light and electron microscopic observations of sporangium formation in *Albugo candida* (Peronosporales; Oomycetes). Can. J. Botany 55:730–739.

KOBAYASHI, Y., and M. ÔKUBO. 1954. On a new genus *Oedogoniomyces* of the Blastocladiaceae. Bull. Natl. Sci. Museum (Tokyo), N.S. 1:59–66.

KUSANO, S. 1930. The life-history and physiology of *Synchytrium fulgens* Schroet., with special reference to its sexuality. Japan. J. Botany 5:35–132.

LINGAPPA, B. T. 1958-a. Sexuality in *Synchytrium brownii*. Mycologia 50:524–537.

———. 1958-b. Development and cytology of the evanescent prosori of *Synchytium brownii* Karling. Am. J. Botany 45:116–123.

———. 1958-c. The cytology of development and germination of resting spores of *Synchytrium brownii*. Am. J. Botany 45:613–620.

MARTIN, R. W., JR., and C. E. MILLER. 1986-a. Ultrastructure of zoosporogenesis in the endoparasite *Olpidiopsis varians*. Mycologia 74:433–459.

———. 1986-b. Ultrastructure of sexual reproduction in *Olpidiopsis varians*. Mycologia 78:359–370.

McKEEN, W. E. 1975. Electron microscopy studies of a developing *Pythium* oogonium. Can. J. Botany 53:2354–2360.

MIDDLETON, J. T. 1943. The taxonomy, host range and geographic distribution of the genus *Pythium*. Mem. Torrey Botan. Club. 20:1–171.

MILLER, C. E. 1976. Substrate-influenced morphological variations and taxonomic problems in freshwater, posteriorly uniflagellate phycomycetes. *In:* Jones, E. B. G., Ed. Recent advances in aquatic mycology. Wiley, New York, pp. 469–487.

———. 1977. A developmental study with SEM of sexual reproduction in *Chytriomyces hyalinus*. Bull. Soc. Botan. France 124:281–289.

MOORE, E. D., and C. E. MILLER. 1973. Resting body formation by rhizoidal fusion in *Chytriomyces hyalinus*. Mycologia 65:145–154.

MULLINS, J. T. 1961. The life cycle and development of *Dictyomorpha* gen. nov. (formerly *Pringsheimella*), a genus of the aquatic fungi. Am. J. Botany 48:377–387.

OLSON, L. W. 1973. The meiospore of *Allomyces*. Protoplasma 78:113–127.

ROANE, M. K. 1973. Two new chytrids from the Appalachian highlands. Mycologia 65:531–538.

SANSOME, E., and B. J. HARRIS. 1962. Use of camphor-induced polyploidy to determine the place of meiosis in fungi. Nature 196:291–292.

SCOTT, W. W. 1961. A monograph of the genus *Aphanomyces*. Virginia Agr. Expt. Sta. Tech. Bull. 151:1–95.

SKUCAS, G. P. 1966. Structure and composition of zoosporangial discharge papillae in the fungus *Allomyces*. Am. J. Botany 53:1006–1011.

—— 1967. Structure and composition of the resistant sporangial wall in the fungus *Allomyces*. Am. J. Botany 54:1152–1158.

SPARROW, F. K. 1960. Aquatic phycomycetes, 2nd ed. University of Michigan Press, Ann Arbor, Mich., 1187 pp.

——. 1973-a. Mastigomycotina (zoosporic fungi). *In:* Ainsworth, G. C., F. K. Sparrow, and A. S. Sussman, Eds. The fungi: an advanced treatise. Academic Press, New York, Vol. 4B, pp. 61–73.

——. 1973-b. Chytridiomycetes, Hyphochytridiomycetes. *In:* Ainsworth, G. C., F. K. Sparrow, and A. S. Sussman, Eds. The fungi: an advanced treatise. Academic Press, New York, Vol. 4B, pp. 85–110.

——. 1973-c. Lagenidiales. *In:* Ainsworth, G. C., F. K. Sparrow, and A. S. Sussman, Eds. The fungi: an advanced treatise. Academic Press, New York, Vol. 4B, pp. 159–163.

——. 1976. The present status of classification in biflagellate fungi. *In:* Jones, E. B. G., Ed. Recent advances in aquatic mycology. Wiley, New York, pp. 213–222.

TOMMERUP, I. C., and D. S. INGRAM. 1971. The life-cycle of *Plasmodiophora brassicae* Woron. in *Brassica* tissue cultures and in intact roots. New Phytologist 70:327–332.

TRUESDELL, L. C., and E. C. CANTINO. 1970. Decay of γ particles in germinating zoospores of *Blastocladiella emersonii*. Arch. Mikrobiol. 70:378–392.

WATERHOUSE, G. M. 1973-a. Plasmodiophoromycetes. *In:* Ainsworth, G. C., F. K. Sparrow, and A. S. Sussman, Eds. The fungi: an advanced treatise. Academic Press, New York, Vol. 4B, pp. 75–82.

——. 1973-b. Peronosporales. *In:* Ainsworth, G. C., F. K. Sparrow, and A. S. Sussman, Eds. The fungi: an advanced treatise. Academic Press, New York, Vol. 4B, pp. 165–183.

WHISLER, H. C. 1979. The fungi versus the arthropods. *In:* Batra, L. R., Ed. Insect-fungus symbiosis: nutrition, mutualism, and commensalism. Allanheld, Osmun, Totowa, N.J., pp. 1–32.

WILLIAMS, P. H. 1966. A cytochemical study of hypertrophy in clubroot of cabbage. Phytopathology 56:521–524.

——, and S. S. McNABOLA. 1967. Fine structure of *Plasmodiophora brassicae* in sporogenesis. Can. J. Botany 215:1665–1669.

YUKAWA, Y., P. H. WILLIAMS, and J. O. STRANDBERG. 1966. Ultrastructural studies on *Plasmodiophora brassicae*. Phytopathology 56:907.

Additional Recommended References

BARR, D. J. S. 1975. Morphology and zoospore discharge in single-pored, epibiotic Chytridiales. Can. J. Botany 53:164–178.

BEAKES, G. W. 1980. Electron microscopic study of oospore maturation and germination in an emasculate isolate of *Saprolegnia ferax*. 4. Nuclear cytology. Can. J. Botany 58:228–240.

BLAND, C. E., and H. V. AMERSON. 1973. Observations on *Lagenidium callinectes:* isolation and sporangial development. Mycologia 65:310–320 (Lagenidiales).

ELLIOTT, C. G., and D. MacINTYRE. 1973. Genetical evidence on the life-history of *Phytophthora*. Trans. Brit. Mycol. Soc. 60:311–316.

FULLER, M. S. 1977. The zoospore, hallmark of the aquatic fungi. Mycologia 69:1–20.

——, and R. REICHLE. 1965. The zoospore and early development of *Rhizidiomyces apophysatus*. Mycologia 57:946–961 (Hyphochytridiales).

HATCH, W. R. 1938. Conjugation and zygote formation in *Allomyces arbuscula*. Am. J. Botany 2:583–614 (Blastocladiales).

HELD, A. A. 1973. Development of endoparasitic, zoosporic fungi. Bull. Torrey Botan. Club 100:203–216.

JOHNS, R. M. 1964. A new *Polyphagus* in algal culture. Mycologia 56:441–451 (Chytridiales).

KOLE, A. P. 1965. Flagella. *In:* Ainsworth, G. C., and A. S. Sussman, Eds. The fungi: an advanced treatise. Academic Press, New York, Vol. 1, pp. 77–93.

——, and A. J. GIELINK. 1963. The significance of the zoosporangial stage in the life cycle of the Plasmodiophorales. Neth. J. Plant Pathol. 69:258–262.

MARCHANT, R. 1968. An ultrastructural study of sexual reproduction in *Pythium ultimum.* New Phytologist 67:167–171 (Peronosporales).

PERROTT, P. E. T. 1955. The genus *Monoblepharis.* Trans. Brit. Mycol. Soc. 38:247–282 (Monoblepharidales).

REICHLE, R. E., and M. S. FULLER. 1967. The fine structure of *Blastocladiella emersonii* zoospores. Am. J. Botany 54:81–92 (Blastocladiales).

SANSOME, E. 1963. Meiosis in *Pythium debaryanum* Hesse and its significance in the life-history of the biflagellatae. Trans. Brit. Mycol. Soc. 46:63–72 (Peronosporales).

——. 1965. Meiosis in diploid and polyploid sex organs of *Phytophthora* and *Achlya.* Cytologia 30:103–117 (Peronosporales and Saprolegniales).

SPARROW, F. K. 1933. The Monoblepharidales. Ann. Botany (London) 47:517–542.

THIRUMALACHAR, M. J., M. D. WHITEHEAD, and J. S. BOYLE. 1949. Gametogenesis and oospore formation in *Cystopus (Albugo) evolvuli.* Botan. Gaz. 110:487–491 (Peronosporales).

TRAVLAND, L. B., and H. C. WHISLER. 1971. Ultrastructure of *Harpochytrium hedinii.* Mycologia 63:767–789 (Harpochytriales).

UMPHLETT, C. J. 1962. Morphological and cytological observations on the mycelium of *Coelomyces.* Mycologia 54:540–554 (Blastocladiales).

——. 1964. Development of the resting sporangia of two species of *Coelomyces.* Mycologia 56:488–497 (Blastocladiales).

VUJIČIČ, R. 1971. An ultrastructural study of sexual reproduction in *Phytophthora palmivora.* Trans. Brit. Mycol. Soc. 57:525–530 (Peronosporales).

WILSON, C. M. 1952. Meiosis in *Allomyces.* Bull. Torrey Botan. Club 79:139–160 (Blastocladiales).

3

Division Amastigomycota— Subdivision Zygomycotina

The remaining groups of fungi are in the Division Amastigomycota. Unlike the fungi in the previous taxon, the Division Mastigomycota, the members of the Amastigomycota lack a motile stage and usually are not adapted to the aquatic habitat.

The fungi in the Subdivision Zygomycotina are characterized by their sexual reproduction, which culminates in *zygospore* formation. In this type of reproduction, two gametangia fuse and subsequently develop into the zygote, which then forms thick walls to become a zygospore. Details of this type of reproduction are discussed in the section on the Order Mucorales, pages 79–82.

The typical nonsexual spore formed by members of the Subdivision Zygomycotina is the *sporangiospore*. Sporangiospores are nonmotile spores that are formed endogenously within a saclike structure, the *sporangium* (Fig. 3–1). Although most sporangia produce several sporangiospores, there are some that produce only a few or even a single one. These reduced sporangia are often called *sporangioles*. In some monosporous sporangioles, the sporangial wall remains distinct and can separate from the sporangiospore at maturity. In other sporangioles, this wall is fused with that of the sporangiospore at maturity. In still other sporangioles, the wall of the sporangiole is fused with that of a sporangiospore, which may also appear to develop exogenously, like most conidia. Although many mycologists call the latter spore type a conidium, the view has been generally supported that the monosporous sporangioles are all homologous structures (e.g., Hawker et al., 1970). An electron-microscopic study of representatives of both types of monosporous sporangioles supports this view, as it showed that in each case the wall layers are deposited *within* the surrounding sporangiole wall and that the total number of wall layers is equivalent (Khan and Talbot, 1975). The sporangia are borne on specialized stalks, the *sporangiophores*.

Members of the Zygomycotina occur on or within a variety of substrata.

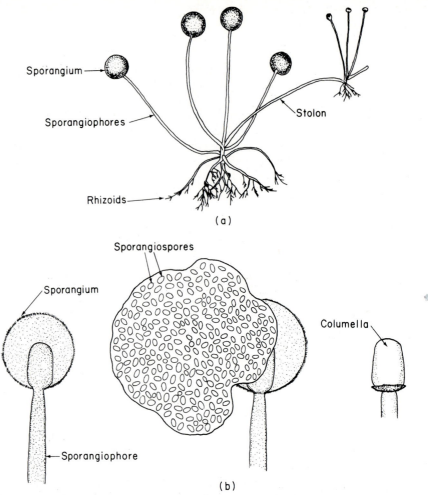

Figure 3-1 Nonsexual reproductive structures in the Mucorales: (a) growth habit of *Rhizopus stolonifer*, showing the production of sporangiophores bearing a single multispored sporangium; (b) single multispored sporangium of *Mucor mucedo* (left), which ruptures to release numerous sporangiospores (center) and finally leaves only the columella (right). [(a) From G. F. Atkinson, 1908, *in:* H. M. Fitzpatrick, 1930, The lower fungi, McGraw-Hill Book Company, New York, 331 pp.; (b) from O. Brefeld, 1872, Botanische Untersuchungen über Schimmelpilze, Heft 1, 1-64, Leipzig.]

The two classes may be distinguished on the basis of the relationship of the fungi to their substratum. The Class Zygomycetes has large numbers of saprophytes and parasites within or upon a variety of hosts. The Class Trichomycetes contains fungi that occur primarily in the gut of arthropods as commensals. We shall use the following classification (Ainsworth, 1973):

> Subdivision Zygomycotina
>   Class Zygomycetes
>   Order Mucorales
>   Order Entomophthorales
>   Order Zoopagales
>   Class Trichomycetes
>   Order Harpellales
>   Order Asellariales
>   Order Eccrinales
>   Order Amoebidiales

ORDER MUCORALES

The majority of the Mucorales occur as saprophytes on plant debris or, to a lesser extent, on animal debris in soil. Some members cause plant or animal diseases. A few of the parasitic ones are obligate parasites on other fungi (Chapter 14). When fungi are isolated from the soil, air, or dung by using agar culture dishes, members of the Mucorales are among the most commonly isolated fungi. They are often notable in culture because of their extremely fast growth rate, a characteristic important in their ability to compete in the soil. Some members not only are common soil saprophytes but also are important or are involved in our lives in other ways. Three such genera are *Mucor, Rhizopus,* and *Absidia.* Members of these genera may cause spoilage of stored grain and diseases known as muromycosis in humans and animals. *Rhizopus stolonifer* is called the ''black bread mold'' because it grows on moist, old bread. Some members of *Rhizopus* and *Absidia* may also cause rotting of vegetables such as sweet potatoes that are stored after harvest. On the positive side, some members of *Rhizopus* and *Mucor* as well as some other members of the Mucorales are used to produce commercially important products. These include organic acids, pigments, fermented Oriental foods, alcohols, and modified steroids. This commercial utilization is discussed further in Chapter 16. *Phycomyces* has served as an important experimental organism because of its response to light and nutritional requirements (Chapter 8). *Pilobolus* has an interesting spore discharge mechanism (Chapter 11).

These fungi have a well-developed vegetative mycelium. The mycelium is multinucleate and nonseptate when first produced, but later it becomes septate at the delimitation of reproductive structures or when aging. In members of two families, septa are regularly formed in the aerial hyphae and sporangiophores. These septa appear Y-shaped in longitudinal sections of the hyphae, with the arms of the Y pointing towards a central pore in the septum and enclosing a lens-shaped area. The pore is plugged with an electron-dense material which is apparently lipid (Saikawa, 1977). The mycelium is usually light in color and is typically submerged in the substratum with only the reproductive portions emerging. In other cases, the mycelium may form a dense, cottony growth on the surface of the substratum.

Sexual Reproduction.   The vegetative mycelium is haploid. Meiosis occurs after zygospore initiation but before or upon germination of the zygospore. Therefore the life cycle is one in which the haploid stage is emphasized (Fig. 3–2).

Individual species may be homothallic, requiring only one thallus type for sexual reproduction (in this group, a single thallus only), or it may be heterothallic, requiring two thalli of opposite mating types. These mating types may be designated ''plus'' and ''minus.'' (See Chapter 7 for additional information on this topic.) The majority of the Mucorales are heterothallic.

Differentiation of the sexual organs is initiated when adjacent branches meet, if homothallic, or when plus and minus hyphae touch. This differentiation is controlled by hormones, as discussed further in Chapter 10. Numerous light microscopic studies have been supplemented by recent ultrastructural studies (e.g., Hawker and Gooday, 1969; O'Donnell et al., 1978). Generally development proceeds in the following manner, although there may be minor morphological variations. Those hyphae that meet to initiate the sexual development, called *progametangia*, are multinucleate with dense cytoplasm and numerous lipid droplets. When the progametangia meet, the walls making contact fuse into a single wall, and the progametangia increase in size and gradually push apart the hyphae from which they have arisen [Fig. 3–3(a)]. The progametangia

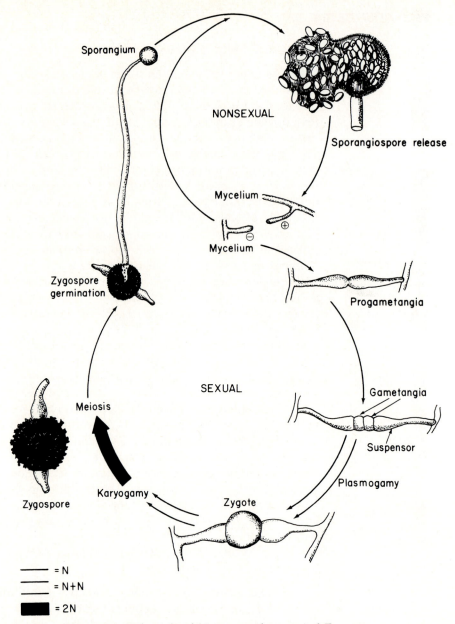

Sporangium

NONSEXUAL

Sporangiospore release

Mycelium

⊕

Mycelium

⊖

Zygospore
germination

Progametangia

SEXUAL

Gametangia

Meiosis

Suspensor

Zygospore

Karyogamy

Plasmogamy

Zygote

———— = N

===== = N+N

▅▅▅▅ = 2N

Figure 3–2 Life cycle of *Mucor mucedo*, a typical Zygomycete.

will become separated by a septum into the *suspensor*, the tapered cell adjoining the vegetative hyphae, and the terminal gametangium. Vesicles, probably arising from the endoplasmic reticulum, may be involved in formation of the wall separating the gametangium from its suspensor. The vesicles fuse, forming the plasmalemma of each cell, and wall material is laid down between the adjacent plasmalemmas (Hawker and Gooday, 1967). Cytoplasmic strands (plasmodesmata) may extend through the wall that separates the suspensor from its gametangium and may provide a means for food to be transported into the gametangium (Hawker and Gooday, 1968). The gametangia are usually equal in size [Fig. 3–3(b)], but gametangia and their supporting suspensors are unequal in some members. The common wall separating the gametangia is then digested enzymatically, beginning at the center. This allows plasmogamy to occur, and the cytoplasm from the two gametangia mingles in the common cell just formed, the *zygote*. The wall of the zygote is at first thin, as it was derived from the gametangia, but as the zygote develops into the zygospore, a thick wall is

(a)

(b)

(c)

Figure 3–3 Stages in the development of the zygospore of *Gilbertella persicaria:* (a) progametangia from opposing hyphae have come into contact, × 810; (b) the zygote has been formed and the outer wall is beginning to rupture from the expansion of the developing zygospore, × 850; (c) nearly mature zygospores with remnants of the outer wall, × 835. Note the prominent markings and suspensors. [Courtesy K. L. O'Donnell, J. J. Ellis, C. W. Hesseltine, and G. R. Hooper, 1977, Can. J. Botany 55:662–675.]

formed inside the original thin wall. Patches of new wall material form and gradually coalesce to form the zygospore wall ornamentations. As the zygospore expands, the thin outer wall is sloughed off [Fig. 3–3(c)]. The wall of the zygospore may have ornate markings such as warts or starfish-shaped plates when mature (Schipper et al., 1975). While the above processes are taking place, a few fungi such as *Phycomyces* produce ornate projections that arise from the suspensors and then surround the zygospore (Fig. 3–4).

The zygospore undergoes a period of rest before germinating. Germination typically occurs through production of a germ tube which is rather short and terminates in a sporangium. Often the sporangiospores within a single sporangium are of the same mating type, indicating that some of the products of meiosis disintegrate.

Cytological events occurring after plasmogamy follow different patterns. In *Mucor*, the nuclei within the zygote fuse in pairs and meiosis follows immediately before the zygospore becomes dormant (Fig. 3–2). In *Rhizopus*, karyogamy occurs before the onset of dormancy and meiosis occurs upon zygospore germination, while in *Phycomyces* both karyogamy and meiosis take place when the zygospore germinates. In some genera, karyogamy does not take place but development of the zygospore is apomictic.

Under some conditions, plasmogamy may fail to take place normally and

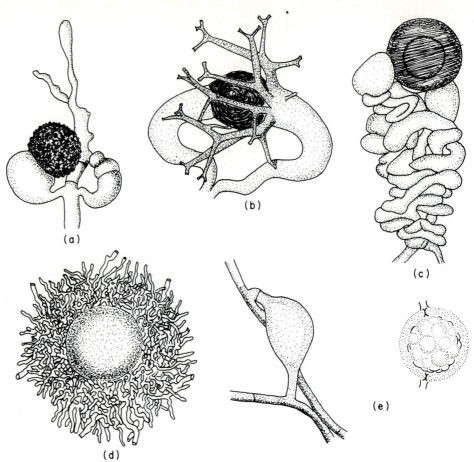

Figure 3–4 Variation in the sexual structures of the Mucorales: (a) unequal suspensors arising from the same hypha in *Zygorhynchus macrocarpus;* (b) spines arising from the tong-shaped suspensors in *Phycomyces microsporus;* (c) twined suspensors of *Choanephora conjuncta;* (d) covering of sterile hyphae around zygospore in *Mortierella rostafinskii;* (e) morphologically simple suspensors and hyaline, smooth zygospore of *Coemansia mojavensis.* [(a) From M. Ling-Young, 1930, Rev. Gen. Bot. 42:147–158; (b) from G. A. Christenberry, 1940, J. Elisha Mitchell Sci. Soc. 56:333–366; (c) from J. N. Couch, 1925, J. Elisha Mitchell Sci. Soc. 41:141–150; (d) from O. Brefeld, 1877, Botanische Untersuchungen über Schimmelpilze, Heft 4, Teil 5, 81–96, Leipzig; (e) from R. K. Benjamin, 1958, Aliso 4:149–169.]

a gametangium may develop parthenogenetically into an *azygospore.* Azygospores are morphologically similar to zygospores but are smaller. They may be either single or double, depending on whether one or both of the copulating gametangia developed parthenogenetically. Sometimes an isolated, noncopulated gametangium will develop into an azygospore. Azygospores are formed infrequently as a result of unfavorable cultural conditions, or they may be induced by forcing a mating between plus and minus strains of different species or genera (Benjamin and Mehrotra, 1963). In a few species, azygospores but not zygospores are routinely formed (O'Donnell et al., 1977).

Nonsexual Reproduction. Nonsexual reproductive spore types formed by the Mucorales are sporangiospores and chlamydospores. Unlike the sporangiospores, which are formed in a special sac and are shed, the chlamydospores are nondeciduous spores that are thick-walled and are formed by the rounding up of vegetative cells. Chlamydospores may be formed in either a terminal or intercalary position and sometimes occur in chains.

The most commonly occurring sporangia are of the type that occurs in the Mucoraceae. In these fungi, the sporangium is large and multispored [Fig. 3–4(a,b)]. A *columella* may be present, which is a domelike extension of the sporangiophore lying within the sporangium. Details of sporangiospore formation within the sporangium are similar to those for zoospore formation. Let us consider a representative, *Gilbertella persicaria* (Bracker, 1968). The sporangium is at first a multinucleate structure with continuous cytoplasm. Vesicles are formed and are at first scattered throughout the cytoplasm [Fig. 3–5(a)]. These vesicles apparently arise from the cisternae of the endoplasmic reticulum. The membranes of these vesicles fuse and they become transformed into tubules. The tubules expand and branch and eventually extend throughout the cytoplasm as membranous cleavage furrows [Fig. 3–5(b)]. Uninucleate segments of protoplasm are cleaved out by these membranes and will become the sporangiospores. Meanwhile the columella is being delimited by a similar mechanism. Immediately after cleavage, the sporangiospore initials appear as naked protoplasts [Fig. 3–5(c)] and are surrounded by an amorphous material derived from the matrix of the cleavage apparatus. The plasmalemma of the sporangiospore is derived from the membrane of the cleavage apparatus, and the wall of the sporangiospore is then deposited outside the plasmalemma. In the Mucorales in general, the sporangial wall may either deliquesce or be persistent, requiring an outside force to rupture it. The sporangiospores are liberated when the sporangial wall breaks down by either mechanism.

Sporangioles are produced by some members of the Mucorales. These reduced sporangia are much smaller than the sporangia described above, lack a columella, and produce only one or a few sporangiospores (Fig. 3–6). In its initial stages, the sporangiole contains continuous cytoplasm. Cleavage of a few sporangiospores in a sporangiole may occur in essentially the way described above in *Thamnidium elegans* (Fletcher, 1973). In monosporous sporangioles in *Mycotypha* and *Cunninghamella*, cleavage is absent and the cytoplasm of the sporangiole becomes that of the sporangiospore when the sporangiospore wall is deposited around it (Khan and Talbot, 1975). Sporangiola may be borne singly on lateral branches of the sporangiophore or in clusters on swollen apices of the sporangiophore. They may be sessile on the sporangiophore or borne on a short, peglike stalk.

Elongated, rodlike sporangiola are formed by some genera and are sometimes called *merosporangia* (Fig. 3–7) (Benjamin, 1959, 1966). These usually contain approximately 10 to 15 sporangiospores arranged in a linear sequence, but some of these contain only one or two spores. The sporangiola may be borne in whorls from a swollen terminal head [Fig. 3–7(a,c)] and may be either sessile on that head or borne on short, peglike projections. The most unusual arrangement is found in the family Kickxellaceae, in which the sporangiola are borne laterally on a modified sporangiophore, much like teeth on a comb [Fig. 3–7(b)]. The sporangiophore branches may break apart at a line of dehiscence to shed the sporangiola, or the sporangiola may be deciduous and drop away from the sporangiophore. In other species, the sporangiola are not shed but the sporangiolum wall fragments, leaving behind single-spored segments, which remain temporarily in chains. Finally, the wall of the sporangiolum may simply be evanescent at maturity. Some species form a slime drop that encloses either the shed sporangiola or the single-spored fragments, while others remain dry.

Development of the elongated sporangiola has been studied ultrastructurally (e.g., Benny and Aldrich, 1975; Fletcher, 1972), and there is some variation in the manner in which development occurs. Let us consider a representative, *Syncephalis sphaerica* (Fig. 3–8) (Baker et al., 1977). In *S. sphaerica*, buds appear simultaneously over the surface of the swollen apex of the sporangiophore. The buds then elongate to form the sporangiola. Three sporangiospores are cleaved out simultaneously when the plasmalemma invaginates, moving

(a)　　　　　　　　　　　　　　　　　　　(b)

(c)　　　　　　　　　　　　　　　　　　　(d)

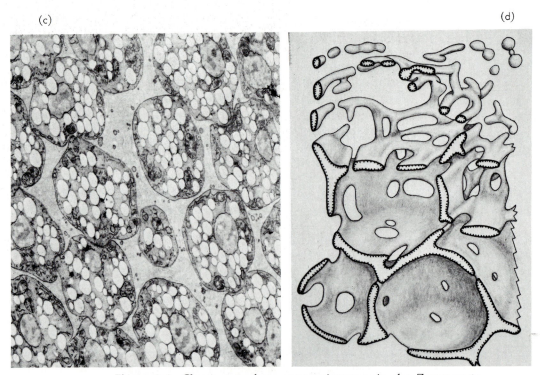

Figure 3–5 Cleavage to form sporangiospores in the Zygomycete, *Gilbertella persicaria:* (a) portion of sporangium before cleavage begins, × 4000; (b) midcleavage, parts of cleavage furrows are evident, × 6200; (c) postcleavage, the spore initials are present as unwalled, individual cells; walls form later, × 1800; (d) diagrammatic model of cleavage, resulting from fusion of vesicles in a sporangium: I, early cleavage; II, midcleavage; III, late cleavage. [Courtesy C. E. Bracker, 1968, Mycologia 60:1016–1067.]

Figure 3–6 Variation in the sporangiophores and sporangia of the Mucorales (compare with Fig. 3–1): (a) sporangia with columellae on terminal and lateral branches of *Actinomucor elegans*; (b) sporangiophore with swollen tips and sporangioles bearing three spores each in *Choanephora trispora*; (c) swollen tips of the sporangiophore bear single-spored sporangia in *Choanephora cucurbitarum*. [From C. T. Ingold and M. H. Zoberi, 1963, Trans. Brit. Mycol. Soc. 46:115–134.]

Figure 3–7 Sporangiophores bearing elongated sporangioles, each with a larger detail of a cluster of sporangioles: (a) *Dispira cornuta*, which has both sterile and fertile branches; (b) *Coemansia majovensis*, which produces monosporous sporangioles; (c) *Piptocephalis lepidula*. Magnifications: (a) × 30 and × 1080, (b) × 30 and × 400, (c) × 180 and × 1700. [(a, c)] From R. K. Benjamin, 1958, Aliso 4:321–433; (b) from R. K. Benjamin, 1958, Aliso 4:149–169.]

(a)   (b)   (c)

Figure 3-8 *Syncephalis sphaerica:* (a) swollen tip of sporangiophore bearing chains of sporangiospores after maturation is completed, × 1700; (b) early stage in which the plasmalemma is invaginating within the sporangiolum to delimit sporangiospores, × 19,250; (c) completed cleavage of sporangiospores by deposition of wall material between the plasmalemma; the wall of the sporangiolum is still present (arrow), × 19,250. [Courtesy K. L. Baker, G. R. Hooper, and E. S. Beneke, 1977, Can. J. Botany 55:2207-2215.]

from the periphery of the sporangiolum to the center and forming the plasmalemma of each spore. Fibrillar wall material is deposited between the plasmalemma of the sporangiospore and the merosporangial wall, thereby forming the walls of the sporangiospores. Meanwhile an electron-dense abscission zone is formed between each spore. At maturity, the wall of the sporangiolum ruptures to leave a chain of single sporangiospores.

The sporangia are borne on sporangiophores. The sporangiophores are typically stout aerial structures, usually terminating in a sporangium or a cluster of sporangioles. Sporangiophores range from relatively simple forms in which a single unbranched stalk terminates in a single sporangium (e.g., *Mucor* or *Rhizopus*) to elaborate branched types (Figs. 3-1 and 3-7). If branched, there may be a main axis with smaller lateral branches, dichotomous branching, or the formation of umbellike clusters of essentially equal branches. The branches usually bear a large single sporangium or perhaps smaller sporangioles at each end, but sometimes some of the branches are sterile. The main axis of the sporangiophore may terminate in a large multispored sporangium, and clusters of sporangioles may occur on the side branches (e.g., in *Thamnidium*). In some species, a prominent rhizoid anchors the sporangiophore to the substratum. A distinct lateral hypha (the *stolon*) may extend from the base of one sporangiophore to another.

The Mucorales may be divided into 14 families, primarily on the basis of the morphology of their nonsexual structures and habitat (Hesseltine and Ellis, 1973).

Endogonaceae. An important family in the Mucorales is the Endogonaceae. These fungi have a symbiotic association with roots of higher plants and have not been cultured in the laboratory on artificial media.

Some members of the Endogonaceae may be readily collected because they produce macroscopic sporocarps (Fig. 3-9). The sporocarps may occur above the surface of the soil, where they may be collected from litter, wood, decaying mushrooms, or the tips of moss gametophytes. Most species produce hypo-

Figure 3–9 Sporocarps of *Endogone*. Top: surface and cross-sectional views of *E. lactiflua*, × 3.2; bottom: *E. flammicorona* attached to roots of a plant, × 1.5. [(Courtesy J. W. Gerdemann and J. M. Trappe, 1974, Mycol. Mem. 5:1–76.]

geous sporocarps, that is, they occur beneath the surface of the soil. The sporocarps sometimes are found connected by hyphae to the roots with which they are associated (Fig. 3–9). The sporocarps range in diameter from approximately 1 to 25 millimeters, are sometimes brightly colored, and may or may not have a distinct covering. The shape varies from globose to ellipsoid or lobed and irregular, and they may be hollow or solid. The sporocarps contain zygospores [Fig. 3–10(a)], chlamydospores [Fig. 3–10(b)], or sporangia which are embedded

(a)

(b)

Figure 3–10 Zygospore (a) and chlamydospores (b) of members of the Endogonaceae. × 625. [Courtesy J. W. Gerdemann and J. M. Trappe, 1974, Mycol. Mem. 5:1–76.]

within or surrounded by sterile hyphae. Only one type of spore is found within a single sporocarp (Gerdemann and Trappe, 1974; Godfrey, 1957). Not all species form sporocarps, and the spores may be found on isolated hyphae in the soil. These spores may be collected by a technique involving the decanting and sieving of a soil suspension.

Species names and descriptions are based on collections of similar isolated spores or sporocarps with one type of spore. It is not known whether these spore types may represent different stages of the same life cycle in some cases or whether they are indeed unrelated fungi. For example, *Endogone* is one of the best known genera, and it usually produces either sporocarps with zygospores or chlamydospores or isolated chlamydospores. Each species of *Endogone* is described as forming only one type of spore, for example, zygospores (Gerdemann and Trappe, 1974; Nicholson and Gerdemann, 1968).

Zygospore formation is known in some species of *Endogone*. In *Endogone*, the gametangia fuse at their tips and then a bud grows out from the point of union or from the larger gametangium. This bud then matures into the zygospore (Fig. 3–11). Two distinct walls are visible, the outer resulting from the expansion of the gametangial wall and the inner developing later. Typically, the gametangia and suspensors eventually disintegrate, leaving the zygospores free within the sporocarp.

Figure 3–11 Successive stages in zygospore formation in *Endogone incrassata*. The zygospore forms above the point of union of the two gametangia. [From J. W. Gerdemann and J. M. Trappe, 1970, Mycologia 62:1204–1208.]

Azygospores are formed free in the soil by some species. Development of azygospores was studied by light and electron microscopy by Mosse (1970-a,b,c) in a species that was later identified as *Acaulospora laevis* (Fig. 3–12) (Gerdemann and Trappe, 1974). A globose vesicle appears as an enlargement at the end of a stout, funnel-shaped supporting hypha. Later an azygospore is initiated as a bud from the side of the supporting hypha, and the contents of the vesicle are transported into the developing azygospore. The vesicle then collapses. Unusual membrane-bounded pigment granules containing melanin are prominent

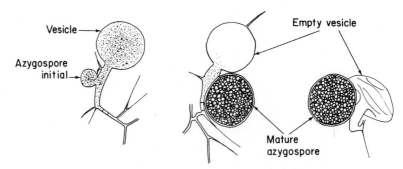

Figure 3–12 Successive stages in azygospore development of *Acaulospora laevis*. [From J. W. Gerdemann and J. M. Trappe, 1974, Mycol. Mem. 5:1–76.]

in the vesicle and azygospore but disappear as the wall of the azygospore develops, which suggests that they play a role in wall formation. The azygospore wall is thick, consisting of six layers, and contains chitin and cellulose in separate parts of the wall. The azygospore undergoes a period of rest before germinating. Upon germination, the peripheral portion of the cytoplasm is cleaved by radial walls into multinucleate segments, and each segment germinates by extending a germ tube through newly formed splits in the azygospore wall.

The chlamydospores are produced in terminal positions on undifferentiated hyphae. Although superficially similar to the zygospores or azygospores, they may be distinguished because they remain attached to the hyphae, which do not resemble suspensors and form only a single wall layer. The chlamydospores are often ovate to spherical in shape and at maturity have oily yellow contents and thick walls. They germinate by forming a germ tube. Sporangia are thin-walled and resemble immature chlamydospores. A large number of sporangiospores may be formed within a sporangium.

Fungi with vesicles resembling the immature chlamydospores or sporangia may be found within or upon roots. These fungi have a symbiotic association, which is described as the vesicular-arbuscular type of mycorrhiza and is discussed further in Chapter 15. At least some members of the Endogonaceae, including some species of *Endogone,* can form these mycorrhizae when their azygospores, zygospores, or chlamydospores are added to the soil of potted plants. No such connection has been established for the species that produce sporangiospores. The fungi within the roots have not been isolated in pure culture, so it is not known whether all these morphologically similar fungi belong to the Endogonaceae (Gerdemann, 1968; Mosse and Bowen, 1968).

ORDER ENTOMOPHTHORALES

The members of the Entomophthorales are predominantly parasites of insects and other arthropods. Members of the genera *Entomophthora* and *Massospora* are insect parasites, and these are discussed further in Chapter 13. Those not parasitic on insects may be parasitic on desmids or fern prothalli, while some species may be saprophytes on plant debris. *Basidiobolus haptosporus* causes a disease of man in which tumorlike enlargements develop in the subcutaneous tissue, while *Entomophthora coronata* is parasitic on humans and horses as well as on insects.

The vegetative stages may consist of a coenocytic mycelium, but often the growth of a true mycelium is sparse and limited. In almost all species, the mycelium later fragments into a number of short segments, the *hyphal bodies,* which separate from each other. The hyphal bodies may serve as propagative units by budding and undergoing fission. Rhizoids may be formed by some species. The rhizoids consist of hyphae that extend from the parasitized insect and adhere to the substratum. The rhizoids may be simple or branched, with each terminating in a holdfast that secretes a viscous, sticky substance. The mycelium or hyphal bodies may produce chlamydospores by forming a thick wall, with or without enlargement of the vegetative cell.

Sexual reproduction is initiated when two specialized mycelial endings (from the same or different hyphae) or hyphal bodies function as gametangia and unite (Fig. 3–13). The gametangia may be alike or dissimilar in size and the zygospore may form either within one of the gametangia, within the conjugated cell formed by both gametangia, or in a bubblelike growth extending from one of the gametangia. Some species (such as *Entomophthora grylli*) do not reproduce sexually but instead produce azygospores that are morphologically similar to zygospores. Azygospores may arise by the parthogenetic development of hyphal bodies, by direct budding of chlamydospores, or at the tips of hyphae arising from hyphal bodies of chlamydospores. Neither zygospores nor azy-

Figure 3–13 *Basidiobolus magnus*, a member of the Entomophthorales: (a) sporangiophore with sporangioles; (b, c) successive stages in gametangial fusion; (d) mature zygospore with a single nucleus. [From C. Drechsler, 1964, Am. J. Botany 51:770–777.]

gospores are known in some species. The thick-walled zygospores, azygospores, and chlamydospores are sometimes termed *resting spores*, as they aid the fungus to survive unfavorable conditions.

Nonsexual reproduction involves the formation of unbranched or comparatively simple branched sporangiophores. The sporangiophores usually bear ovoid to spherical sporangioles, but other shapes occur. The sporangioles are monosporous, except that in *Basidiobolus* more than one sporangiospore may be formed in a sporangiole. The sporangiospores are uni- or multinucleate. Except in the case of *Massospora,* the sporangiole is discharged with great force. Mechanisms responsible for forcible discharge are considered in Chapter 11. The sporangiophores are phototropic in many species. Germination occurs by production of germ tubes or if conditions are not favorable for subsequent growth, by production of another sporangiophore and sporangiole. The shape of the secondary sporangiolum may be either like that of the primary sporangiolum or dissimilar (e.g., globose or fusiform). This sporangiole may be discharged and again germinate by producing another sporangiophore and sporangiole. Each generation becomes progressively smaller, and this process may continue until either the vitality of the fungal structures is exhausted or a suitable substratum is encountered.

A particularly interesting member of this order is *Basidiobolus ranarum,* which has an unusual life cycle. *B. ranarum* grows on excrement of frogs and lizards. Beetles crawl about over the excrement and devour the sporangia. The beetles in turn may be eaten by a lizard or frog, and while in the stomach the sporangiole cleaves into a number of sporangiospores that will undergo no further development until liberated with the feces. Once these are exposed to the air, mycelium formation follows.

ORDER ZOOPAGALES

All members of the Zoopagales are predacious fungi, which means that they trap their prey. (Other predacious fungi are discussed in Chapter 13.) There are at least 65 species in the Zoopagales and they are of widespread geographical distribution (Duddington, 1973). The filamentous thallus is often at first devoid of septa, but septa are later formed and delimit degenerating portions of the mycelium from that part which is physiologically active and from the reproductive structures. Sexual reproduction does not always occur, but if it does, it involves the pairing of similar gametangia followed by zygospore formation. Nonsexual reproduction occurs through the formation of sporangiospores ranging in shape from filiform to spherical. The sporangiospores may be borne on long aerial sporangiophores or they may be nearly sessile on short *sterigmata* (peglike projections). Sporangiospores may be borne singly along the length of the sporangiophore or in long chains.

The Zoopagales occur predominantly on amoebae and other protozoa, although a few species are large enough to trap nematodes. These fungi are apparently obligate predators, requiring a living animal in order to satisfy their nutritional requirements. Often they are host specific and are restricted to a certain kind of animal. These fungi may either develop within their host (endogenously) or exist primarily outside their host (exogenously).

Endogenous Zoopagales include members of the genera *Cochlonema* (Fig. 3–14) and *Endocochlus,* which are somewhat similar. As an example of this group, we examine *Cochlonema verrucosum,* a predator on *Amoeba* spp. Infection is initiated by a fusiform sporangiospore which is attached to the ectoplasm of

(a)

(b)

Scale-μ

Figure 3–14 *Cochlonema verrucosum:* (a) conidia arising from endogenous thallus within dead amoeba; (b) zygospores formed on disintegrating amoeba. [From C. Drechsler, 1935, Mycologia 27:6–40.]

Figure 3–15 Portion of hypha of *Stylopage rhynchospora* on which five amoebae are attached. Amoeba *e* has been captured so recently that it has not been penetrated; amoebae *a* and *d* have each been invaded by a haustorium; and amoebae *b* and *c* have been depleted of their contents. [From C. Drechsler, 1939, Mycologia 31:338–415.]

the amoeba, and penetration is accomplished by a germ tube. A globular swelling develops within the amoeba at the tip of the germ tube, and both the sporangiospore and germ tube fall away, leaving only this swelling within the animal. The globular swelling develops into a minute elongate coiled thallus about 6 micrometers in length. During the early stages of infection, the amoeba leads a normal life, apparently unhindered by the fungus infection. As the fungus continues to grow and to feed on the endoplasm of the amoeba, however, the amoeba becomes increasingly sluggish and finally dies. The fungus continues to feed on the contents of the amoeba until only the shriveled ectoplasm remains (Fig. 3–14). Chains of sporangiospores are formed on the exterior of the amoeba before its death. Sexual reproduction occurs after the death of the host when slender club-shaped gametangia emerge from the remains of the amoeba and fuse, leading to zygospore formation.

The predacious habit is more obvious in a fungus such as *Stylopage rhynchospora*, an example of an exogenous member of the Zoopagales (Drechsler, 1959). Like the majority of the species of *Stylopage* that prey upon protozoa, *S. rhynchospora* preys upon amoebae. The sparingly septate, delicate mycelium adheres to any passing amoeba, probably by forming a sticky secretion. At the point of contact, the fungus forms a number of haustoria that penetrate the amoeba and absorb its contents. The amoeba is at first relatively unharmed, as cytoplasmic streaming continues to occur and the contractile vacuole continues to function. With further depletion of the endoplasm, the amoeba becomes rounded, all movements cease, and death follows (Fig. 3–15). The fungus forms sporangiospores and thick-walled zygospores with walls marked by hemispherical warts.

CLASS TRICHOMYCETES

Members of the Class Trichomycetes inhabit a variety of arthropods, including insect larvae, crayfish, millipedes, beetles, and isopods (Lichtwardt, 1976, 1985). The arthropod hosts may live on land or in fresh water or marine waters. A particular member of the Trichomycetes may be restricted to a narrow range of host species, perhaps all within a single family. There is also a specificity for parts of the host, for example, a given fungus may be found in the rectum. The Trichomycetes do not grow independently of their hosts in nature.

The fungi almost always occur within the gut, and may be detected if the thallus protrudes from the anus of the host or if the host is dissected. Members of one genus, *Amoebidium*, occur on the outer surface. The thalli are attached to the host by a holdfast, which penetrates the lining of the gut or exoskeleton only and does not penetrate into the living tissues of the host. When the arthropod molts, the lining of the gut and exoskeleton are shed and the fungus is also expelled. These fungi are not parasitic on their hosts and live as commensals, apparently neither harming nor benefiting them (Moss, 1979). The fungi obtain their nutrients from the surrounding gut contents or water. In experiments in which spores of the fungus *Smittium* were fed to bacteria-free mosquito larvae, it was found that the fungal spores could germinate only in the gut of the larva, although the thallus could be cultured alone in defined medium. Therefore, it was concluded that the spores of this Trichomycete require components of the gut contents produced by the larvae and not by bacteria for spore germination (Williams and Lichtwardt, 1972). Only a few of these fungi have been isolated and grown in pure culture; mostly these are species of *Smittium* or *Amoebidium*. Therefore, our understanding of the life cycle usually results from the dissection of numerous host specimens from a heavily infected population.

The thallus of the Trichomycetes is comparatively simple (Fig. 3–16). It is anchored by a holdfast, which may be a simple secretion or may have a more elaborate shape such as that of a bulb. The thallus may have one of two basic forms: (1) unbranched and coenocytic, or (2) branched with septa. The septa are similar to those described for members of the Kickxellaceae in the Mucorales, consisting of a bifurcated wall with a plug [Fig. 3–18(b)]. Similarity of the septa

Figure 3–16 Thalli of two members of the Trichomycetes. Left: part of a sporulating thallus of *Smittium culisetae* showing the basipetal development of sporangioles, × 900; right: generative cells (arrows) and sporangioles of *Genistellospora homothallica*, × 720. [Courtesy R. W. Lichtwardt; photograph of *G. homothallica* from 1972, Mycologia 64:167–197.]

in the Kickxellaceae and Trichomycetes has been used to argue for their close phylogenetic relationship (Chapter 7).

There are four orders in the Trichomycetes. These orders are distinguished primarily by the morphology of nonsexual reproductive structures. *Arthrospores* are produced by the Order Asellariales. Arthrospores are formed when the vegetative thallus fragments into sporelike segments, which then fall apart from each other. Sporangiospores are produced by the remaining orders. In the Eccrinales, the unbranched coenocytic thallus cleaves into sporangia, beginning at the apex and proceeding downward. Each sporangium then forms a single sporangiospore, which then escapes. Thin-walled, multinucleate sporangiospores germinate in the gut, while the thicker-walled, uninucleate sporangiospores are passed from the gut and are apparently able to withstand unfavorable conditions. In the Order Amoebidiales, naked amoeboid sporangiospores are formed within the thallus, which assumes the role of a sporangium. In the last order, the Harpellales, sporangioles bearing one to several filaments are formed and disseminated (Fig. 3–17). In all cases, the spores escape from the host when the host defecates, molts, or dies.

Figure 3–17 Scanning electron micrographs of sporangioles of *Smittium culicus.* Left: generative cell bearing a collar and a terminal sporangiole, × 2600; right: released sporangiole with its collar and uncoiling single appendage, × 2650. [Courtesy S. T. Moss.]

We will discuss the Order Harpellales further as it is the best-known order in the Trichomycetes and is also the only order in which sexual reproduction is known. Additional details about remaining orders may be seen in Table 3–1.

ORDER HARPELLALES

Members of this order occur in larvae of insects that inhabit fresh water, most usually the water of rapidly flowing streams. Some species form unbranched, coenocytic thalli in the midgut, while others form branched thalli in the hindgut (Fig. 3–16).

The unbranched coenocytic thalli become entirely separated into single cells, the *generative cells,* while the branched thalli have distinct terminal generative cells. The generative cells support the sporangioles externally (Figs. 3–16 and 3–17). The sporangioles may be borne directly upon the generative cells or may be borne on short, nonseptate lateral branches extending from the generative cells. Each sporangiole contains a single uninucleate sporangiospore and is typically elongate in shape. During development of the sporangiole in *Genistellospora homothallica* (Figs. 3–16 and 3–18), the young sporangiole is initiated as

Chap. 3 / Division Amastigomycota—Subdivision Zygomycotina

TABLE 3-1
Characteristics of the Trichomycete Orders

Order	Host	Thallus	Nonsexual Reproduction	Sexual Reproduction
Harpellales	Hindgut lining or peritrophic membrane of immature insects	Branched or unbranched	Sporangiospores	Zygospores
Asellariales	Hindgut lining of isopods or insects	Branched and septate	Arthrospores	Unknown
Eccrinales	Hindgut or foregut of insects, millipedes, isopods, amphipods, decapods	Unbranched or branched only at the base	Sporangiospores	Unknown
Amoebidiales	External parts of aquatic crustacea and insects	Unbranched, entire thallus becoming a sporangium	Amoeboid cells	Unknown

(a)   (b)

Figure 3-18 Ultrastructural details of the point of attachment of the sporangiole to the generative cell in *Genistellospora homothallica*. In both electron micrographs, the sporangiole is on top while the generative cell is on the bottom. (a) Note the continuity of the outermost wall layer between the sporangiole and the generative cell. An additional wall layer is present inside the sporangiole. An appendage is visible at the arrow. The septum appears continuous because this is a nonmedian section. × 7500. (b) Septum separating the sporangiole from the generative cell in a median section. Note the characteristic forking of the septum. × 14,500. [Courtesy S. T. Moss and R. W. Lichtwardt, 1976, Can. J. Botany 54:2346-2364.]

an extension of its generative cell, from which it is separated by a septum. The wall of the sporangiole and that of the generative cell are continuous and electron-opaque (Fig. 3–18). The sporangiole wall expands, and meanwhile an inner electron-transparent wall layer, representing the wall of the sporangiospore, is deposited immediately inside the sporangiole wall. Vesicles are involved in the deposition of wall material in the expanding outer wall and newly forming inner wall. The appendages are formed in *G. homothallica* as in other Trichomycetes studied, that is, they are extensions of the sporangial wall which form within the generative cell but external to the plasmalemma (Fig. 3–18). Vesicles are also involved in the deposition of the appendages (Moss and Lichtwardt, 1976). At maturity, sporangioles are shed from the generative cell by its rupture either at the base of the sporangiole or at the base of the supporting branch (Fig. 3–17). A minute collar consisting of the remnant of the generative cell is left behind. As the sporangiole moves away from the thallus, the appendages unwind and trail behind. They are nonmotile but may be adaptive in preventing the sporangiole from floating away from a population of potential hosts and also in entangling the sporangiole in debris that might be consumed by the host. A second adaptive feature is that germination cannot occur in the water but only in the gut of the host, where the thallus can develop. Once within the host, the sporangiospore germinates. In *Smittium,* germination may occur when

(a)   (b)

Figure 3–19 Zygospores of *Trichozygospora chironomidarum,* a Trichomycete: (a) the conjugating cells (*cj*) have produced a zygosporophore (*zs*) and an immature zygospore (*z*); (b) mature released zygospore with collar and numerous appendages. [(a) Courtesy S. T. Moss and R. W. Lichtwardt, 1977, Can. J. Botany 55:3099–3110; (b) courtesy R. W. Lichtwardt, 1976; *in*: Jones, E. B. G., Ed., Recent advances in aquatic mycology, John Wiley & Sons, New York.]

the sporangiospore forms a germ tube that ruptures the sporangiole wall or when the entire sporangiospore protoplast and wall escape from the sporangiole. Once the sporangiospore germinates, it attaches itself to the cuticle of the host by a holdfast. In studies of *Smittium*, germination and attachment were observed to occur within a half hour of ingestion of the spores (Williams and Lichtwardt, 1972). Fairly rapid germination and holdfast production help prevent the sporangiospores or young thallus from being expelled from the gut.

Many members of the Harpellales reproduce sexually by producing zygospores (Figs. 3–19 and 3–20). Conjugation may occur between different thalli if the thalli are unbranched and between cells of the same thallus or different thalli in the branched forms. Conjugation occurs between hyphae, and then a specialized branch (the *zygosporophore*) develops from one of the conjugated cells (Fig. 3–19) in all genera except *Glotzia,* where it arises from a conjugation tube. The zygospore develops as an outgrowth of the zygosporophore. At maturity, the zygospore is an elongated uninucleate cell terminating in two conelike ends

(a)

(b)

Figure 3–20 Ultrastructural details of zygospores of *Trichozygospora chironomidarum:* (a) zygospore borne on the zygosporophore; both the apical thickenings (*ta*) and appendages (*ap*) of the zygospores are visible, × 2001; (b) detail of the numerous folded zygospore appendages that lie to the outside of the plasmalemma of the zygosporophore but are within the collar region, × 3680. [Courtesy S. T. Moss and R. W. Lichtwardt, 1977, Can. J. Botany 55:3099–3110.]

with thickened walls (Figs. 3–19 and 3–20). The zygospore may be perpendicular to the zygosporophore or may occur in other spatial relationships to it. Cytological aspects of sexual reproduction are incompletely known, as neither karyogamy nor meiosis has been observed.

Conjugation does not occur in *Genistellospora homothallica,* and parthenogenetic azygospores are formed. The azygospores are released by rupture of the zygosporophore, and appendages similar to those formed on the sporangioles unwind and trail behind as they are dispersed (Moss and Lichtwardt, 1977).

REFERENCES CITED

AINSWORTH, G. C. 1973. Introduction and keys to higher taxa. *In:* Ainsworth, G. C., F. K. Sparrow, and A. S. Sussman, Eds. The fungi: an advanced treatise. Academic Press, New York, Vol. 4B, pp. 1–7.

BAKER, K. L., G. R. HOOPER, and E. S. BENEKE. 1977. Ultrastructural development of merosporangia in the mycoparasite *Syncephalis sphaerica* (Mucorales). Can. J. Botany 55:2207–2215.

BENJAMIN, R. K. 1959. The merosporangiferous Mucorales. Aliso 4:149–169.

——. 1966. The merosporangium. Mycologia 58:1–42.

——, and B. S. MEHROTRA. 1963. Obligate azygospore formation in two species of *Mucor* (Mucorales). Aliso 5:235–245.

BENNY, G. L., and H. C. ALDRICH. 1975. Ultrastructural observations on septal and merosporangial ontogeny in *Linderina pennispora* (Kickxellales; Zygomycetes). Can. J. Botany 53:2325–2335.

BRACKER, C. E. 1968. The ultrastructure and development of sporangia in *Gilbertella persicaria.* Mycologia 60:1016–1067.

DRECHSLER, C. 1959. Several Zoopagaceae subsisting on a nematode and on some terricolous amoebae. Mycologia 51:787–823.

DUDDINGTON, C. L. 1973. Zoopagales. *In:* Ainsworth, G. C., F. K. Sparrow, and A. S. Sussman, Eds. The fungi: an advanced treatise. Academic Press, New York, Vol. 4B, pp. 231–234.

FLETCHER, J. 1972. Fine structure of developing merosporangia and sporangiospores of *Syncephalastrum racemosum.* Arch. Mikrobiol. 87:269–284.

——. 1973. Ultrastructural changes associated with spore formation in sporangia and sporangiola of *Thamnidium elegans* Link. Ann. Botany (London) 37:963–971.

GERDEMANN, J. W. 1968. Vesicular-arbuscular mycorrhizae and plant growth. Ann. Rev. Phytopathol. 6:397–418.

——, and J. M. TRAPPE. 1974. The Endogonaceae in the Pacific Northwest. Mycol. Mem. 5:1–76.

GODFREY, R. M. 1957. Studies of British species of *Endogone.* I. Morphology and taxonomy. Trans. Brit. Mycol. Soc. 40:1127–1135.

HAWKER, L. E., and M. A. GOODAY. 1967. Delimitation of the gametangia of *Rhizopus sexualis* (Smith) Callen: an electron microscope study of septum formation. J. Gen. Microbiol. 49:371–376.

——, and ——. 1968. Development of the zygospore wall in *Rhizopus sexualis* (Smith) Callen. J. Gen. Microbiol. 54:13–20.

——, and ——. 1969. Fusion, subsequent swelling and final dissolution of the apical walls of the progametangia of *Rhizopus sexualis* (Smith) Callen: an electron microscope study. New Phytologist 68:133–140.

——, B. THOMAS, and A. BECKETT. 1970. An electron microscope study of structure and germination of conidia of *Cunninghamella elegans* Lendner. J. Gen. Microbiol. 60:181–189.

HESSELTINE, C. W., and J. J. ELLIS. 1973. Mucorales. *In:* Ainsworth, G. C., F. S. Sparrow, and A. S. Sussman, Eds. The fungi: an advanced treatise. Academic Press, New York, Vol. 4B, pp. 187–217.

KHAN, S. R., and P. H. B. TALBOT. 1975. Monosporous sporangiola in *Mycotypha* and *Cunninghamella*. Trans. Brit. Mycol. Soc. 65:29–39.

LICHTWARDT, R. W. 1976. Trichomycetes. *In:* Jones, E. B. G., Ed. Recent advances in aquatic mycology. Wiley, New York, pp. 651–671.

——. 1985. The Trichomycetes: fungal associates of arthropods. Springer-Verlag, New York, 343 pp.

MOSS, S. T. 1979. Commensalism of the Trichomycetes. *In:* Batra, L. R., Ed. Insect-fungus symbiosis: nutrition, mutualism, and commensalism. Allanheld, Osmun, Totowa, N.J., pp. 175–227.

——, and R. W. LICHTWARDT. 1976. Development of trichospores and their appendages in *Genistellospora homothallica* and other Harpellales and fine-structural evidence for the sporangial nature of trichospores. Can. J. Botany 54:2346–2364.

——, and ——. 1977. Zygospores of the Harpellales: an ultrastructural study. Can. J. Botany 55:3099–3110.

MOSSE, B. 1970-a. Honey-coloured, sessile *Endogone* spores. I. Life history. Arch. Mikrobiol. 70:167–175.

——. 1970-b. Honey-coloured, sessile *Endogone* spores. II. Changes in fine structure during spore development. Arch. Mikrobiol. 74:129–145.

——. 1970-c. Honey-coloured, sessile *Endogone* spores. III. Wall structure. Arch. Mikrobiol. 74:146–159.

——, and G. D. BOWEN. 1968. A key to the recognition of some *Endogone* spore types. Trans. Brit. Mycol. Soc. 51:469–483.

NICHOLSON, T. H., and J. W. GERDEMANN. 1968. Mycorrhizal *Endogone* species. Mycologia 60:313–325.

O'DONNELL, K. L., J. J. ELLIS, C. W. HESSELTINE, and G. R. HOOPER. 1977. Azygosporogenesis in *Mucor azygosporus*. Can. J. Botany 55:2712–2720.

——, S. L. FLEGLER, and G. R. HOOPER. 1978. Zygosporangium and zygospore formation in *Phycomyces nitens*. Can. J. Botany 56:91–100.

SAIKAWA, M. 1977. Ultrastructure of septa of two species of Dimargaritaceae (Mucorales). J. Japan. Botany 52:200–203.

SCHIPPER, M. A. A., R. A. SAMSON, and J. A. STALPERS. 1975. Zygospore ornamentation in the genera *Mucor* and *Zygorhynchus*. Persoonia 8:321–328.

WILLIAMS, M. C., and R. W. LICHTWARDT. 1972. Infection of *Aedes aegyptii* larvae by axenic cultures of the fungal genus *Smittium* (Trichomycetes). Am. J. Botany 59:189–193.

Additional Recommended References

CUTTER, V. M. 1942-a. Nuclear behavior in the Mucorales. I. The *Mucor* pattern. Bull. Torrey Botan. Club 69:480–508.

——. 1942-b. Nuclear behavior in the Mucorales. II. The *Rhizopus, Phycomyces,* and *Sporodinia* patterns. Bull. Torrey Botan. Club 69:592–616.

DRECHSLER, C. 1961. Two species of *Conidiobolus* often forming zygospores adjacent to antheridium-like distentions. Mycologia 53:278–303 (Entomophthorales).

GAUGER, W. L. 1961. The germination of zygospores of *Rhizopus stolonifer*. Am. J. Botany 48:427–429 (Mucorales).

——. 1965. The germination of zygospores of *Mucor hiemalis*. Mycologia 57:634–641.

GERDEMANN, J. W., and J. M. TRAPPE. 1975. Taxonomy of the Endogonaceae. *In:* Sanders, F. E., B. Mosse, and P. B. Tinker, Eds. Endomycorrhizas. Academic Press, New York, pp. 35–51 (Mucorales).

GIBSON, J. L., J. W. KIMBROUGH, and G. L. BENNY. 1986. Ultrastructural observations on Endogonaceae (Zygomycetes): *Endogone pisiformis*. Mycologia 78:543–553 (Mucorales).

KING, D. S. 1977. Systematics of *Conidiobolus* (Entomopthorales) using numerical taxonomy. Can. J. Botany 55:718–729.

——. 1979. Systematics of fungi causing Entomophthoramycosis. Mycologia 71:731–745 (Entomophthorales).

MOSS, S. T. 1975. Septal structure in the Trichomycetes with special reference to *Astreponema gammari* (Eccrinales). Trans. Brit. Mycol. Soc. 65:115–127.

SASSEN, M. M. A. 1962. Breakdown of cell wall in zygote formation of *Phycomyces blakesleeanus*. Proc. Koninkl. Ned. Akad. Wetenschap. Ser. C 65:447–452 (Mucorales).

YOUNG, T. K. W. 1968. Electron microscopic study of asexual spores in Kickxellaceae. New Phytologist 67:823–836 (Mucorales).

——. 1973. Ultrastructure of the sporangiospore of *Coemansia reversa* (Mucorales). Trans. Brit. Mycol. Soc. 60:57–63.

4

Division Amastigomycota—
Subdivision Ascomycotina

Members of the Subdivision Ascomycotina are often called "ascomycetes," a name known to all mycologists but without taxonomic rank. This is the largest fungal group, containing almost 2000 genera. These include a wide range of diverse organisms such as the yeasts, powdery mildews, cup fungi, and the edible delicacies, the morels and truffles. Members of the Ascomycotina are predominantly terrestrial, although a large number live in fresh or marine waters. The majority are saprophytes on decaying plant debris. Many saprophytic members are highly specialized and are capable of growing only on decaying parts of a certain host species; they may even be restricted to a particular part of the host, such as the petiole of the leaf. Other specialized saprophytic types include those fungi that form sporocarps only where a fire recently occurred or on the dung of a certain kind of animal. Many ascomycetes are parasites on plants and, less commonly, on insects or other animals (Chapter 13).

Fungi in the Subdivision Ascomycotina are of great importance, in either a destructive or beneficial sense. Devastating plant diseases such as chestnut blight, apple scab, blackspot of rose, Dutch elm disease, and peach leaf curl are caused by ascomycetes. Beneficial members include the yeasts used in commercial production of alcohol and bread and *Claviceps purpurea*, or ergot, used as a source of medicinal agents. Manufacture of these products is discussed further in Chapter 16.

THE ASCUS AND SEXUAL REPRODUCTION

One characteristic shared by these fungi is that their sexual spores are borne within an *ascus* (plural, *asci*). The ascus is a cell that at first contains a diploid nucleus resulting from karyogamy, and this nucleus undergoes meiosis. Haploid ascospores with thick walls are formed within the ascus (Fig. 4-1).

Figure 4–1 Asci and ascospores. Note the variation in form and numbers of the ascospores. Paraphyses, discussed on page 106, are illustrated by two of the asci. [From A. Engler and K. Prantl, 1897, Die natürlichen Pflanzenfamilien, Engelmann, Leipzig; and H. M. Fitzpatrick, 1920, Mycologia 12:206–267.]

To understand the role of the ascus in the life cycle, we begin with the ascospore. The haploid ascospore germinates, producing either individual vegetative cells or a monokaryotic mycelium. These cells or mycelia may exist for an extended period of time in the assimilative or vegetative state, perhaps reproducing nonsexually.

Plasmogamy

The sexual cycle is usually initiated when plasmogamy takes place. Plasmogamy may occur by the isogamous union of two similar cells or hyphae or by the heterogamous union of dissimilar cells, a differentiated female gametangium with a detached male cell (a *spermatium*) or a female gametangium with a male gametangium (the antheridium). The gametangia may be simple and rather similar single-celled structures, or they may be quite complex (Figs. 4–2 and 4–3). The female gametangium (the *archicarp*) may be divided into a number of cells. Proceeding from the apex downward, these are (1) the *trichogyne*, (2) the *ascogonium*, and (3) the basal cells that support the remainder of the archicarp. The

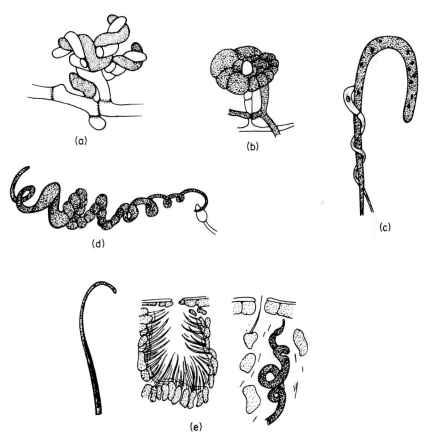

Figure 4–2 Examples of reproductive structures in the Ascomycotina (the archicarp is shaded and the male elements are unshaded): (a) Discomycete, *Ascodesmis nigricans*; (b) Discomycete, *Ascobolus magnificus*; (c) Plectomycete, *Penicillium wortmanni*; (d) Discomycete, *Ascobolus carbonarius*; (e) Pyrenomycete, *Polystigma rubrum*. The elongated cells in the sporocarp at the left are probably spermatia. [Adapted from: (a) P. Claussen, 1905, Z. Botan. 63:1–27; (b) B. O. Dodge, 1920, Mycologia 12:115–134; (c) P. A. Dangeard, 1907, Botaniste 10:1–385; (d) B. O. Dodge, 1912, Bull. Torrey Botan. Club 39:139–197; (e) V. H. Blackman and E. J. Welsford, 1912, Ann. Botany (London) 26:761–767.]

Figure 4–3 Scanning electron micrograph of the gametangia of *Pyronema domesticum*. The ascogonia are the balloon-like structures, and each is producing a tubular trichogyne that is in contact with a club-shaped antheridium. × 1000. [Courtesy K. L. O'Donnell.]

trichogyne is the terminal cell that initially receives the male protoplasm and allows it to pass to the ascogonium. The ascogonium consists of either a single cell or a chain of cells and is the final recipient of the male protoplasm in plasmogamy (Fig. 4–4). Many ascomycetes are apogamous and the male nuclei are not transferred to the archicarp. Also, many lack sexual organs. Any exchange of nuclei must occur by hyphal anastomosis in the latter cases.

Following plasmogamy there are two basic types of behavior:

1. Karyogamy may follow immediately after plasmogamy, transforming those cells that functioned as gametes into zygotes. Either the newly formed zygote or its vegetative progeny becomes the ascus.
2. Alternatively, karyogamy may be delayed and the dikaryotic condition initiated after plasmogamy is prolonged for several cell generations. This may be accomplished by forming a succession of dikaryotic cells (as in the Taphrinales, pages 118–119) or by forming a special hyphal system, the ascogenous hyphae.

The Ascogenous Hyphae

Ascogenous hyphae are formed by the majority of the members of the Ascomycotina. The ascogenous hyphae arise from the ascogonium (Fig. 4–4) or from other cells if the ascogonium is absent. Usually the ascogenous hyphae contain haploid nuclei of both male and female origin and are therefore dikaryotic. These nuclei divide repeatedly, and the cells in the ascogenous system contain several nuclei. The ascogenous hyphae repeatedly branch, and finally, the individual tips reach the site where the asci will be produced. Usually a *crozier* is formed when the tip of the ascogenous hypha recurves and forms a hooked cell. Two nuclei are present in the crozier. These two nuclei divide simultaneously, with their division figures parallel to the long axis of the crozier. Two septa form across the spindle and form two cells from the crozier. These are the apical cell,

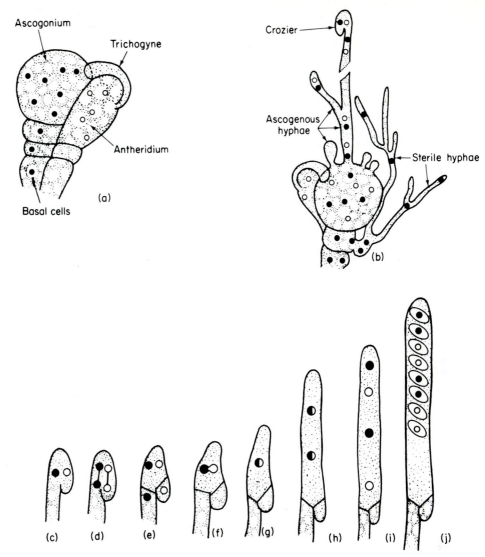

Figure 4–4 Sexual organs, plasmogamy, and ascus formation in *Pyronema domesticum*: (a) the archicarp and antheridium are undergoing plasmogamy; note the continuity between the trichogyne and antheridium; (b) ascogenous hyphae are arising from the ascogonium, and sterile hyphae are arising from the basal cells of the archicarp; (c) binucleate crozier; (d) mitotic division of nuclei; (e) formation of septa to delimit penultimate cell; (f) karyogamy; (g) diploid nucleus in young ascus; (h) binucleate ascus after first division of meiosis: (i) four-nucleate ascus after second divison of meiosis; (j) mature ascus with eight ascospores after final mitotic division. [(a, b) Adapted from E. J. Moore, 1963, Am. J. Botany 50:37–44.]

the recurved portion of the hyphal tip, and the second (penultimate) domelike cell, which will become the ascus. Sister nuclei are separated by the septa, which isolate two nuclei in the penultimate cell and a third nucleus in the apical cell; the fourth nucleus remains in the ascogenous hypha. The two nuclei in the penultimate cell fuse and form a diploid nucleus, and this cell is now a young ascus (Fig. 4–4). In those fungi forming an ascogenous system, the ascus is the only diploid stage in the life cycle, and the haploid stage is quickly regained in the meiotic division that follows.

Ascus Formation

Beginning with the young diploid ascus, ascospore formation is similar in all the Ascomycotina. The diploid nucleus in the ascus undergoes a meiotic division, and the ascus acquires two nuclei as a result of meiosis I and four nuclei as a result of meiosis II. In some ascomycetes the asci will have no further divisions, while in most species a single mitotic division follows and the four nuclei divide to form eight nuclei. The resulting nuclei are then incorporated into ascospores (Fig. 4-4). Ascospore development has been studied extensively at the ultra-structural level, and with the exception of members of the Hemiascomycetes (page 113), virtually all members of the Ascomycotina conform to the following general pattern (e.g., Rosing, 1982; Stiers, 1974, 1976; Tyson and Griffiths, 1976). Two continuous unit membranes form a sac near the periphery of the ascus and lie parallel to each other and the ascus wall [Fig. 4-5(a)]. These membranes comprise the *ascus vesicle*. The origin of the ascus vesicle is difficult to determine but the membranes apparently develop from a preexisting membrane system such as the plasmalemma or nuclear envelope. The two membranes then move inward, meanwhile pushing organelles into new positions near the nuclei [Fig. 4-5(b)]. Each nucleus is eventually isolated, along with its associated organelles and surrounding cytoplasm, from the remainder of the cytoplasm by two concentric membranes. Each cell isolated in this manner becomes an ascospore. The inner membrane of the concentric pair becomes the plasmalemma of the ascospore, and ascospore wall material is deposited between the two membranes [Fig. 4-5(c)]. The wall is at first uniform in appearance but as it matures, distinct layers become apparent. The number of wall layers and the sequence in which they differentiate vary from species to species (e.g., Merkus, 1976). Some cytoplasm remains in the ascus around the ascospore but it disappears as the ascus matures [Fig. 4-5(c,d)]. Although 8 ascospores are usually produced in each ascus, the number of ascospores may be 4 if the third division does not occur or it may be 16, 32, or any multiple of 8, depending on the number of mitotic divisions taking place (Fig. 4-1). As many as 7000 ascospores occur in *Trichobolus* (Kimbrough and Korf, 1967). Asci may or may not be produced within or upon complex multicellular sporocarps, the *ascocarps*.

Sterile Tissues in an Ascocarp

The sequence from plasmogamy through ascospore formation represents only the sexual phase of development in a typical Ascomycotina member that forms an ascocarp. In addition to the sexual organs, ascogenous hyphae, and asci, an ascocarp usually contains a large quantity of sterile tissue. The sterile tissue arises as hyphae, usually from the basal cells supporting the gametangia, and then envelopes the sexual structures and becomes organized into tissues. In some instances, the sterile elements may be represented by only a few hyphae incompletely surrounding the developing sexual system. The sterile hyphae are not dikaryotic, as the ascogenous hyphae frequently are, but rather are monokaryotic, as they arise from mycelium or cells that are not the products of plasmogamy (Fig. 4-6).

The hyphae are intertwined so that distinct tissues are formed, and the ascocarp is often divided into anatomical regions characterized by a distinctive type of tissue. Hyphae from these tissues may terminate in sterile, hairlike processes. These are (1) *paraphyses*, which lie among the asci and are free at their apices; (2) *pseudoparaphyses*, which grow downward from the upward surface of the ascocarp cavity and form a vertical palisade of hairs, usually becoming attached at the bottom by interweaving to form a matlike plexus; and (3) *periphyses*, which are small hairs that line a canal in the necklike extension of some ascocarps (perithecia, see below) (Fig. 4-7).

There are fundamental differences in the morphology of the ascocarp and in the origin of the surrounding tissues. Some ascomycetes form a *stroma*, a pad of vegetative tissue. *Locules* (cavities) may form within the stroma. Asci are then formed within the locules and are not surrounded by any sterile tissue that originates from the archicarp or that is not sexually stimulated. A second major type of ascocarp is the *perithecium*, which may be immersed within a stroma or free on the substratum and is characterized by an enclosed cavity bearing asci. This cavity is lined by a wall of sterile cells arising from the archicarp. The *perithecium* may be of various shapes, most often flask-shaped, and opens by forming a slit or ostiole. A third type of ascocarp is the *cleistothecium*, which is globular and without a regular means of opening. Finally, an *apothecium* is an open ascocarp that bears asci on an exposed surface. These ascocarp types are illustrated in Fig. 4–8.

The number of asci borne by an ascocarp varies from one to several. They may be borne randomly within the ascocarp cavity, or they may arise as a clustered *fascicle* from the base of the ascocarp cavity. Most often, they are borne in a palisadelike layer, the *hymenium*, which lines the base of the ascocarp cavity or an exposed part of the apothecium.

Ascus Types

Two fundamentally different types of asci occur. These are the bitunicate and unitunicate asci. The *bitunicate ascus* has a rigid outer wall and an inner wall which is flexible and can become considerably extended. When ascospore discharge takes place, the outer wall splits and may sometimes slip downward to form a collar, and the inner wall protrudes through the opening and expands (Fig. 4–9). The ascospores are expelled in succession through the opening of the inner ascus wall. The *unitunicate* ascus possesses a single ascus wall, which may be laminated. The unitunicate ascus wall either may be uniform in thickness throughout [Fig. 4–9(a)] or may have a thickened apex. The wall may rupture to allow ascospore discharge to take place, or a pore may occur in the apex. Ascospore discharge occurs simultaneously.

NONSEXUAL REPRODUCTION

Many members of the Ascomycotina have nonsexual reproduction (Fig. 4–10). Therefore their life cycles consist of two alternating phases, nonsexual and sexual. This entire life cycle including both nonsexual and sexual cycles is sometimes termed the *holomorph*. The nonsexual phase of the cycle may be referred to as the *anamorph* while the sexual phase is termed the *telomorph*. (Until recently, the anamorph was called the ''imperfect'' stage and the telomorph the ''perfect'' stage: you may encounter these terms as you read other literature.) At its simplest, nonsexual reproduction may occur by budding of cells in the yeasts. (This is discussed further on page 115.) Nonsexual reproduction by means of conidia is very common throughout the group. Conidia are borne on specialized hyphae, the *conidiophores*. The conidiophores may be scattered loosely over the mycelium, or they may be organized into groups or into sporocarps that closely resemble ascocarps. The conidial stage (anamorph) may follow germination of ascospores. Often the conidia are borne on a mycelium separate from that producing the sexual stage, which gives a spatial separation of the anamorph and telomorph. In many ascomycete life cycles, the conidia will be produced in abundance throughout the spring and summer, with the ascocarp beginning its development in the late autumn and maturing the following spring, which culminates in ascospore discharge.

An ascomycete may propagate so constantly and persistently through

(a)

(b)

Figure 4–5 Stages in ascospore delimitation: (a) longitudinal section of a young ascus showing the membrane sac forming, × 9900; (b) the membrane sac is at an advanced stage of invagination, and the ascospore initials are almost completely delimited, × 8500; (c) production of wall material (clear areas) has begun, × 4900; (d) nearly mature reticulated ascospore in ascus; note that the surrounding protoplasm has disappeared, × 12,600. [Courtesy G. C. Carroll; (a, b) 1967, J. Cell Biol. 33:218–224; (c, d) 1966, Thesis, University of Texas.]

(c)

(d)

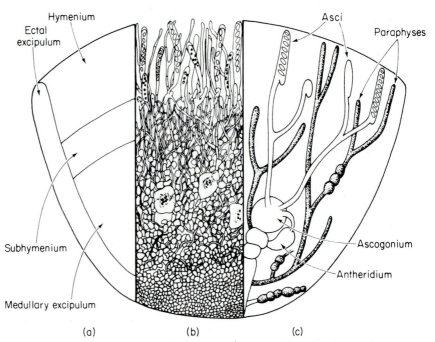

Figure 4–6 Longitudinal section through an apothecium of *Pyronema domesticum:* (a) division into major tissue zones; (b) natural appearance of the tissues; (c) diagrammatic representation of the origin of sexual structures (unshaded) and sterile structures (shaded) that intertwine to form the tissues. [(b) From E. J. Moore, 1962. Thesis, Cornell University.]

Figure 4–7 Diagrammatic longitudinal sections through perithecia to show the position and nature of sterile filaments.

Figure 4–8 Ascocarp types.

Figure 4–9 Ascus types: (a) unitunicate ascus; (b) bitunicate ascus. From left to right, a mature bitunicate ascus before spore discharge begins; the inner wall expanded in the process of spore ejaculation; and a collapsed ascus showing the thickened inner wall and the thin, rolled-back outer wall which forms the collar. [(a) From A. Engler and K. Prantl, 1897, Die natürlichen Pflanzenfamilien, Engelmann, Leipzig; (b) from E. S. Luttrell, 1960, Mycologia 52:64–79.]

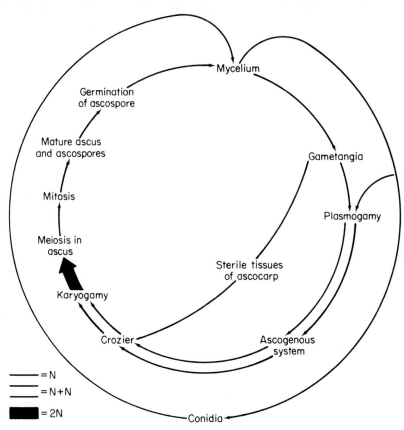

Figure 4–10 Life cycle of a typical member of the Ascomycotina that forms a multicellular ascocarp, showing both sexual and nonsexual reproduction.

its nonsexual stage that mycologists are unaware that a particular conidia-producing fungus is the anamorph of an ascomycete. They may discover this relationship when they germinate ascospores of a known ascomycete and find that a conidial fungus is produced which under certain conditions produces an ascocarp. For further information about the morphology and development of the nonsexual structures, consult Chapter 6.

CLASSIFICATION

Taxonomy of the Subdivision Ascomycotina is based largely on gross morphology, anatomy, and life history of the individual organisms. To a lesser extent, physiological characters are used and are especially important in the taxonomy of the yeasts. This subdivision may be divided into six classes (Ainsworth, 1973) based primarily on the presence or absence of an ascocarp and its morphology. Structure of the ascus also plays an important role in classification. Within the following classes, the orders are those used by Kreger-Van Rij (1973), Kramer (1973), Fennell (1973), Müller and Von Arx (1973), Korf (1973), Benjamin (1973), and Von Arx and Müller (1975), respectively. Partly because of the large number of organisms in this group and their great diversity, not all mycologists agree on how they should be classified. An outline of the classification system used here follows:

Subdivision Ascomycotina
Class Hemiascomycetes
Order Endomycetales

Order Protomycetales (not treated further)
Order Taphrinales
Class Plectomycetes
Order Eurotiales
Class Pyrenomycetes
Order Erysiphales
Order Meliolales
Order Coronophorales
Order Sphaeriales
Class Discomycetes
Order Medeolariales (not treated further)
Order Cyttariales (not treated further)
Order Phacidiales
Order Ostropales
Order Helotiales
Order Pezizales
Order Tuberales
Class Laboulbeniomycetes
Order Laboulbeniales
Class Loculoascomycetes
Order Dothideales

Class Hemiascomycetes

The Class Hemiascomycetes contains simple reduced members of the Ascomycotina which are either unicellular or form a limited mycelium. The unicellular organisms are commonly called "yeasts" although there is often no sharp dividing line between the unicellular and mycelial forms. The Hemiascomycetes are especially characterized by the absence of an ascocarp and of ascogenous hyphae. Either a zygotic cell, the diploid progeny developing nonsexually from that cell, or a usually isolated dikaryotic cell is transformed directly into an ascus bearing ascospores. Structurally, the ascus containing ascospores may superficially resemble a sporangium or the fused gametangia of the Zygomycotina, but this cell functions as an ascus and, after karyogamy, the nucleus undergoes two or three successive divisions with no resting stage. Ascospore formation shares many similarities with that outlined on page 106 but differs in one important feature (Ashton and Moens, 1979; Black and Gorman, 1971; Moens, 1971; Syrop and Beckett, 1976). A continuous ascus vesicle is not formed near the periphery of the ascus. Instead, discontinuous membranes arise in the vicinity of each nucleus, perhaps originating as invaginations of the plasmalemma or from the endoplasmic reticulum. The membranes enlarge and surround a nucleus and some associated cytoplasm and cytoplasmic organelles. The nucleus is isolated by the membranes, and development of the ascospore walls proceeds as described previously (page 106). Usually either four or eight ascospores are formed, but the number is variable if some nuclei degenerate after the nuclear divisions occur. In some species, more or fewer than eight ascospores may be formed. The ascospores are usually oval or spherical, but may be elliptical, reniform, needlelike, or shaped like a hat with a brim. Although the ascospores are usually smooth, some yeasts produce rough, warted ascospores.

Fungi resembling the unicellular yeasts also occur in other taxonomic groups (the Basidiomycotina and Deuteromycotina, Chapters 5 and 6). Unlike the yeasts in the Hemiascomycetes, the yeastlike fungi in other groups do not form asci and ascospores. In addition, the DNA base composition differs (e.g., the mole percent of guanine and cytosine combined is lower in those Hemiascomycetes tested than in the yeastlike fungi in the Basidiomycotina). Other differences include cell-wall composition and the ultrastructure of the cell wall.

The Class Hemiascomycetes may be divided into the Orders Endomycetales, Protomycetales, and Taphrinales. The Protomycetales is a very small group of specialized plant parasites and will not be treated further.

ORDER ENDOMYCETALES

Members of the Endomycetales include those fungi that form an ascus directly from a zygote or from the diploid progeny of a zygote. The approximately 150 species include the well-known yeast of commerce *Saccharomyces cerevisiae* (see Chapter 16), in addition to many others of little or no economic importance. The yeasts are well-known for their ability to ferment sugars (Chapter 8), and the ability or inability to ferment particular sugars is useful in delimiting species and genera in this group. In nature, the yeasts are frequently found growing saprophytically in sugary materials. Yeasts and their allies may be found in nectaries of flowers; on leaves and fruits; in plant exudates; in association with insects; in soil; in fresh and marine waters; in sewage; and in foods such as syrups, brines, meats, and milk products. Some parasitic yeasts cause diseases of plants and animals. Yeasts may be distributed by being carried in the air and are probably carried on dust particles. The unicellular yeasts have been widely studied from both the biological and commercial viewpoints owing to their extensive and varying physiology, their potential contribution to basic biological problems such as cytology and phylogeny, and their ease of manipulation in culture.

Thallus. The unicellular yeasts vary from round, oval, or elongate shapes to rather peculiar dumbbell-like or triangular configurations. Macroscopically, yeasts cultured on solid media appear as slimy, mucoid colonies which are often brightly colored. These colonies strongly resemble bacterial colonies.

Cytology of the yeast cells has been studied extensively, and much of our knowledge has been derived from studies of *Saccharomyces cerevisiae* and closely related yeasts (Matile et al., 1969). A typical yeast cell will be described. The yeast cell is surrounded by a cell wall which consists of layers: the outermost layer contains mainly mannan-protein and some chitin, the middle layer consists of glucan and appears lighter than the other two layers in electron micrographs, and the innermost layer consists of a proteinaceous material. The plasmalemma exhibits numerous invaginations. The cytoplasmic contents include glycogen, ribosomes, and mitochondria that may be spherical, rod-shaped, or threadlike and branched. Vesicles containing lipid or catalase are distributed within the cytoplasm and may have originated from the endoplasmic reticulum. The endoplasmic reticulum is sometimes attached to the plasmalemma but is continuous with the nuclear envelope. The nucleus has a large, dense cup-shaped nucleolus and more or less translucent dome-shaped nucleoplasm occupying the remainder of the nucleus (both the nucleolus and nucleoplasm may be seen with a phase contrast microscope). The most conspicuous organelle is a vacuole that is usually so large that it is visible with the light microscope. This vacuole, which has a diameter of approximately 0.3 to 3 micrometers, is bounded by a single membrane and has watery contents. Under certain conditions (e.g., anaerobic conditions), the vacuole may have granular inclusions. By analyzing the contents of isolated vacuoles it has been determined that they contain high levels of hydrolytic enzymes, including proteases, ribonuclease, and esterase. The vacuole apparently functions as a lysosome, representing an isolated compartment of the cell where macromolecules such as proteins and nucleic acids may be degraded by the hydrolytic enzymes before their components are incorporated into newly synthesized macromolecules. In addition to functioning as a lysosome, the vacuole serves as a storage place for cellular components such as amino acids or potassium ions that may be utilized later.

Cellular division in the unicellular yeasts occurs when nuclear division is accompanied by *fission* or *budding.* Fission is the division of the mother cell into two daughter cells and involves the formation of new wall material to divide two essentially equal protoplasts. In budding, a new cell arises as a protrusion on a narrow base from the mother cell, increases in size, and eventually breaks away [Fig. 4–11(b)]. Several cellular changes are involved in budding. In a cell preparing to form a bud, the endoplasmic reticulum fuses and forms a nearly closed envelope containing the nucleus and vacuole and subsequently produces vesicles near the site where the bud will form. The vesicles are thought to contain enzymes involved in the localized weakening of the wall that is followed by pore formation. The vesicles may also contain proteinaceous wall precursors. Production of a pore in the mother cell is accompanied by the simultaneous production of new wall materials appearing in the daughter cell. The nucleus originally lies between the vacuole and the developing bud. Nuclear division

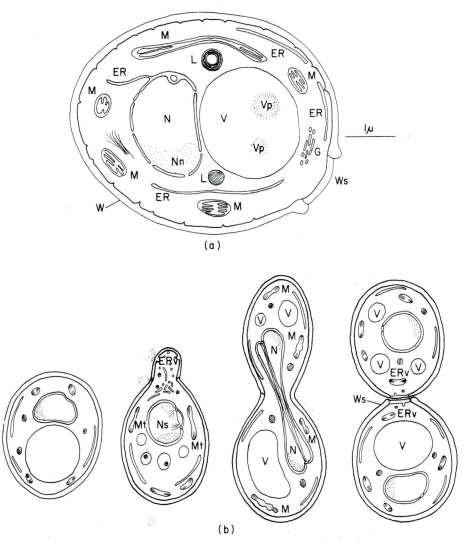

Figure 4–11 Yeast *Saccharomyces cerevisiae.* (a) Diagram of a resting cell: *ER,* endoplasmic reticulum; *ERv,* vesicles derived from endoplasmic reticulum; *G,* Golgi apparatus; *L,* lipid granule; *M,* mitochondrion; *N,* nucleus; *Nm,* nucleolus; *Ns,* spindle apparatus; *V,* vacuole; *Vp,* polymetaphosphate granule; *W,* cell wall; *Ws,* bud scar; (b) successive stages in budding. [Adapted from P. Matile, H. Moor, and C. F. Robinow, 1969, *in:* Rose, A. H., and J. S. Harrison, Eds., The yeasts, Academic Press, New York, Vol. 1, pp. 219–320.]

occurs by elongation and constriction and without loss of the nuclear envelope. One half of the elongated nucleus enters the bud in an amoeboid manner, temporarily leaving a long, narrow neck between it and the other half of the nucleus, which remains in the mother cell. Eventually, division of the nucleus is completed. The vacuole fragments and the products of the fragmentation are distributed between the mother and daughter cells. The plasmalemma constricts inward to separate the protoplast of the bud from that of the mother cell, and deposition of two walls occurs at the point of junction. Finally, the bud separates from the mother cell, leaving a prominent scar on the mother cell and a smaller almost invisible scar on the daughter cell where the walls were connected. As many as 24 bud scars have been detected on a single yeast cell, corresponding to the number of daughter cells formed by that cell. In some yeasts, especially in older cultures, new cells formed by budding may remain together in simple or branched chains which superficially resemble a mycelium. Other genera, especially the yeast allies, form a true but limited mycelium through cell division. More than one thallus type may occur in a single species, depending partially on age and cultural conditions.

Nonsexual Reproduction. In the unicellular yeasts, cell division or budding comprises nonsexual reproduction. Mycelial allies may reproduce nonsexually by forming yeastlike conidia, or the mycelium may fragment into a number of *oidia*. Conidia and oidia may propagate by budding or dividing for a period of time. These cells are often indistinguishable from the cells of a unicellular yeast, and they may form colonies of a similar appearance. A reversion to the mycelial type occurs eventually.

Sexual Reproduction. Sexual reproduction occurs by conjugation of either similar or dissimilar cells or gametangia. Morphology of the reproductive structures and events involved in plasmogamy vary widely. There may be the isogamous union of haploid vegetative cells or of haploid ascospores to yield a zygote. A heterogamous union may occur between a large and small vegetative cell or between morphologically distinct gametangia. In addition to the morphological variation in reproductive structures, there is variation in the overall life cycle pattern. There may be a long haploid stage with a transient diploid stage, a long diploid stage with a transient haploid stage, or a haploid and diploid stage of approximately equal length (alternation of generations). To explore at least a little of this variation, we examine sexual reproduction and life cycles of three yeasts.

In *Saccharomyces cerevisiae*, similar haploid vegetative cells fuse to form a zygote. The diploid zygote typically buds and produces several generations of diploid cells. Eventually, meiosis takes place in one of these cells, and the cell is converted into an ascus in which the four haploid nuclei are incorporated into ascospores. Upon their release, the haploid ascospores bud and produce several generations of haploid vegetative cells. Later, some of these again function as gametes to initiate the cycle again. In this case, isogamous union of vegetative cells is coupled with a life cycle with both extended haploid and diploid stages [Fig. 4–12(a)].

In *Saccharomycodes ludwigii*, isogamous union between ascospores takes place. There are four ascospores in each ascus, and they become separated into two pairs which lie at opposite ends of the ascus. The ascospores swell and form small beaks. The beaks extend to meet each other and meanwhile broaden to form a canal that eventually provides cytoplasmic continuity between the ascospores. The nuclei from the ascospores migrate into the canal and undergo karyogamy. The canal now penetrates through the ascus and forms a number of diploid vegetative cells. The cycle is completed when meiosis takes place in a diploid vegetative cell, converting it into an ascus. This life cycle combines isoga-

Figure 4–12 Life cycles of: (a) *Saccharomyces cerevisiae;* (b) *Saccharomycodes ludwigii;* (c) *Dipodascopsis aggregatus.* [(b) Figures from A. Guilliermond, 1905, Rev. Gen. Botan. 17:337–376; (c) figures from L. R. Batra, 1959, Mycologia 51:320–355.]

mous union of ascospores with a long diploid phase and a short haploid phase [Fig. 4–12(b)].

Dipodascopsis aggregatus is an example of the mycelial yeast allies that form specialized gametangia [Fig. 4–12(c)]. The vegetative mycelium is haploid. Sexual reproduction is initiated when two adjacent cells in a filament form protrusions which will become an ascogonium and antheridium, respectively. In *D. aggregatus,* the ascogonium is considerably larger than the antheridium. The gametangia fuse at their apex and are separated from their parental cells by septa. Since only the tip fuses, two stalklike structures, or "feet," remain behind. Karyogamy takes place, and the ascogonium elongates into an ascus with two footlike projections. While the ascus is elongating, meiosis takes place, followed by a number of further divisions. The mature ascus is filled with numerous haploid ascospores that form a haploid mycelium upon germination. This life cycle combines heterogamous gametangial copulation with a long haploid phase and a short diploid phase. Although this fungus is a mycelial yeast ally, similar cycles occur in the unicellular yeasts.

ORDER TAPHRINALES

Members of the Taphrinales are plant parasites that cause localized deformation and swelling of the host tissues. A severe disease, peach leaf curl, is caused by *Taphrina deformans,* a member of this group. Peach leaf curl is characterized by pocketlike bulges and curled areas on the leaves. The Taphrinales differ from the Endomycetales principally in that (1) the asci are organized into a palisadelike layer on the surface of the host, and (2) the ascus develops from a dikaryotic cell, not from the zygote. Otherwise, the Taphrinales share many similarities with the yeasts and their allies.

The assimilative structures of the Taphrinales are found within the host tissue, usually as a limited mycelium with dikaryotic cells in which the nuclei divide conjugately (Mix, 1935). The mycelium may occur intercellularly or subcuticularly, or it may penetrate the walls of the host where it develops. In addition to the mycelium, the vegetative structures may consist partially or completely of binucleate, separate cells distributed among host cells.

In most species of *Taphrina,* the vegetative mycelium fragments into binucleate ascogenous cells, usually immediately beneath the upper cuticle of the leaf. These binucleate cells form the asci that penetrate the leaf surface. Less commonly, the cells of the vegetative hyphae may give rise to asci without separating, or if a mycelium is not formed, the yeastlike cells will simply differentiate into asci. Karyogamy takes place in the binucleate cell, and the cell elongates and forms the ascus. The diploid fusion nucleus usually undergoes a single mitotic division; one of the daughter nuclei of this division remains in the basal portion of the developing ascus, which becomes separated as a stalk cell, and the second nucleus migrates into the tip of the young ascus. If the species does not form a stalk cell, this single mitotic division is lacking. Meiosis follows, and the single nucleus in the young ascus divides to give either four or eight ascospore nuclei (Fig. 4–13).

While in the ascus or after their release, the ascospores produce a large number of uninucleate conidia by budding in a yeastlike fashion. In culture, yeastlike colonies consist of these cells. Conidia are capable of overwintering and of initiating new infections in the spring. Apparently, mycelia are incapable of overwintering. In some species (including *T. epiphylla*), conidia regularly conjugate to reestablish the dikaryon before penetration of the host occurs, but in *T. deformans* the single haploid nucleus in the conidium divides to establish a dikaryon that invades the host.

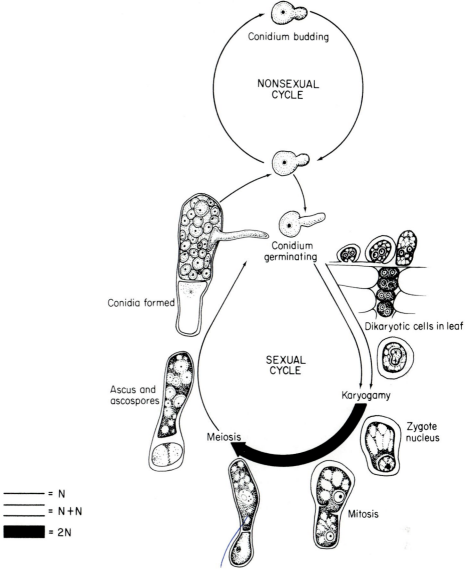

Figure 4–13 Life cycle of *Taphrina deformans.* [Figures in sexual cycle from C. L. Kramer, 1960, Mycologia 52:295–320.]

Class Plectomycetes

ORDER EUROTIALES

The Class Plectomycetes includes the simplest members of the ascocarp-forming ascomycetes. The members of this class occur in a single order, the Eurotiales, and share a combination of characteristics (Fennell, 1973). The asco-carp is a cleistothecium, an enclosed ascocarp lacking an ostiole or other regular means of opening (Figs. 4–14 and 4–15). Paraphyses are lacking and asci are scattered randomly within the cleistothecium. Typically the asci are rounded in shape, contain eight ascospores, and have thin walls that break down at maturity to release the ascospores within the cleistothecium. The ascospores are unicellular and lack germ pores or slits. The cleistothecium opens from exposure to the weather or by pressure exerted by maturation of the contents.

Figure 4–14 Simple cleistothecium of *Emericellopsis microspora*. The asci are surrounded by a thin, transparent peridium. × 850. [Courtesy M. P. Backus and P. A. Orpurt, 1961, Mycologia 53:64–83.]

Figure 4–15 Two simple members of the Eurotiales; *Myxotrichum chartarum* (left, × 125) and *Ctenomyces serratus* (right, × 175). Each cleistothecium bears elaborate appendages. [Courtesy R. K. Benjamin.]

Many of the Plectomycetes produce a conidial stage. This anamorph may be denoted by the name of a conidial genus. This grouping is a *form-genus* because genetically diverse (distantly related) fungi may have similar anamorphs. Therefore fungi assigned to the same form-genus may or may not be related. For a further discussion of taxonomy of conidial fungi, see Chapter 6. The conidial stages of the Plectomycetes are morphologically diverse and have been assigned to a variety of form-genera.

The majority of the Eurotiales live as saprophytes in the soil or on substrata such as dead insects, wood, dung, hair, bones, feathers, horns, or hooves. Skin diseases of humans and other animals may be caused by parasitic species of the form-genera *Trichophyton* and *Microsporum*, which are the conidial stages with telomorphs in *Arthroderma* and *Nannizzia*, respectively. Like those Eurotiales that occur saprophytically on hair, feathers, horns, or hooves, these skin-inhabiting parasites are able to live saprophytically on such materials because they also utilize keratin as a nutrient. The causal agent of a serious and often fatal disease of humans, histoplasmosis, is caused by *Histoplasma capsulatum*, the anamorph of *Emmonsiella capsulata* (Kwon-Chung, 1973).

The form-genera *Penicillium* and *Aspergillus* are among the most commercially important of the conidial fungi. Sometimes a conidial culture of either of these fungi will produce cleistothecia that may be assigned to several genera in the Eurotiales (for example, *Aspergillus* may have cleistothecia in *Eurotium*, while *Penicillium* may have cleistothecia in *Talaromyces* (Benjamin, 1955; Raper and Fennell, 1965; Malloch and Cain, 1972). The telomorphs for many species of *Aspergillus* and *Penicillium* are unknown, however, and therefore these species are considered to be members of the Deuteromycotina (Chapter 6). In general, *Penicillium* and *Aspergillus* are among the most common of soil fungi (Chapter 12) and many species are important in human affairs (Chapter 16). Species of *Penicillium* are used commercially as a source of the antibiotic penicillin and in the production of Roquefort cheese. Species of *Aspergillus* are used commercially to produce citric acid, to biochemically alter steroids that will be used medically, and to manufacture certain Oriental foods. Species of both genera may be responsible for the destruction of stored foods, including grains. *Aspergillus flavus* is especially destructive as it produces aflatoxins, animal carcinogens, in stored grains. *Aspergillus* may also produce cellulolytic enzymes that destroy cloth.

The cleistothecia of some members may be collected directly on their substrata in nature. For example, species of *Onygena* form stalked cleistothecia on weathered hooves or feathers. Some keratinophilic (keratin-utilizing) species may be obtained by placing an appropriate "bait" such as hair in or on soil. After a short period of time, the hair will be invaded by mycelium that subsequently forms cleistothecia. The conidial stages of many members of the Eurotiales are more easily obtained than are the sexual stages but may produce cleistothecia in culture. For example, the conidial stage of *Toxotrichum cancellatum* was isolated from frozen blueberry pastries and subsequently produced cleistothecia in culture (Orr and Kuehn, 1964).

The cleistothecia are typically globose in shape, although stalked cleistothecia may be formed by a few genera (e.g., *Onygena*). The cleistothecia are comparatively small (usually only 5 millimeters or less in diameter), and may be hyaline or some shade of yellow, orange, red, brown, or black. Most species produce isolated cleistothecia surrounded only by the cleistothecial wall (the *peridium*). A few species will produce confluent cleistothecia, or cleistothecia embedded within a stroma. The peridium shows a great deal of variation, ranging from a netlike weft of hyphae that barely covers the asci in some species of *Gymnoascus* to a compact covering consisting of interwoven hyphae or pseudoparenchymatous cells (e.g., in *Eurotium*). Some species have characteristic appendages protruding from the cleistothecium (Fig. 4–15). Among the various types of appendages are some that are spinelike and unbranched, some that have numerous small lateral branches so that each resembles a comb, and some that are unbranched and coil at the tips. Hülle cells, chlamydosporelike enlargements formed on some of the hyphae surrounding the cleistothecia, may occur in species of *Emericella*.

Developmental studies of cleistothecia of various species have been made by several mycologists (e.g., Benjamin, 1955; Kuehn, 1957; Wong and Chien, 1986). When cleistothecia are formed from the vegetative mycelium, differentiated gametangia may be lacking. In these cases, it is not known whether plasmogamy occurred between hyphae. Differentiated gametangia are formed by the majority of the species and may be either clavate or coiled (Fig. 4–16). The antheridium is most often a straight, clublike cell. Usually the archicarp forms a flat circular coil in a single plane or a symmetrical cylindrical spiral around the antheridium, or it may become entwined with the antheridium. Fusion occurs when a pore is formed between the trichogyne and the antheridium but has not been observed in some species that form an antheridium. Ascogenous hyphae arise from the ascogonium and grow in various directions and lengths. Some species form croziers prior to ascus formation, while others do not. In those species that do not form croziers, the ascogenous hyphae may become septate and the separate cells enlarge and differentiate into the asci. Sometimes the asci are borne in short chains or are formed singly on short lateral branches. The asci

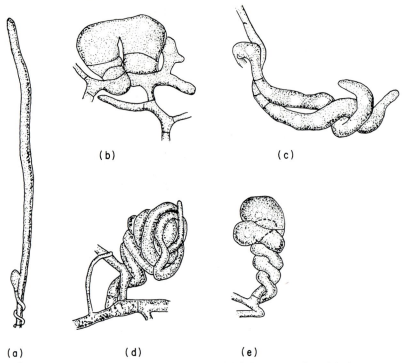

Figure 4–16 Representative gametangia of the Eurotiales: (a) elongate clavate ascogonium and spiraled antheridium of *Penicillium vermiculatum;* the ascogonium may reach a length of 250 micrometers; (b) similar clavate gametangia of *Myxotrichum conjugatum* have fused; (c) similar clavate gametangia of *Arachniotus reticulatus* are loosely entwined; fusion has not been observed between these gametangia; (d) the archicarp of *Myxotrichum emmonsii* has formed a tight coil around a clavate antheridium; (e) the gametangia of *Myxotrichum uncinatum* are coiled together. (b–e) × 1300. [(a) from C. W. Emmons, 1935, Mycologia 27:128–150; (b) from H. H. Kuehn, 1955, Mycologia 47:878–890; (c) from H. H. Kuehn, 1957, Mycologia 49:55–67; (d, e) from H. H. Kuehn, 1955, Mycologia 47:533–545.]

occur randomly within the cleistothecium. Early during the above developmental sequence the gametangia become surrounded by hyphae arising from the basal cells of the archicarp, from the vegetative mycelium, or from both. These sterile hyphae will form the peridium (Fig. 4–17).

Class Pyrenomycetes

The Pyrenomycetes comprise the largest fungal class, more than 6000 representatives being known. Not all authorities agree on which fungi should be included in this class or on the manner in which they should be divided into orders. Müller and Von Arx's (1973) treatment will be followed here, and the Pyrenomycetes divided into the orders Erysiphales, Meliolales, Coronophorales, and Sphaeriales.

In the Pyrenomycetes, the ascocarp is an enclosed structure. The ascocarp usually is a perithecium having a necklike extension opening by a canal, the ostiole. Periphyses line the ostiole. Some Pyrenomycetes form a cleistothecium without an ostiole or an ascostroma with locules. The asci are typically arranged in a basal layer, the *hymenium,* within the ascocarp (in contrast to the random arrangement in the Plectomycetes). The asci are unitunicate.

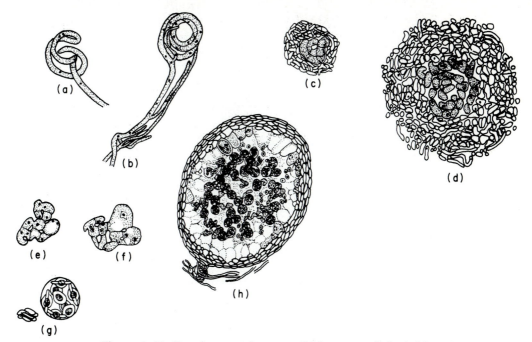

Figure 4–17 Developmental stages of *Neosartorya fischeri*: (a) young, nonseptate ascogonial coil; (b) septate ascogonial coil giving rise to vegetative hyphae that will form the cleistothecial wall; (c) enlarged ascogonial coil surrounded by vegetative hyphae; (d) further enlargement of young cleistothecium with ascogonium giving rise to ascogenous hyphae; (e) crozier formation; (f) ascus formation; (g) mature ascus with characteristic ringed ascospores; (h) mature cleistothecium. [From L. S. Olive, 1944, *Mycologia* 36:266–275.]

ORDER ERYSIPHALES

The Erysiphales contain a single family, the Erysiphaceae (Yarwood, 1957, 1973). There are at least 90 species in this group, and all of them are obligate parasites, causing the diseases known as the powdery mildews. The diseased plant has a whitish, powdery appearance owing to the superficial mycelium and conidia growing on its surface. Powdery mildew diseases differ greatly in their damage to the host (Fig. 4–18).

Infection may be initiated by overwintered ascospores in the early spring or by conidia during the late spring and summer. The mycelium is rather delicate, consisting of white or colorless septate hyphae that branch and interweave freely. The mycelium is usually superficial, occurring only on the surface of the leaf, and at intervals it sends a projection, a *haustorium*, into the epidermal cells of the leaf. The haustoria anchor the mycelium to the leaf and absorb food from it. The form of the haustorium varies and may be peglike, globular with a narrow neck connecting it to the hypha, or lobed with many fingerlike projections (Fig. 4–19).

Conidiophores and conidia are produced abundantly on the surface of the leaf in the spring and summer. The conidia are colorless or white and somewhat cylindrical in shape; they may be borne either singly or in chains on the simple, erect conidiophore.

Near the end of the summer, spherical cleistothecia begin to form on the superficial mycelium (Fig. 4–18). The cleistothecia are small and brown or black; they occur superficially on the surface of the leaf and may be easily scraped off. The cleistothecial walls consist of many layers of tightly woven tissue and bear long conspicuous appendages. The appendages may be simple and myceloid,

Figure 4–18 Powdery mildew fungus, *Erysiphe graminis* on a leaf. The dark cleistothecia are immersed within the light mycelium and conidia. Approximately × 10. [Courtesy Plant Pathology Department, Cornell University.]

Figure 4–19 Structures produced by powdery mildew fungi; (a) haustorium with fingerlike projections in host cell; (b) cleistothecium with coiled appendages; (c, d) dichotomously branched appendages; (e) appendage with a bulbous base. [(a) From G. Smith, 1900, Botan. Gaz. 29:153–184; (b–d) from E. S. Salmon, 1900, Mem. Torrey Botan. Club 9:1–292; (e) from D. H. Linder, 1943, Mycologia 35:465–458.]

have a bulbous base, or have either coiled or dichotomously branched tips (Fig. 4–19). The appendage form is one of the criteria used in delimiting genera. There may be one or several asci, and if there are several, they are formed in a group at the base of the cleistothecium. Their organization within the cleistothecium differs sharply from the scattered arrangement found in the Eurotiales.

Specific events involved in cleistothecial formation vary from organism to organism, but they comply with the general type for this group. An ovoid or club-shaped ascogonium pairs with a more elongate and slender antheridium [Fig. 4–20(a)]. These organs may be laterally appressed or they may coil together. Meanwhile cells supporting the gametangia give rise to hyphae which will develop into the wall of the cleistothecium. It is uncertain whether or not plasmogamy occurs in all species [Fig. 4–20(b)] but, in any event, the ascogonium divides into three or more cells. The penultimate cell (the second cell from the apical end) contains two nuclei and the remainder of the cells contain a single nucleus [Fig. 4–20(c)]. If the fungus forms a single ascus [Fig. 4–20(e)], the penultimate cell of the ascogonium enlarges and the two nuclei then fuse and undergo three successive divisions, thus converting that cell into an ascus [Fig. 4–20(d)]. If many asci form, the central ascogonial cell gives rise to the ascogenous hyphae that form asci without crozier formation. The asci form an irregular layer across the middle of the perithecium. Owing to the tremendous expansion of the asci, some of the surrounding cells are crushed and form the cavity that will be totally occupied by the asci. The cleistothecium lacks an ostiole and opens at maturity when the wall splits. The asci are uniformly thin-walled and discharge their ascospores forcibly through a slit that forms in the ascus apex.

ORDER MELIOLALES

The Order Meliolales contains approximately 1000 known species, called the black mildews because of their dark superficial growth on the surface of their hosts (Doidge, 1942; Hansford, 1961, 1963). These fungi occur primarily in moist

(a) (b) (c)

(d) (e)

Figure 4–20 Sexual organs, plasmogamy, and stages in the development of the single ascus of *Sphaerotheca castagnei*. See the text for an explanation. [From R. A. Harper, 1895, Ber. Deut. Botan. Ges. 13:475–481.]

tropical regions but may also occur in the warmer regions of the temperate zone. They are obligate parasites on plants and occur primarily on the leaves but may also occur on twigs and petioles. Although these fungi may be parasitic on commercially important plants, they are not considered to be serious pests. The parasites may cause a reduction in the chlorophyll content of the host cells immediately beneath the colony but rarely kill the cells.

The dark brown to black superficial mycelium is thick-walled, is of uniform diameter, has cells of a uniform length, and may have a characteristic undulation and branching typical for a given species. Specialized branches, the *hyphopodia*, are produced from the hyphae [Fig. 4–21(a)]. Some of the hyphopodia consist of a single flask-shaped cell oriented at right angles to the leaf, standing erect. The function of these flask-shaped hyphopodia is not known. A second type of hyphopodium is that with two cells. The terminal cell is usually swollen and pressed against the surface of the leaf and gives rise to a simple ovate, one-celled haustorium that penetrates an epidermal cell of the host. In the genus *Meliola*, bristlelike setae may project from the mycelium or occur at the base of the ascocarp. The setae are usually straight and unbranched but may have various forms. Usually the colonies formed by the mycelium, hyphopodia, and setae can be traced to growth from one ascospore that germinated on the surface of the leaf. The colonies may have thin, sparse growth spreading over the entire surface of the leaf, or they may be comparatively dense, more or less circular, and sharply delimited. In some species, the hyphal cells may be so short and thickly packed that they form a continuous solid, thalluslike plate. Sometimes the colonies appear to be woolly or velvety, especially if the setae are numerous.

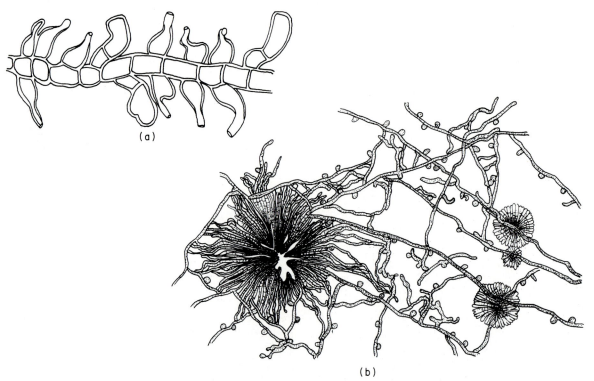

(a)

(b)

Figure 4–21 Structures formed by members of the Meliolales: (a) segment of a hypha produced by *Meliola ptaeroxyli* showing the two types of hyphopodia; (b) portion of a colony of *Asterina ferruginosa* showing a cleistothecium that has opened in a stellate manner. × 460. [Extracts from Bothalia 4:193–217 and 273–420, reproduced under Copyright Authority 6684 of 5-10-80 of the Government Printer of the Republic of South Africa.]

The ascocarps are either true cleistothecia, lacking an ostiole, or are perithecia and have a rudimentary neck with an ostiole. These ascocarps are dark brown to black and are generally globose, but they may be flattened and disklike. The peridium is composed of thick-walled, relatively large, dark cells. In a few genera, the ascocarps may possess appendages of various forms. Paraphyses may be present or absent within the ascocarps. The asci usually contain two or, rarely, four ascospores (more ascospores are initiated but fail to mature) and have evanescent walls that disintegrate at maturity. The dark brown ascospores are usually divided into four or five cells. The ascocarp becomes irregularly broken at the apex at maturity [Fig. 4–21(b)], allowing the ascospores to escape.

Details concerning sexuality and development of the ascocarps have not been extensively studied and are poorly known. From his study of the development of *Meliola circinans*, Graff (1932) concluded that there are many similarities to the development of members of the Erysiphales. However, a conidial stage is lacking, unlike species in the Erysiphales.

ORDER CORONOPHORALES

The Order Coronophorales is a comparatively small order containing only about 20 known species, many first described by Fitzpatrick (1923). These fungi occur as saprophytes on wood, forming their ascocarps on the surface of weathered old wood or within the bark (Fig. 4–22). A few species are parasitic on other Pyrenomycetes.

The mycelium is typically buried within the substratum although a weblike *subiculum* of superficial hyphae may be visible beneath the ascocarp. The superficial hyphae are dark brown to black and have a characteristic metallic iridescence. The ascocarps are ascostromata containing locules. They are carbonaceous, dark, and surrounded by a distinct peridium. The ascostromata are globose or turbinate in shape and are often solitary and free from each other. In some species, several ascostromata form an aggregate and may be seated upon a common subiculum or a compact stroma. The ascostromata are originally solid structures, and locules are formed by breakdown of some of the internal tissue. Asci then develop and eventually fill the locule. The asci are numerous and are irregularly distributed throughout the cavity. The asci have thin walls, are unitunicate, and have long, slender stalks. The ascospores may be hyaline or brown and may have one to four cells. The asci are evanescent, disintegrating to release the ascospores into the locule at maturity. The ascostromata are closed throughout most of their development as they lack ostioles. Eventually the ascostromata open when the apices disintegrate. Disintegration is accomplished when a specialized pad of tissue at the apex undergoes gelatinization. At maturity, the ascostromata collapse to become cupulate.

ORDER SPHAERIALES

The Sphaeriales constitutes the largest group of the Pyrenomycetes, as there are at least 5000 known members of this order. The fungi are extremely diverse morphologically. Because of the large numbers and morphological diversity, the classification of this group has undergone many changes and usually the fungi have been divided into at least four orders. However, there has been little agreement on the disposition of many borderline fungi into one order or another. Luttrell (1951) traces many of the changes in the classification of some of the fungi. Müller and Von Arx (1973) maintain that there are numerous intergrading fungi that span the more traditional orders, and also that sometimes fungi belonging to the same genus differ so much from each other that they would have to be classified in more than one of the traditional orders if certain fundamental characters such as ascus and ascospore morphology are consid-

Figure 4–22 *Coronophora angustata:* long-stalked, multispored asci; (b) appearance on wood; (c) diagram through longitudinal section of ascostromata on wood; (d) enlarged view of ascostromata. [Plate from A. N. Berlese, 1894, Icones fungorum, vol. 3.]

ered. Therefore, Müller and Von Arx claim that these fungi cannot be divided into discrete orders and instead place them into a single large order. We shall follow this treatment here. See Table 4–1 for a synopsis of some of the families.

Many of the Sphaeriales are saprophytes and occur on a wide variety of substrata such as soil, dung, wood, and decaying leaves or petioles. Some occur in unusual environments, for example, on wood submerged in marine water. Many of the Sphaeriales are parasites on a wide range of organisms, including the marine red algae, lichens, other fungi, insects, and higher plants. Many of the most devastating plant diseases are caused by members of this order. *Cerato-cystis ulmi* causes Dutch elm disease, which has resulted in the loss of large numbers of ornamental elm trees in the United States. Similarly, *Cryphonectria* (= *Endothia*) *parasitica* is responsible for the chestnut blight that almost elimi-nated chestnut trees in the United States. Some species of *Ceratocystis* cause a blue stain of the wood in which they grow, greatly decreasing its commercial value. Some members of the Sphaeriales are symbionts with algae, forming li-chens (Chapter 15). Scientists use species of *Neurospora* as a research tool to investigate genetic problems.

TABLE 4–1

Main Features of Some of the Families of the Sphaeriales as Delimited by Müller and Von Arx (1973)

Family	Ascocarp	Asci and Ascospores	Representative Genera
Melanosporaceae	Cleistothecia or perithecia, usually not in a stroma	Asci evanescent, freeing ascospores in ascocarp, clavate, obovate, or spherical	Thielavia Chaetomium
Ophiostomataceae	Cleistothecia or perithecia; perithecia may have a long beak and be partially embedded in a stroma	Asci spherical, evanescent, and freeing ascospores in ascocarp	Ceratocystis
Coryneliaceae	Ascostromata with locules; dark, opening usually clefts	Asci evanescent, usually stalked and clavate, sometimes forming only two ascospores	Corynelia
Sordariaceae	Cleistothecia or perithecia; dark, lacking a stroma	Asci persistent, clavate, simple apex; ascospores usually dark with germ pores or slits and often with mucous sheaths or ornamented	Sordaria Neurospora Gelasinospora
Polystigmataceae	Perithecia immersed in host tissue or stroma; mostly parasitic on higher plants	Asci persistent with simple apex; ascospores lacking germ pores or slits, often forming appressoria upon germination	Glomerella Phyllachora Polystigma
Hypocreaceae	Perithecia often seated on a stroma; brightly colored and fleshy	Asci persistent, simple apex, clavate; ascospores lacking germ pores or slits	Nectria Gibberella Hypomyces Pseudomeliola
Sphaeriaceae	Perithecia often upon or within a stroma; brown or black and carbonaceous	Asci persistent, simple apex, clavate; ascospores lacking germ pores or slits	
Diaporthaceae	Perithecia embedded in host tissue or stroma, sometimes covered by a shieldlike clypeus; stroma light or dark	Asci sometimes loosening from the ascogenous hyphae and forming a slimy mass; apex of ascus with an apical ring usually visible as two refractive bodies	Gnomonia Cryphonectria Diaporthe
Clavicipitaceae	Perithecia usually immersed in a stroma, stroma bright or dark, fleshy or horny, in various forms including crustlike, cushion-shaped, or stalked	Elongate persistent asci, usually with a slightly thickened apical cap perforated by a narrow canal, often plugged with a chitinoid body; ascospores needlelike, multicellular	Epichloë Claviceps Balansia Cordyceps
Diatrypaceae	Perithecia usually immersed in a large stroma, sometimes solitary or arranged in groups	Asci persistent, clavate with long stalks; thickened apex with simple amyloid ring; ascospores one-celled and allantoid	Diatrype Eutypa
Xylariaceae	Cleistothecia or perithecia; free or immersed in host tissues or stroma; stroma dark throughout or with a dark crust, carbonaceous; stroma in various forms such as crustlike, cushion-shaped, or upright	Asci either persistent or evanescent, large with amyloid ring or system of rings, often with chitinoid pulvillus stained by cotton blue; ascospores one-celled, dark, usually with germ slit	Xylaria Hypoxylon Daldinia

The Perithecium. Ascocarps may be ascostromata with locules, cleistothecia, or perithecia. Only a few members, *Corynelia* and its allied genera (in the Coryneliaceae), form ascostromata with locules. The vast majority of the members form a true perithecium, the typical ascocarp for this group (Fig. 4–23). Other than lacking an opening, the cleistothecial members are similar to many perithecial members, sharing such important characters as ascus and ascospore morphology or developmental features (Malloch and Cain, 1973; Von Arx, 1975). Experimental evidence confirms that the cleistothecial forms are closely allied to the perithecial forms in this order. For example, *Chaetomium deceptivum* normally forms perithecia, but if cultured on certain media, will form cleistothecia that closely resemble those formed by the cleistothecial genus *Thielavia* (Malloch and Benny, 1973). A strain of *Gelasinospora calospora* underwent spontaneous mutation and subsequently formed cleistothecia rather than perithecia in culture. If the cleistothecial strain was mated with a normal perithecial strain, the ascospores segregated in approximately a 1:1 ratio in their ability to form cleistothecia or perithecia (Maniotis, 1965).

The perithecium is usually flask-shaped and has a neck surmounted by a pore. Periphyses line the ostiole within the neck. The perithecial neck varies greatly in its length and may range from extremely short, barely protruded forms to those that are so long that they resemble a bristle. In some species, the neck may be lacking and the pore seated directly within the main part of the

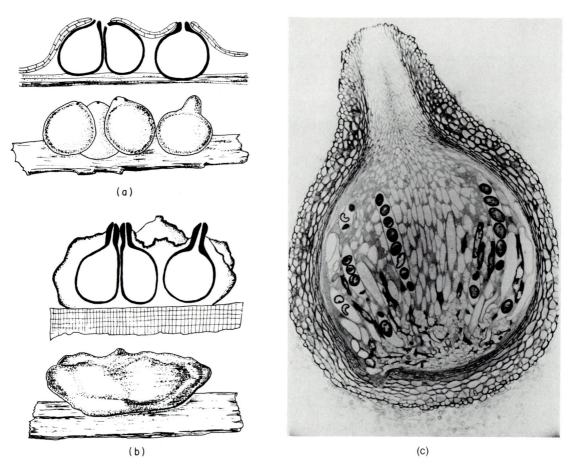

Figure 4–23 Perithecia: (a) free perithecia without a stroma; a longitudinal section is shown above and a habit sketch showing the entire perithecium is below; (b) similar diagrams of perithecia embedded in a stroma; note the prominent walls separating the perithecia from the stroma; (c) longitudinal section through a perithecium of *Gelasinospora reticulospora*. [(a, b) From A. N. Berlese, 1900, Icones fungorum, Vol. 3; (c) from J. D. Jensen, 1982, Mycologia 74:724–737, The New York Botanical Garden.]

perithecial wall. Although most perithecia are flask-shaped, the perithecium may be somewhat globular. The wall of the perithecium may be either brightly colored or dark and may be somewhat fleshy, membranous, or carbonaceous in texture. Usually, bright colors are associated with a fleshy texture while the darker colors are associated with the drier textures, but this correlation is not absolute. Perithecia may occur free on the surface of the substratum or alternatively may be seated upon a pad of stroma, immersed within the host tissue, or partially or completely embedded within a stroma consisting of fungal tissue. If the perithecia are embedded within a stroma, the stroma may brightly colored (a shade of yellow, pink, orange, red, blue, or violet) or subdued (usually dark brown or black). Although there are some exceptions, the brightly colored stromata are fleshy in consistency while the dark ones are dry and carbonaceous. There is considerable variation in the manner in which the perithecia are arranged in the stroma (Fig. 4–24). Within the perithecium, hymenial paraphyses or pseudoparaphyses may be present in addition to the asci. Some species lack both paraphyses or pseudoparaphyses, some have only paraphyses, and others have only pseudoparaphyses.

The Asci. The asci, in combination with the pseudoparaphyses or paraphyses, more or less fill the cavity that occurs within the perithecial wall, and

Figure 4–24 Longitudinal sections of perithecia (heavy outline) and stromata of the Sphaeriales: (a) solitary, free on surface of substratum; (b) similar, but immersed in substratum; (c) immersed in host with a shieldlike layer of stromal tissue; (d) long-necked perithecium immersed in substratum; (e) similar, but on surface of substratum; (f) perithecia immersed in a single layer of a crustlike stroma; (g) perithecium in a leaf, surrounded by a circular stroma; (h) perithecia grouped between a thin layer of stroma tissue and the substratum; (k–m) stromata assuming various shapes. All with perithecia embedded in a single layer except in (j). [From A. N. Berlese, 1894, Icones fungorum, Vol. 1.]

these elements within the cavity are collectively termed the *centrum*. The asci are variable in shape and may be spherical but are usually clavate or cylindrical and are almost always borne within a distinct hymenium. In some members (e.g., *Ceratocystis ulmi* and *Chaetomium*) the asci will deliquesce and the ascospores will ooze out of the perithecium in a slime. These deliquescent asci have simple walls that are uniformly thin and lack a distinct pore. The remainder of the fungi in the Sphaeriales have persistent asci that remain attached to the perithecium and forcibly discharge their ascospores through the ostiole (Chapter 11). There is considerable morphological variation of the ascus apex among those species with persistent asci (Fig. 4–25). For example, the apical pore may be simple and lie within a thickened apical region of the ascus wall. Alternatively, the apical pore may be surrounded by distinct ringlike structures that lie within the ascus wall and are refractive or become blue when stained with iodine; the apical pore may contain a plug that blues with iodine; or the apical

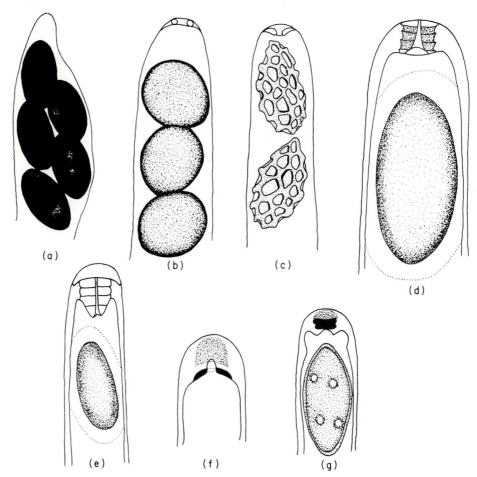

Figure 4–25 Variation in ascus apices among members of the Sphaeriales: (a) simple apex; (b) slightly thickened apex with a single ring; (c) funnel-like identation, surrounded by a thickened ring; (d) thickened apex with three rings in a cylinder, becoming blue in iodine; (e) cylindrical apex with four rings, some of which project downward into the ascus; the rings stain very faintly with iodine; (f) apex with pulvillus (stippled) and a ring that blues in iodine; (g) apex with pulvillus and two rings, blueing in iodine. [(a) From R. F. Cain, 1962, Can. J. Botany 40:447–490; (b) from R. F. Cain, 1957, Can. J. Botany 35:255–268; (c) from R. F. Cain, 1950, Can. J. Res. C28:566–576; (d) adapted from J. C. Krug and R. F. Cain, 1974, Can. J. Botany 52:589–605; (e) from J. C. Krug and R. F. Cain, 1974, Can. J. Botany 52:809–843; (f, g) from R. A. Shoemaker et al., 1966, Can. J. Botany 44:247–254.]

pore may be recessed within a funnel-shaped depression formed by the ascus apex. There are additional variations that are visible with the light microscope. The ascus apices of several species have been studied with the electron microscope (Griffiths, 1973; Reeves, 1971; Stiers, 1977). Basically, the apices consist of an electron-transparent ground material within which electron-dense granules are arranged as a ring or plate; the granules occur in combination with longitudinally oriented fibrils in similar configurations. The pore is delimited by these rings of somewhat different materials. The bottom edge of the apex, bounded by the inner wall layer of the ascus, projects downward as a shoulderlike ring into the cytoplasm of the ascus (Fig. 4–26). Electron-dense material or fibrils may be in this ring. According to Stiers (1977), some of the variation seen with the light microscope might be attributed to the degree to which the shoulderlike ring extends downward into the ascus lumen, its width, and the extent to which it is flared. It is generally difficult to precisely correlate the image of an

Figure 4–26 Diagrammatic longitudinal section of ascus apex of *Hypoxylon fragiforme*, prepared from an electron micrograph. *a*, annulus of electron-dense material; *c*, annular region of slightly electron-dense wall material; *i*, internal wall layer; *o*, external wall layer; *p*, electron-transparent apical region of ascus wall. [From G. N. Greenhalgh and L. V. Evans, 1967, Trans. Brit. Mycol. Soc. 50:183–188.]

apex seen with the light microscope with that seen with the electron microscope. However, differences among the ascus apices are considered to be extremely important in establishing relationships within the group (Müller and Von Arx, 1973). Usually eight ascospores are formed within each ascus. Ascospore morphology is also varied and considered to be important taxonomically. Germ pores or slits may be present or absent, the ascospores may be hyaline or colored and uni- to multicellular, and they may vary in shape from spherical to elongated, needlelike forms. Ornate appendages occur on the ascospores of a few species.

Ascocarp Development. Perithecium formation is often initiated by formation of an archicarp (Figs. 4–27 and 4–28). Typically the archicarp is coiled and may be either unicellular or multicellular. There are four patterns into which sexuality may fall. First, in many of the species investigated, no evidence exists for formation of male gametes or gametangia, plasmogamy is lacking, and further development of the ascogenous system is apogamous. This type of development is extremely widespread and has been observed in representatives of

Figure 4–27 Sexual structures of *Gelasinospora adjuncta*. All × 900. This fungus is heterothallic and requires two genetically compatible cultures to produce a fertile cross. (a) Branched spermatiophore with short cells and small projections from which the spermatia are budded; (b) coiled archicarps; (c) unfertilized archicarp within a young perithecium. This will not develop further without fertilization. Plasmogamy occurs by transfer of spermatia to archicarps of a second, genetically compatible strain. Spermatia are sometimes absent, and hyphal fusion between compatible cultures will take place and produce fertile perithecia. [From R. F. Cain, 1950, Can. J. Res. C28:566–576.]

several families [e.g., it has been described in species of *Chaetomium* by White-side (1961, 1962)]. Second, many fungi produce *spermatia*, discrete sporelike cells that function as male gametes (Fig. 4–27). The spermatia, like spores, are formed on specialized hyphae, can be disseminated from the thallus producing them to other thalli, and may be responsible for cross or outbreeding. After dissemination, a spermatium fuses with a trichogyne of the receptive thallus, and its nucleus travels to the ascogonium. Spermatization occurs in several fungi, including species of *Neurospora* and *Gelasinospora* (Fig. 4–27) (Nelson and Backus, 1968; Dodge, 1932; Goos, 1959). Third, an antheridium may be formed that fuses with the trichogyne. This apparently occurs infrequently but has been reported in *Gelasinospora calospora* (Ellis, 1960). Fourth, gametangia have not been observed in some species, and in these plasmogamy presumably occurs by hyphal fusion. Sometimes hyphal fusion may occur in fungi which form an archicarp that normally participates in plasmogamy (e.g., *Gelasinospora* and *Glomerella*) [Cain (1950) and McGahen and Wheeler (1951), respectively].

After the archicarps complete their maturation and possibly plasmogamy, they produce ascogenous hyphae. In most species, the ascogenous hyphae form typical croziers from which asci develop (croziers are absent in a few of the fungi, including some species of *Chaetomium*) (Whiteside, 1961). The asci are usually borne in a hymenium near the base of the perithecial cavity (Fig. 4–28).

The sterile hyphae that will comprise the perithecial wall usually start to appear after the archicarp has been differentiated. If the perithecium will be embedded within a stroma, the archicarp typically appears within the stroma and sterile hyphae are then formed from the basal cells of the archicarp. These sterile hyphae will develop into a discrete perithecial wall. At maturity, such a perithecial wall originating from the archicarp or from a sexual stimulus is distinct from the tissue of the stroma, which originated by modification of vegetative hyphae. Whether the perithecium is borne free or within a stroma, the hyphae initiating wall formation typically enclose the gametangia and continue to enlarge and differentiate into a distinct wall (often finally consisting of pseudoparenchymatous tissue). Meanwhile the centrum differentiates within the expanding perithecial wall. Centrum development is a critical part of development, as it may proceed in different ways. Closely related species share a similar type of development, and therefore the mode of centrum development is important in establishing relationships among these fungi. Luttrell (1951) described developmental patterns that occur among this group, including the *Diaporthe*, *Xylaria*, and *Nectria* types examined below (Fig. 4–29). The *Sordaria* type was added as result of subsequent research (Huang, 1976). Each type is named for an important genus in which it occurs but actually occurs in several genera. The types are as described below.

1. *Diaporthe*-type centrum: *Diaporthe* and its allies lack both paraphyses and pseudoparaphyses. The centrum is at first filled with pseudoparenchyma. Formation of the perithecial cavity is accomplished by expansion and disintegration of the pseudoparenchymatous cells. While the pseudoparenchyma is disintegrating, the growing asci push into it as they elongate. Eventually the asci form a layer lining the base of the perithecial cavity. Of the representative fungi that we consider on pages 137–139 this type of development also occurs in *Gnomonia* and *Hypomyces* (Luttrell, 1951).

2. *Xylaria*-type centrum: These fungi form paraphyses but not pseudoparaphyses. After the perithecial wall is initiated and encloses the gametangia, paraphyses begin to grow from the inner surface of the wall at both the base and sides. The paraphyses grow upward and inward, exerting pressure by pressing against those growing in the opposite direction, and finally creating a central cavity. The developing asci

Figure 4–28 Developmental stages of *Chaetomium globosum*. The asco-gonium undergoes parthenogentic development, and the centrum has the *Xylaria*-type of development. (a) Various configurations of the coiled archicarp; (b) archicarp (arrow) is becoming surrounded by sterile hyphae arising from its supporting hypha; (c) archicarp (arrow) is sur-rounded by sterile tissue originating from the hyphae in (b), and which will differentiate into the perithecial wall and centrum; (d) ascogenous hyphae showing the absence of croziers and two young asci; (e) section through a perithecium at the time that the cavity and asci are begining to form; (f) mature perithecium extruding ascospores released by deli-quescence of the asci. (a–d) × 1000; (e–f) × 170. [From W. C. White-side, 1961, Mycologia 53:512–523.]

push their way up through the paraphyses. In some species, the para-physes are evanescent and will later disappear. Of those fungi dis-cussed on pages 137–139, *Xylaria*, *Claviceps* (Luttrell, 1951), *Neurospora* (Nelson and Backus, 1968), and *Chaetomium* (Whiteside, 1961) have this type of development.

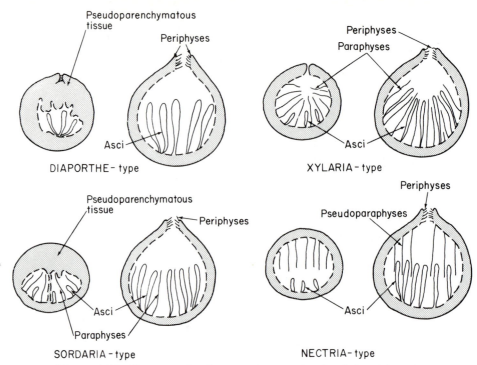

Figure 4–29 Centrum developmental types in the Sphaeriales. An early stage with an enclosed perithecium and young asci is on the left, while a more mature, ostiolate perithecium is on the right.

3. *Sordaria*-type centrum: These, like the *Xylaria* type, have paraphyses and lack pseudoparaphyses. However, development proceeds more or less as in the *Diaporthe* type, as the centrum is filled with pseudoparenchyma. The paraphyses differentiate from some of the pseudoparenchymatous cells in the centrum (Huang, 1976). This type also occurs in *Gelasinospora* (Mai, 1977).

4. *Nectria*-type centrum: Species of *Nectria* and other fungi having this type centrum do not form basal paraphyses but rather have pseudoparaphyses. After the perithecial wall is initiated, there is a downward growth of pseudoparaphyses from a specialized meristematic area lying in the inner wall near the apex. The perithecial cavity is created from a combination of the pressures exerted by growing pseudoparaphyses and by further growth of the perithecial wall. The pseudoparaphyses remain free at their lower ends (Fig. 4–30). The asci form in a concave layer at the base of the cavity, and grow upward among the pseudoparaphyses. This centrum type occurs in *Nectria* (Hanlin, 1961), in *Hypomyces* (Fig. 4–30), discussed further on page 138 (Hanlin, 1964; Samuels, 1973) as well as many other allies of *Nectria* (Rogerson, 1970).

Anamorphs. Conidial states are known for many of the Sphaeriales. Most often, the conidial states belong to the Form-Order Moniliales, but sometimes they belong to the Form-Orders Melanconiales or Sphaeropsidales. Usually the conidia have the blastic type of development. Chlamydospores may be formed by some species. Further clarification of these form-orders and the blastic type of development may be found in Chapter 6.

Representative Members. We shall examine some members of the Sphaeriales so that some of the variation within this group may be explored.

Figure 4–30 Perithecia of *Hypomyces trichothecoides* having the *Nectria*-type of centrum development. Left: young stage showing the downward growth of a layer of pseudoparaphyses (arrow), × 480; right; mature perithecium showing asci with the ascospores (dark) interspersed among the pseudoparaphyses. Note that the ostiole has elongated also. × 410. [Courtesy R. Hanlin.]

Many members of the Sphaeriales have dark perithecial walls and possibly dark stromata. One of the simplest members is *Thielavia*, which forms small isolated cleistothecia on the surface of plant roots on which it lives parasitically. *Chaetomium* is similar to *Thielavia*, but differs in having an almost globose perithecium opening by an ostiole and in bearing extremely long olive brown to black bristles on its outer surface. Perithecia of *Chaetomium* may be found on decaying plant material or dung in nature. Another common dung inhabitant is *Sordaria*, which forms black perithecia with an elongated neck so that the overall form is pear-shaped (Fig. 4–23). The isolated perithecia may be found partially immersed in dung in nature. *Neurospora* forms perithecia much like those found in *Sordaria* when cultured, but in nature the conidial state may be found on plant material that has been charred by fire.

Unlike the above representatives that lack a stroma, two common genera with a stroma are *Hypoxylon* and *Xylaria*. In *Hypoxylon* (Fig. 4–31), the globose to flask-shaped perithecia are embedded within a common carbonaceous stroma, which is black, brown, or reddish in color and may vary in shape from a hemispherical cushion to a thin spreading crust. The stromata are found on dead wood or bark. Species of *Xylaria* form an erect, stalked, club-shaped or cylindrical stroma. The stromatal tissue is white at first but as it matures, the surface becomes crustlike and black. The perithecia are embedded in a single layer beneath the surface, and their ostioles protrude through the black surface crust. One common species, *X. polymorpha*, is known as ''dead man's fingers'' because the cluster of stromata arises from the base of tree stumps and has an eerie resemblance to fingers.

Among the fleshy members of the Sphaeriales, we find some species of

(a)

(b)

Figure 4–31 Two species of *Hypoxylon* (Sphaeriales). The perithecia are embedded within the hard, black stromata. (a) Approximately × 2; (b) approximately natural size. [(a) Courtesy Plant Pathology Department, Cornell University; (b) from E. J. Moore and V. N. Rockcastle, 1963, Fungi, Cornell Science Leaflet.]

Balansia that form black or yellow stromata (Fig. 4–32). Species of *Balansia* occur as parasites on grasses or sedges. The remaining representatives with fleshy walls or stromata that we will discuss are brightly colored. Members of *Hypomyces* often parasitize other fungi, occurring on the basidiocarps of bracket fungi or mushrooms (Chapter 14). The yellow to red perithecia of *Hypomyces* may occur, densely crowded but separate, in a common web on the surface of the substratum (decaying basidiocarps) or the ground where infected basidiocarps have

Figure 4–32 Fleshy stalked stromata of *Balansia claviceps.* × 9. [Courtesy U.S. Department of Agriculture.]

disintegrated, or they may be embedded within the host tissues of living basidiocarps. Some species of *Nectria* may also form scattered perithecia on the wood or bark of their host plants, but many form yellow, coral, or red stroma that are cushion-shaped and break through the bark or other superficial tissues of the host. The perithecia occur in clusters upon these stromata and may be partially immersed in the stromal tissue. Some species of *Nectria* are weakly parasitic and cause dieback disease. *Cryphonectria parasitica* (causal agent of chestnut blight) similarly has perithecia embedded within a reddish stroma in the form of a cushion or elongated crust erumpent through cracks of the bark of the host plant. Both *Cordyceps* and *Claviceps* produce comparatively large stromata having an elongate, upright sterile stalk surmounted by a larger head in which the perithecia are immersed. Species of *Cordyceps* may be found as parasites on subterranean fungi or on insects (Chapter 13). The stromata of *Cordyceps militaris* are up to 5 centimeters in height, are orange-red to red, and have a club-shaped head merging with the stem. The stromata of *Claviceps purpurea* are somewhat smaller, are pale purple, and have a globose head about 2 millimeters in diameter. *C. purpurea* causes a disease of rye plants in which the developing rye grain is converted into a hardened black body (the *ergot*) that is actually a sclerotium and in turn produces the stromata after overwintering. Consumption of the ergots may result in poisoning (Chapter 16).

Class Discomycetes

In most species of the Discomycetes, the hymenium partially covers the surface of an open ascocarp, the apothecium. This group is commonly called the "cup fungi" because many apothecia are cup-shaped with the hymenium lining the inner surface of the cup. Many apothecia are not cupulate but may be flat, convex, or stalked, with an enlarged, club- or saddle-shaped fertile apex partially covered by a hymenium. Apothecia occur in other forms but typically share the common characteristic of having a hymenium lining a surface that is exposed at maturity. Apothecia range in color and include shades of white, buff, pink, orange, red, green, brown, and black. Characteristically they are fleshy and moist, but some form a hardened stroma resembling that of some of the Pyrenomycetes. Their size varies considerably, and they may be barely visible to the naked eye or up to several centimeters in diameter.

Apothecia are divided into distinct anatomical regions that are readily visible in longitudinal sections. The regions often differ in the structure of the tissue (Fig. 4–33). The hymenium bears a palisade layer of asci and paraphyses and is most often on the upper surface of the apothecium. Beneath the hymenium is the *subhymenium*, a relatively thin zone of tissue giving rise to the asci and paraphyses. The remaining tissues in an apothecium comprise the excipulum. The excipulum may be divided into a more or less distinct *ectal excipulum*, the outermost sterile layer of the apothecium, and the *medullary excipulum*, the central enclosed mass of sterile tissue (Fig. 4–34). In some fungi, parts of the excipulum may appear jellylike owing to the secretion of mucilage, which then accumulates among the hyphae and absorbs water [Fig. 1–15(e)].

The Ascus. Asci in species of the Discomycetes are usually cylindrical or clavate but occasionally may be spherical. Ultrastructural studies show that the ascus wall is laminated, usually having two or three lamina. Unlike the bitunicate ascus, which has two distinct walls, these lamina comprise only a single wall and remain permanently attached to each other. When mounted in a reagent containing iodine, the asci of some species will turn entirely blue or turn blue at the apex only. This blueing reaction is apparently due to the reaction of a minute covering of mucilaginous material (Samuelson, 1978). One of the most fundamental differences among asci is whether they are *operculate* or *inoperculate* (Fig. 4–35).

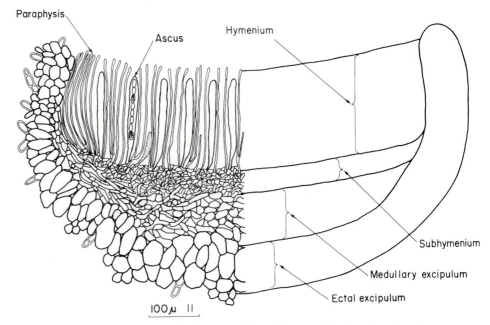

Figure 4–33 Longitudinal section through an apothecium showing the anatomical zones. [Adapted from K. S. Thind and K. S. Waraitch, 1971, Res. Bull. Panjab Univ., N. S. 22:109–123.]

Figure 4–34 Scanning electron micrograph of an apothecium of the Discomycete *Pyronema domesticum*. This apothecium is pulvinate in form, and the visible surface tissue is the ectal excipulum. The tips of the paraphyses are visible at the top. × 560. [Courtesy K. L. O'Donnell.]

Figure 4–35 Types of ascus openings: (a) no regular means of opening; (b) pore in tip of the ascus; both (a) and (b) are inoperculate; (c) operculate ascus which opens by throwing back a lidlike operculum.

Figure 4–36 Scanning electron micrograph of the apex of an ascus showing the operculum, which has been thrown aside from the opening. × 11,250. [Courtesy K. L. O'Donnell.]

The operculate asci open by throwing back a lidlike *operculum,* which may remain attached on one side, or ejecting a lidlike operculum from the ascus (Fig. 4–36). Usually the operculum is centered at the apex of the ascus, but in some species, it is obliquely located on one side of the apex. The ascus wall consists of at least two layers, a thin outer layer and a thicker, less rigid inner layer (Fig. 4–37). The wall near the apex of the ascus may be thinner (Samuelson, 1978) or

Figure 4–37 Diagrams of electron micrographs of operculate asci, both before and after opening. (a) *Pyronema omphalodes:* the operculum is not sharply delimited, and there is no indentation or ring. Irregular tearing of the operculum is due to the absence of strengthening. (b) *Thecotheus* sp. illustrates the most usual type of ascus apex. The large operculum is sharply delimited both by strengthening and by an indented ring. Right: detail of wall; center: entire ascus tip when closed; bottom: opened ascus. (c) *Sarcoscypha coccinea* has an extremely thick operculum that is sharply delimited, especially in the inner wall. The inner layer is greatly enlarged, forming an opercular plug, which will be ejected. C, cleavage of ascus wall; F, line of fracturing; I, indentation of ascus wall; o, operculum; OP, opercular plug; SL, sublayering of ascus wall. [Adapted from J. Van Brummelen, 1978, Persoonia 10:113–128.]

thicker than the lateral walls (Van Brummelen, 1978). Parts of the wall near the apex or at the apex may become strengthened and rigid, possibly by undergoing localized thickening or the addition of wall layers. Other parts of the adjacent wall may become weakened, perhaps by a ringlike indentation of the wall form. In the mature ascus, a swollen or indented ring or rings and possibly conspicuously thickened apices may be present. The weakened area is the dehiscent zone, which breaks open by fracturing, undergoing wall disintegration, or gelatinizing (Van Brummelen, 1978). The dehiscent zone frequently will show different staining reactions than the remainder of the apex (Samuelson, 1975, 1978). The operculum is that part of the apex enclosed within the dehiscent zone. It may be thrown off entirely or may remain attached to the ascus by one or more wall layers. Van Brummelen (1978) distinguishes several types of operculate asci; some are illustrated in Fig. 4–37.

The inoperculate ascus may have no definite means of opening and may rely upon animals to break it and allow the spores to escape or may open by a slit or pore at maturity. The apices of the inoperculate asci may be uniformly thin or have various degrees of thickening. A distinct plug may be present in the thickened apex. When mounted in a reagent containing iodine, there may be a blueing of the entire ascus apex, of a ring around the plug, or of the plug itself. An ultrastructural study was made of the ascus of *Ciboria acerina*, which forms a cylindrical pore with a plug in an apical thickening (Corlett and Elliott, 1974). In this fungus, the outer ascus wall appears to curve inward to line the cylindrical pore. The pore is filled with a granular, amorphous material separated from the cytoplasm of the ascus by the inner wall layer but is not bounded on the outer surface. A similar amorphous material forms the thickened apex by lying between the two wall layers of the apex. The ascospores exit through the pore, and after spore discharge, the pore is empty and the outer wall that lined the cylinder is everted and pressed against the top of the ascus (Fig. 4–38). Asci of other inoperculate Discomycetes were studied by Bellemere (1977) and Schoknecht (1975) and the results are generally similar to those reported for Pyrenomycetes (page 130) and above.

Ascocarp Development.   An archicarp and fertilizing male elements have not been observed in the majority of the Discomycete species and are thought to be absent. Some of the Discomycetes having an archicarp develop apogamously. If sexual organs are present, however, they are dissimilar, making sexual reproduction heterogamous. The male element may be either a spermatium or antheridium. *Pyronema domesticum*, used to illustrate sexual reproduction in the ascomycetes (Figs. 4–3 and 4–4), is a member of the Pezizales, a Discomycete order.

Once plasmogamy has been accomplished, if it occurs, the ascogenous hy-

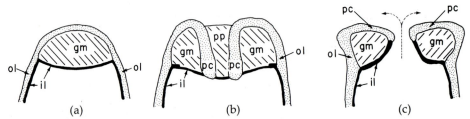

Figure 4–38 Development of the ascus apex of *Ciboria acerina:* (a) early stage in which the ground material (*gm*) has accumulated at the ascus apex, pushing the inner layer (*il*) downward; (b) the outer wall of the ascus has invaginated to form a pore cylinder (*pc*), which is filled with the granular material (*pp*); (c) mature ascus in which the granular material has lysed and the pore cylinder has everted. [From M. Corlett and M. E. Elliott, 1974, Can. J. Botany 52:1459–1463.]

phae develop. The sterile hyphae develop simultaneously, and are intertwined with ascogenous hyphae. The tissues formed by the sterile hyphae and ascogenous hyphae in combination comprise most, if not all, of the medullary excipulum and the subhymenium. The paraphyses are formed by the sterile endings of some of these hyphae. Usually the paraphyses appear before the asci, and the croziers lie among the bases of the paraphyses. The developing asci push their way up through the paraphyses to form the hymenium. There is some variation among members of the Discomycetes in regard to the exposure of the hymenium during development. Van Brummelen (1967) distinguishes two basic types of apothecia on this basis. These types are the *cleistohymenial* and *gymnohymenial*. In the cleistohymenial type of development, the hymenium is enclosed within the excipulum of the developing ascocarp. This enclosure may be permanent in some species that form cleistothecia, but in most the hymenium will be exposed when the developing asci and paraphyses rupture the excipulum. The hymenium may become exposed after the paraphyses are formed but before croziers develop or at any point up to that at which mature ascospores are present. In contrast to the cleistohymenial development, the hymenium is exposed throughout its development in the gymnohymenial type (Fig. 4–39).

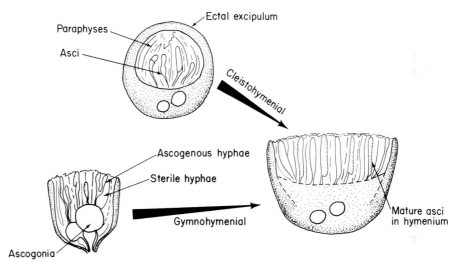

Figure 4–39 Two major developmental types in the Discomycetes.

Vegetative Structures. In addition to forming ascocarps, some Discomycete species produce vegetative structures. These may be *sclerotia* or *stromata*. Both a sclerotium and stroma have a dark outer rind surrounding a pale inner medulla and may include host tissue in addition to fungus tissue. A sclerotium is a hardened, resistant structure capable of overwintering or surviving unfavorable conditions. The sclerotium is determinant in size and is usually rounded in shape (Fig. 4–40). Upon germination, the sclerotium may give rise to vegetative hyphae or to an apothecium. Like the sclerotium, the Discomycete stroma is a resistant structure of approximately the same internal structure, but it differs primarily in that it is somewhat indeterminant in size, it may be irregular in shape, and it is often flattened or padlike. If a stroma is formed in a leaf or wood, a stromatized area of host cells together with mycelium is delimited by a border of darkened cells. A stroma may form within floral parts and fruits, and the entire plant part may become invaded by the fungal hyphae and converted into a stroma. The fruit becomes dry, hardened, and shriveled (Fig. 4–41). A stroma gives rise to an apothecium directly; this apothecium either may arise above the stroma or may be embedded within it.

Figure 4–40 Section through a sclerotium of a Discomycete *Pyronema domesticum.* × 440. [From E. J. Moore, 1962, Mycologia 54:312–316.]

Figure 4–41 Apothecia of *Monilinia fructicola* arising from peach fruits that have been converted to stromata. [Courtesy L. R. Batra.]

Taxonomy. Seven orders are represented in this group (Korf, 1973). Two of these orders have only a single genus, each a rather unique plant parasite group. These orders, the Medeolariales and Cyttariales, will not be discussed further. The remaining orders, which we shall consider further, are the Phacidiales, Ostropales, Helotiales, Pezizales, and Tuberales.

ORDER PHACIDIALES

Members of the Phacidiales are parasitic on plants, usually occurring on conifer leaves (*needles*), on which they cause diseases known as needle blights. Species of *Rhytisma* may invade leaves of deciduous plants, causing black, shiny spots to develop that gives the disease the common name of tar spot (Fig. 4–42).

Generally, these parasites may be divided into two broad groups. First, many of the fungi are weakly parasitic and invade injured or weakened needles, perhaps near the end of the season, and cause them to drop from the trees. These particular fungi are also called "needle-cast" fungi and include members of the largest genus, *Lophodermium*. The needle-cast fungi continue to live sapro-

Figure 4–42 Tar-spot fungus, *Rhytisma andromedae,* on its host. × 1.5 [Courtesy U.S. Department of Agriculture.]

phytically on the fallen leaves and produce their ascocarps during the following spring or summer. The second group includes the majority of the members of this order and are strong parasites. They invade relatively healthy needles and do not cause them to be shed but instead form their ascocarps within the needles while they are still on the tree. They also require at least a year for the maturation of the ascocarp (Darker, 1967).

The ascocarp consists of a hymenium embedded in a stroma (Fig. 4–43). The stroma is embedded within the superficial tissues of the leaf (just beneath the cuticle, epidermal, or hypodermal tissues). The leaf tissue may become discolored and inseparable from the stroma because of the penetration by mycelium extending from the stroma. The stroma is typically elongate and is usually somewhat linear or boat-shaped. Its size varies from a fraction of a millimeter to 4 centimeters or more. These stromata are usually black but may be brown to almost colorless. According to Darker (1967), the color is related to water supply. He noted in a collection of *Lophodermium pinastri* that needles caught in a small depression filled with water were almost colorless, while normally the needles that are on the drier forest floor have black stromata. In addition, the texture of the stroma varied with the availability of water. The substance that makes the

Figure 4–43 *Phacidium nigrum,* a member of the Phacidiales. Left: stromata of *P. nigrum* on its host. × 1.25; right: ruptures of host epidermis reveal the hymenium. × 10. [From E. K. Cash, 1942, Mycologia 34:59–63; courtesy U.S. Department of Agriculture.]

stroma dark seems to originate within the hyphae and then eventually lies in the interhyphal spaces (Silverberg and Morgan-Jones, 1974).

Each stroma contains at least one hymenium (Fig. 4–44). The hymenium contains both paraphyses which are free at the apices, and asci. The asci have an apical thickening, and sometimes the plug or surrounding pore becomes blue in iodine. The asci contain four or eight ovoid to filiform ascospores. The ascospores have a gelatinous sheath and may be single-celled or multiseptate. The hymenium is seated upon a relatively thin pad of pseudoparenchymatous tissue, which in turn is in contact with the host tissues. The cells of the host may be discolored beneath this fungal tissue. According to Uecker and Staley (1973), who studied *Lophodermella morbida,* and Campbell and Syrop (1975), who studied *Lophodermella sulcigena,* only the basal pseudoparenchymatous layer and the covering black shield are present in the early stages of development. The pseudoparenchymatous basal layer grows more rapidly than the covering shield, causing the basal layer to push deeper into the host tissue and meanwhile creating a cavity. Both layers of the stroma remain permanently attached to each other at the edges. Archicarps are lacking, and meanwhile some of the pseudoparenchymatous cells give rise to the ascogenous hyphae while others give rise to the paraphyses. The developing hymenium occupies the cavity created by differential growth.

(a)

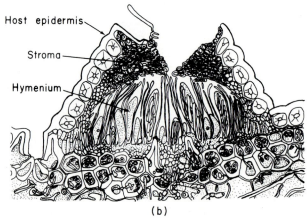

Host epidermis

Stroma

Hymenium

(b)

Figure 4–44 *Bifusella striiformis* on a conifer: (a) macroscopic appearance of boatlike apothecia on host; (b) longitudinal section through an apothecium. [From G. D. Darker, 1932, Contrib. Arnold Arboretum Harvard Univ. 1:1–131.]

After the asci have matured, the dark stromatal shield ruptures. The split may be lengthwise or stellate. If more than one hymenium is formed, there will be a split above each hymenium. The splitting sometimes occurs along the line of greatest stress, resulting perhaps from the pressure of the underlying hymenium. In some species, there is a differentiated dehiscence zone. This may consist in some cases of a band of cells within the stroma that rupture easily or in other cases of cells that become gelatinous to form the slit. The gelatinous cells may also seal off the hymenium under adverse conditions. Some species have a thickened, liplike band of tissue at the edge of the opening. After the stroma splits, the edges may only form a narrow opening or they may fold back to form most or all of the hymenium. As in the typical apothecium formed by the Discomycetes, the hymenium is exposed at maturity. Unlike the usually dark stroma, the hymenium may be light in color.

These fungi form a nonsexual stage. Conidia are formed in *acervuli* or *pycnidia*, more or less rounded or elongate dark structures formed in the superficial layers of the leaf. The anamorph is usually intermixed with the ascocarps, but the acervuli or pycnidia mature earlier than the ascocarps. Some investigators believe that these conidia may function as spermatia and be responsible for fertilization of archicarps found in some species (Darker, 1967). Some species that form conidia do not form archicarps (Uecker and Staley, 1973) while other species may form neither archicarps nor conidia (e.g., Thyr and Shaw, 1966; Woo and Partridge, 1969).

ORDER OSTROPALES

The Ostropales contain a single small family, characterized by their ascus and ascospore structure. The asci are elongate, rather cylindrical in form, and have a thickened apex penetrated by a pore. The ascospores are elongate and filiform or threadlike, and at maturity they break crosswise into a number of short, cylindrical segments. The apothecia are only a few millimeters in diameter and light but drab in color. Their form varies from a stipitate apothecium with a convex pileus or a sessile, shallow cup in *Vibrissea* to a partly submerged shallow cup in *Stictis*. These fungi are saprophytes and are often found on woody substrata submerged in a quickly running stream or semiaquatic habitats. The ascospores are discharged while submerged and have a tendency to adhere to the apothecium after their discharge. The cluster of ascospores vibrates or oscillates about in the water, which has given rise to the generic name *Vibrissea* (Latin, *vibrare*, to oscillate or vibrate) (Sánchez and Korf, 1966).

ORDER HELOTIALES

The Helotiales are a very diverse group of fungi, containing over 1500 species. Important characteristics of the group are found in the asci and ascospores. The asci are inoperculate and have an apical pore through which ascospore discharge will occur. There may or may not be a blueing reaction with iodine. The asci are cylindrical to clavate in form. The ascospores are short and ellipsoid to needlelike and do not fragment at maturity (unlike those of the Ostropales). The apothecium may be seated within the host tissue or entirely superficial. Unlike the members of the Phacidiales, the apothecium is not immersed within a stroma. Some members of the Helotiales form a sclerotium that gives rise to the apothecium. Nonsexual stages are common and are usually Hyphomycetes.

The majority of these organisms are saprophytes on plant debris and may be collected from dead plant parts such as grass blades, leaves, and twigs. Comparatively few occur directly on soil or dung. The fungi occur in abundance in relatively moist environments such as the forest floor, the edge of streams, or in bogs. Some are virulent and important plant parasites. The plant diseases may be caused either by the apothecial stage (as in the needle blight diseases) or by the nonsexual stage (as in the brown rot of peaches, discussed below).

The majority of the apothecia formed by members of the Helotiales are comparatively small, often only a few millimeters in width, and are typically cupulate or pulvinate. The apothecium bears the hymenium on its upper surface, which may be concave or convex. The cup may be seated directly on the substratum, or it may be supported by a stemlike portion, the *stipe*. The texture of the apothecium may be soft and fleshy, somewhat waxy, or gelatinous. Colors range from pallid earth colors to bright yellow, orange, red, or green. Common genera include *Dasyscyphus*, which forms small stipitate apothecia (about 1 to 2 millimeters in diameter) with hairs on the excipulum (Fig. 4–45), and *Bisporella*. *Bisporella citrina* forms bright, citron yellow apothecia that are only up to 3 millimeters in diameter but are relatively noticeable because they occur in dense swarms on the surface of logs.

Some of the Helotiales are parasitic on conifers, where they cause a needle blight disease similar to that caused by members of the Phacidiales. The fungi are members of the Hemiphacidiaceae, and they form relatively simple fleshy, usually orange to yellow-brown apothecia beneath the epidermis or hypodermis of the host plant. The apothecia form on either living or recently killed leaves that are still attached to the plant. Unlike those of the Phacidiales, the apothecia are not covered by fungal tissue but are covered simply by the host tissue, which eventually tears to expose the apothecium (Korf, 1962).

Perhaps the best known and most important group of this order belongs to the family Sclerotiniaceae, which consists almost entirely of parasitic fungi. Important plant parasites include *Sclerotinia whetzelii*, which occurs on eastern poplar, *Sclerotinia sclerotiorum* (Fig. 4–45), which occurs on vegetables such as lettuce, *Monilinia fructicola*, which causes brown rot of peaches, and *Botryotina fuckeliana*, which occurs on grapes.

These fungi have a long stalk with a cuplike or funnel-like fertile region, are usually somewhat waxy in texture, and are some shade of yellow, tan, or

(a)   (b)

Figure 4–45 Cupulate representatives of the Helotiales: (a) *Dasyscyphus* on dead herbaceous stems collected under leaves, approximately × 2; (b) *Sclerotinia sclerotiorum*; this species forms yellowish-brown cups borne on a downy stalk, approximately natural size. [(a) Courtesy Plant Pathology Department, Cornell University; (b) from E. J. Moore and V. N. Rockcastle, 1963, Fungi, Cornell Science Leaflet.]

brown. The apothecia are larger than those previously described, as the fertile region may be about 10 to 20 millimeters in diameter.

The life history of many members of the Sclerotiniaceae is known much better than is that of other members of the order. Spermatia are common in the Sclerotiniaceae but are probably either rare or absent in the remainder of the Helotiales. The spermatia have been demonstrated to be functional in some members, uniting with the apothecial fundaments (the female elements) arising from the sclerotia or stromata. After plasmogamy occurs, the apothecial fundaments complete their development and form the mature apothecia.

Let us briefly examine the life history of *Monilinia fructicola* (Honey, 1936). *Monilinia* overwinters as a mycelium within a hardened and blackened fruit (a stroma or ''mummy'') produced by the diseased host (Fig. 4–41). In the spring, spermatia are produced from short, flask-shaped spermatiophores. The pillar-shaped fundaments of the apothecium arise from the mummy and continue their development into apothecia, presumably after fertilization by spermatia. The apothecia discharge ascospores that infect the young flowers or leaves of the peach tree. During the spring and summer, hyaline chains of conidia are produced on the leaves and young twigs. These conidia are responsible for initiating many secondary infections throughout the summer.

Some of the most conspicuous members of this group are known as ''earthtongues.'' These are members of *Geoglossum* and closely related fungi. They occur on the ground in forests and bogs. These fungi are tall (about 2 to 8 centimeters) and have an elongate stipe bearing a slightly expanded, somewhat tonguelike region at the apex which is covered by the hymenium. The earthtongues may be dark or shades of bright orange or red. A rubbery gelatinous fungus, *Leotia,* is closely related to the earthtongues but differs in having a cap-like structure at the apex of the stipe (Fig. 4–46). The hymenium covers the upper surface of the cap only.

ORDER PEZIZALES

At least 500 species comprise the Pezizales. All members of the Discomycetes with operculate asci belong to this group. Some of the fungi have asci with subapical opercula, and coupled with this character, apothecia with a carti-

Figure 4–46 Apothecia of *Leotia stipitata*. Approximately two-thirds natural size. [Courtesy Plant Pathology Department, Cornell University.]

laginous or leathery texture. The remaining members have centrally located opercula, and typically form brittle and rather fleshy apothecia.

These fungi occur in diverse habitats as saprophytes on dung, soil, wood, or plant debris. They may be readily collected on cut banks of roads and in forests. Some members will form apothecia only on recently burned substrata (e.g., *Pyronema*, whose name literally means "fire thread"). A few members form their apothecia beneath the soil's surface.

There is a great deal of morphological variation in this group. Some of the dung-inhabiting members simply form a mycelial mat or tuft (e.g., species of *Ascodesmis*) or a small cleistothecium. Only one ascus may occur in some of these cleistothecial dung-inhabitants, such as *Lasiobolus monascus* (Kimbrough, 1974).

The majority of the fungi in the Pezizales produce a larger, more complex apothecium. In some, such as *Pyronema domesticum*, the apothecium may be only a few millimeters in diameter, but in the majority it is over 1 centimeter and apothecia as large as 10 centimeters are common. Most members, including species of *Peziza* and *Sarcoscypha* (Fig. 4-47) form a cupulate apothecium, with the hymenium lining the inside of the cup. Pulvinate apothecia with a convex hymenium are also common (Fig. 4-34). The apothecia may be sessile or have a stipe of varying lengths. Many members of the Pezizales form apothecia that are noncupulate but bear fertile regions of varying forms at the apex of the stipe. The fertile region may be saddle-shaped as in *Helvella* species, spoon-shaped as in *Wynnea* species, funnel-shaped as in *Urnula* species, or deeply convoluted as in the edible morels (*Morchella* species) (Fig. 4-48).

Developmental studies have been made on numerous members of this group and have been especially numerous on species of *Pyronema* (illustrated in Figs. 4-3 and 4-4 to show many general features for the ascomycetes). *Pyronema* has received so much attention because it forms unusually large club-shaped antheridia and globose ascogonia, which may be up to 50 micrometers in diameter, and also because it will develop in culture (Moore, 1963). As indicated, the antheridium is usually functional and fuses with the trichogyne in *Pyronema*. However, even though some of the antheridia are functional, ascogonia in the same culture may not undergo fusion and may produce ascogenous hyphae without apparent fertilization (Moore-Landecker, 1975). Some other members of the Pezizales form antheridia and ascogonia, but their morphology varies greatly. For example, the ascogonium is a multicellular coil fertilized by a thread-like antheridium in *Anthracobia melaloma* (Rosinski, 1956). Some of the Pezizales

Figure 4–47 Apothecia of *Sarcoscypha coccinea,* the scarlet elf cup, a member of the Pezizales. These apothecia are conspicuously colored, with an organge-red hymenium and a white exterior. About natural size. [Courtesy Plant Pathology Department, Cornell University.]

Chap. 4 / Division Amastigomycota—Subdivision Ascomycotina

Figure 4–48 Noncupulate members of the Pezizales: (a) *Urnula geaster;* (b) *Helvella infula;* (c) *Wynnea americana;* (d) *Morchella esculenta.* [(a) Courtesy U.S. Department of Agriculture; (b–d) courtesy Plant Pathology Department, Cornell University.]

have archicarps fertilized by spermatia; an example is *Ascobolus stercorarius*, which forms a coiled archicarp and spermatia. (These are discussed further in Chapter 10.) Some species form archicarps but not spermatia or antheridia (e.g., *Peziza quelepidotia* described by O'Donnell and Hooper, 1974). Sexual structures are unknown for the vast majority of the Pezizales. Subsequent development of the ascogenous system and sterile hyphae generally occurs as described on pages 104–106.

ORDER TUBERALES

The Tuberales is a small order containing about 140 species (Gilkey, 1939, 1961). These fungi are commonly known as "truffles" and many species have been widely esteemed as a delicacy since ancient times, when they were imported from North Africa by the Romans (Trappe, 1971). The early Greeks not only relished the truffles but speculated about their nature (Chapter 16). In nature, the ascocarps are often associated with vascular plants, with which they establish a mycorrhizal relationship (Chapter 15). Efforts to culture the truffles as a food crop have been futile, perhaps because of their intricate biological requirements, met by their mycorrhizal association but not easily duplicated in cultivation (Hawker, 1955).

The apothecia usually occur beneath the soil's surface at varying depths up to approximately 8 to 10 centimeters, but they also may be partially buried in the litter on the forest floor. Macroscopically, the apothecium may appear to be cupulate or globose (Figs. 4–49 and 4–50). The exterior is often convoluted or irregularly lobed. The size of the apothecia ranges from a few millimeters to more than 10 centimeters, and they may be white, yellow, brown, black, rose, or purple in color.

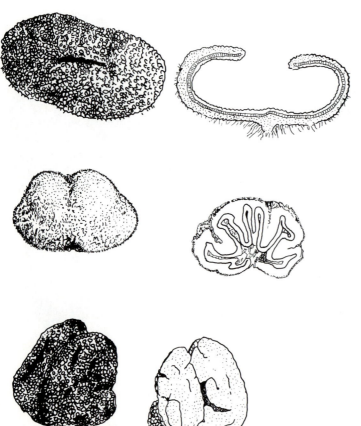

Figure 4–49 Representative ascocarp forms in the Tuberales. A habit sketch is on the left, accompanied by a longitudinal section on the right. [From L. E. Hawker, 1954, Phil. Trans. Roy. Soc. London, Ser. B 237:429–546.]

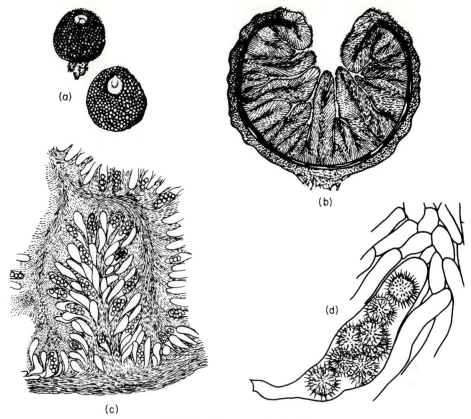

Figure 4–50 Member of the Tuberales, *Pachyphloeus:* (a) ascocarps as they would appear in nature; (b) longitudinal section through the ascocarp; (c) detail of the ascocarp as it would appear near the wall; (d) ascus. [From A. Engler and K. Prantl, 1897, Die natürlichen Pflanzenfamilien, Engelmann, Leipzig.]

The simplest members form hollow, cupulate apothecia opening by a pore or slit. In these, the hymenium occurs as a single layer lining the inside of the cup (Fig. 4–49). Unlike members of the Pezizales, the paraphyses extend beyond the asci and their tips fuse to form a tissue (the *epithecium*) that encloses the asci. In more complex forms, there is a folding inward of the excipulum to form shelf- or lobelike projections extending into the central cavity. These projections may even anastomose when meeting an opposing fold. The particular degree and pattern of infolding is important in distinguishing genera and species within this group (Fig. 4–49). Apothecia with a great deal of infolding tend to have extremely small cavities and are essentially solid. Individual chambers may be isolated within the apothecium by the folds, or alternatively, there may be a network of canals that wind through the apothecium. These chambers and canals become lined with a hymenium containing both asci and paraphyses. In some species, the chambers or canals become filled with loose hyphae. The individual chambers may eventually open to the outside. Some species do not form a hymenium but rather form asci scattered within the sterile tissue.

Members of the Tuberales do not forcibly discharge their ascospores. The ascospores may be released when the ascocarp decays or when the apothecium is broken or carried away by an animal. Usually the ripe apothecium emits a strong, sometimes unpleasant odor attracting animals that dig them up for food. If you wish to collect truffles, search for woodland sites where it appears that a small animal may have been digging recently, and then search carefully through the surrounding layer of soil.

ORDER LABOULBENIALES

Members of the Laboulbeniales are minute insect parasites, ranging in size from approximately 0.1 millimeter to slightly over 1 millimeter. At least 1500 of these tiny fungi have been described (Tavares, 1985; Thaxter, 1896, 1908, 1924, 1926, 1931). They penetrate the chitinous exoskeleton of the insect and form their thalli outside the insect's body. The thallus appears to the naked eye as a dark or yellowish bristle, and the insect will appear furry if it is heavily infected. The insects are apparently either unharmed or only slightly harmed by these fungi. The infections are contagious and may be spread by contact of one insect with another; infections are especially severe in dense insect populations (Benjamin, 1973). The vast majority of the Laboulbeniales are highly host-specific: they live on only certain insect species and often inhabit a certain part of the insect, such as the mouth parts, the anterior legs, or the wings (Fig. 4–51). In addition, the fungus may be restricted to the right or left side, to the upper or lower surfaces, and perhaps even to male or female insects (Benjamin and Shanor, 1952). Some specificity for a particular position may be correlated with mating behavior or other behavioral patterns accounting for contact of certain parts of the insect's body (Whisler, 1968). A few species have comparatively broad host ranges. These fungi have not been successfully cultured apart from the host.

Figure 4–51 Two thalli of a member of the Laboulbeniales (*Stigmatomyces* sp.) growing on the femur of an insect. × 500. [Courtesy W. Rossi.]

The thallus is attached to its host by a darkened foot cell that typically forms a small haustorium penetrating the host's integument. The haustorium absorbs nutrients from the insect, as a mycelium is absent. The foot cell is immediately beneath the main body of the thallus, the *receptacle*. The receptacle consists of a column of cells that may be more or less definitely arranged in one or more rows or may be somewhat parenchymatous. The receptacles may bear one or more sterile hairlike appendages, which may be unicellular or multicellular and possibly branched. The receptacle also bears the male and female reproductive structures, as well as a perithecium. Some species are monoecious and form both male and female reproductive structures on the same thallus. Other species are dioecious and form separate male and female thalli occurring in close proximity to each other in pairs (Fig. 4–52).

The male gametes are spermatia. Spermatia may be borne exogenously as cells formed from some appendages similar to sterile appendages. Usually the spermatia are formed endogenously within a flask-shaped cell, the antheridium,

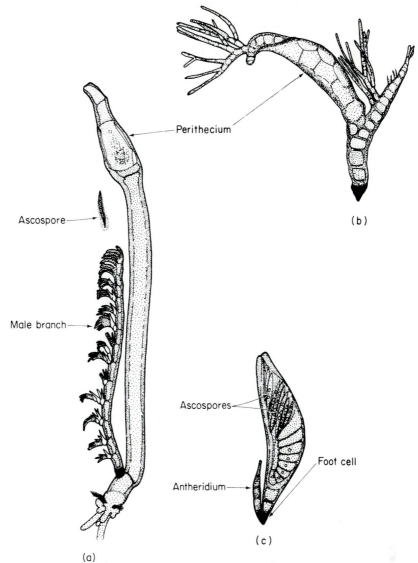

Figure 4–52 Members of the Laboulbeniales: (a) monoecious species, *Rhizomyces ctenophorus*; (b) monoecious species, *Ceratomyces contortus*; (c) dioecious species, *Amorphomyces falagriae*. The male and female thalli developed from the individual members of an ascospore pair which were formed in the same ascus, stuck together, and germinated side by side. [From R. Thaxter, 1896, Mem. Am. Acad. Arts Sci. 12:195–429.]

and escape through a neck region [Fig. 4–53(b,c)]. The antheridium is most often borne individually on the side of an appendage but may also be formed from one or more cells of the appendage itself. In some species, a compound structure consisting of fused antheridia is formed and the spermatia are discharged into a common cavity before being finally released.

The perithecium and its contents are derived from a single cell of the receptacle or one of its branches [Fig. 4–53(b)]. At maturity, the perithecium consists of three basal cells supporting a cellular perithecial wall. The perithecial wall consists of a double layer of cells regularly arranged in both horizontal and vertical rows. Usually there are four vertical rows and either four or five horizontal rows. The perithecium lacks paraphyses, pseudoparaphyses, and periphyses but contains the ascogenous system.

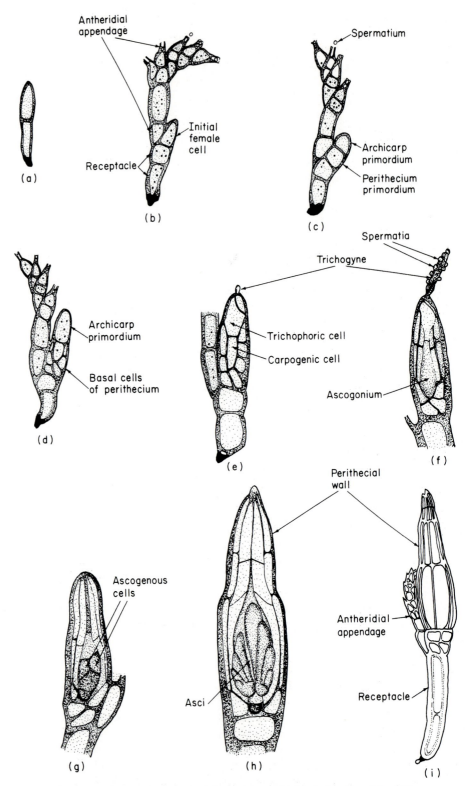

Figure 4–53 Development of the ascocarp in the Laboulbeniales (*Stigmatomyces baeri*). See the text for an explanation. [From R. Thaxter, 1896, Mem. Am. Acad. Arts Sci. 12:195–429.]

Development of the Perithecium.   The ontogeny of some species was first described by Thaxter (1896), and some later studies include those of Blackwell (1980), Benjamin and Shanor (1950), Shanor (1952), and Tavares (1966, 1985). A typical sequence follows. Development begins when the two-celled ascospore lands on the surface of the host and germinates. The ascospore becomes attached by a blackened extremity that forms from the lower cell of the ascospore [Fig. 4–53(a)]. This blackened structure is the foot cell. After the foot cell is formed, the remainder of the ascospore differentiates into the receptacle. Development is characterized by apical growth of an elongating column of cells, all formed in a very precise manner. The antheridia and spermatia are differentiated before the female structures are and arise from the receptacle. The perithecium and archicarp primordia develop from a single initial cell [Fig. 4–53(b)] that divides, the lower cell giving rise to the surrounding perithecium while the upper cell gives rise to the archicarp [Fig. 4–53(b,c)].

During development of the perithecial wall, the lower cell resulting from the division of the basal cell again divides by forming an oblique septum, becoming two cells. Each of these two cells will remain as stalk cells supporting the perithecium. In turn, the stalk cells produce a total of three cells that will become the basal cells of the perithecium [Fig. 4–53(d)]. The basal cells produce a total of eight upgrowths. Four of these are external and give rise to the outer perithecial wall, while four are internal and give rise to the inner perithecial wall. The continued terminal growth and septation of the cells will result in formation of the sheathlike perithecium which envelopes the asci at maturity [Fig. 4–53(h,i)].

Meanwhile the sexual system continues its development. Let us return to the division of the single initial cell [Fig. 4–53(b)] and recall that the upper cell resulting from this division is the archicarp primordium [Fig. 4–53(c)]. The archicarp primordium divides to form three uninucleate cells; proceeding from the bottom and progressing upward, these are the *carpogenic* cell, an intermediate cell (the *trichophoric* cell), and the trichogyne in the terminal position [Fig. 4–53(e)]. In some species, the trichogyne may consist of more than one cell. The developing perithecium surrounds the basal two cells (the carpogenic and intermediate cells), while the trichogyne is either partially or wholly exposed. Spermatization occurs at about this time [Fig. 4–53(f)]. The spermatia come into contact with the trichogyne, possibly owing to the insect's movement or to the presence of a water film. The trichogyne and the trichophoric cell eventually disappear, leaving only the carpogenic cell. The carpogenic cell divides again to form an ascogonium and supporting cells [Fig. 4–53(f)]. The ascogonium undergoes a division, and sequential divisions may occur, resulting in a precise number and arrangement of cells that is constant for a particular species. These divisions result in the production of ascogenous cells [Fig. 4–53(g)]. Each ascogenous cell produces asci budding from it both in an upward and downward direction and will produce a succession of asci [Fig. 4–53(h)]. Cytological observations of karyogamy and meiosis are lacking for most species. The asci generally become free in the cavity of the perithecium as their supporting cells disintegrate. Also, the inner wall of the perithecium disappears to provide room for the expanding asci.

Ascospores are formed with the typical pattern described for the ascomycetes (Hill, 1977). Usually, four ascospores remain after the disintegration of four nuclei. Eventually the ascus wall disintegrates, freeing the ascospores within the perithecium. The ascospores are embedded in a slime that may aid in lubricating their exit from the perithecium. The slime originates in some cells of the perithecium and is secreted by vesicles into the perithecial cavity. The ascospores tend to cling together in pairs owing to their sticky sheaths. In this fashion, ascospores land together, adhere to a surface, and germinate side by side. In dioecious species, typically one member of the ascospore pair develops into the male plant while the other, larger ascospore develops into the female plant.

Class Loculoascomycetes

Members of the Class Loculoascomycetes are characterized by: (1) the possession of a bitunicate ascus, and (2) the production of a stroma within which asci are produced. This is the only group of fungi having a bitunicate ascus. Ascostromata with asci also occur in the Coronophorales and Coryneliaceae but are formed in conjunction with a unitunicate ascus. Von Arx and Müller (1975) recognize a single order, the Dothideales, for which they delimit 34 families (Table 4-2).

ORDER DOTHIDEALES

Approximately 2000 species are known. Numerous members occur as saprophytes on dead leaves, herbaceous stems, wood, plant debris, and dung. Species of *Myriangium* are parasitic on scale insects, which in turn occur on host plants. Other fungi are parasitic on other fungi, lichens, mosses, or higher plants. Some of the plant parasitic fungi will form ascostromata on or immersed within the living plant, while others will invade the living plant, kill it, and then form ascostromata on or within dead plant parts (Figs. 4-54 and 4-55). Such fungi are responsible for a wide variety of plant diseases usually characterized by fruit rots or leaf spots. An example is *Mycosphaerella tulipiferae*, which causes leaf spot and some defoliation of the tulip poplar in the Middle Atlantic and Gulf states. The mycelium kills the host cells, and conidiophores and conidia develop on the lower surface of the leaf. Development of the ascocarp is initiated in the autumn in leaves that are still attached to the tree but is not completed until the following spring in the fallen leaves. *Venturia inaequalis*, the causal agent of apple scab which results in leaf spots and fruit rot of the apple, undergoes a similar sequence of events.

Figure 4-54 Pseudothecium of a member of the Dothideales, *Mycosphaerella* sp. in the leaf of its host. [Courtesy Plant Pathology Department, Cornell University.]

TABLE 4–2
Some Important Characteristics of Representative Families of the Dothideales (as Delimited by Von Arx and Müller, 1975)

Family	Ascocarp	Asci and Ascospores	Representative Genera
Dothideaceae	Locules in stromata; stromata small to large, often erumpent, shapes include crustlike, pulvinate, and spherical; locules opening by fissures or dehiscence; pseudoparaphyses absent	Asci in fascicles or parallel at base, sometimes on column of tissue; clavate to cylindrical, with 8 or more spores; ascospores uni- to multiseptate	*Dothidea* *Plowrightia*
Myriangiaceae	Often parasitic on scale insects; stroma of various shapes (pulvinate, discoid, spherical, crustlike); small to medium; pseudoparaphyses absent	Asci borne singly in small locules, often at different levels, spherical to broadly clavate; ascospores usually both longitudinally and transversely septate	*Myriangium* *Elsinoë*
Hysteriaceae	Saprophytic on wood; stroma superficial or immersed at the base, elongated, elliptical, linear, or boat-shaped; opening by a longitudinal cleft; pseudoparaphyses often branched	Asci parallel, clavate, or nearly cylindrical; apical thickening of ascus; ascospores sometimes with many cells, relatively wide	*Hysterium*
Pleosporaceae	Ascocarp usually a pseudothecium, conical, spherical, or flask-shaped; sometimes immersed in host tissue or forming locules in an erumpent stroma; medium to large; pseudoparaphyses present	Asci formed in a parallel array at the base of the locule, cylindrical or baglike, thickened apically	*Pleospora*
Mycosphaerellaceae	Pseudothecia small, spherical, and smooth, immersed in host tissue or forming locules in a stroma which may be erumpent; pseudoparaphyses absent	Asci in a fascicle or parallel at the base of the locule, ovate to cylindrical with thickened apex; ascospores elongated, narrow, uni- to triseptate	*Mycosphaerella*
Capnodiaceae	Epiphytic, forming superficial dark mycelium ("sooty molds"); ascocarps usually spherical; immersed in a subiculum or covered with hyphae; pseudoparaphyses absent	Asci ovoid or baglike; ascospores usually many septate	*Capnodium*
Microthyriaceae	Superficial, often shield-shaped with a covering wall of radially arranged cells; pseudoparaphyses present	Asci arranged in a ring with their tips converging to a pore; cylindrical or baglike; ascospores usually uniseptate	*Microthyrium*

159

Figure 4–55 Flattened shieldlike ascocarps of *Platypeltella angustispora*, a member of the Dothideales. × 100. [Courtesy M. L. Farr and F. G. Pollack, 1969, Mycologia 61:191–195.]

Although many species of *Leptosphaeria* cause a variety of leaf spot diseases, one species, *L. sengalensis*, is responsible for a disease of humans (El-Ani, 1966). The so-called "sooty molds" occur as harmless epiphytes on plants, forming their dark superficial mycelium, conidia, and ascostromata on the surface of plants. Actually the sooty molds are not parasitizing the plants but feeding on the honeydew secreted from the alimentary canal of scale insects. The honeydew contains a variety of nutrients, including sugars, amino acids, proteins, and minerals (Hughes, 1976). Some members form lichens (Chapter 15).

The Ascocarp. The ascostromata form locules that are not separated by a distinct wall from the stromal tissue. This character separates these fungi from the majority of the species in the Sphaeriales, in which a distinct wall is visible. The asci may exist individually in a small space that is essentially an individual locule (Fig. 4-56), but usually larger locules are formed in which several asci occur (Figs. 4-54 and 4-55). These larger locules resemble the cavities of the perithecia described on pages 107–111. Each ascostroma may have from one to several locules. In some species, there is a single locule surrounded by only a small amount of stroma tissue; such an ascostroma superficially resembles a perithecium and is sometimes called a *pseudothecium* (Fig. 4-54). Sometimes when mature pseudothecia and perithecia are compared, it is difficult to distinguish between the two ascocarp types. They do differ, however, in their mode of development and ascus type (unitunicate or bitunicate). Other than having a peritheciumlike form, the ascocarp shows numerous morphological variations. The shape may be lenslike, owing to the formation of upper and lower plates of radiate tissue (superficially resembling the ascocarps of the Meliolales) (Fig. 4-55), elongated and boatlike and also opening by a slit, or superficially resembling an apothecium (Fig. 4-56). The ascocarps range in size from comparatively small (40 to 150 micrometers) to quite large (1 to 5 millimeters) but are solitary. Even larger crusts may be formed if the ascocarps fuse together at their sides. The ascocarps may be brightly colored or dark. In some species, the ascocarp may be gelatinous or dissolve into a slimy mass.

Asci are usually borne at the base of the locule in a group. They may occur in a fanlike cluster (fascicle) or in a basal layer. Asci range in shape from globose to elongated cylindrical or clavate types. Unlike the ascus apex of many fungi with unitunicate asci, that of the bitunicate ascus is comparatively simple and more or less invariable and does not blue with iodine. The ascospores are usu-

Figure 4–56 Longitudinal section through the stroma of *Diplotheca* showing scattered locules, each containing a single ascus with septate ascospores. Details of the ascospores are at the right. [From B. O. Dodge, 1939, Mycologia 31:96–108.]

ally septate and may show a number of other variations, such as a gelatinous sheath or appendages. The asci may be the only contents of the locule or they may be interspersed among pseudoparaphyses (Fig. 4–57). Hymenial paraphyses are not formed in this group. The pseudoparaphyses are hairlike struc-

Figure 4–57 Centrum of *Sporomiella australis,* a member of the Pleosporales. Note the thin, threadlike pseudoparaphyses among the wider asci. [Courtesy R. Blanchard, 1972, Am. J. Botany 59:537–548.]

tures growing downward from the apex of the locule, and almost always become attached at the bottom of the locule when their tips recurve and become entangled in the resulting mass of tissue or push into the stromal tissue at the base of the locule (Luttrell, 1965). Growth of the pseudoparaphyses is largely intercalary, and they may also be branched (Corlett, 1973). Periphyses may be formed in those species which form a neck with an ostiole.

Ascocarp Development. In tracing the development of the ascostroma, we may begin with a relatively small initial consisting of sterile hyphae, the *stromal initial.* The stromal initial will continue to grow and will develop into all sterile elements of the ascocarp.

Apparently most species form differentiated archicarps. One or more archicarps may be formed in each stromal initial and lie embedded within the stromal tissue. The archicarps are often coiled and multicellular. In some species, plasmogamy may occur between the archicarp and spermatia: this occurs in species of *Mycosphaerella* (Fig. 4–58) (Higgins, 1936) and *Cochliobolus sativus* (Shoemaker, 1955). In others, including *Venturia inaequalis,* the archicarp undergoes plasmogamy with a differentiated antheridium (Killian, 1917). Alternatively, male gametes or gametangia may be lacking and the archicarp is then not involved in plasmogamy. In observations on *Trichometosphaeria turica,* there was no morphological evidence for plasmogamy involving the archicarp, although genetic evidence indicated that plasmogamy had occurred, perhaps involving hyphal fusion (Luttrell, 1964). Some species, including *Pleospora herbarum,* lack differentiated gametangia. In *P. herbarum,* two undifferentiated hyphae undergo fusion prior to pseudothecium formation.

Luttrell (1951) describes three major types of development (Fig. 4–59) which he considers to be typical for orders that he recognizes (1973). Because we are recognizing only a single order, each developmental type will be designated by the name of the genus on which the name of Luttrell's order is based. These types are described below.

1. *Myriangium* type: The globose asci occur individually in the tissue of the stroma and may occur throughout or in restricted fertile regions of those stromata superficially resembling apothecia. One or more archicarps are formed and produce ascogenous hyphae that penetrate through the stromal tissue, producing asci at various intervals. Although the stromal tissue is not greatly altered, it may sometimes be slightly compressed and disintegrated by the pressure of the expanding asci. When the asci are to discharge their ascospores, they penetrate individually through the overlying stromal tissue, which has become softened or crumbles.

2. *Dothidea* type: Species of *Dothidea* and their allies produce asci in fascicles in small perithecium-shaped locules in ascostromata and do not form pseudoparaphyses. This developmental sequence begins when ascogonia are formed within the stromal initial, which continues to enlarge and differentiate further. While embedded in the pseudoparenchymatous tissue, the ascogonium produces a short, radiating mass of ascogenous hyphae. This particular configuration of ascogenous hyphae ultimately produces the fanlike fascicle of asci. The elongating asci push up into the pseudoparenchyma, which becomes crushed and disintegrates to form the locule. At maturity, the locule is an unwalled cavity that is almost completely filled by asci. In some species, partially disintegrated pseudoparenchymatous tissue may remain behind as strands. A pore is formed when some tissue at the apex dissolves. This type of development also occurs in *Mycosphaerella* (Luttrell, 1951) illus-

Figure 4–58 Sexual cycle of *Mycosphaerella tulipiferae* which has spermatization and the *Dothidea*-type of development: (a) mature spermagonium within which spermatia are produced; the spermagonium, like the pseudothecium, is formed within the leaf tissue; (b) young stromal initial developing, with an archicarp at the lower center and with a trichogyne growing toward the surface; (c) part of the threadlike trichogyne extending above the leaf surface with a spermatium attached to its tip; (d) portion of an archicarp in which the male nucleus is migrating down the trichogyne; the base of the archicarp is swollen, and the tip of the trichogyne is not shown; (e) later stage in which the ascogonium is binucleate near the base, and the trichogyne is disintegrating; note the further development of the surrounding stroma; (f) ascogonium with eight pairs of conjugate nuclei, each pair held in a mass of denser protoplasm; (g) latter stage of pseudothecium with branched ascogonium; (h) tip of ascogenous hyphae with a crozier (left) and a young ascus (right); (i) nearly mature pseudothecium with asci in various stages of development and ascogonium and ascogenous hyphae still visible; note the formation of the neck with an ostiole and periphyses, and also the relatively conspicuous locule; (j) mature pseudothecium with asci and ascospores. [Adapted from B. B. Higgins, 1936, Am. J. Botany 23:598–602.]

Classification

MYRIANGIUM-type

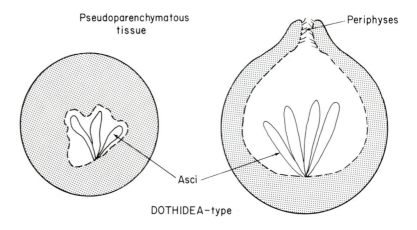

Pseudoparenchymatous tissue

Periphyses

Asci

DOTHIDEA-type

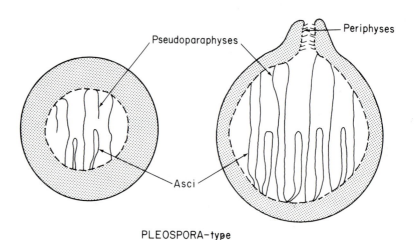

Pseudoparaphyses

Periphyses

Asci

PLEOSPORA-type

Figure 4–59 Centrum development in the Loculoascomycetes. A younger, nonostiolate pseudothecium is on the left, showing young asci. The more mature stage with an ostiole is on the right.

trated in Fig. 4–58, and also in a few species of sooty molds (Corlett, 1970; Reynolds, 1975).

3. *Pleospora* type: Pseudoparaphyses are formed in this type of development. *Pleospora* and its allies form a pseudoparenchymatous stroma in which the ascogonia are embedded. Pseudoparaphyses appear in the vicinity of the ascogonia, grow downward, and become attached at the bottom of the locule, meanwhile forming a locule by exerting outward pressure. Elongation of the pseudoparaphyses produces a flask-shaped locule within the stroma. Asci develop among the pseudoparaphyses, forming a concave layer (hymenium) across the base of the locule. Finally a pore is dissolved in the stroma at the apex of the locule. This type also occurs in *Leptosphaeria avenaria* (Corlett, 1966) and *Didymosphaeria sadasivanii* (Luttrell, 1975).

AINSWORTH, G. C. 1973. Introduction and keys to higher taxa. *In:* Ainsworth, G. C., F. K. Sparrow, and A. S. Sussman, Eds. The fungi: an advanced treatise. Academic Press, New York, Vol. 4B, pp. 1–7.

ASHTON, M. L., and P. B. MOENS. 1979. Ultrastructure of sporulation in the Hemiascomycetes *Ascoidea corymbosa, A. rubescens, Cephaloascus fragrans,* and *Saccharomycopsis capsularis.* Can. J. Botany 57:1259–1284.

BELLEMERE, A. 1977. L'appareil apical d l'ascue chez quelques discomycetes: Étude ultrastructurale comparative. Rev. Mycol. 41:233–264.

BENJAMIN, C. R. 1955. Ascocarps of *Aspergillus* and *Penicillium.* Mycologia 47:669–687.

BENJAMIN, R. K. 1973. Laboulbeniomycetes. *In:* Ainsworth, G. C., F. K. Sparrow, and A. S. Sussman, Eds. The fungi: an advanced treatise. Academic Press, New York, Vol. 4A, pp. 223–246.

——, and L. SHANOR. 1950. The development of male and female individuals in the dioecious species *Laboulbenia formicarum* Thaxter. Am. J. Botany 37:471–476.

——, and ——. 1952. Sex of host specificity and position of certain species of *Laboulbenia* on *Bembidion picipes.* Am. J. Botany 39:125–131.

BLACK, S. H., and C. GORMAN. 1971. The cytology of *Hansenula.* III. Nuclear segregation and envelopment during ascosporogenesis in *Hansenula wingei.* Arch. Mikrobiol. 79:231–248.

BLACKWELL, M. 1980. Developmental morphology and taxonomic characters of *Arthrorhynchus nycteribiae* and *A. eucampsipodae* (Laboulbeniomycetes). Mycologia 72:159–168.

CAIN, R. F. 1950. Studies of coprophilous ascomycetes. I. *Gelasinospora.* Can. J. Res. Sect. C28:566–576.

CAMPBELL, R., and M. SYROP. 1975. Light and electron microscope study of hysterothecium development in *Lophodermella sulcigena.* Trans. Brit. Mycol. Soc. 64:209–214.

CORLETT, M. 1966. Perithecium development in *Leptosphaeria avenaria.* Can. J. Botany 44:1141–1149.

——. 1970. Ascocarp development of two species of sooty molds. Can. J. Botany 48:991–995.

——. 1973. Observations and comments on the *Pleospora* centrum type. Nova Hedwigia 24:347–360.

——, and M. E. ELLIOTT. 1974. The ascus apex of *Ciboria acerina.* Can. J. Botany 52:1459–1463.

DARKER, G. D. 1967. A revision of the genera of the Hypodermataceae. Can. J. Botany 45:1399–1444.

DODGE, B. O. 1932. The non-sexual and the sexual functions of microconidia of *Neurospora.* Bull. Torrey Botan. Club 59:347–360.

DOIDGE, E. M. 1942. A revision of the South African Microthryriaceae. Bothalia 4:273–420.

EL-ANI, A. S. 1966. A new species of *Leptosphaeria,* an etiologic agent of mycetoma. Mycologia 58:406–411.

ELLIS, J. J. 1960. Plasmogamy and ascocarp development in *Gelasinospora calospora.* Mycologia 52:557–573.

FENNELL, D. I. 1973. Plectomycetes; Eurotiales. *In:* Ainsworth, G. C., F. K. Sparrow, and A. S. Sussman, Eds. The fungi: an advanced treatise. Academic Press, New York, Vol. 4A, pp. 45–68.

FITZPATRICK, H. M. 1923. Monograph of the Nitschkieae. Mycologia 15:45–67.

GILKEY, H. M. 1939. Tuberales of North America. Oregon State Monogr. Stud. Botan. 1:1–63.

——. 1961. New species and revisions in the order Tuberales. Mycologia 53:215–220.

GOOS, R. D. 1959. Spermatium–trichogyne relationship in *Gelasinospora calospora* var. *autosteira.* Mycologia 51:416–428.

GRAFF, P. W. 1932. The morphological and cytological development of *Meliola circinans.* Bull. Torrey Botan. Club 59:241–266.

GRIFFITHS, H. B. 1973. Fine structure of seven unitunicate Pyrenomycete asci. Trans. Brit. Mycol. Soc. 60:261–271.

HANLIN, R. T. 1961. Studies in the genus *Nectria*. II. Morphology of *N. gliocladiodes*. Am. J. Botany 48:900–908.

———. 1964. Morphology of *Hypomyces trichothecoides*. Am. J. Botany 51:201–208.

HANSFORD, C. G. 1961. The Meliolineae: a monograph. Beih. Sydowia Ann. Mycol. Ser. 2, 2:1–806.

———. 1963. Icongraphia Meliolinearum. Beih. Sydowia Ann. Mycol. Ser. II, 5: plates I–CCLXXXV.

HAWKER, L. E. 1955. Hypogeous fungi. Biol. Rev. Cambridge Phil. Soc. 30:127–158.

HIGGINS, B. B. 1936. Morphology and life history of some ascomycetes with special reference to the presence and function of spermatia. III. Am. J. Botany 23:598–602.

HILL, T. W. 1977. Ascocarp ultrastructure of *Herpomyces* sp. (Laboulbeniales) and its phylogenetic implications. Can. J. Botany 55:2015–2032.

HONEY, E. E. 1936. North American species of *Monilinia*. I. Occurrence, grouping, and life-histories. Am. J. Botany 23:100–106.

HUANG, L. H. 1976. Developmental morphology of *Triangularia backusii* (Sordariaceae). Can. J. Botany 54:250–267.

HUGHES, S. J. 1976. Sooty moulds. Mycologia 68:693–820.

KILLIAN, K. 1917. Über die Sexualität von *Venturia inaequalis* (Cooke) Ad. Z. Botan. 9:353–398.

KIMBROUGH, J. W. 1974. Structure and development of *Lasiobolus monascus,* a new species of Pezizales, Ascomycetes. Mycologia 66:907–918.

———, and R. P. KORF. 1967. A synopsis of the genera and species of the tribe Theleboleae (= Pseudoascoboleae). Am. J. Botany 54:9–23.

KORF, R. P. 1962. A synopsis of the Hemiphacidiaceae, a family of the Helotiales (Discomycetes) causing needle-blights of conifers. Mycologia 54:12–33.

———. 1973. Discomycetes and Tuberales. *In:* Ainsworth, G. C., F. K. Sparrow, and A. S. Sussman, Eds. The fungi: an advanced treatise. Academic Press, New York, Vol. 4A, pp. 249–319.

KRAMER, C. L. 1973. Protomycetales and Taphrinales. *In:* Ainsworth, G. C., F. K. Sparrow, and A. S. Sussman, Eds. The fungi: an advanced treatise. Academic Press, New York, Vol. 4A, pp. 33–41.

KREGER-VAN RIJ, N. J. W. 1973. Endomycetales, basidiomycetous yeasts, and related fungi. *In:* Ainsworth, G. C., F. K. Sparrow, and A. S. Sussman, Eds. The fungi: an advanced treatise. Academic Press, New York, Vol. 4A, pp. 11–32.

KUEHN, H. H. 1957. Observations on the Gymnoasceae. V. Developmental morphology of two species representing a new genus of the Gymnoasceae. Mycologia 49:694–706.

KWON-CHUNG, K. J. 1973. Studies on *Emmonsiella capsulata*. I. Heterothallism and development of the ascocarp. Mycologia 65:109–121.

LUTTRELL, E. S. 1951. Taxonomy of the Pyrenomycetes. Univ. Missouri Stud. 24(3):1–20.

———. 1964. Morphology of *Trichometasphaeria turica*. Am. J. Botany 51:213–219.

———. 1965. Paraphysoids, pseudoparaphyses, and apical paraphyses. Trans. Brit. Mycol. Soc. 48:135–144.

———. 1973. Loculoascomycetes. *In:* Ainsworth, G. C., F. K. Sparrow, and A. S. Sussman, Eds. The fungi: an advanced treatise. Academic Press, New York, Vol. 4A, pp. 135–219.

———. 1975. Centrum development in *Didymosphaeria sadasivanii* (Pleosporales). Am. J. Botany 62:186–190.

MAI, S. H. 1977. Morphological studies in *Sordaria fimicola* and *Gelasinospora longispora*. Am. J. Botany 64:489–495.

MALLOCH, D., and G. L. BENNY. 1973. California Ascomycetes: four new species and a new record. Mycologia 65:648–660 (Eurotiales).

———, and R. F. CAIN. 1972. The Trichocomataceae: Ascomycetes with *Aspergillus, Paecilomyces,* and *Penicillium* imperfect states. Can. J. Botany 50:2613–2628.

———, and ———. 1973. The genus *Thielavia*. Mycologia 65:1055–1077.

MANIOTIS, J. 1965. A cleistothecial mutant of the perithecial fungus *Gelasinospora calospora*. Mycologia 57:23–35.

MATILE, P., H. MOOR, and C. F. ROBINOW. 1969. Yeast cytology. *In:* Rose, A. H., and J. S. Harrison, Eds. The yeasts. Academic Press, New York, Vol. 1, pp. 219–302.

MCGAHEN, J. W., and H. E. WHEELER. 1951. Genetics of *Glomerella*. IX. Perithecial development and plasmogamy. Am. J. Botany 38:610–617.

MERKUS, E. 1976. Ultrastructure of the ascospore wall in Pezizales (Ascomycetes). IV. Morchellaceae, Helvellaceae, Rhiziniaceae, Thelebolaceae, and Sarcoscyphaceae. General discussion. Persoonia 9:1–38.

MIX, A. J. 1935. The life history of *Taphrina deformans*. Phytopathology 25:41–66.

MOENS, P. B. 1971. Fine structure of ascospore development in the yeast *Saccharomyces cerevisiae*. Can. J. Microbiol. 17:507–510.

MOORE, E. J. 1963. The ontogeny of the apothecia of *Pyronema domesticum*. Am. J. Botany 50:37–44.

MOORE-LANDECKER, E. 1975. A new pattern of reproduction in *Pyronema domesticum*. Mycologia 67:1119–1127.

MÜLLER, E., and J. A. VON ARX. 1973. Pyrenomycetes, Meliolales, Coronophorales, Sphaeriales. *In:* Ainsworth, G. C., F. K. Sparrow, and A. S. Sussman, Eds. The fungi: an advanced treatise. Academic Press, New York, Vol. 4A, pp. 87–132.

NELSON, A. C., and M. P. BACKUS. 1968. Ascocarp development in two homothallic *Neurosporas*. Mycologia 60:16–28.

O'DONNELL, K. L., and G. R. HOOPER. 1974. Scanning ultrastructural ontogeny of paragymnohymenial apothecia in the operculate Discomycete *Peziza quelpidotia*. Can. J. Botany 52:873–876.

ORR, G. F., and H. H. KUEHN. 1964. A re-evaluation of *Myxotrichum spinosum* and *M. cancellatum*. Mycologia 56:473–481.

RAPER, K. B., and D. I. FENNELL. 1965. The genus *Aspergillus*. Williams & Wilkins, Baltimore, 686 pp.

REEVES, F., JR. 1971. The structure of the ascus apex in *Sordaria fimicola*. Mycologia 63:204–212.

REYNOLDS, D. R. 1975. The centrum of the sooty mold ascomycete *Limacinula samoensis*. Am. J. Botany 62:775–779.

ROGERSON, C. T. 1970. The hypocrealean fungi (Ascomycetes, Hypocreales). Mycologia 62:865–910.

ROSING, W. C. 1982. Ultrastructure of ascus and ascospore development in *Chaetomium brasiliense*. Mycologia 74:960–974.

ROSINSKI, M. A. 1956. Development of the ascocarp of *Anthracobia melaloma*. Mycologia 48:506–533.

———. 1961. Development of the ascocarp of *Ceratocystis ulmi*. Am. J. Botany 48:285–293.

SAMUELS, G. J. 1973. Perithecial development in *Hypomyces aurantius*. Am. J. Botany 60:268–276.

SAMUELSON, D. A. 1975. The apical apparatus of the suboperculate ascus. Can. J. Botany 53:2660–2679.

———. 1978. Asci of the Pezizales. VI. The apical apparatus of *Morchella esculenta, Helvella crispa,* and *Rhizina undulata*. General discussion. Can. J. Botany 56:3069–3082.

SÁNCHEZ, A., and R. P. KORF. 1966. The genus *Vibressea,* and the generic names *Leptosporium, Apostemium, Apostemidium, Goroniceps* and *Ophiogloea*. Mycologia 58:722–737.

SCHOKNECHT, J. D. 1975. Structure of the ascus apex and ascospore dispersal mechanisms in *Sclerotinia tuberosa*. Trans. Brit. Mycol. Soc. 64:358–362.

SHANOR, L. 1952. The characteristics and morphology of a new genus of the Laboulbeniales on an earwig. Am. J. Botany 39:498–504.

SHOEMAKER, R. A. 1955. Biology, cytology, and taxonomy of *Cochliobolus sativus*. Can. J. Botany 33:562–576.

SILVERBERG, B. A., and J. F. MORGAN-JONES. 1974. *Lophodermium pinastri:* certain fine-structural features of ascocarp morphology. Can. J. Botany 52:1993–1995.

STIERS, D. L. 1974. Fine structure of ascospore formation in *Poronia punctata*. Can. J. Botany 52:999–1003.

——. 1976. The fine structure of ascospore formation in *Ceratocystis fimbriata*. Can. J. Botany 54:1714–1723.

——. 1977. The fine structure of the ascus apex in *Hypoxylon serpens, Poronia punctata, Rosellinia aquila,* and *R. mammiformis*. Cytologia 42:697–702.

SYROP, M., and A. BECKETT. 1976. Leaf curl disease of almond caused by *Taphrina deformans*. III. Ultrastructural cytology of the pathogen. Can. J. Botany 54:293–305.

TAVARES, I. I. 1966. Structure and development of *Herpomyces stylopygae* (Laboulbeniales). Am. J. Botany 53:311–318.

——. 1985. Laboulbeniales (Fungi, Ascomycetes). Mycologia Mem. 9:1–627.

THAXTER, R. 1896. Contribution towards a monograph of the Laboulbeniaceae. Mem. Am. Acad. Arts Sci. 12:187–429, Plates I–XXVI.

——. 1908. Contribution towards a monograph of the Laboulbeniaceae. Part II. Mem. Am. Acad. Arts Sci. 13:217–469, Plates XXVIII–LXXI.

——. 1924. Contribution towards a monograph of the Laboulbeniaceae. Part III. Mem. Am. Acad. Arts Sci. 14:309–426.

——. 1926. Contribution towards a monograph of the Laboulbeniaceae. Part IV. Mem. Am. Acad. Arts Sci. 15:427–580.

——. 1931. Contribution towards a monograph of the Laboulbeniaceae. Part V. Mem. Am. Acad. Arts Sci. 16:1–435.

THYR, B. D., and C. G. SHAW. 1966. Ontogeny of the needle cast fungus, *Hypodermella arcuata*. Mycologia 58:192–200.

TRAPPE, J. M. 1971. A synopsis of the Carbomycetaceae and Terfeziaceae (Tuberales). Trans. Brit. Mycol. Soc. 57:85–92.

TYSON, K., and D. A. GRIFFITHS. 1976. Developmental morphology and fine structure of *Placoasterella baileyi*. II. Ascus and ascospore development. Trans. Brit. Mycol. Soc. 66:263–279.

UECKER, F. A. 1976. Development and cytology of *Sordaria humana*. Mycologia 68:30–46.

——, and J. M. STALEY. 1973. Development of the ascocarp and cytology of *Lophodermella morbida*. Mycologia 65:1015–1027.

VAN BRUMMELEN, J. 1967. A world-monograph of the genera *Ascobolus* and *Saccobolus*. Persoonia Suppl. 1:1–260.

——. 1978. The operculate ascus and allied forms. Persoonia 10:113–128.

VON ARX, J. A. 1975. On *Thielavia* and some similar genera of Ascomycetes. Studies Mycol. 8:1–29.

——, and E. MÜLLER. 1975. A re-evaluation of the bitunicate ascomycetes with keys to families and genera. Studies Mycol. 9:1–159.

WHISLER, H. C. 1968. Experimental studies with a new species of *Stigmatomyces* (Laboulbeniales). Mycologia 60:65–75.

WHITESIDE, W. C. 1961. Morphological studies in the Chaetomiaceae. I. Mycologia 53:512–523.

——. 1962. Morphological studies in the Chaetomiaceae. II. Mycologia 54:152–159.

WONG, H.-C., and C.-Y. CHIEN. 1986. Ultrastructure of sexual reproduction of *Monascus purpureus*. Mycologia 78:713–721.

WOO, J. Y., and A. D. PARTRIDGE. 1969. The life history and cytology of *Rhytisma punctatum* on bigleaf maple. Mycologia 61:1085–1095.

YARWOOD, C. E. 1957. Powdery mildews. Botan. Rev. 23:235–300.

——. 1973. Pyrenomycetes: Erysiphales. *In:* Ainsworth, G. C., F. K. Sparrow, and A. S. Sussman, Eds. The fungi: an advanced treatise. Academic Press, New York, Vol. 4A, pp. 71–86.

Additional Recommended References

BENJAMIN, R. K. 1965. A new genus of the Gymnoascaceae with a review of the other genera. Aliso 3:310–328 (Eurotiales).

BUTLER, G. M. 1966. Vegetative structures. *In:* Ainsworth, G. C., and A. S. Sussman, Eds. The fungi: an advanced treatise. Academic Press, New York, Vol. 2, pp. 83–112.

DENNIS, R. W. G. 1978. British Ascomycetes. J. Cramer, Springäckern, West Germany, 585 pp.

FUNK, A., and R. A. SHOEMAKER. 1967. Layered structure in the bitnunicate ascus. Can. J. Botany 45:1265–1266.

GOOS, R. D. 1974. A scanning electron microscope and *in vitro* study of *Meliola palmicola*. Proc. Iowa Acad. Sci. 81:23–27 (Meliolales).

HAGLER, A. N., and D. G. AHEARN. 1987. Ecology of aquatic yeasts. *In:* Rose, A. H., and J. S. Harrison, Eds. Biology of yeasts, 2nd ed. Academic Press, New York, Vol. 1, pp. 181–205.

KAMAT, M. N., and A. PANDE-CHIPLONKAR. 1971. Patterns of spermatization in the ascomycetes. Rev. Mycol. 36:257–276.

KIMBROUGH, J. W. 1963. The development of *Pleochaeta polychaeta* (Erysiphaceae). Mycologia 55:608–626 (Erysiphales).

——. 1970. Current trends in the classification of Discomycetes. Botan. Rev. 36:91–161.

KORF, R. P. 1958. Japanese Discomycete notes I–VIII. Sci. Rept. Yokohama Natl. Univ. Sect. II 7:7–35 (Discomycete anatomy).

KRAMER, C. L. 1960. Morphological development and nuclear behavior in the genus *Taphrina*. Mycologia 52:295–320 (Taphrinales).

KREGER-VAN RIJ, N. J. W. 1987. Classification of yeasts. *In:* Rose, A. H., and J. S. Harrison, Eds. The yeasts, 2nd ed. Academic Press, New York, Vol. 1, pp. 5–61.

LODDER, J., and N. J. W. KREGER-VAN RIJ. 1952. The yeasts: a taxonomic study. Wiley, New York, 713 pp.

McCLARY, D. O. 1964. The cytology of yeasts. Botan. Rev. 30:168–225 (Endomycetales).

MOORE, E. J. 1965. Ontogeny of gelatinous fungi. Mycologia 57:114–130 (Helotiales).

NANNFELDT, J. A. 1976. Iodine reactions in ascus plugs and their taxonomic significance. Trans. Brit. Mycol. Soc. 67:283–287.

PHAFF, H. J., and W. T. STARMER. 1987. Yeasts associated with plants, insects and soil. *In:* Rose, A. H., and J. S. Harrison, Eds. Biology of yeasts, 2nd edition. Academic Press, New York, Vol. 1, pp. 123–180.

VAN BRUMMELEN, J. 1975. Light and electron microscopic studies of the ascus top in *Sarcoscypha coccinea*. Persoonia 8:259–271 (Pezizales).

VON ARX, J. A., L. R. DEMIRANDA, M. T. SMITH, and D. YARROW. 1977. The genera of the yeasts and yeast-like fungi. Studies Mycol. 14:1–42.

5

Division Amastigomycota— Subdivision Basidiomycotina

The Subdivision Basidiomycotina contains about 15,000 known species. We may refer to these fungi informally as the ''basidiomycetes,'' a term which is commonly used but without taxonomic status. Fungi such as the jelly fungi, rusts, smuts, bracket fungi, mushrooms, and puffballs belong to this group. Many of the common names refer to the visible part of the fungus, which is a conspicuous reproductive body, the *basidiocarp*. The basidiocarp is supported by an extensive assimilative mycelium penetrating the soil or plant material and thus deriving the nutrients that supply the basidiocarp.

THE BASIDIUM AND SEXUAL REPRODUCTION

All members of the Basidiomycotina produce *basidia* (singular, *basidium*). Basidia are organs homologous to and presumably derived from the ascus. Like the typical ascus, the basidium usually originates as a binucleate structure and is the site of karyogamy and meiosis. Unlike the ascus, the basidium bears its basidiospores on the outside of the basidium.

In order to understand the role of the basidium in the Basidiomycotina life cycle, we now explore a typical life cycle, which is characterized by an extremely short diploid stage, occurring in the basidium only, and a prolonged dikaryotic stage. With the exception of the Uredinales and Ustilaginales, almost all basidiomycetes have virtually the same life cycle, which follows (Fig. 5–1). Some variations are described on pages 174–175.

The basidiospore usually contains a single haploid nucleus. When the basidiospore germinates, it initiates a haploid monokaryotic mycelium. The monokaryon is at first nonseptate but later becomes divided into a number of uninucleate cells. This phase of the life cycle is usually of short duration.

Plasmogamy then takes place. Sexual organs are lacking in the basidiomy-

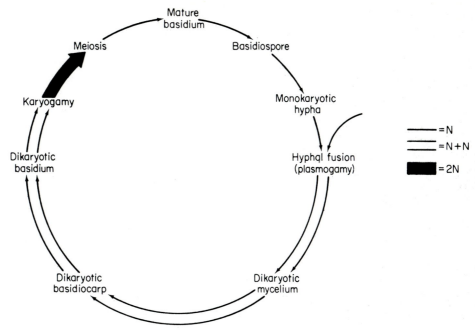

Figure 5–1 Generalized life cycle of a basidiomycete in which basidio-carp formation occurs.

cetes (except in the Uredinales), and plasmogamy is usually accomplished by the fusion of two monokaryotic hyphae. When the monokaryotic hyphae fuse, the nuclei of one hypha flow into the other. Usually this exchange of nuclei is reciprocal between the fusing hyphae. Invading nuclei may travel many millimeters or centimeters before stopping. The hyphae now have nuclei of two genetic types and are dikaryotic. Each cell of this dikaryon has two haploid nuclei (*N* + *N*).

The dikaryon formed after plasmogamy continues to proliferate, meanwhile maintaining the binucleate, dikaryotic condition. Usually, *clamp connections* are formed to maintain the binucleate condition (Fig. 5–2). The young

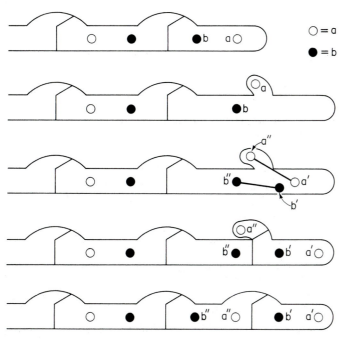

Figure 5–2 Steps in the formation of a clamp connection. See the text for an explanation.

clamp begins to form as a bulging pocket from the hyphal wall between the two nuclei (*a* and *b*) of the terminal dikaryotic cell. Both nuclei divide mitotically. The division figure of the terminal nucleus *a* lies obliquely along the main axis of the clamp and, at the completion of its division, one daughter nucleus *a"* remains in the clamp, while the second daughter nucleus *a'* remains in a terminal position. The division figure of the second nucleus, *b*, lies along the main axis of the hypha, and its one daughter nucleus *b"* remains in a posterior position, while the second daughter nucleus *b'* is in a terminal position near *a'*. A transverse septum is laid down and separates the new terminal binucleate cell (*a'b'*). The cell remaining behind still has only one nucleus (*b"*) and the other nucleus (*a"*) is still isolated in the lateral clamp. The clamp recurves and comes into contact with the uninucleate cell, and the walls between are dissolved. The nucleus from the clamp migrates into the hyphal cell, reconstituting the dikaryotic condition (*a"b"*). The clamp remains as a permanent part of the hypha, and it is often used as morphological evidence that the hypha is dikaryotic (Figs. 5–2 and 5–3).

Figure 5–3 Clamp connection of *Coprinus cinereus*. Note the continuity between the clamp and the main body of the hypha, and also the two septa. × 25,000. [Courtesy D. J. McLaughlin.]

The dikaryotic mycelium serves as the assimilative mycelium which is concealed deep within the substratum. When conditions are favorable for reproduction, some of the dikaryotic mycelium undergoes a complex morphogenesis to form a basidiocarp (such as the visible mushroom). All the mycelium within the basidiocarp is of the same type, in contrast to some ascomyetes in which two distinct mycelial systems take part in the formation of an ascocarp. Within the basidiocarp, some of the cells in sterile tissues may remain binucleate while others become multinucleate, depending upon the species.

Some of the cells in a specialized fertile region of the basidiocarp are the young basidia (Fig. 5–4). The basidia are at first dikaryotic and have two haploid nuclei (Fig. 5–5). The nuclei fuse to form a diploid nucleus, which then undergoes meiosis. Usually four uninucleate basidiospores are produced on each basidium. Although basidiospores may be borne directly upon the surface of the basidium, they are usually produced upon elongate projections, the *sterigmata*.

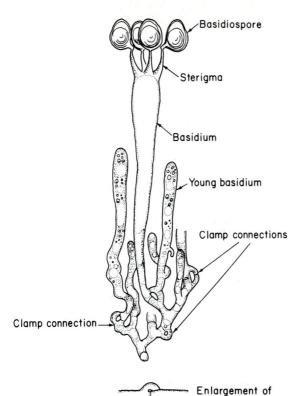

Basidiospore

Sterigma

Basidium

Young basidium

Clamp connections

Clamp connection

Enlargement of
clamp connection

Figure 5–4 Hyphae with clamp connections producing basidia. [From E. J. H. Corner, 1950, Ann. Botan. Mem. 1:1–74.]

Each sterigma bears a basidiospore at its apex. If a basidium is to form its basidiospores upon sterigmata (the usual situation), the sterigmata originate as projections from the apex of the basidium. The wall at the tip of the sterigma is at first thin and swells into a more or less globose apex; the swelling then elongates and enlarges asymmetrically. Meanwhile, additional wall layers are produced around the basidiospore initial and then gradually thicken. Cytoplasmic changes accompany maturation of the basidiospore initial: These may include the presence of many vesicles during the early stages and the appearance of mitochondria and other organelles during the later stages. A nucleus migrates into each basidiospore initial. In *Schizophyllum commune*, the nuclei and cyto-

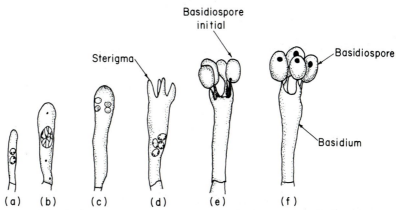

Basidiospore
initial

Sterigma

Basidiospore

Basidium

(a) (b) (c) (d) (e) (f)

Figure 5–5 Development of a typical basidium; (a) young dikaryotic basidium; (b) diploid basidium; (c) basidium with four nuclei resulting from meiosis; (d) basidium after development of sterigmata; (e) migration of nuclei into basidiospore initials; (f) mature basidium with basidiospores. [Adapted from A. H. Smith, 1934, Mycologia 26:305–331.]

The Basidium and Sexual Reproduction

plasm are apparently pushed through the sterigmata by the increasing vacuolation of the basidium (Wells, 1965), while in *Coprinus cinereus,* cytoplasmic microtubules may be responsible for nuclear migration as well as for the migration of other organelles. Finally, the cytoplasm within the basidiospore is separated from that of the sterigma by the formation of a plug at their juncture. At that point of juncture is the *hilar appendix* which consists primarily of a kneelike bulge enclosed by a wall continuous with that of the sterigma and containing an electron-light zone. The bulging is initiated after the nucleus migrates into the basidiospore initial, and the hilar appendix becomes increasingly distended while the basidiospores are maturing. The hilar appendix is extremely important in effecting basidiospore discharge and is discussed further in Chapter 11. In some species, a specialized electron-dense body is associated with the inner wall of the hilar appendix in its early developmental stages and may possibly contribute to its thickening (McLaughlin, 1977).

Variations of the General Pattern

The vast majority of the members of the Basidiomycotina have the characteristics outlined in the general life cycle on pages 170–174. There are, however, some that do not comply with one or more aspects of the general pattern. We will examine some of these deviations.

Most basidiomycete species are heterothallic, requiring fusion of two genetically different but compatible hyphae to produce a fertile cross resulting in a basidiocarp. (Heterothallism is discussed further in Chapter 7.) Some basidiomycete species are homothallic and do not require plasmogamy between genetically different hyphae to produce a basidiocarp. An example is the edible straw mushroom, *Volvariella volvacea,* considered a delicacy in China. Although hyphal fusion may occur between monokaryotic hyphae and nuclear migration may occur, the nuclei are randomly distributed in the multinucleate mycelium and there is neither a tendency for the nuclei to occur in pairs nor clamp connections. Both karyogamy and meiosis occur in the basidium (Chang and Ling, 1970).

Clamp connections may also be lacking in heterothallic basidiomycete species. This absence of clamp connections may be sporadic in some species but occurs regularly in others. Some species that normally form clamps may become clampless if there is inadequate aeration. Unlike the multinucleate cells in the homothallic species, the cells of the mycelium are typically binucleate and dikaryotic. The dividing nuclei are found in the terminal cell of the hypha, which may be either binucleate or multinucleate. As in those cells forming a clamp connection, paired nuclei undergo simultaneous divisions. In this case, the spindles are parallel to each other, thus maintaining the close association of nuclear pairs (Kühner, 1977).

In heterothallic species, fusion of the primary mycelium to form the dikaryotic state may occur at various points in the life cycle. Fusion usually occurs during mycelial growth, but in some species it occurs at the time that the basidiocarp is initiated, or rarely, even in the subhymenium or hymenium. Binucleate, dikaryotic basidiospores may be regularly produced by some species. When these dikaryotic basidiospores germinate, they give rise directly to a dikaryotic mycelium that does not need to undergo plasmogamy to produce a basidiocarp.

The cells of the sterile tissues of basidiocarps may maintain the regularly binucleate condition or may become multinucleate, depending on the species.

Although the vast majority of the species have only a meiotic division in the basidium and subsequently produce four uninucleate basidiospores, there are deviations from this pattern. In some species, a mitotic division will follow meiosis, resulting in an eight-nucleate basidium; or alternatively, an additional nuclear division occurs in the sterigma or young basidiospore. Other possibilities are that some of the nuclei within the basidium may not be incorporated into basidiospores but instead degenerate, or that more than one nucleus will be incorporated in a basidiospore. Various combinations of these possible nuclear events sometimes result in basidia that form up to eight basidia (e.g., two, three, or five) or have basidiospores that are binucleate or contain up to eight nuclei (Kühner, 1977).

Variations in Basidia

We have just seen that two important events take place in the basidium: these are karyogamy and meiosis. There is some variation in both basidial structure (discussed further on pages 190–191) and the site at which karyogamy and meiosis occur. We may describe that cell in which karyogamy occurs as the *probasidium* and that cell in which meiosis occurs as the *metabasidium* (Talbot, 1973-a).*

The most common situation is that in which both karyogamy and meiosis occur in essentially the same location, and are separated only by time. This is the type used to illustrate the general life cycle on pages 170–174. Functionally, the young basidium at first is the probasidium and is the site of karyogamy, but it is entirely transformed into the metabasidium when meiosis occurs. Morphologically, there are no remnants of the probasidial stage. At maturity, the basidium consists of a metabasidium, possibly with sterigmata (Figs. 5–4 to 5–6).

In the second type, karyogamy and meiosis are separated both in time and in space. In its earliest stages, the basidium consists entirely of a probasidium, which may be a thick-walled resting spore. Karyogamy occurs, and the probasidium produces a thin-walled tubular extension into which the diploid nucleus migrates. The thin-walled extension, together with any remnants of the probasidium, functions as a single cell, the metabasidium. Meiosis occurs within the metabasidium. In some species, the metabasidium becomes divided by transverse septa after meiosis. The mature basidium consists of the metabasidium (possibly including remnants of the probasidium), and sterigmata, if formed (Fig. 5–23).

NONSEXUAL REPRODUCTION

Nonsexual reproduction occurs commonly in members of the Basidiomycotina, although much less frequently than in members of the Division Mastigomycota or the Subdivisions Zygomycotina and Ascomycotina. Nonsexual reproductive structures that are formed include buds (similar to those produced by yeasts), conidia, chlamydospores, and oidia.

Oidia commonly originate from a monokaryotic mycelium when a branch fragments into its component cells. The haploid oidia then germinate, yielding a haploid mycelium. Such oidial mycelia may unite with each other to form a dikaryotic mycelium, or an oidial hypha may unite with a haploid mycelium derived from a germinated basidiospore (Fig. 5–7). The most important role of oidia is apparently to increase the possibility that mating will take place. Nonsexual structures may be produced directly from the basidium, from the basidiospores, from the mycelium, and among the basidia in a hymenium. In a few cases, a separate sporocarp bearing the nonsexual spores may be produced.

CLASSIFICATION

We shall divide the Subdivision Basidiomycotina into three classes, the Teliomycetes, Hymenomycetes, and Gasteromycetes (Ainsworth, 1973). The members

---

*The terms *probasidium* and *metabasidium* may be used differently in other mycological works. See Talbot (1973-a) for clarification if required.

(a)

(b)

Figure 5–6 Basidia with basidiospores: (a) electon micrograph of a longitudinal section of a basidium at interphase II of meiosis (two of the four nuclei are visible); one of the four basidiospores is attached to its sterigma while a second basidiospore lies above the basidium, x 5,000; (b) scanning electron micrograph of a hymenium; three of the four basidiospores are present on the basidium in the center, x 7,600. [(a) Courtesy D. J. McLaughlin; (b) Courtesy K. S. Yoon and D. J. McLaughlin.]

of the Class Teliomycetes do not form a basidiocarp but produce resting spores that function as probasidia. Neither characteristic is found in the remaining two classes, as the majority of members produce a well-developed basidiocarp and the basidium does not originate from a resting spore. Members of the Class Hymenomycetes produce an exposed hymenium, while in the Class Gasteromycetes a hymenium is usually lacking but if present is formed in an enclosed basidiocarp. A further outline of the classification of this group follows:

Subdivision Basidiomycotina
Class Teliomycetes
Order Uredinales
Order Ustilaginales

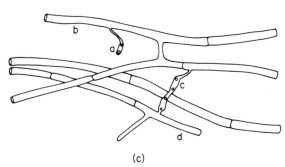

(a) (b) (c)

Figure 5–7 Nonsexual reproduction by means of oidia in *Coprinus lago-pus:* (a) modified hypha bearing oidia in a drop of liquid; (b) the hypha and oidia in (a) as it would appear if immersed in a film of water; (c) fusion of oidial germ tubes of the oidia *a* and *c* of one mating type with the hyphae *b* and *d* of a mycelium derived from a basidiospore of the opposite mating type. [From H. J. Brodie, 1931, *Ann. Botany* (London) 45:315–344.]

Class Hymenomycetes
   Subclass Phragmobasidiomycetidae
      Order Tremellales
      Order Auriculariales
      Order Septobasidiales
   Subclass Holobasidiomycetidae
      Order Exobasidiales
      Order Brachybasidiales
      Order Dacrymycetales
      Order Tulasnellales
      Order Aphyllophorales
      Order Agaricales
Class Gasteromycetes
      Order Hymenogastrales
      Order Melanogastrales
      Order Gautieriales
      Order Phallales
      Order Lycoperdales
      Order Tulostomatales
      Order Sclerodermatales
      Order Nidulariales

Class Teliomycetes

Members of the Teliomycetes are plant parasites that do not form a basidiocarp. The probasidia are thick-walled resting spores that often overwinter before producing metabasidia. The resting spores are formed in a loose cluster, the *sorus*.

    The Teliomycetes may be divided into two orders, the Uredinales and Ustilaginales (Durán, 1973; Laundon, 1973).

ORDER UREDINALES

The Uredinales are the "rusts," a group of almost 5000 specialized parasites occurring on ferns, gymnosperms, and angiosperms. The rust fungi are host-specific, and a given species or genetic strain of the fungus is known to occur only on certain genera, species, or genetic strains of the host (for example,

Puccinia graminis var. *tritici* occurs on wheat while *Puccinia graminis* var. *secalis* occurs on rye). These fungi are known to occur only as parasites in nature although they have been recently cultured in the laboratory (Laundon, 1973; Petersen, 1974). Rust diseases result in widespread physiological damage to the host, causing an increase in transpiration rate and a decrease in photosynthesis and respiration rates. Sometimes abnormal enlargements are induced to form. Rust diseases have been known since ancient times when it was thought that they were inflicted by a revengeful deity.

The mycelium occurs between cells of the host plant and penetrates them by forming haustoria. Clamp connections are not formed. The spores are usually borne in a loose cluster, the sorus, which often becomes exposed by rupture of the host epidermis (Fig. 5–8). Some sori formed by the rust fungi are orange or red giving the characteristic rustlike appearance to the diseased areas.

Although morphologically simple, some rusts have the most complex life cycle found in the fungi, in addition to possessing a high degree of host specificity. As many as five separate spore stages may be formed in succession in the life cycle of a rust. These spore stages may be defined according to their role in the life cycle (Hiratsuka, 1973). The spore stages are described below.

Stage 0. Pycnium bearing pycniospores. Pycnia are monokaryotic, haploid structures that produce *pycniospores.* The pycniospores are haploid, uni-

(a) (b)

Figure 5–8 Sori in the Uredinales: (a) uredinial stage of *Puccinia sorghi* on maize; note that the mass of urediniospores in each sorus causes the host epidermis to rupture; (b) scanning electron micrograph of the telial stage of *Puccinia podophylli;* the teliospores have ruptured the host epidermis, x 425. [(a) Courtesy Plant Pathology Department, Cornell University; (b) courtesy K. L. O'Donnell.]

nucleate structures that function as spermatia when transferred to another pycnium. The pycnia are hermaphroditic structures, containing the female receptive structures as well as the male pycniospores. The dikaryotic state is most often established by the transfer of pycniospores from one pycnium to the receptive hyphae in a second pycnium.

The pycnia vary in shape and are frequently flask-shaped but may also be flat or globose. They occur beneath or within the epidermis, beneath the cuticle, or within the cortex of the host plant (Fig. 5–9).

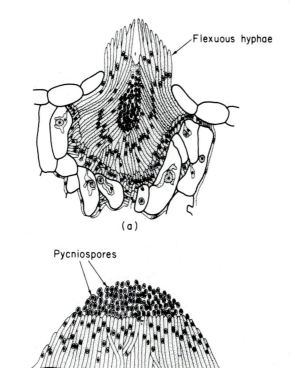

Figure 5–9 Pycnia in the Uredinales: (a) flask shape pycnium, x 510; (b) cushionlike superficial pycnium, × 850. [From P. Sappin-Trouffy, 1896, Botaniste 5:59–244.

Stage I. Aecium bearing aeciospores. The *aecia* are usually associated with a pycnium that has functioned in plasmogamy to establish the dikaryotic state, and they usually bear the first dikaryotic spores, the *aeciospores*, in the life cycle. Upon germination, the aeciospores establish a dikaryotic mycelium.

A typical aecium occurs as an open cup, bearing the binucleate aeciospores in chains. The cupulate aecium is surrounded by a peridium, which encloses the aecium during its early stages and separates it from the host tissues in which the aecium is embedded. The peridium later ruptures and curves backward to form an irregular collar. The aeciospores are yellow to orange. A second common aecial form is that in which an elongated hornlike protrusion is formed and is surrounded by a peridium that opens by tearing into shreds (Fig. 5–10).

Stage II. Uredinium bearing urediniospores. The *urediniospores* are also binucleate and dikaryotic, but unlike the aeciospores, they represent the "repeating" stage. This means that a urediniospore may germinate, forming a dikaryotic mycelium that again gives rise to uredinia. The urediniospores are actually equivalent to an anamorph, as they are responsible for nonsexual reproduction.

Figure 5–10 Aeica in the Uredinales: (a) aecium of *Puccinia graminis* showing typical cupulate form, × 510; an enlarged aeciospore is also shown; (b) a hornlike aecium of *Gymnosporangium juniperinum* (× 80) and its appearance on the host plant. [From P. Sappin-Trouffy, 1896, Botaniste 5:59–244.]

Typically, the *uredinium* is a sorus that erupts through the host epidermis and contains one-celled urediniospores on short stalks. Germ pores are almost always visible. The urediniospore very often has spines protruding from its middle layer and penetrating the thin outer wall layer. The uredinia sometimes have a peridium surrounding the edge or contain scattered sterile hyphal tips among the urediniospores (Fig. 5–11).

Stage III. Telium bearing teliospores. The *telium* is a sorus bearing *teliospores.* The teliospores function as probasidia and are the site of karyogamy; they subsequently produce a metabasidium within which meiosis occurs. Ultimately, the teliospores bear basidiospores. Teliospores are usually dark, thick-walled spores with two or more cells. They may be sessile or borne

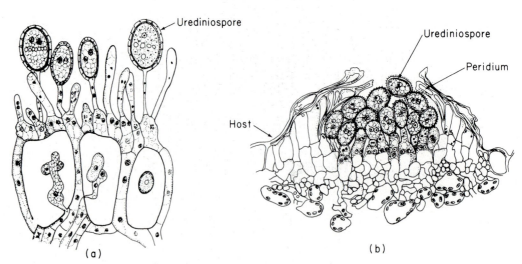

Figure 5–11 Uredinia in the Uredinales: (a) portion of an uredinium and urediniospores of *Puccinia graminis,* × 700; (b) uredinium of *Cronartium ribicola* that has a peridium and is buried within the host tissue, × 500. [(a) From P. Sappin-Trouffy, 1896, Botaniste 5:59–244; (b) from R. H. Colley, 1918, J. Agr. Res. 15:619–659.]

on a short stalk. The teliospores usually occur in a sorus embedded in the host mesophyll or in the epidermal region. In some species, the teliospores occur on gelatinous strands or hornlike structures (the telial column) that extend from the host (Fig. 5–12). Other morphological forms of both teliospores and telia occur.

Stage IV. Basidiospores. The haploid basidiospores are produced on the metabasidium and are capable of reinfecting a host. Because this reinfection usually results in the formation of pycnia, the life cycle is initiated again.

Some rusts must have all the above five spore stages in their life cycle, occurring in the order presented. Such a rust is termed *long-cycled*, in contrast to those that are *short-cycled* and have only some of the spore stages (see page 183). In addition, rusts differ in whether they require one or two hosts to complete their life cycle. An *autoecious* rust completes its entire life cycle on a single host plant. A *heteroecious* rust requires two hosts to complete its life cycle. The haploid phase (stages 0 and I) occur on one host species, while the dikaryotic and diploid stages (stages II, III, and IV) occur on a second host species. The first host is often termed the *alternate* host because it is not the host on which basidiospores are formed. An example of a heteroecious, long-cycled rust is *Cronartium ribicola*, the cause of white pine blister rust, which produces its pycnia and aecia on white pines (the alternate host) and its uredinia and telia on species of *Ribes* such as currants [Figs. 5–11(b) and 5–12(b)]. A second example of a heteroecious, long-cycled rust is *Puccinia graminis*, which we discuss below.

Puccinia graminis. To clarify the life cycle of a typical long-cycled, heteroecious rust, we examine the life cycle of *Puccinia graminis*, the causal agent of

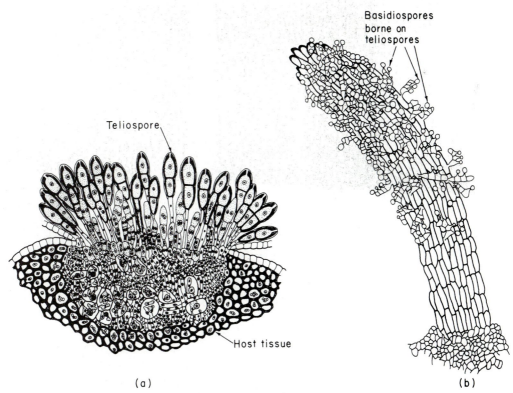

Teliospore

Host tissue

Basidiospores
borne on
teliospores

(a)

(b)

Figure 5–12 Telia in the Uredinales: (a) telium and teliospores of *Puccinia graminis*, × 450; (b) telial horn formed by *Cronartium ribicola*, × 170. [(a) From P. Sappin-Trouffy, 1896, Botaniste 5:59–244; (b) from R. H. Colley, 1918, J. Agr. Res. 15:619–659.]

the black stem rust of wheat and other cereals (Fig. 5–13). In the spring, infection of the first host, the barberry, occurs when a basidiospore lands on the surface of the leaf. The basidiospore germinates, and a monokaryotic hypha invades the leaf. A flask-shaped pycnium which is immersed in the surface leaf tissue is formed. The pycnium has paraphyses and nonseptate flexuous hyphae. Pycni-

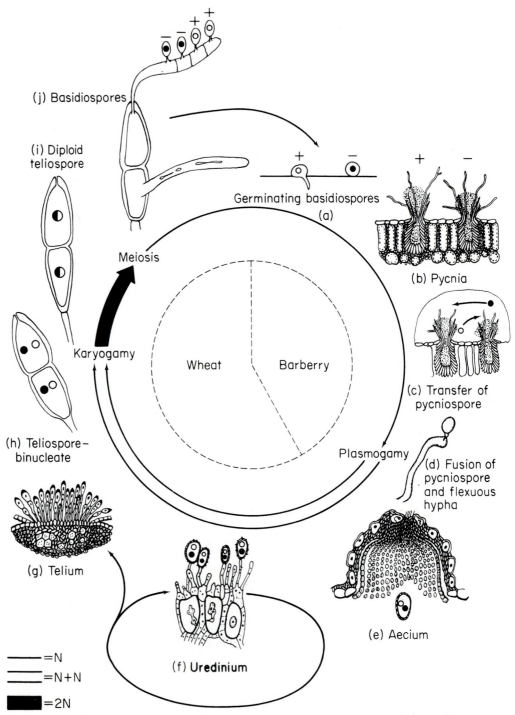

(j) Basidiospores

(i) Diploid teliospore

Germinating basidiospores
(a)

Meiosis

(b) Pycnia

(c) Transfer of pycniospore

Karyogamy

Wheat   Barberry

(d) Fusion of pycniospore and flexuous hypha

Plasmogamy

(h) Teliospore–binucleate

(g) Telium

(e) Aecium

=N

=N+N

=2N

(f) Uredinium

Figure 5–13 Life cycle of a typical heteroecious, long-cycled rust, *Puccinia graminis*. [Figures (b), (c), and (d) from A. H. R. Buller, *Researches on fungi*, Vol. 7, by permission of Royal Society of Canada and University of Toronto Press. Figures (e), (f), and (g) from P. Sappin-Trouffy, 1896, Botaniste 5:59–244.]

ospores are produced within the pycnium and are exuded in a drop of nectar. These pycniospores function as spermatia and fuse with the flexuous hyphae (the female element) of a pycnium of another mating type. (See Chapter 7 for additional information on mating types.) The nucleus of the pycniospore migrates into the flexuous hyphae, and its progeny make their way down through the haploid hyphae to the fundaments of the aecium, establishing the dikaryon. The aecial fundament, now dikaryotic, forms the cupulate aecium which bears binucleate, thin-walled aeciospores in chains. The aecia are formed on the underneath surface of the barberry plant.

The aeciospores are dispersed by air currents and may land on the second host, a cereal plant. The aeciospores germinate, and the dikaryotic hyphae penetrate the plant. The hyphae within the cereal plant eventually form pustules on the surface of the host. These pustules are uredinia and are reddish in appearance owing to the formation of thick-walled red urediniospores, each borne on a short stalk, within them. Each urediniospore is binucleate, preserving the dikaryotic condition. The urediniospores are capable of initiating new infections on the cereal plants, and this particular stage may repeat numerous times during the summer. This stage is responsible for the huge buildup of spores which is a major factor in making this fungus a devastating parasite. As autumn approaches, a new type of sorus appears on the wheat plant. This is the telium which bears teliospores. The teliospores are composed of two binucleate cells and have thick, dark walls making the telium appear black to the naked eye. The teliospore overwinters and, in the spring, each of its cells functions as a probasidium. Karyogamy takes place in each cell of the teliospore, and a metabasidium is produced from each cell. Meiosis occurs in the metabasidium, which is then divided by transverse septa into four cells, each containing a single nucleus. Each cell of the metabasidium forms a sterigma and the nucleus is included in the basidiospore which is then formed. This basidiospore is capable of initiating infection again on the barberry plant.

Short-Cycled Rusts.   In contrast to the long-cycled rusts, many rusts are short-cycled, as one or more of the spore stages does not occur in its life cycle. For example, a short-cycled rust may have all stages except one, possibly deleting the pycnial or the uredinial stages (that is, it would have spore stages I, II, III, and IV or 0, I, III, and IV, respectively). These rusts may be either heteroecious or autoecious. If the pycnial stage is lacking, plasmogamy frequently occurs by hyphal fusion within the host tissue, thus establishing the dikaryotic stage. Those rusts lacking the uredinial stage do not have a repeating stage.

Even shorter life cycles may be found, as more than one spore stage may be deleted. For example, both the aecial and uredinial stages may be absent, leaving only stages 0, III, and IV. One of the rusts with the simplest life cycle is *Puccinia malvacearum*, which has only stages III and IV and is autoecious, occurring on mallow and its allies. In *P. malvacearum*, monokaryotic mycelium is established in the host after the basidiospore germinates. A telium is formed, and plasmogamy occurs in some of the hyphae of the telium, thereby establishing the dikaryon. The dikaryotic hyphae produce the teliospores, which form the metabasidium and basidiospores without undergoing a rest period. The basidiospores become binucleate when their single haploid nucleus divides mitotically. Several cycles may occur in a single growing season.

In some life cycles, a particular spore stage may have the morphology of one spore stage but carry out the role of another (Hiratsuka, 1973). For example, let us examine *Endophyllum sempervivi*, in which the uninucleate basidiospore lands on the host and the mycelium overwinters in the host. In the spring, pycnia are formed but are nonfunctional. Plasmogamy occurs at the base of the telium, which is associated with the pycnium. Morphologically, this telium is indistinguishable from a typical aecium as it is cupulate, has a peridium, and produces binucleate spores in chains. Functionally, the sorus is a telium and its spores are teliospores. Karyogamy occurs in the teliospores, which then germinate, forming a metabasidium in which meiosis occurs, and finally, uninucleated basidiospores are produced. This life cycle may be designated as 0, III, IV.

ORDER USTILAGINALES

Unlike the Uredinales, the members of the Ustilaginales produce a yeastlike unicellular stage that may generally be cultured readily and is presumably capable of existing saprophytically in nature. A few of the genera include nonparasitic yeastlike fungi that complete their entire life cycles in the absence of a host (Lodder, 1970). One serious human disease, cryptococcus, is caused by the yeastlike nonsexual stage of *Filobasidiella neoformans* (Kwon-Chung, 1976). The vast majority of the members of the Ustilaginales are plant parasites and apparently require the plant for completion of their life cycles in nature. These fungi are the "smut" fungi.

The smut fungi number approximately 850 species. They are parasites on fungi and angiosperms, especially on grasses and sedges. The smut fungi derive their name from the sooty appearance of their sori, filled loosely with dark brown or black teliospores that form on the affected plant part (Fig. 5–14). They most commonly attack the reproductive structures of the plant, and depending upon the species, cause abnormalities of the ovaries or anthers, or they may invade the embryo in the developing seed. One particularly interesting abnormality is called "induced hermaphoditism" or alternatively, "parasitic castration." This phenomenon is caused by *Ustilago violacea*, which attacks members of the Carophyllaceae (e.g., the pinks and chickweed), plants that normally bear the male parts (stamens and anthers) in flowers separate from those bearing the female parts (ovaries). Infection by *U. violacea* in the male plant results in replacement of the pollen grains in the anthers by teliospores (the only site where teliospores are produced) and may induce formation of ovaries. Conversely, infection in the female plant induces formation of sterile anthers. As a

(a)   (b)

Figure 5–14 Smut fungus, *Urocystis cepulae*, on onion (a), and balls of teliospores from the onion (b). [Courtesy Plant Pathology Department, Cornell University.]

result of infection by *U. violaceae,* the appearance of the teliospore-laden anthers so closely resembles that of normal pollen-laden anthers that insects visit the anthers as usual and are probably responsible for transferring teliospores to other flowers. Other smut fungi may cause sterility, abnormal enlargement, or other modification of the reproductive structures. A common smut disease is boil smut of corn, caused by *Ustilago maydis* and in which the most conspicuous symptom is the massive enlargement of the developing grain to form a large, fleshy, pearly gray gall consisting of host tissue and mycelium and filled with coal dustlike teliospores at maturity (Fig. 13–19). Smuts may also invade the stems, leaves, and occasionally the roots. An example of such a smut is *Tilletia controversa,* cause of dwarf bunt of wheat. Spores of *T. controversa* in the soil invade seedlings and the resulting infection results in a marked shortening of the stem and mottling of the leaves. Some smuts produce their sori on vegetative organs (e.g., onion smut, Fig. 5–14).

The smut life cycle is considerably less complicated than that of the rusts, although these organisms share certain features. The smut mycelium that initiates the infection is usually dikaryotic, although in *Ustilago maydis* the infecting mycelium is haploid in its initial stages and the dikaryotic condition is soon established by mycelial fusion. The slender, septate mycelium develops intercellularly. Clamp connections and haustoria may be present. The mycelium in the sorus eventually breaks apart into dikaryotic cells which differentiate into teliospores. The teliospores have dark, thick walls which may be smooth, spiny, or reticulated. In some instances, a number of teliospores remain together in a cluster of definite pattern and are called *spore balls* [Fig. 5–14(b)]. As the binucleate teliospore nears maturity, karyogamy occurs and the diploid teliospore gives rise to a metabasidium. Unlike its regularly triseptate counterpart in the rusts, the smut metabasidium may be nonseptate or, if septate, there may be any number of septa formed. The majority of the septate species form a three- or four-celled metabasidium. Meiosis occurs, and the metabasidium usually gives rise to uninucleate haploid basidiospores. Unlike the Uredinales, these basidiospores are not borne on sterigmata but are sessile on the metabasidium. If the metabasidium is septate, the basidiospores may be formed from each cell; and if the metabasidium is nonseptate, the basidiospores are typically formed at the apex of the metabasidium. An indefinite number of basidiospores are formed. They may give rise to additional spores and in culture will produce a yeastlike colony of spores. The haploid phase of the life cycle is very short, and plasmogamy typically follows with little delay. The majority of the smuts require different mating types to accomplish plasmogamy (see Chapter 7). Depending on the species, plasmogamy may occur between (1) metabasidia, which either may or may not have produced basidiospores; (2) basidiospores or spores produced from them; (3) haploid mycelium during the early infection stages; or (4) any combination of these structures. The dikaryotic hypha produced after plasmogamy is typically the structure that is capable of penetrating the plant. A typical life cycle of a smut is given in Fig. 5–15.

Class Hymenomycetes

The Class Hymenomycetes contains some of the most conspicuous and commonly known fungi, such as the jelly fungi, mushrooms, coral fungi, and bracket fungi. The majority of these fungi are saprophytes on wood or plant material in the soil, and the mycelium plays an important role in breaking down the plant remains and in producing humus (see Chapter 12). Many Hymenomycete species are partners in the mycorrhizal relationship where they are symbionts with the roots of higher plants (see Chapter 15). Many of these fungi are serious plant parasites, and one member, *Schizophyllum commune,* may cause a

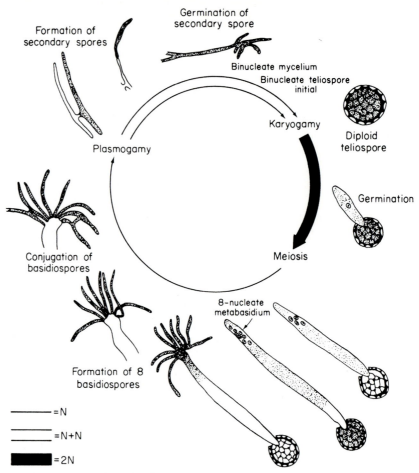

Formation of
secondary spores

Germination of
secondary spore

Binucleate mycelium

Binucleate teliospore
initial

Karyogamy

Diploid
teliospore

Germination

Plasmogamy

Meiosis

Conjugation of
basidiospores

8-nucleate
metabasidium

Formation of 8
basidiospores

———— = N

════ = N+N

▬▬▬ = 2N

Figure 5–15   Life cycle of a smut, *Tilletia eleusines.* The teliospores occur primarily in the ovaries of the host, replacing the seeds but occurring within the pericarp that normally surrounds the seeds. The binucleate mycelium formed by germination of the secondary spore is capable of invading the host. [Figures from N. C. Joshi, 1960, *Mycologia* 52:829–836.]

disease in humans. Species in the Hymenomycetes share the important characteristic of forming a hymenium that is exposed at maturity.

The Basidiocarp and the Hymenium.   In almost all these fungi, the hymenium is borne on a basidiocarp (Fig. 5–16). The basidiocarp is generally a macroscopic structure and may be several centimeters in diameter or height. From species to species, there is a great deal of internal variation of the mature basidiocarp, both in arrangement and cellular morphology of the tissue zones. Generally at least three major anatomical zones may be distinguished: These are the hymenium, subhymenium, and sterile tissues. The hymenium is a layer of basidia, basidiospores, and often sterile elements that may cover the entire exposed surface of the basidiocarp or only part of the surface. The morphologically distinct tissue from which the hymenium arises is the subhymenium. Sterile tissue may separate adjacent hymenia or may comprise the bulk of the basidiocarp and is frequently called the *trama.* Portions of a basidiocarp not covered by a hymenium are usually covered by a skinlike layer, the *cuticle.* The cuticle may or may not be markedly differentiated from the trama tissue. (We discuss the cuticle further on page 205.)

Typically, the early developmental stages of either the whole basidiocarp or of an enlarging portion of the basidiocarp consist of thin-walled, tubular es-

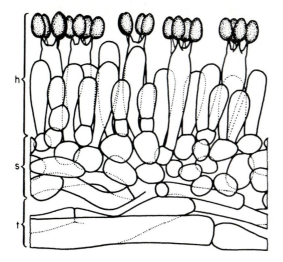

Figure 5–16 Part of a cross section of a gill of *Agaricus campestris* showing a typical hymenium (*h*) with basidia in various stages of development. Some of the elements of intermediate length are probably basidioles. The subhymenium (*s*) gives rise to the cells in the hymenium, and the sterile tissue or trama (*t*) bears both the subhymenium and hymenium. × 400. [From A. H. R. Buller, 1922, Researches on fungi, Vol. 2, Longmans, Green & Co. Ltd., London.]

sentially undifferentiated hyphae that may be interwoven or cohere to form a tissue. Sometimes meristematic regions are active and are centers of growth in the more highly differentiated basidiocarp (Reinjnders, 1977; Kennedy, 1972; Kennedy and Wong, 1978). These nonspecialized, thin-walled, and nonpigmented hyphae are called the *generative* hyphae. The generative hyphae grow by undergoing cell division, cell enlargement, extension at their tips, or a combination of these processes. When mature, the resulting tissue may consist entirely of generative hyphae, or alternatively the generative hyphae may give rise to other hyphal types or to discrete cells. When the generative hyphae produce other hyphal types, they may undergo a structural modification, become pigmented, or serve as depositories for metabolic by-products (Smith, 1966). Contents of these cells may be fats, glycogen, resins, tannins, and other materials (Lentz, 1954). Clamp connections may be either present or absent in the generative hyphae and in hyphae derived from them.

Examples of thick-walled hyphae include unbranched *skeletal* hyphae, which may make the basidiocarp texture somewhat tough, and the highly branched, relatively limited *binding* hyphae, which seem to weave other hyphal types together. The skeletal and binding hyphae are especially common in basidiocarps that are perennial or tough and leathery in texture. In these basidiocarps, generative hyphae may be intermixed with either skeletal or binding hyphae or with both types (Fig. 5–17). Other types of thick-walled hyphae may occur in fleshy basidiocarps. An example of a hyphal system accumulating a metabolic waste is provided by the *lactiferous* hyphae. The lactiferous hyphae form an undulating or anastomosing hyphal system filled with a milky fluid that appears as a turbid emulsion of droplets. A fungus with a lactiferous system will exude the milky fluid when broken, at first the fluid is usually white or clear but it may turn to other colors such as pink or yellow upon exposure to the air. Other hyphae which may be similar to the lactiferous hyphae in appearance may contain fats, aromatic oils, or a blood-colored fluid which may be exuded when the basidiocarp is damaged. The hyphae in some tissues will either break down to form a gel, yielding a soft watery tissue, or the hyphae will secrete a mucilaginous material, forming a sticky coating (Smith, 1966).

The generative hyphae may become divided by additional septa to form a number of separate cells. These cells sometimes remain tubular and elongated, but often they undergo inflation, becoming both wider and longer. This inflation contributes greatly to the overall increase in size of many basidiocarps. Cells in basidiocarps may become globose, barrel-shaped, sausage-shaped, or irregularly branched. Enlarged cells may comprise a tissue entirely, or they may be

Figure 5–17 Representative hyphae and cell types in the Hymenomycetes: (a) thin-walled generative hyphae with clamps; (b) thick-walled generative hyphae without clamps; (c) skeletal hyphae; (d) binding hyphae; (e) two types of encrusted hyphae; (f) hyphae with comparatively short, inflated cells; (g) chains of subglobose to globose cells; (h) greatly inflated sphaerocysts. [(a–d) From S. Domanski, 1965, Fungi, U.S. Department of Agriculture Translation; (e, left) from F. L. Lombard and R. L. Gilbertson, 1966, Mycologia 58:827–845; remaining figures adapted from D. Largent, D. Johnson, and R. Watling, 1977, How to identify mushrooms to genus III: microscopic features, Mad River Press, Eureka, Calif., 148 pp.]

intermixed with elongated hyphae. The enlarged cells usually retain their thin walls, but in some cases the cell walls thicken. Soft, fleshy or brittle basidiocarps characteristically have an abundance of inflated, thin-walled cells in their tissues.

The hyphae often terminate as morphologically distinct entities. Most important are the basidia and *basidioles,* which terminate generative hyphae and occur in the hymenium. The basidiole may be a basidium which has not yet produced sterigmata, or it may be a cell morphologically identical to an immature basidium but which never produces basidiospores. Some of the basidioles become broad, forming a pavement of cells which helps support the basidia that produce basidiospores. *Cystidia* may occur in the hymenium as well as in the trama or sterile surfaces of the basidiocarp. A cystidium is morphologically distinct from an immature basidium but the morphology of the cystidia may vary greatly. Some of them are smooth and thin-walled like the basidia but have a different shape [Fig. 5–18(a)]. Others may have either colored or thickened walls and may be partially covered with incrustations (often calcium oxalate) [Fig. 5–18(b)]. Many cystidia contain distinctive contents that are oily, amorphous, and refractive or that stain deeply with specific stains [Fig. 5–18(c)]. Cystidia of

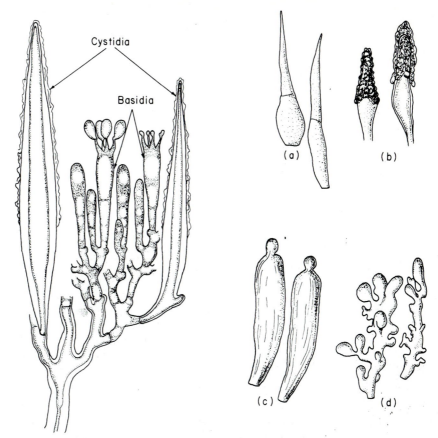

Figure 5–18 Left: cystidia arising from hypha that also produces basidia; right: types of cystidia that might occur in the hymenium. See the text. × 1000. [Left from E. J. H. Corner, 1950, Ann. Botan. Mem. 1:1–740; right from D. N. Pelger, 1966, Persoonia 4:73–124.]

this type often terminate hyphae having similar contents. The last cystidial type is that which is peculiarly branched [Fig. 5–18(d)].

The function of the cystidia apparently varies. In a few species, cystidia play a mechanical role in holding opposing hymenia apart (Fig. 5–34). The branched cystidia are thought to trap air and humidity in pockets around developing basidia, thereby maintaining favorable conditions for their further development. Many cystidia apparently provide an exit point for metabolic wastes accumulated in the hyphae. For example, some cystidia may secrete mucilage from a thin apex, and this secreted mucilage subsequently encrusts on the outer surface (Lentz, 1954; Smith, 1966). In an ultrastructural study of cystidia of *Suillus americanus*, Clémençon (1975) observed that a dark resin first accumulates within the endoplasmic reticulum but is then deposited between the plasmalemma and the cell wall. As the wall thickens, the dark material is first incorporated in patches in the cell wall but finally forms a granular coating on the surface of the cystidium and becomes interspersed with a slime that may be derived from the cell wall itself.

Frequently specific cells or tissues of the basidiocarp or basidiospores will react to chemical treatments and give characteristic color changes. In some species, entire structures will turn either green or blue when moistened with a reagent containing iodine. Moistening with KOH (potassium hydroxide) may induce a brown, yellow, red, or violet color of the tissue or spores of certain species (the particular color induced is constant for a given species). Lactiferous hyphae will sometimes give a positive reaction with the sulfovanillin test. Unfortunately, the chemical basis for many of these reactions is unknown (Watling,

1971). These and similar color reactions are of great value in making identifications of basidiocarps collected in the field (Baroni, 1978).

Vegetative structures produced by some species are *rhizomorphs* and sclerotia. Rhizomorphs are elongated, often hardened strands consisting essentially of bundles of parallel hyphae. Rhizomorphs may be divided into concentric tissue zones and elongate from the apex. They resemble roots and often may be found between the wood and bark of fallen trees or stumps [Fig. 5–19(a)]. Rhizomorphs are a multicellular link between the food supply and the basidiocarp, and they have a role in the conduction or storage of nutrients. Sclerotia are resistant structures which may give rise to mycelium or to a basidiocarp, depending on the species [Fig. 5–19(b)].

(a) (b)

Figure 5–19 Vegetative structures produced by some basidiomycetes: (a) rhizomorphs of *Armillaria mellea* on the surface of a log from which the bark has been removed, approximately × ½; (b) basidiocarp of *Typhula* arising from sclerotium. [Courtesy Plant Pathology Department, Cornell University.]

Basidium Types and Classification of the Hymenomycetes. At maturity, the morphology of the basidium may vary considerably from one genus to another. The most commonly occurring basidium is a single-celled structure, the *holobasidium.** The holobasidium is the type used to illustrate the general life cycle (pages 172–174). The holobasidium is nonseptate and generally clavate in shape, but it may also be deeply divided by a central cleft or may be somewhat spheroidal. Sterigmata may be present or absent. If present, the sterigmata may vary from comparatively short spicules as illustrated in Fig. 5–6 to elongate tubular or balloonlike structures (Fig. 5–20). A second major type of the basidium is a *phragmobasidium,* a basidium that is divided into separate cells by either transverse or longitudinal septa, which are formed after meiosis and separate the daughter nuclei from each other. The phragmobasidium is typical of the Subclass Phragmobasidiomycetidae, while the holobasidium is found in the Subclass Holobasidiomycetidae as well as in the Class Gasteromycetes.

---

*This term is used differently by some authors. See Talbot (1973-a) for a discussion of its usage.

Holobasidia   Phragmobasidia

Figure 5–20 Different types of basidia in the Hymenomycetes. Mature basidia are represented; shown are the metabasidium (*m*), serigmata (*st*), and basidiospores (*bs*). (a) A typical holobasidium found in the Agaricales that is not divided or cleft; (b) A typical holobasidium that bears only two sterigmata and basidiospores (*Dacrymyces*); (c) holobasidium with swollen sterigmata (*Tulasnella*); (d) phragmobasidium with longitudinal septa and with extremely long sterigmata (*Tremella*); (e) phragmobasidium with transverse septa separating individual cells of the metabasidium (*Auricularia*).

SUBCLASS PHRAGMOBASIDIOMYCETIDAE

In this subclass, the basidiocarps may be waxy, fleshy, gelatinous, or dry and leatherlike in consistency. Those species which form a gelatinous basidiocarp are commonly known as "jelly fungi" (Fig. 5–21). The basidiocarp is not gelatinous early in its development, but later the hyphae in some tissues disintegrate and leave a gelatinous mass (Fig. 5–22) (Moore, 1965). These fungi are able to survive long periods of drought by drying to a horny texture. When moisture is again available, they absorb water and again become gelatinous. Spore formation and discharge can occur after the moisture is absorbed. Buller (1922) found that basidiocarps of a jelly fungus could revive and produce spores after being kept dry for 8 months. Many of these fungi are brightly colored and are some shade of orange, yellow, or red, while others are white or brown.

Figure 5–21 Basidiocarps of *Tremella fibulifera*, a jelly fungus. The larger basidiocarp is about 4 centimeters in diameter. [Courtesy J. S. Furtado.]

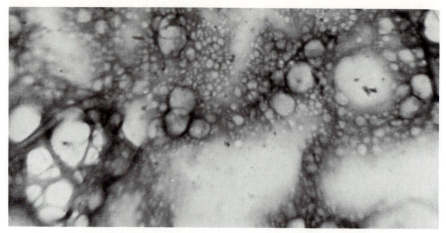

Figure 5–22 Gelatinous tissue from the center of the basidiocarp of a jelly fungus, *Pseudohydnum gelatinosum.* Note the apparent absence of hyphae. × 828. [From E. J. Moore, 1965, Am. J. Botany 52:389–395.

Basidiocarp form varies considerably. Some basidiocarps are resupinate, lying flattened on the substratum with the hymenium on the outer side. Others have a pustulate or cushionlike form, an erect hornlike form that may be either branched or unbranched, or the shape of a funnel or of a cap on a stalk. The hymenium may occur on either one or two sides and is exposed throughout development.

Phragmobasidia and possibly cystidia occur in the hymenium. The phragmobasidia may be buried within a gelatinous layer (Fig. 5–23). An important characteristic shared by these organisms is that basidiospores may frequently germinate by repetition, a process in which a sterigma and a new basidiospore form from the original basidiospore. Alternatively, basidiospores sometimes germinate by forming conidia. Nonsexual spores (oidia) are occasionally formed in some species in a separate sporocarp from that in which the basidiospores occur.

The Subclass Phragmobasidiomycetidae may be divided into three orders, which are distinguished primarily by the morphology of the basidium. These are the Orders Tremellales, Auriculariales, and Septobasidiales (McNabb, 1973).

Order Tremellales. Members of the Order Tremellales usually have *cruciately divided* metabasidia. That means that there are longitudinally oriented septa dividing the metabasidium into four cells. In a transverse section of a metabasidium, the septa appear in the form of a cross, as the septa are perpendicular to each other. One of the septa is formed immediately after meiosis is completed, and the second is formed shortly thereafter. As the majority of these fungi are gelatinous, the metabasidium is embedded within a layer of gel and produces elongate sterigmata that pass through this layer so that the basidiospores are borne on the surface.

Members of the Tremellales occur in diverse forms and are white, yellow, brown, or almost black in color. Most of these fungi are saprophytes on wood or the ground.

Order Auriculariales. These fungi form an elongate metabasidium that is divided by transverse septa, usually into four cells. The probasidial wall may or may not remain permanently distinct. Typically each cell of the metabasidium produces an elongate sterigma that bears a basidiospore. Some of these fungi are saprophytes on wood. An example is *Auricularia auricula,* which often forms brown, ear-shaped basidiocarps on trees, earning it the common name "fungus ears." It is eaten by Oriental people. Many species are parasitic on other fungi,

Figure 5–23 Basidia in Tremellales. Top: developing phragmobasidia of *Exidia nucleata:* (a) dikaryotic probasidium; (b) diploid probasidium; (c) segmented metabasidium after meiosis and the segmentation of the probasidium; (d) sterigmata developing from the metabasidium; (e) nuclear migration through the sterigma to the basidiospore initial; (f) mature basidiospore borne on sterigma and metabasidium. Bottom: Basidia in various stages of development in hymenium within gel of *Tremella mesenterica.* Note that the basidiospores are formed above the surface of the gel. [Top series adapted from K. Wells, 1964, Am. J. Botany 51:360–370; bottom from P. A. Dangeard, 1895, Botaniste 4:119–181.]

mosses, ferns, or higher plants. As in the Tremellales, a variety of basidiocarp forms are produced.

Order Septobasidiales. The phragmobasidium of the Order Septobasidiales is essentially similar to that of the Order Auriculariales. There is a persistent probasidial wall and an elongate metabasidium that is divided into as many as four cells. The basidiospores are not usually repetitive. These fungi are parasitic on insects, with which they establish a unique relationship favorable to some insects in the colony (Chapter 15). The basidiocarp is resupinate and crustlike and is somewhat leathery in texture. Only a single genus, *Septobasidium,* is represented in this order.

SUBCLASS HOLOBASIDIOMYCETIDAE

The members of the Subclass Holobasidiomycetidae include those Hymenomycetes that form a nonseptate basidium or a holobasidium. This group includes diverse types of fungi, including simple plant parasites, jelly fungi, and the mushrooms and bracket fungi with which you are undoubtedly familiar.

The simplest of the plant parasites are found in the Order Exobasidiales. They are parasites on stems, leaves, or flower buds of vascular plants and cause leaf spots or abnormal enlargements to occur. The mycelium occurs inside the leaf tissue and forms basidia which penetrate through epidermal cells or stomates of the host. The basidia are typical cylindrical to clavate holobasidia and may occur singly, in fascicles, or in a hymenium on the surface of the host. Basidiocarps are not formed. The few members of a second order, the Brachybasidiales, are somewhat similar plant parasites. These fungi form an extremely small pustule or disk-shaped basidiocarp which emerges through the epidermis or stomate and also have a persistent probasidial wall from which the undivided metabasidium emerges (Fig. 5–24).

The remaining orders in this subclass are the Dacrymycetales, Tulasnellales, Aphyllophorales, and Agaricales.

Orders Dacrymycetales and Tulasnellales. In many respects, these orders are similar to those in the Subclass Phragmobasidiomycetidae, and have been classified with them previously or by other authors. Some of these fungi form gelatinous basidiocarps and are also known as jelly fungi. The orders are separated primarily on the basis of basidial morphology (McNabb and Talbot, 1973).

In the Order Dacrymycetales, the typical basidium is a deeply divided form

(a)

(b)

Figure 5–24 Member of the Brachybasidiales, *Ceraceosorus bombacis,* × 1125: (1) developmental stages of the basidium, proceeding from the thick-walled probasidium (left) to the metabasidium (right); the metabasidium supports two sterigmata and basidiospores and retains the thickened wall that was present in the probasidium; (b) portion of the hymenium, showing the persistent walls of the probasidia at various levels. [From J. L. Cunningham, B. K. Bakshi, P. L. Lentz, and M. S. Gilliam, 1976, Mycologia 68:640–650.]

but without septa. The metabasidium is divided into two elongated extensions, each bearing a sterigma and basidiospore. This basidium type resembles a tuning fork, and is sometimes called a tuning-fork basidium. The basidiospores may be nonseptate but are usually septate. The septa often do not appear until after the basidiospores are discharged, when one to seven septa may be formed. The basidiocarps may range greatly in form, as may those of the Subclass Phragmobasidiomycetidae. They are usually gelatinous but may be waxy or dry and form a thin sheet or horny mass on drying. These fungi are often brightly colored and may be shades of yellow or orange as well as deep brown. They occur on wood as saprophytes. Oidia sometimes occur in separate sporocarps similar to those which bear the basidia.

In the Tulasnellales, the basidium bears balloonlike sterigmata. The sterigmata may be sporelike and deciduous or elongated and nondeciduous. The basidiocarp occurs as a flat layer which varies from a filmy, cobweblike covering to a crust. The consistency of the basidiocarp may also vary, both gelatinous and dry forms being known. The colors are soft hues of gray, pink, violet, or lilac. Members of the Tulasnellales are saprophytes on wood or old fungi or may occur as parasites on plants, especially plant parts near the soil. The anamorph of some species may form mycorrhizae with orchids (Chapter 15) (McNabb and Talbot, 1973). Also, some species trap nematodes.

Order Aphyllophorales. The Aphyllophorales is a large order containing about 2000 known species. Many of these are the bracket and coral fungi. The vast majority of these fungi are saprophytic on plant debris. Many species are significant in decomposing plant remains, as they are able to digest cellulose or lignin that occurs in plant cell walls. Also, many members of the Aphyllophorales cause destruction of timber and wood products or decay wood in standing trees (Chapter 12).

Many of these fungi have been involved in everyday human affairs. A few were used medicinally by the Greeks and Romans as a remedy for many complaints, including colic, fractured limbs, and bruises. Later, in the fifth century, a bracket fungus was immersed in saltpeter and dried and then used to cauterize surgical wounds. Other bracket fungi have been used as curry combs for horses, as snuff, as razor strops, and as a source of dye for clothing (Rolfe and Rolfe, 1925). A few species are edible.

There is a great deal of morphological variation in this group. Those characters unifying the group are the possession of a holobasidium in an exposed hymenium borne on a dry or comparatively dry basidiocarp. The texture of the basidiocarp may be similar to that of cork, wood, leather, paper, or cartilage. Unlike the basidiocarps of the Order Agaricales, the basidiocarps of the Aphyllophorales are *not* fleshy and moist.

The simplest basidiocarp forms are crustlike and resupinate, lying appressed to the surface of the substratum and with a hymenium covering the entire exposed surface. Other basidiocarps may be similarly flat and appressed to the substratum but have a portion of the basidiocarp reflexed or turned away from the substratum (in these, the hymenium would occur on one surface only). Upright basidiocarps also occur and may assume various forms such as a funnel (with the hymenium on the outer surface) or upright clubs or stalks, which may be branched or unbranched (Fig. 5–27). Some of the basidiocarps form a definite cap, the *pileus*, which bears the hymenium on the lower surface. The pileus may be bilaterally symmetrical, resembling a shell or shelf, and may be attached directly to the substratum at its side (*sessile*) (Fig. 5–32) or it may be attached by a stalk (the *stipe*). A few of the species that form a pileus have one that is radially symmetrical, and is supported by a stipe attached to its center. Geotropism can be demonstrated in the upright or pileal forms (see Chapter 11).

We have seen that these basidiocarps range from somewhat cartilaginous forms to those that are hard and woody. The trama is generally comprised of elongate hyphae. Generative hyphae, sometimes with inflated segments or cells, predominate in the softer forms. The hard, woody forms often have skeletal or binding hyphae or both in addition to the generative hyphae. We use the term *monomitic* to refer to a tissue having only generative hyphae, *dimitic* for a tissue having generative hyphae and one additional hyphal type (either skeletal or binding hyphae), and *trimitic* for a tissue with all three hyphal types (Fig. 5–25).

The form of the hymenium is most often determined by that of the trama immediately beneath it. Because this trama bears the hymenium and determines its shape, it is called the *hymenophore*. The hymenophore may be smooth or in the form of warts or spines; it may line the interior of tubes or cover the surface of *gills* (leaflike lamellae that resemble the pages of a half-opened book). In terms of development of the hymenium, two distinct types may be distinguished, the *catahymenium* and *euhymenium*. The catahymenium consists at first of a layer of sterile hyphal elements; the developing basidia then push into the catahymenium and gradually extend through and beyond it. In contrast, the euhymenium lacks such a layer, but the first hymenial elements to appear are the basidia intermixed with a few cystidia. Sometimes the euhymenium may become progressively thicker as it ages. If the euhymenium thickens, it will produce successive layers of basidia, with the older basidia and cystidia buried beneath a layer of younger basidia. As each layer of basidia reaches maturity and then disintegrates, it is penetrated by hyphae that arise from the subhymenium and form new basidia on the surface. The euhymenium does not thicken in many species, and if it does, all basidia in it will mature simultaneously and will not be replaced by younger basidia.

Division of the members of the Aphyllophorales into genera was originally made on the basis of gross morphology of the basidiocarp and hymenium. In the more modern systems, a complex of characters is used. These include many

(a) (b) (c)

Figure 5–25 (a) Monomitic tissue consisting of generative hyphae only; (b) dimitic tissue consisting of generative hyphae (light) and skeletal hyphae (dark); (c) trimitic tissue consisting of generative, skeletal, and binding hyphae. [From E. J. H. Corner, 1953, Phytomorphology 3:152–167.]

microscopical and anatomical features, as well as the more conspicuous ones of gross morphology (Harrison, 1973; Pegler, 1973; Petersen, 1973; Talbot, 1973-b). Donk (1964) recognizes 22 families in this order, some of which are shown in Table 5–1. We shall briefly consider some representatives of this order.

Corticium. Probably the simplest basidiocarps are those formed by species of *Corticium.* These are thin, flattened resupinate basidiocarps that are completely appressed to the wood or bark on which they occur. The hymenium is flat and covers the entire surface of the basidiocarp. When fresh, the basidiocarp is somewhat membranous or crusty, but when dry it may crack in a manner similar to the old paint that it resembles. Basidiocarps of species of *Stereum* are similar except that they have a leathery texture and are not completely resupinate but have a recurved edge (Fig. 5–26).

Figure 5–26 Basidiocarps of *Stereum* having a smooth hymenophore. [Courtesy Plant Pathology Department, Cornell University.]

Clavaria and Allies. Upright basidiocarps with a smooth hymenium occur in members of several genera, including *Typhula* [Fig. 5–19(b)] and *Clavaria.* *Typhula* species form a club-shaped unbranched basidiocarp arising from a sclerotium. The outer surface of this basidiocarp is covered by a smooth hymenium, and the texture is fleshy to waxy. Many species of *Clavaria* and allied genera also form unbranched basidiocarps, but this group of fungi is especially notable for producing repeatedly branched basidiocarps that often resemble antlers or coral, hence their common name, the "coral fungi" (Fig. 5–27). This branching is often quite regular, and may occur in a dichotomous or symmetrical radial pattern. These fungi are often colorful and may be yellow, red, purple, or violet as well as the more sedate shades of white, cream, gray, or brown. Large specimens may weigh several pounds. These fungi usually form their basidiocarps directly on the ground, usually in humus. The hymenium is smooth and covers virtually the entire basidiocarp. In members of some genera, basidia bear only two basidiospores each. The trama has generative hyphae that may be inflated but lacks binding hyphae. Clamp connections may be present or absent, depending upon the species. Cystidia are frequently present in the hymenium.

Cantharellus and Allies. These fungi form a fleshy basidiocarp which is usually trumpet- or funnel-shaped. They are most often some shade of yellow or orange but may occur in other colors and usually occur on the ground in autumn. The hymenium is borne on the lower surface of the pileus but not on the stipe. In members of the genus *Craterellus,* as well as in some species of *Cantharellus,* the hymenium is smooth (Fig. 5–28). In most species of *Cantharellus,* the hymenium is folded into shallow, rounded ridges which are longitudinally oriented or repeatedly merge and separate from each other in a reticulate pattern. According to Corner (1966), the ridges originate when numerous new

TABLE 5-1

Some Characteristics of Selected Families in the Aphyllophorales as Delimited by Donk (1964)

Family	Basidiocarp	Tissue/Cell Types	Hymenophore/ Hymenium	Representative Genera
Cantharellaceae	Tubular to funnel-shaped or with central stipe and pileus; fleshy to tough and membranous	Monomitic with thin-walled inflated hyphae; clamps ±; cystidia rare	Smooth or with shallow gills; hymenium thickening; 2–8 spores/basidium	*Cantharellus*
Clavariaceae	Erect, simple, or branched; usually fleshy, waxy, or gelatinous	Monomitic with inflating hyphae; dimitic with skeletal hyphae; clamps ±; cystidia rare	Covering all surfaces; smooth or becoming wrinkled; often thickening; 2–8 spores/basidium	*Clavaria*
Corticiaceae	Flat, usually web-like to membranous, sometimes waxy or gelatinous	Monomitic usually, rarely dimitic with skeletal hyphae; clamps ±; cystidia ±	Smooth, hymenium may or may not thicken; 2–4 spores/basidium	*Corticium*
Hericiaceae	Pileate, upright and branching, or flat; fleshy to membranous	Monomitic, sometimes with inflating hyphae, or dimitic with skeletal hyphae; clamps present; cystidia with distinctive contents in hymenium	Smooth or toothed; nonthickening hymenium; 4 spores/basidium	*Hericium*
Hydnaceae	Stipe and pileus; fleshy	Monomitic, hyphae thin-walled and inflating; clamps present; cystidia absent	Teeth or pointed spines; nonthickening hymenium; 2–6 spores/basidium; basidia often aborting	*Hydnum*
Hymenochaetaceae	Highly variable, including flat, one-sided sessile and those with a stalk and pileus; fibrous to woody; annual or perennial	Monomitic or dimitic with skeletal hyphae; hyphae almost always colored, becoming red and then permanently dark in KOH; clamps absent; prominent pointed, dark cystidia almost always present	Smooth, tubular, or with teeth; nonthickening hymenium; 2–4 spores/basidium	*Hymenochaete*
Polyporaceae	Highly variable, including flat, one-sided sessile and those with a central stipe and pileus; texture variable but often woody and hard; annual or perennial	Mono-, di-, or trimitic with skeletal or binding hyphae; clamps ±; cystidia ±	Typically tubular, but also with gills or other forms; nonthickening hymenium; 2–4 spores/basidium	*Polyporus Fomes*
Thelephoraceae	Highly variable, includes flat, those with a stipe and pileus, and many other forms	Monomitic, sometimes with inflating hyphae; becomes greenish in KOH; clamps ±; cystidia rare	Smooth, warted, toothed, poroid, or folded; hymenium thickening; 2–4 spores/basidium	*Thelephora*

Figure 5–27 Basidiocarp of a coral fungus. The smooth hymenophore covers the upright branches. The basidiocarp is about 10 centimeters tall. [From E. J. Moore and V. N. Rockcastle, 1963, Fungi, Cornell Science Leaflet.]

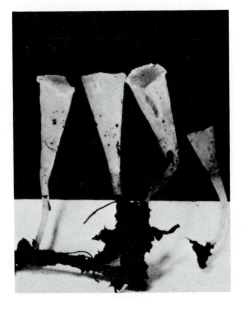

Figure 5–28 Basidiocarps of *Craterellus taxophilus*. The smooth hymenophore lines the outer surface. These basidiocaps are about 5 to 10 centimeters in height. [Courtesy Plant Pathology Department, Cornell University.]

basidia are formed among those already existing, thereby causing the hymenium to buckle and stretch the underlying hyphae of the trama. This stretching may continue until there is only a loose weft of hyphae remaining. Also there may be further thickening of the gill fold by the addition of new basidia. The trama consists of generative hyphae only, and cystidia are usually lacking.

Hydnum and Allies. In members of the genus *Hydnum* and its allies, the hymenium covers spines or teeth. There is great variety in the form of the basidiocarps. They may be resupinate and have the hymenium covering the entire surface or have a pileus. The pileus, which has the hymenium on the underside, may be either laterally attached or borne on a stipe (Fig. 5–29). The hedgehog fungus, *Hericium americanum* (Fig. 5–30) forms a basidiocarp consisting principally of a fleshy mass of pendant teeth hanging down from a mass of fungal

Figure 5–29 Lower surface of a basidiocarp of *Echniodontium tinctorium* showing the toothed hymenophore. [From J. R. Weir and E. E. Hubert, 1918, U.S. Department of Agriculture Bull. 722, 39 pp. Courtesy U.S. Department of Agriculture.]

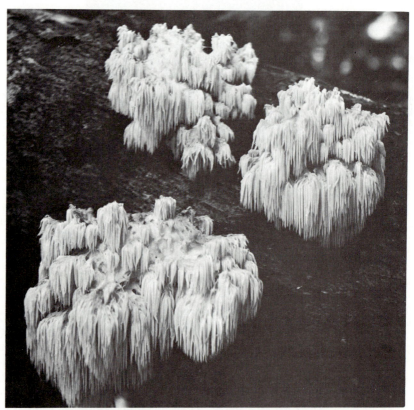

Figure 5–30 Basidiocarps of the hedgehog fungus *Hericium americanum*. The hymenophore covers the downward-hanging teeth. The basidiocarps are among the largest in the Aphyllophorales. [From E. J. Moore and V. N. Rockcastle, 1963, Fungi, Cornell Science Leaflet.]

tissue. The size of these tooth fungi varies considerably, as well as their textures, which may be leathery, woody, waxy, or membranous. They are most often drab in color and are usually some shade of brown. The majority occur as saprophytes on wood, while others occur on the ground in forests.

Polyporus and Allies (the Polypores). These fungi occur predominantly as saprophytes on dead woody plant parts such as tree trunks, stumps, and branches. A few occur on living trees, where they cause heartrot, and others occur on soil.

Species of *Polyporus* and allied genera form tubes lined with the hymenium (Figs. 5–31 to 5–33). If these tubes are examined from the end (or from the lower surface of the basidiocarp) one sees a surface punctuated by many pores. This

Figure 5–31 Resupinate basidiocarp of *Junghuhnia nitida.* Note the tiny pores in the hymenophore, which covers the entire surface. × 2. [Courtesy Plant Pathology Department, Cornell University.]

Figure 5–32 Basidiocarps of the bracket fungus *Daedaleopsis.* Note the irregularly shaped pores. Basidiocarps of this species range from 3 to 15 centimeters at their widest point. [From E. J. Moore and V. N. Rockcastle, 1963, Fungi, Cornell Science Leaflet.]

Classification

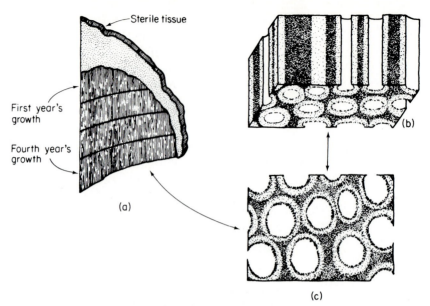

Figure 5–33 (a) Longitudinal section through a perennial bracket fungus; each year a new layer of tubes is formed; (b) three-dimensional view of hymenium-lined tubes; (c) cross section through tubes.

may be easily visualized if one examines a handful of parallel drinking straws from the end. The generic name is derived from this poroid appearance (Greek: *poly* = many, *poros* = passage) and provides a common name for these fungi, the "polypores." Although the pores are usually circular (Fig. 5–33), they may be elongated and labyrinthine (Fig. 5–32), elongated and hexagonal, or elongated and almost gill-like.

The basidiocarps may be fleshy when they are young, but become dry and like wood, cork, or leather as they mature. The basidiocarps are often quite large, sometimes reaching 25 centimeters or more in diameter. Some of the basidiocarps are resupinate and have the hymenial tubes covering the surface (Fig. 5–31). Others form a pileus that is attached by a broad surface to the substratum and thus resemble a shelf. These shelflike fungi are often called "bracket fungi." Still other basidiocarps may have a stalk, extending from either the side or the center of the pileus. The majority of the fungi are annual and the mycelium produces new basidiocarps every year. Species of *Fomes*, however, are perennial and the laterally attached basidiocarps are viable for many years. In these perennial polypores, a new layer of tubes develops each year (Fig. 5–33). The new layer of tubes is formed beneath the one that was produced the previous season, and only the most recently formed layer of tubes produces basidiospores. The basidiocarps are usually cream, gray, brown, or black in color but are sometimes shades of yellow, orange, or red. In those species forming a pileus, the color of the upper, sterile surface may be the same or different from that of the tubular region.

Order Agaricales. The Agaricales include the familiar mushrooms and toadstools, well known for their edibility or deadly effects (Chapter 16). Popularly, "mushroom" is used to refer to edible members of this group, while "toadstool" refers to poisonous members. The word toadstool is a corruption of the German name "Todestuhl," literally meaning "death's chair." The hallucinogenic fungi, used in many tribal religious rites, also belong to this group. Other members include the "boletes."

In nature, the members of the Agaricales occur predominantly as saprophytes on plant debris in the soil, on wood, or on dung. Some species are parasitic on plants, while others are involved in symbiotic mycorrhizal associations with vascular plants (Chapter 15). Members of the Agaricales may be found in diverse habitats, such as pastures, lawns, meadows, old fields, bogs, on wood or trees, or in humus on the forest floor. They are often found in city environments, where they may be associated with stumps or roots left from a tree that had been removed. They are important in decomposing plant materials in nature (see Chapter 12).

More than 5000 species are known in this order. The feature that unifies this group is their formation of a fleshy, moist basidiocarp bearing a hymenium that is exposed at maturity. The hymenium may be borne in tubes on the lower surface of a pileus. Those that bear tubes are members of the family Boletaceae and are often informally called "boletes." The majority of the Agaricales bear the hymenium on gills that hang downward from the pileus. The gill-bearing members of this order are classified in 13 families (Smith, 1973) and are often informally called "agarics." The characteristics of some families of the Agaricales are given in Table 5–2.

Basidiocarp Structure and Development. The Agaricales possess a pileus that bears the hymenium on its lower surface on either a tube or gill-shaped hymenophore (Figs. 5–34 to 5–36). The pileus may or may not be supported by

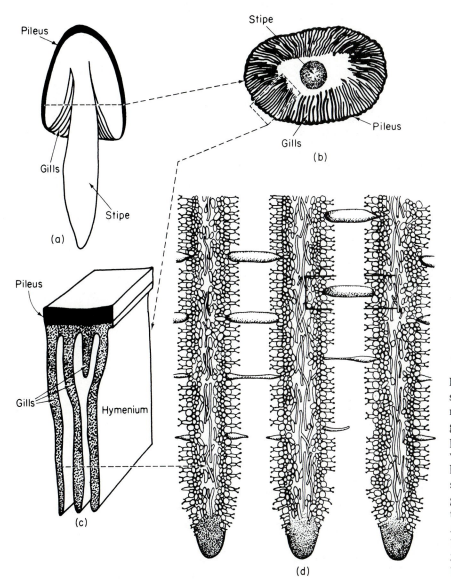

Figure 5–34 (a) Longitudinal section through *Coprinus*, a member of the Agaricales with gills; (b) cross section through pileus; (c) three-dimensional view of a section of the gills. The hymenium covers the sheetlike surface; (d) section through the gills; note the cystidia that extend between the gills. [(a, c, d) Adapted from A. H. R. Buller, 1924, Researches on fungi, Vol. 3, Longmans, Green & Co. Ltd., London.]

TABLE 5–2
Selected Characteristics of Representative Families of the Agaricales Which Are Recognized by Smith (1973)

Family	General Features	Accessory Structures	Basidiospores	Representative Genera
Boletaceae	Pileus on central stipe; fleshy; hymenium in tubes; associated with trees	Annulus and veil present or absent	Basidiospore deposit various colors	*Boletus*
Hygrophoraceae	Pileus on central stipe; gills thick, edge sharp, waxy in texture, decurrent; pileus brightly colored; on duff or ground	Absent	Spore print white; spores smooth	*Hygrophorus*
Tricholomataceae	Stipe present or absent; pileus may be seated eccentrically or centrally on stipe; gills attached but not decurrent; may occur on wood	Annulus present or absent	Spore print white or yellow to pink	*Marasmius* *Pleurotus* *Clitocybe*
Amanitaceae	Central stipe, separating cleanly from pileus; gills free; on ground in forests	Volva and annulus present; scales on pileus present or absent	Spore print white; spores smooth, thin-walled	*Amanita*
Lepiotaceae	Central stipe, separating cleanly from pileus; gills free; in grassy areas or humus, litter	Annulus and scales on pileus present	Spore print white to yellow or green	*Lepiota* *Chlorophyllum*
Agaricaceae	Central stipe, separating cleanly from pileus; gills gray to pink when young, becoming brown at maturity; cuticle of the pileus usually filamentous; on humus, grassy areas, duff	Annulus and veil present	Spore print chocolate brown; spores subglobose to elliptical; pore at apex	*Agaricus*
Strophariaceae	Stipe eccentric or central; stipe and pileus confluent; gills attached; pileus cuticle filamentous, often gelatinous; in grassy areas or plant debris such as mulch or wood chips	Annulus and veil present or absent	Spore print violet to brown; spores elliptical with pore at apex	*Stropharia* *Psilocybe*
Coprinaceae	Central stipe; basidiocarps thin and fragile; gills usually attached, may undergo autodigestion; pileus cuticle cellular; on dung or other organic waste	Annulus and veil present or absent	Spore print purple to brown or black; spores elliptical with pore at apex	*Coprinus* *Psathyrella*
Russulaceae	Central stipe; trama of pileus and stipe brittle, composed of nests of large rounded cells bounded by connective hyphae; latex sometimes present; on ground associated with trees (mycorrhizal)	Absent	Spore print white to yellow; markings on spore in form of warts or ridges which become blue in iodine.	*Russula* *Lactarius*

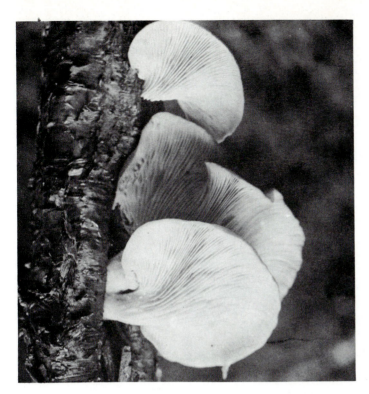

Figure 5–35 Laterally attached basidiocarps of *Pleurotus,* a member of the Agaricales. Note the gills on the underneath surface. [Courtesy Plant Pathology Department, Cornell University.]

Figure 5–36 Bolete, *Suillus granulatus.* About natural size. [Courtesy A. H. Smith.]

a stipe. We may recognize the following anatomical divisions of the basidiocarp: the hymenium, subhymenium, trama of the hymenophore, trama of the pileus, trama of the stipe, and cuticle.

The cuticle consists of one to three layers and is a skinlike covering of the sterile portions of the basidiocarp which sometimes may be peeled off easily. Most often the cuticle consists of interwoven hyphae (Fig. 5–37), but in some species the outer layer may consist of separate cells. The cuticle may be poorly differentiated from the trama. In some cases, the outermost layer of the cuticle may form a palisade layer resembling the hymenium, pseudoparenchymatous

(a)

(b)

Figure 5–37 Scanning electron micrographs of (a) the pileus cuticle and (b) trama of the pileus in two members of the Agaricales. (a) × 1650, (b) × 475. [Courtesy of R. L. Homola and J. Kimball.]

cells, greatly inflated cells, or other modifications (Fig. 5–38). The cuticle of some species may become viscous or sticky, owing to the gelatinization of some of its components. A cuticle that becomes gelatinous makes the surface feel slippery or sticky. In still other cases, some of the cells of the cuticle may separate from each other and form a powdery coating on the basidiocarp's surface.

The trama of the pileus and stipe generally consist primarily of generative hyphae or of thin-walled hyphal types or cells derived from the generative hyphae. Some species have lactiferous or other specialized hyphal types. The structure of the pileus trama and stipe trama often differ in a given basidiocarp. The gill trama consists of hyphae projecting downward from the pileus trama. Within the gill, the tramal hyphae may lie parallel to each other, may be interwoven more or less at random, or may diverge when approaching the subhymen-

Figure 5–38 Some modifications of the cuticle found in the Agaricales. In each drawing, the cuticle is shown with some of the underlying trama. (a) Cuticle is of a pseudoparenchymatous layer and bears a layer of cystidia; (b) cuticle is a single layer of greatly enlarged cells; (c) cuticle has undergone gelatinization; (d) cuticle is an upright palisade of cells, each with a hairlike projection. (a-c) × 525, (d) × 1500. [From A. H. Smith, 1949, Mushrooms in their natural habitat, Hafner Press, New York, 626 pp. Copyright GAF Corporation 1949.]

ium (Fig. 5–39). The texture of the hymenophore is generally moist, but in *Hygrophorus* the gills have a distinct waxy texture.

In addition to the pileus, stipe, and hymenophore, other structures may be found on mature basidiocarps. These include a skirtlike ring around the stipe, the *annulus;* a veil on the margin of the pileus; a cuplike membrane surrounding the base of the stipe, the *volva;* and scales which may adhere to either the stipe or the pileus (Figs. 5–40 and 5–41). These accessory structures result from the presence of specialized *veils* during the early stages of development. The veils were torn through the expansion of the basidiocarp, resulting in the accessory structures. Any combination of accessory structures may occur, depending on the particular developmental pattern that has been followed. Three developmental patterns may be distinguished, which are described below.

1. *Gymnocarpic development.* The basidiocarp always begins as a primordium of generative hyphae formed on the mycelium. This primordium becomes recognizable as the rounded ''button'' stage. In gymnocarpic

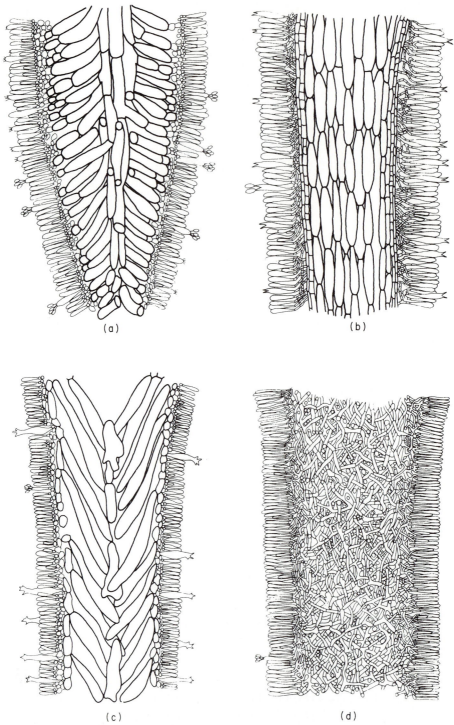

Figure 5–39 Semidiagrammatic drawings of cross sections of gills showing the four important types of arrangement of the hyphae: (a) divergent type; (b) parallel type; (c) convergent type; (d) interwoven type. [From A. H. Smith, 1949, Mushrooms in their natural habitat, Hafner Press, New York, 626 pp. Copyright GAF Corporation 1949.]

Figure 5–40 Agaric which forms a universal veil. Left: A young basidiocarp is just beginning to rupture the universal veil; right: the universal veil is left as a volva at the base of an older basidiocarp. Note that a cortina and an annulus are lacking. [Courtesy Plant Pathology Department, Cornell University.]

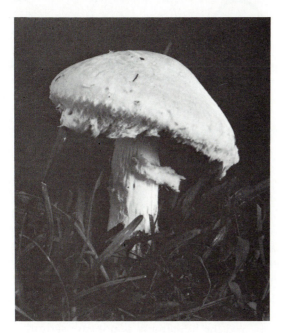

Figure 5–41 Field mushroom, *Agaricus brunnescens*, with a veil and an annulus remaining after the recent rupture of the partial veil. [Courtesy Plant Pathology Department, Cornell University.]

development, the button continues to enlarge to form the basidiocarp, with the apical portion expanding to form the pileus. The hymenium differentiates from a region on the lower side of the developing pileus and is never covered by veils at any point in development. The mature basidiocarp typically lacks accessory structures (Fig. 5–42) (some gymnocarpic species have scales on the pileus but they do not originate from veils).

2. *Pseudoangiocarpic development.* Development of the primordium more or less proceeds as in the gymnocarpic type, but the enlarging pileus recurves and the edge of the young pileus becomes rejoined to the stipe by a loose weft of fundamental hyphae, the *partial veil*. A small cavity is enclosed, and the tubes or gills form within the cavity. As the pileus

GYMNOCARPIC DEVELOPMENT

PSEUDOANGIOCARPIC DEVELOPMENT

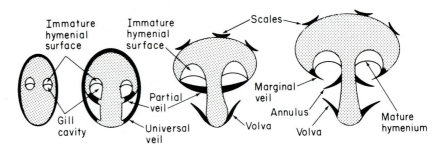

HEMIANGIOCARPIC DEVELOPMENT

Figure 5–42 Developmental patterns in the Agaricales.

expands, the partial veil is torn and may be left as the collarlike annulus around the stipe, or remnants may remain on the edge of the pileus as a *marginal* veil, scales, or flaps of tissue. Sometimes the remaining veil is cobweblike and called a *cortina*. Therefore any basidiocarp which underwent this development may have the annulus alone; marginal veil, scales, or cortina alone; or both the annulus and the marginal veil, scales, or cortina (Fig. 5–42).

3. *Hemiangiocarpic development.* Unlike the previous types, the hymenium is formed endogenously within gill cavities in the tissue of the young basidiocarp, and the hymenium is not exposed until late in the developmental sequence. The outer layer of the primordium differentiates into a distinct surrounding layer of tissue, the *universal veil.* A partial veil consisting of loose tissue joining the margin of the expanding pileus and stipe is also formed. As the basidiocarp enlarges, the universal veil ruptures. If the universal veil was poorly differentiated, it may remain behind in the substratum, gelatinize, or form scales on the stipe or pileus. If the universal veil was a well-defined layer, it leaves a cuplike volva at the base of the stipe. After rupture, the partial veil may remain as scales on the stipe or as an annulus (Fig. 5–42).

Any combination of structures may remain in the mature basidiocarp as a result of hemiangiocarpic development. For example, both the universal and partial veils may remain only as scales on the stipe and pileus. Many, including the cultivated mushroom *Agaricus brunnes-*

cens, have only an annulus remaining from the partial veil. Some may have a volva with remnants of the partial veil remaining as patches on the stem. The deadly species of *Amanita* have a volva, annulus, and scales.

Morphological Variation. As we have just read, there may be considerable variation in the presence or absence of accessory structures such as the volva, scales, or annulus. There are many other macroscopic variations that are important not only for recognizing species in the field but also for understanding relationships. A small number of these variations are examined below.

Basidiocarps may range from those that are extremely small and perhaps also delicate to those that are large and perhaps also massive. The smallest species may have a pileus that is only about 2 millimeters in diameter, and there may be so few cells in the pileus that one may read print through the trama. An example of a small, delicate agaric is *Marasmius rotula*, often called the ''horse-hair fungus'' because of the extremely long black stalks which support the delicate white pileus. This fungus occurs in dense populations, and a dried basidiocarp will revive when moisture becomes available. In contrast to such delicate agarics, some may have a pileus which is 30 centimeters or more in diameter. The most massive basidiocarps are often found among the boletes, which often have very thick, bowl-like pilei with a great deal of trama and the stout stipes required to support such a massive pileus.

The stipe may be either present or absent. If present, the stipe may have the same diameter throughout its length, or it may be somewhat enlarged and bulbous near the base. In the boletes, the stipe often has a reticulated pattern or glandular areas on its surface. The stipe may be attached to the pileus at its side, at its center, or in an intermediate position.

The pileus shapes include those which are asymmetrical and somewhat shell-shaped. These are typically associated with either the absence or the noncentral insertion of the stipe. In the majority of the species, the pileus shape is symmetrical and the stipe is centrally inserted. At maturity, the pileus shapes include those which are conical and more or less closed, as well as some which are bell-shaped, flattened, or even funnel-shaped. During maturation, the pileus usually changes from a more enclosed, somewhat globose or conical form to one that is more open and bell-like or flattened. A fully open pileus can easily discharge its basidiospores into open space, while a partially closed, conical form is not as well adapted for basidiospore discharge. Other variations of the pileus include the nature of the margin, which may be flared, curved, or of some other form. Toward the middle of the pileus, there may be an indentation or a knob.

If we examine the underneath surface of the pileus, we might find that the gills may be distantly spaced from each other or so close that they almost touch. There is some variation in the shape of the gills as well as in their margin. At the point where the hymenophores approach the stipe, there may be a free space where these structures are not attached and only a short section where they are attached to the stipe, or they may be attached to the stipe throughout their entire length. In some species, the gills are decurrent, running down the stipe and gradually merging with it. In most species, the edges of the gills are wedge- or knife-shaped, and the basidiospores will mature randomly over the entire surface of the hymenophore. In contrast, in species of *Coprinus* the basidiospores begin to mature at the bottom of the gill and the process then progresses upward. The basidiospores of *Coprinus* cannot fall out of the pileus readily because of the nontapering crowded gills, thickened gill margin, and the somewhat closed, conical shape of the pileus. Instead, the gills of some species undergo autolysis beginning at the base and progressing upwards as the basidiospores mature. The resulting black fluid drips away, carrying the basidiospores

(a)   (b)

Figure 5–43 Basidiocarps of the inky cap mushroom. *Coprinus:* (a) unopened basidiocarps with conical pileus; (b) autolysis of the pileus results in the dripping of a black fluid containing basidiospores. [From E. J. Moore and V. N. Rockcastle, 1963, Fungi, Cornell Science Leaflet.]

with it. For this reason, these fungi have been called the ''inky caps'' and the black fluid has been collected and used as ink (Fig. 5–43).

Basidiocarps may occur in virtually any color, ranging from somber shades such as white, cream, tan, gray, or brown to brightly colored shades of yellow, orange, red, violet, or green. The pigments responsible for these colors may occur within the cytoplasm of hyphae, in hyphal walls, or as encrustations such as crystals or spirals on the outer surface of hyphal walls. In some species, damaged tramal tissue will turn a distinctive color such as blue or red. The colors of the pileus, hymenophore, and stipe may be similar or dissimilar in a given basidiocarp. Further, the color of the gills or tubes may or may not correspond exactly to the color of the basidiospores that they bear. When viewed in quantity in a spore deposit, basidiospores may appear to be shades of white, pink, purple, ocher, or black.

Bioluminescence. Some agarics are bioluminescent, emitting light of a green, greenish-white, or bluish-green color. This light is sometimes quite bright and may be bright enough to read by or to be used as the sole light source in prolonged exposure of photographic plates. In the past, this phosphorescent material was known as ''fox fire'' and chunks were used to mark trails or on hats to illuminate the pathway at night.

The luminescent portion of the fungus varies: for some species of *Collybia,* it is the germinating sclerotium; for *Omphalotus olearius,* it is both the mycelium and basidiocarp; and for *Armillaria mellea,* it is both the mycelium and rhizomorphs.

Omphalotus olearius is perhaps the most conspicuous of these fungi, as this species forms yellow-orange basidiocarps that occur in clusters containing up to 50 individuals. Each basidiocarp is extremely large: the diameter of the pileus and length of the stipe may be up to 20 centimeters. Basidiocarps appear around old stumps in the autumn and this species is called "jack-o'-lantern" fungus. Bioluminescence may be demonstrated by collecting the luminescent part of the fungus and then examining it in a dark room after allowing sufficient time for the eyes to become accustomed to the dark. Wood penetrated by the mycelium of *Armillaria mellea* is good material for this purpose. Such wood may be found by locating the basidiocarps of *A. mellea* or the long, dark rhizomorphs which grow beneath the bark of logs or stumps (Fig. 5–19).

Class Gasteromycetes

The Gasteromycetes include the puffballs, false truffles, stinkhorn fungi, bird's nest fungi, and earthstars. These fungi usually occur on the surface of the ground, but some members occur on logs or dung or beneath the surface of the soil. The Gasteromycetes are especially prevalent in the drier regions of the world and in sandy, dry habitats. The unifying characteristic of this class is that its members form basidiocarps which enclose the basidiospores until maturity is reached.

The basidiocarp is surrounded by a peridium, an outer covering consisting of one or more layers (Fig. 5–44). The peridium may be thick and leatherlike, papery, scaly, or membranous. Often the peridium will persist and remain intact throughout development, but in some species it will flake away or dehisce rather early. The peridium surrounds the fertile portions of the basidiocarp in which the basidiospores are borne. This internal, fertile tissue is the *gleba*. The gleba consists of sterile tissue, the trama, in combination with the basidia and basidiospores. The basidia may be scattered throughout the gleba or alternatively may occur in a hymenium lining elongated or branched chambers within the gleba. Most of the basidia are somewhat globose in shape and often bear sessile basidiospores seated directly on the basidium rather than on sterigmata. If sterigmata are present, the entire sterigma breaks off and remains attached to its basidiospore. Basidiospores are not forcibly ejected (see Chapter 11). At maturity, the gleba may undergo autodigestion to become slimy or may become powdery and dry. In addition to the gleba and peridium, the basidiocarp may have a columella as well as a stipe. The columella is a central column of sterile tissue which occurs within the gleba. It may occur only near the base of the basidiocarp or may transverse the entire gleba. Some species have an external stipe upon which the peridium and gleba are seated, and this stipe may or may not extend into the columella.

(a)   (b)   (c)

Figure 5–44 Gasteromycete, *Truncocolumella citrina* (Hymenogastrales): (a) external view of a basidiocarp; (b) median section through a basidiocarp; (c) section of the peridium and gleba. [From S. M. Zeller, 1939, Mycologia 31:1–32.]

Development of many Gasteromycetes has been studied and various developmental patterns described. One is the *lacunar* type, in which the gleba is at first a fleshy mass of hyphae. During development, the tissues pull apart to form various lacunae (cavities) which later become lined by the hymenium or filled with basidia. In other developmental types, plates or branches of trama tissue originate from the columella and grow upward or originate in undifferentiated tissue near the apex of the basidiocarp and grow downward. To a certain extent, the course of development controls the final morphology of the gleba (Fig. 5–45).

Figure 5–45 Diagrams of longitudinal sections of basidiocarps in the Gasteromycetes showing different arrangements of the gleba derived from distinctive developmental patterns. The lacunar form is at the left. [Adapted from E. Fischer, 1933, in Engler, A., and K. Prantl, (Eds.), Die natürlichen Pflanzenfamilien, Vol. 7A, W. Engelmann, Leipzig.]

The Gasteromycetes may reproduce nonsexually by producing conidia. The conidiophores and conidia may be borne in the hymenium with the basidia or in a separate sporocarp. If they are in a separate sporocarp, that sporocarp superficially resembles the basidiocarp but bears conidia rather than basidiospores. Among the Gasteromycetes there are important differences, such as whether a hymenium is present or absent and differences in the texture of the gleba and the nature of the stipe. The orders of the Gasteromycetes, as recognized by Dring (1973), follow.

ORDER HYMENOGASTRALES

In this heterogenous order, the peridium is connected to plates of trama which divide the interior of the basidiocarp into chambers. These chambers are often elongated or irregular (Fig. 5–44) but may also be boxlike. A distinct hymenium lines the surface of the chambers. These chambers are usually empty early in development but later may become partially or completely filled with basidiospores. Two to eight basidiospores are borne on each basidium. The gleba is cartilaginous to fleshy and usually retains this structure until maturity. In some members, the gleba becomes powdery at maturity.

Some of the members of this order strikingly resemble members of the Agaricales and may be an evolutionary link between these two groups. These fungi are comparatively large and may occur below or above the ground. They have a stipe and a pileus (Fig. 5–46). The peridium forms the surface tissue of the pileus and extends into plates of trama which resemble the gills of the agarics or tubes of the boletes. The tramal plates may anastomose and crumple to form irregular cavities. The stipe extends as a columella through the fertile portion of the basidiocarp. At maturity, the peridium breaks away from the stipe or may crack in a longitudinal manner. In some cases, there may be further expansion of the basidiocarp. The morphology of these fungi has been compared with that of an agaric that failed to open during its development and had subsequent internal maturation of its basidiospores (Smith, 1973).

The remaining members of this group superficially resemble truffles and are sometimes called "false truffles" (Figs. 5–44 and 5–47). They often occur

Figure 5–46 Member of the Hymenogastrales, *Gyrophragmium delilei,* which forms both a stipe and a pileus. Left: intact basidiocarp; note the rupturing of the peridium; right: a longitudinal section through the basidiocarp showing the gill-like tramal plates in the gleba. [Courtesy U.S. Department of Agriculture.]

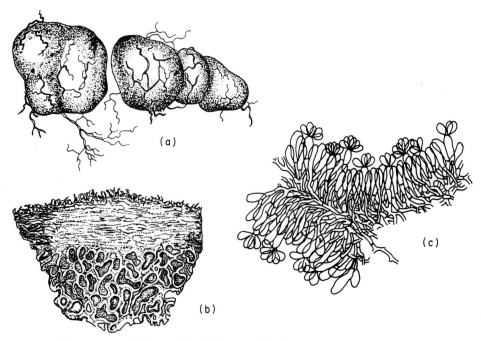

Figure 5–47 False truffle, *Rhizopogon luteolus*: (a) mature basidiocarps with rhizomorphs (natural size); (b) portion of the periphery of the basidiocarp showing the peridium (top) and gleba (bottom), × 450; (c) portion of the gleba, × 450. [From C. Tulasne and H. Rehsteiner., *in:* Gäumann, E. A. and C. W. Dodge, 1928, Comparative morphology of fungi, McGraw-Hill Book Company, New York, 701 pp.]

underground or may occur in the duff on the forest floor. The basidiocarp is usually somewhat globose or pear-shaped and may be lobed. It is often connected to fairly prominent rhizomorphs, and the size is typically quite small, the basidiocarps being often only a few millimeters or centimeters in diameter. These basidiocarps have a fleshy gleba that disintegrates along with the basidiocarp at maturity. The gleba often develops a strong odor at maturity, which helps attract rodents and insects that eat the basidiocarps.

An important genus of the false truffles is *Hymenogaster*. The development of several species of *Hymenogaster* has been studied, and although the developmental pattern is not necessarily the same, many of them have the lacunar type of development (Fig. 5–48). Generally, the peridium and trama tissues of the gleba are poorly differentiated from each other in the young stages and may undergo no further differentiation in some species. As development proceeds in other species, the outer hyphae become large and form a coarse outer layer that gradually merges into the gleba. The cavities develop by splitting and tearing in the upper portion of the basidiocarp, leaving a hemispherical to conical sterile base that may persist to maturity. When the cavities expand above, the lower ones next to the sterile base are stretched longitudinally so that they appear to radiate from the sterile base. This expansion causes the plates of trama to become gradually thinner (Dodge and Zeller, 1934). Other developmental patterns may occur within the Hymenogastrales (Hawker, 1954, 1955).

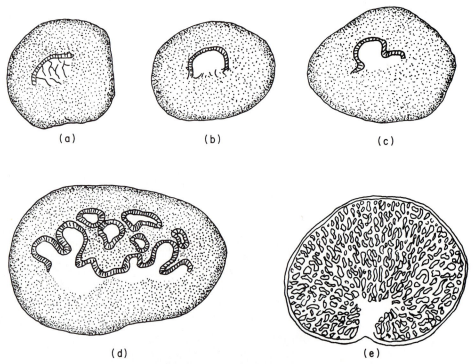

(a) (b) (c)

(d) (e)

Figure 5–48 Development of the basidiocarp of *Hymenogaster tener*, with all figures shown in longitudinal section: (a) very young basidiocarp, showing a cavity formed by stretching and tearing of hyphae; hymenium (shaded zone) is already forming at upper margin of cavity; (b,c) slightly older basidiocarps in which cavity is clear of torn hyphae and wrinkling of hymenium is beginning; (d) older basidiocarp in which hymenium is greatly wrinkled and infolded; (e) mature basidiocarp showing the sterile base and elongated, more or less radially arranged chambers. [From L. E. Hawker, 1954, Phil Trans. Roy. Soc. London Ser. B237:429–546.]

As in the Hymenogastrales, members of the Melanogastrales and Gautieriales are usually hypogeous. Their basidiocarps are similar in their overall form to those of the so-called "false truffles" described in the preceding section.

In the Melanogastrales, the chambers within the hymenium do not become lined with a well-defined hymenium as in the Hymenogastrales. Instead, the basidia and basidiospores are formed more or less randomly within the chambers along with the loose "stuffing" hyphae that make the basidiocarp solid. At maturity, the hyphae and basidia undergo autolysis and fill the cavities with a gelatinous slime in which the basidiospores are distributed.

The basidiocarps in the Order Gautieriales are unique in that they lack a peridium at maturity (Fig. 5–49). The cartilaginous gleba terminates in a number of pileuslike plates of tissue which more or less surround the basidiocarp.

ORDER PHALLALES

The Order Phallales contains approximately 65 species which are especially widespread in the tropical regions of the world, although they are common in the woods of the temperate zone in the late summer and fall. In their early developmental stages, these fungi form globose or ovoid basidiocarps, which often superficially resemble an egg and occur beneath the surface of the soil.

Figure 5–49 *Gautieria* sp., a member of the Gautieriales. Top: entire basidiocarps; bottom: longitudinal section showing the gleba. Note the absence of a peridium. [Courtesy Plant Pathology Department, Cornell University.]

These range in size from approximately 0.5 to 5 centimeters in diameter. The young basidiocarp is covered by a peridium, and underneath the peridium is a well-defined gelatinous layer. The gleba, which is somewhat fleshy in this early stage, may be attached to the peridium at the base only, or it may be attached by platelike extensions which extend through the gelatinous layer. When mature, the gleba undergoes autolysis to form a slime.

Some members such as *Phallogaster* (Fig. 5–50) have a persistent peridium and will remain egg- or pear-shaped permanently. These hypogeous fungi share some similarities with the Hymenogastrales (for example, the chambered gleba) and are apparently transitional forms between that order and the remainder of the Phallales (Dring, 1973).

In the remaining members (the majority), the internal portion of the young basidiocarp is occupied by some spongy, sterile tissue. This tissue is capable of

(a)

(b)

(c)

Figure 5–50 *Phallogaster saccatus,* a member of the Phallales: (a) longitudinal section through a basidiocarp showing the thick, persistent peridium with a gelatinous layer beneath it; the gleba with labyrinthine, hymenium-lined chambers occupies most of the internal space; (b) portion of the gleba is shown in greater detail; (c) mature basidiocarp showing the rupture of the peridium, slimy contents, and conspicuous rhizomorphs. [Courtesy Plant Pathology Department, Cornell University.]

undergoing rapid expansion to form a *receptacle* (the stem), which may fully expand in only a few hours. When the receptacle expands, it ruptures the peridium, which remains behind in the ground as the volva. The receptacle, which is now aboveground, may occur in various forms, such as an unbranched column, a hollow, netlike ball, or arms which may be united at the top or be spread in a raylike fashion (Fig. 5–51). The gleba is carried upward by the expansion of the receptacle, and meanwhile the gleba undergoes autolysis to form a foul-smelling slime in which the basidiospores are embedded. The slimy, putrid gleba is very attractive to flies and attracts them to the basidiocarps. The flies carry the spores to new locations.

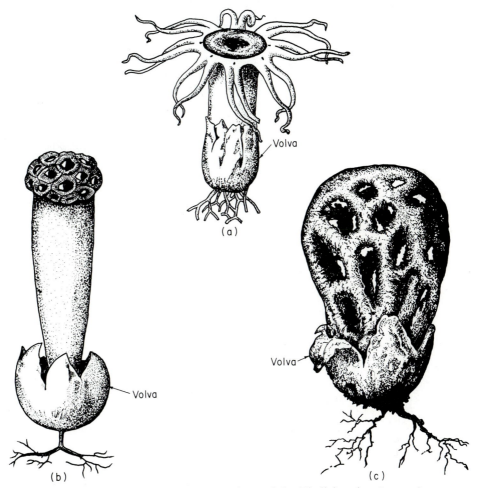

Figure 5–51 Representative members of the Phallales showing variations in the form of the receptacle. Note that a prominent volva is present at the base of the receptacle of each. (a) *Aseröe* sp. with receptacle extended into arms; (b) *Simblum* sp. with receptacle terminating as a pileus like hollow network; (c) *Clathrus* sp. with receptacle as a sessile hollow network. [From M. J. Berkeley, J. Gerard, and V. Fayod, respectively, *in:* Gäumann, E. A., and C. W. Dodge, 1928, Comparative morphology of fungi, McGraw-Hill Book Company, New York, 701 pp.]

North American Phallales are represented principally by the so-called "stinkhorn" fungi, members of the family Phallaceae. The volva is prominent in these fungi, and either cups around or sheaths the elongate, unbranched receptacle (Figs. 5–52 and 5–53). The hollow receptacle may be white, delicate pink, lilac, or orange-red and somewhat spongy in consistency. The receptacle

Figure 5–52 Stinkhorn, *Mutinus caninus.* Early stages, still surrounded by the peridium, are on the right. Mature basidiocarps are on the left. Note the ruptured peridium (the volva) and the slimy apex on the mature basidiocarps. [Courtesy Plant Pathology Department, Cornell University.]

may either have a bell-shaped pileus or may terminate bluntly without such a cap. The terminal portion of the pileus bears the foul-smelling olive-colored slime containing the basidiospores. A skirtlike veil may be present beneath the pileus (Fig. 5–53).

ORDERS LYCOPERDALES AND TULOSTOMATALES

These fungi form basidiocarps that have dry contents at maturity. The gleba is fleshy when young but the generative hyphae undergo autodigestion, resulting in breakdown of the gleba. This autodigestion frees a mass of hyphae, the *capillitium,* and basidiospores. At first, the disintegrated tissue is watery but it eventually dries. Most of the capillitial hyphae are thick-walled, brown, and without septa and are derived from skeletal or binding hyphae. However, some of them are hyaline, thin-walled, and septate and will collapse. These two types of capillitium may occur in the same or different basidiocarps. Further, the capillitium may be branched or unbranched and may sometimes have external markings such as spines. The capillitium is a tangled mass that traps the basidiospores and may also break into even shorter fragments. At maturity, the basidiocarp is filled with a powdery material consisting of fragments of the capillitium and basidiospores. Both the capillitium and basidiospores may escape through an ostiole or fissures in the persistent peridium, or the peridium may scale away during development. If the peridium is struck lightly (perhaps by raindrops) or is broken suddenly (perhaps by being stepped upon by an animal), a powdery cloud of basidiospores is suddenly emitted. For this reason, these fungi are often called ''puffballs.'' Usually, however, the basidiospores are released over a period of time, as they are blown away by breezes but are pre-

Chap. 5 / Division Amastigomycota—Subdivision Basidiomycotina

Figure 5-53 Basidiocarp of *Dictyophora duplicata*, a member of the Phallales. This stinkhorn has a lacy, skirtlike veil. [Courtesy U.S. Department of Agriculture.]

vented from being blown away all at once because they are trapped within the tangled capillitium.

Members of the Tulostomatales have a prominent stalk consisting mostly of hyphae arranged parallel to the long axis. The stalk supports a rounded peridium (Fig. 5-54). In most other respects, the Tulostomatales are similar to the members of the Lycoperdales, which lack such a well-developed stalk (Figs. 5-55 and 5-56).

Members of the Lycoperdales are almost always epigeous, varying in size from a few millimeters to over a meter in diameter. They form a sessile basidiocarp or have a short, somewhat spongy stalk which consists of modified glebal tissue. A columella may also be present. The basidiocarps are often globose, subglobose, or pyriform in shape. The peridium is usually subdivided into two or more layers, ranging from those that are extremely thick to those that are comparatively thin and flake away at maturity. The surface of the peridium may be smooth or may have ornamentations such as spines or warts. The color of the peridium is usually white or some shade of tan or yellow. The color, thickness, and markings of the peridial layers of a basidiocarp may differ among species. During the early developmental stages, glebal chambers are formed that are lined by a well-defined hymenium. Various types of development have been

Figure 5–54 Member of the Tulostomatales, *Tylostoma lysocephalum* (natural size). Note the prominent pore in the peridium. [From W. H. Long, 1944, Mycologia 36:318–339. Courtesy U.S. Department of Agriculture.]

described, including the upward or downward growth of trama tissue as well as the lacunar type. At first, the gleba is light in color (e.g., white or pale yellow) but the contents darken as the basidiocarp matures because of the maturation of the dark-colored basidiospores. The capillitium and intermixed basidiospores often become deep shades of brown or olive. In addition to being dark, the basidiospores are usually globose and warted or spiny.

Figure 5–55 Basidiocarps of a puffball, *Lycoperdon*. [From E. J. Moore and V. N. Rockcastle, 1963, Fungi, Cornell Science Leaflet.]

Figure 5–56 Basidiocarps of the giant puffball, *Calvatia gigantea*. Compare their size with the 36-inch ruler. [Courtesy Plant Pathology Department, Cornell University.]

Some members of the Lycoperdales are known as ''earthstars'' (Fig. 5–57). Earthstars possess a prominent, thickened outer peridium of two or three layers and a thin inner membrane. As maturity is reached, the outer peridium separates from the inner peridium and dehisces in a stellate manner. Segments of the outer peridium recurve and superficially resemble petals of a flower while the inner peridium retains its rounded shape. Some earthstars have been used as hygrometers because of their response to moisture (the peridium closes when the weather is dry and opens when the weather is damp).

Most fungi in these orders have fewer peridial layers and do not have outer peridial layers that recurve as do those of the earthstars. This group is generally called puffballs because of the cloud of capillitial fragments and basidiospores that emerges. Common puffball genera include *Lycoperdon* and *Calvatia*, both in the Lycoperdales. Species of *Lycoperdon* form pear-shaped or subglobose basidio-

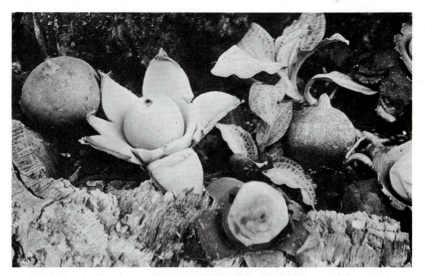

Figure 5–57 Earthstar, *Geastrum*. Note that the peridium of the basidiocarp at the left has not recurved and also note the ostiole in the thin inner peridium of the central basidiocarp. The peridium of the basidiocarp at the right has collapsed after discharge of the capillitium and basidiospores. Approximately × ½. [From E. J. Moore and V. N. Rockcastle, 1963, Fungi, Cornell Science Leaflet.]

Classification

carps that are about 3 to 5 centimeters in diameter. The peridium has two layers and the outer layer forms warts or spines. An ostiole is formed that allows the basidiospores to escape at maturity (Fig. 5–55). *Calvatia gigantea,* the giant puffball, is the best known member of the genus *Calvatia* (Fig. 5–56). *C. gigantea* forms creamy white, globose, or subglobose basidiocarps up to 120 centimeters in diameter. These basidiocarps have a smooth, leathery texture. *C. gigantea* is edible when young while the gleba is still firm and white, and it is often sold in farmers' markets. At maturity, the thin peridium flakes away to expose the capillitium. Large quantities of soft capillitium have been used to dress surgical wounds and to stop bleeding from injuries and was kept on hand in rural households for such purposes (Rolfe and Rolfe, 1925).

ORDER SCLERODERMATALES

This order shares many similarities with the Lycoperdales. These fungi occur in similar habitats and the overall appearance of the basidiocarp is generally similar to that of the sessile puffballs or earthstars.

Unlike the gleba in members of the Lycoperdales, the gleba in this order is divided into numerous chambers by thick or thin plates of trama, a characteristic that this group shares with the Hymenogastrales. Development is of the lacunar type. Basidia may be scattered within the glebal tissue or may occur in poorly defined clusters on the lining of the chambers (unlike the well-defined hymenium in members of the Lycoperdales). At maturity, the peridium contains a dry, powdery material. Most often, this consists only of the disintegrated gleba that is mixed with basidiospores, but in a few species a scant amount of capillitium is formed.

A common genus is *Scleroderma.* Species of *Scleroderma* occur above the ground and form sessile, globose basidiocarps with a thick, tough peridium. At maturity, the outer layer of the peridium breaks into scales. The peridium is usually some shade of yellow or tan while the gleba is a dark purple-brown at maturity.

ORDER NIDULARIALES

Most of the members of the Nidulariales are called "bird's nest fungi," as their basidiocarps resemble miniature bird's nests filled with eggs. The basidiocarps range from several millimeters to a centimeter in diameter. These fungi are frequently found in large numbers in lawns, fields, and gardens (especially where manure has been distributed). In addition, they occur commonly on decaying wood in forests and are especially common in the autumn. The "nest" is actually a funnel- or goblet-shaped peridium with a large open mouth (Fig. 5–58). It is usually cream, tan, or ocher, sometimes has a hairy exterior, and is typically leathery and tough in texture. In some species, the base is immersed within a hyphal cushion (Fig. 5–59). The peridium may have one to three distinct layers. The peridium contains egg-shaped peridioles loosely piled on top of each other. Each peridiole is a membrane-bounded segment of the gleba with a central cavity containing basidiospores. In species of the common genera *Crucibulum* and *Cyathus,* the peridiole is attached to the peridium by a tubular structure that contains a coiled cord consisting of parallel hyphae, the *funiculus* (Fig. 5–59). A funiculus is lacking in some bird's nest fungi as well as in *Sphaerobolus,* another member of this group which forms only a single spherical peridiole in a globose peridium. The bird's nest fungi and *Sphaerobolus* have interesting adaptations for spore discharge which are discussed in Chapter 11.

When the basidiocarps of the bird's nest fungi are first evident, they consist of undifferentiated, hyaline hyphae and are more or less globose in shape. The outer layers of the peridium gradually differentiate, becoming dark or hairy depending upon the species. At first the differentiating peridium surrounds an

Figure 5–58 Bird's nest fungus, *Cyathus striatus,* growing on dead twigs. [Courtesy H. J. Brodie.]

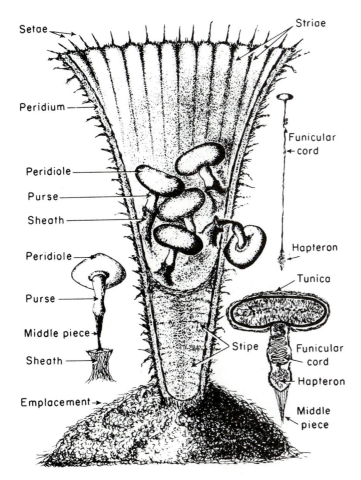

Setae

Striae

Peridium

Funicular cord

Peridiole

Purse

Sheath

Peridiole

Hapteron

Purse

Tunica

Middle piece

Stipe

Funicular cord

Sheath

Hapteron

Emplacement

Middle piece

Figure 5–59 Longitudinal section through a basidiocarp of *Cyathus striatus* showing the attachment of the peridioles to the peridium. When a peridiole is ejected by a raindrop, the purse is torn at its lower end, freeing the funicular cord (funiculus) which bears an adhesive organ (hapteron) that aids in attachment. [Figure reproduced by permission of the National Research Council of Canada from Can. J. Botany, 1951, 29:224–234 and courtesy H. J. Brodie.]

undifferentiated, more or less homogenous ground tissue (the gleba). As the basidiocarp elongates and expands, the peridium gradually becomes somewhat funnel-shaped [Fig. 5–60(d)] and opens at the apex. In a few species, the peridium tears to form the opening of the funnel, but in most species the peridium has a central pore that may contain a drumlike membrane concealing the gleba. At a fairly early stage in the above sequence, the glebal tissue becomes distinct from that of the peridium because of the gelatinization of a thin layer of tissue [Fig. 5–60(a)]. Both the funiculus and peridiole are initiated when some hyphae grow inward from the peridium, penetrating the previously homogenous ground tissue and converging with other hyphae that originate in the ground tissue. The convergent hyphae delimit a peridiole initial [Figs. 5–60(b) and 5–61] that gradually becomes more distinct as it enlarges and layers of tissue differentiate and mature [Fig. 5–60(c)]. Eventually each peridiole is covered by three distinct layers of tissue. Depending on the species, the peridioles mature simultaneously or differentiate successively, beginning from the top and progressing downward. A cavity forms within each peridiole by gelatinization of ground tissue, and both the peridiole and the enclosed cavity enlarge further [Fig. 5–60(d)]. The cavity is lined by basidia when it is only about half its final size, but by the time the peridiole is mature, the cavity is filled with basidiospores that have been released by the collapse and gelatinization of the basidia. While the peridioles are maturing, the ground tissue which surrounds them undergoes gelatinization. By the time that the peridium expands fully and the apical membrane ruptures, the peridioles are fully formed and are embedded in a pool of gelatinous material that eventually dries. Because the peridioles are no longer supported by glebal tissue, they often settle to the bottom of the cup-shaped peridium.

One member of this group, *Cyathus stercoreus,* is tetraploid, having 12 chromosomes which represent the 4N condition. Evidence to support this conclusion is that the haploid nucleus contains similar pairs of chromosomes and that quad-

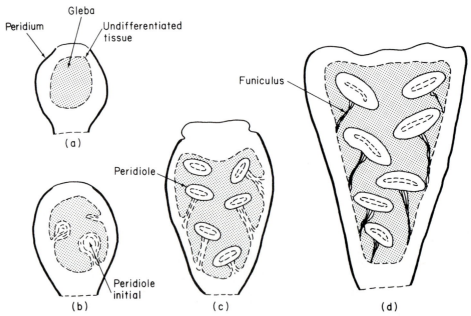

Figure 5–60 Progressive stages in development of a basidiocarp of the Nidulariales: (a) young stage with a homogeneous gleba; (b) peridiole initials are forming within the gleba; (c) further enlargement of the basidiocarp and differentiation of the peridioles; (d) nearly mature basidiocarp just prior to opening. [Adapted from L. B. Walker, 1920, Botan. Gaz. 70:1–24.]

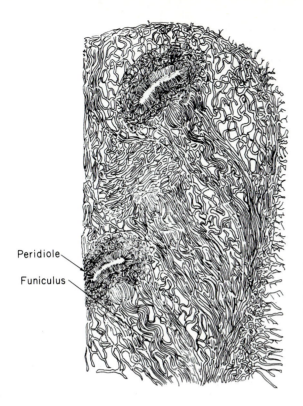

Peridiole

Funiculus

Figure 5–61 Differentiation of the peridioles within the gleba of an immature basidiocarp of *Crucibulum vulgare*, a member of the Nidulariales. Note that the peridioles are attached to the peridium by the parallel hyphae within each funiculus. [From J. Sachs, *in:* E. A. Gäumann and C. W. Dodge, 1928, Comparative morphology of fungi, McGraw-Hill Book Company, New York.]

rivalent configurations are formed in meiosis. Apparently, *C. stercoreus* is the first basidiomycete in which polyploidy has been demonstrated (Lu, 1964).

REFERENCES CITED

AINSWORTH, G. C. 1973. Introduction and keys to higher taxa. *In:* Ainsworth, G. C., F. K. Sparrow, and A. S. Sussman, Eds. The fungi: an advanced treatise. Academic Press, New York, Vol. 4A, pp. 1–7.

BARONI, T. J. 1978. Chemical spot-test reactions: boletes. Mycologia 70:1064–1076.

BULLER, A. H. R. 1922. Researches on fungi. Longmans, Green, London, Vol. 2, 492 pp.

CHANG, S. T., and K. Y. LING. 1970. Nuclear behavior in the basidiomycete, *Volvariella volvaceae*. Am. J. Botany 57:165–171.

CLÉMENÇON, H. 1975. Ultrastructure of hymenial cells in two boletes. Beih. Nova Hedwigia 51:93–98.

CORNER, E. J. H. 1966. A monograph of Cantharelloid fungi. Oxford University Press, London (Ann. Botany Mem. 2), 255 pp.

DODGE, C. W., and S. M. ZELLER. 1934. *Hymenogaster* and related genera. Ann. Missouri Botan. Gardens 21:625–708.

DONK, M. A. 1964. A conspectus of the families of Aphyllophorales. Persoonia 3:199–324.

DRING, D. M. 1973. Gasteromycetes. *In:* Ainsworth, G. C., F. K. Sparrow, and A. S. Sussman, Eds. The fungi: an advanced treatise. Academic Press, New York, Vol. 4B, pp. 451–478.

DURÁN, R. 1973. Ustilaginales. *In:* Ainsworth, G. C., F. K. Sparrow, and A. S. Sussman, Eds. The fungi: an advanced treatise. Academic Press, New York, Vol. 4B, pp. 281–300.

HARRISON, K. A. 1973. Aphyllophorales. III. Hydnaceae and Echinodontiaceae. *In:* Ainsworth, G. C., F. K. Sparrow, and A. S. Sussman, Eds. The fungi: an advanced treatise. Academic Press, New York, Vol. 4B, pp. 351–368.

HAWKER, L. E. 1954. British hypogeous fungi. Phil. Trans. Roy. Soc. London Ser. B 237:429–546.

——. 1955. Hypogeous fungi. Biol. Rev. Cambridge Phil. Soc. 30:127–158.

HIRATSUKA, Y. 1973. The nuclear cycle and the terminology of spore states in Uredinales. Mycologia 65:432–443.

KENNEDY, L. L. 1972. Basidiocarp development in *Calocera cornea*. Can. J. Botany 50:413–417.

——, and W. W. WONG. 1978. *Fomitopsis cajandrei:* development in nature and in culture. Can. J. Botany 56:2319–2327.

KÜHNER, R. 1977. Variation of nuclear behaviour in the Homobasidiomycetes. Trans. Brit. Mycol. Soc. 68:1–16.

KWON-CHUNG, K. J. 1976. Morphogenesis of *Filobasidiella neoformans,* the sexual state of *Cryptococcus neoformans.* Mycologia 68:821–833.

LAUNDON, G. F. 1973. Uridinales (sic). *In:* Ainsworth, G. C., F. K. Sparrow, and A. S. Sussman, Eds. The fungi: an advanced treatise. Academic Press, New York, Vol. 4B, pp. 247–279.

LENTZ, P. L. 1954. Modified hyphae of Hymenomycetes. Botan. Rev. 20:135–199.

LODDER, J. 1970. The yeasts, 2nd ed. North-Holland, Amsterdam, 1485 pp.

LU, B. C. 1964. Polyploidy in the basidiomycete *Cyathus stercoreus.* Am. J. Botany 51:343–347.

MCLAUGHLIN, D. J. 1977. Basidiospore initiation and early development in *Coprinus cinereus.* Am. J. Botany 64:1–16.

MCNABB, R. F. R. 1973. Phragmobasidiomycetidae: Tremellales, Auriculariales, Septobasidiales. *In:* Ainsworth, G. C., F. K. Sparrow, and A. S. Sussman, Eds. The fungi: an advanced treatise. Academic Press, New York, Vol. 4B, pp. 303–325.

——, and P. H. B. TALBOT. 1973. Holobasidiomycetidae: Exobasidiales, Brachybasidiales, Dacrymycetales, Tulasnellales. *In:* Ainsworth, G. C., F. K. Sparrow, and A. S. Sussman, Eds. The fungi: an advanced treatise. Academic Press, New York, Vol. 4B, pp. 317–325.

MOORE, E. J. 1965. Fungal gel tissue ontogenesis. Am. J. Botany 52:389–395.

PEGLER, D. N. 1973. Aphyllophorales. IV. poroid families. *In:* Ainsworth, G. C., F. K. Sparrow, and A. S. Sussman, Eds. The fungi: an advanced treatise. Academic Press, New York, Vol. 4B, pp. 397–420.

PETERSEN, R. H. 1973. Aphyllophorales. II. the clavarioid and cantharelloid basidiomycetes. *In:* Ainsworth, G. C., F. K. Sparrow, and A. S. Sussman, Eds. The fungi: an advanced treatise. Academic Press, New York, Vol. 4B, pp. 351–368.

——. 1974. The rust fungus life cycle. Botan. Rev. 40:453–513.

REINJNDERS, A. F. M. 1977. The histogenesis of bulb and trama tissue of the higher basidiomycetes and its phylogenetic implications. Persoonia 9:329–362.

ROLFE, R. T., and F. W. ROLFE. 1925. The romance of the fungus world. Dover, New York, 301 pp. [Reprint (1974) of 1925 edition published by Chapman & Hall.)]

SMITH, A. H. 1966. The hyphal structure of the basidiocarp. *In:* Ainsworth, G. C., and A. S. Sussman, Eds. The fungi: an advanced treatise. Academic Press, New York, Vol. 2, pp. 151–177.

——. 1973. Agaricales and related secotioid Gasteromycetes. *In:* Ainsworth, G. C., F. K. Sparrow, and A. S. Sussman, Eds. The fungi: an advanced treatise. Academic Press, New York, Vol. 4B, pp. 421–450.

TALBOT, P. H. B. 1973-a. Towards uniformity in basidial terminology. Trans. Brit. Mycol. Soc. 61:497–512.

——. 1973-b. Aphyllophorales. I. General characteristics: thelephoroid and cupuloid families. *In:* Ainsworth, G. C., F. K. Sparrow, and A. S. Sussman, Eds. The fungi: an advanced treatise. Academic Press, New York, Vol. 4B, pp. 327–349.

WATLING, R. 1971. Chemical tests in agaricology. *In:* Booth, C., Ed. Methods in microbiology. Academic Press, New York, Vol. 4, pp. 567–597.

WELLS, K. 1965. Ultrastructural features of developing and mature basidia and basidiospores of *Schizophyllum commune.* Mycologia 57:236–261.

Additional Recommended References

ATKINSON, G. F. 1914-a. The development of *Agaricus arvensis* and *A. comtulus*. Am. J. Botany 1:3–22 (Agaricales).

——. 1914-b. The development of *Lepiota clypeolaria*. Ann. Mycol. 12:346–360 (Agaricales).

BLIZZARD, A. W. 1917. The development of some species of agarics. Am. J. Botany 4:221–240 (Agaricales).

BRODIE, H. J. 1936. The occurrence and function of oidia in the Hymenomycetes. Am. J. Botany 23:309–327.

——. 1975. The bird's nest fungi. University of Toronto Press, Toronto, 199 pp. (Nidulariales).

COKER, W. C., and A. H. BEERS. 1943. The Boletaceae of North Carolina. University of North Carolina Press, Chapel Hill, N.C., 96 pp. (Agaricales).

——, and J. N. COUCH. 1928. The Gasteromycetes of the eastern United States and Canada. University of North Carolina Press, Chapel Hill, N.C., 195 pp.

CORNER, E. J. H. 1970. Supplement to ''A monograph of *Clavaria* and related genera.'' Beih. Nova Hedwigia 33:1–299 (Aphyllophorales).

CRAIGIE, J. H., and G. J. GREEN. 1962. Nuclear behavior leading to conjugate association in haploid infections of *Puccinia graminis*. Can. J. Botany 40:163–178 (Uredinales).

DOUGLAS, G. E. 1916. A study of development in the genus *Cortinarius*. Am. J. Botany 5:36–54 (Agaricales).

——. 1918. The development of some exogenous species of agarics. Am. J. Botany 5:36–54 (Agaricales).

ELROD, R. P., and W. H. SNELL. 1940. Development of the carpophores of certain Boletaceae. Mycologia 32:493–504 (Agaricales).

FISCHER, G. W. 1965. The romance of the smut fungi. Mycologia 57:331–342 (Ustilaginales).

——, and C. S. HOLTON. 1957. Biology and control of the smut fungi. Ronald Press, New York, 622 pp. (Ustilaginales).

FITZPATRICK, H. M. 1913. A comparative study of the development of the fruit body in *Phallogaster, Hysterangium,* and *Gautieria*. Ann. Mycol. 11:119–149 (Gautieriales, Phallales).

FLEGLER, S. L., and G. R. HOOPER. 1978. Ultrastructure of *Cyathus stercoreus*. Mycologia 70:1181–1190 (Nidulariales).

HIRATSUKA, Y., and G. B. CUMMINS. 1963. Morphology of the spermagonia of the rust fungi. Mycologia 55:487–507 (Uredinales).

KENNEDY, L. L. 1972. Basidiocarp development in *Polyporus adustus*. Mycologia 63:69–78 (Aphyllophorales).

LOWE, J. L., and R. L. GILBERTSON. 1961. Synopsis of the Polyporaceae of the western United States and Canada. Mycologia 53:474–511 (Aphyllophorales).

LOWY, B. 1971. Tremellales. Flora Neotropica, Monogr. 6, pp. 1–153.

MARTIN, G. W. 1952. Revision of the north central Tremellales. Univ. Iowa Studies Nat. Hist. 19(3):1–122.

MILLER, O. K., JR. 1972. Mushrooms of North America. E.P. Dutton, New York, 360 pp. (Agaricales).

MOSER, M., translated by S. Plant. 1983. Keys to agarics and boleti. Roger Phillips, London, 535 pp. (Agaricales).

OVERHOLTS, L. O. 1953. The Polyporaceae of the United States, Alaska, and Canada. University of Michigan Press, Ann Arbor, Mich., 466 pp. (Aphyllophorales).

PHILLIPS, R. 1981. Mushrooms and other fungi of Great Britain and Europe. Pan Books, London, 288 pp.

SHAFFER, R. L. 1975. The major groups of basidiomycetes. Mycologia 67:1–18.

SINGER, R. 1962. The Agaricales in modern taxonomy. J. Cramer, Springäckern, West Germany, 915 pp.

——, and E. E. Both. 1977. A new species of *Gastroboletus* and its phylogenetic significance. Mycologia 69:59–72.

Smith, A. H., and H. D. Thiers. 1971. The boletes of Michigan. University of Michigan Press, Ann Arbor, Mich., 418 pp. (Agaricales).

Teixeira, A. R. 1962. The taxonomy of the Polyporaceae. Biol. Rev. Cambridge Phil. Soc. 37:51–81 (Aphyllophorales).

Townsend, B. B. 1954. Morphology and development of fungal rhizomorphs. Trans. Brit. Mycol. Soc. 37:222–233.

Walker, L. B. 1920. Development of *Cyathus fasicularia, C. striatus,* and *Crucicubulum vulgare.* Botan. Gaz. 70:1–24 (Nidulariales).

——. 1940. Development of *Gasterella lutophila.* Mycologia 32:31–42 (Hymenogastrales).

Watling, R., and A. E. Watling. 1980. A literature guide for identifying mushrooms. Mad River Press, Eureka, Calif., 121 pp.

Yoon, K. S., and D. J. McLaughlin. 1979. Formation of the hilar appendix in basidiospores of *Boletus rubinellus.* Am. J. Botany 66:870–873 (Agaricales).

6

Division Amastigomycota—
Form—Subdivision
Deuteromycotina

Members of the Form-Subdivision Deuteromycotina are often known as "deuteromycetes," and also were formerly called the "imperfect fungi," or Fungi Imperfecti, terms that you may see in other literature. This is a group of about 15,000 species. They are combined in a single group because they apparently lack a sexual stage and therefore have no convenient place in the remainder of the fungus classification system, which is based principally on the mode of sexual reproduction. It is possible either that many species never had sexual reproduction in their life cycles or that once-existent sexual reproduction was deleted. It is generally believed that the majority of the deuteromycetes are the nonsexual stages, or anamorphs, of sexually reproducing fungi that belong to the Ascomycotina or Basidiomycotina, with the largest number occurring in the Ascomycotina. Frequently the sexual stage is discovered later, and then both stages of the fungus, telomorph and anamorph, are taxonomically transferred to the group to which the telomorph belongs. This transfer is consistent with the concept that the anamorph and telomorph together constitute the complete life cycle, the holomorph. As an example, *Curvularia intermedia* was classified in the Form-Subdivision Deuteromycotina and placed in the genus *Curvularia* because it possesses the enlarged central cell of the conidium that is typical for this genus [Fig. 6–1(c)]. Its telomorph was later discovered and found to be an ascomycete in the genus *Cochliobolus*. This fungus has since then been considered to belong to the Subdivision Ascomycotina in the genus *Cochliobolus* and is identified as *Cochliobolus intermedius* with the anamorph or conidial stage *Curvularia intermedia*. It is formally designated as *Cochliobolus intermedius* (= *Curvularia intermedia*) (Nelson, 1960). It is evident from this discussion that the Form-Subdivision Deuteromycotina represents a "wastebasket" assemblage of fungi classified together for the sake of convenience and, as a group, lacks any common phylogenetic origin or relationship. This taxon is designated as a form-subdivision rather than as a subdivision because of its artificial composition.

231

Figure 6–1 Some common members of the deuteromycetes; (a) *Alternaria;* (b) *Cladosporium;* (c) *Curvularia;* (d) *Fusarium;* (e) *Gonatobotrys;* (f) *Helicosporium;* (g) *Nigrospora;* (h) *Penicillium;* (i) *Periconia;* (j) *Septonema;* (k) *Thielaviopsis;* (l) *Trichoderma;* (m) *Verticillium.* [From drawings by H. H. Swart.]

Although members of the deuteromycetes as a group are unrelated, it may be assumed that there are some natural phylogenetic groups within the overall assemblage and also that speciation may occur even in the absence of sexual reproduction. The required genetic variation for speciation occurs through mutation and the parasexual cycle, in which genetic recombination occurs without sexual reproduction (Chapter 7).

Members of the deuteromycetes invade virtually every possible habitat and type of substratum. Large numbers are saprophytic inhabitants of soil, and many are plant and animal parasites. Some occur in swiftly running streams, forming tetraradiate (four-armed) conidia that may be collected in foam formed on a turbulent stream (Fig. 11–18). Some commonly known members of the deuteromycetes include the "molds," which usually may be easily cultured on artificial media and yield a prolific growth of mycelium. The molds have wide application as research tools for genetic (Chapter 7), nutritional, and biochemical studies (Chapters 8 and 9). Some species produce metabolic by-products that are commercially useful and have, in essence, supplied the foundation for industrial mycology. For example, these fungi are used to produce cheese, antibiotics, and various organic acids (Chapter 16). Some members of the deuteromycetes that are important elsewhere in this book are illustrated in Fig. 6–1.

SPORE TYPES AND MORPHOLOGY

Some members of the deuteromycetes are single-celled, yeastlike organisms that reproduce by budding. The vast majority are mycelial. The mycelium may be of the ascomycete type, in which septa with simple pores are formed, or of the basidiomycete type, in which dolipore septa and clamp connections are formed. The structure of the mycelium is useful in determining the natural affinities of these fungi.

Reproduction may occur by cellular division or by germination of mycelial fragments or spores of various types (Fig. 6–2). The spores may be *chlamydospores* or conidia. Chlamydospores are formed by transformation of an intercalary vegetative cell or group of cells by rounding up of the cells and deposition of a thick wall. Chlamydospores are not shed but are released when the parent

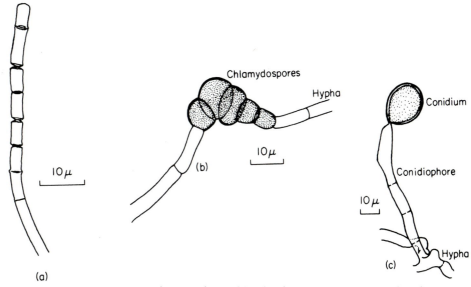

Figure 6–2 Types of spores formed by the deuteromycetes: (a) oidia of *Geotrichum*; (b) chlamydospores of *Peyronellaea*; (c) conidium of *Acremoniella*. [From drawings by H. H. Swart.]

Chlamydospores
vs
Conidia

hypha disintegrates. In contrast, conidia are produced from a specialized cell that is not part of the vegetative mycelium; when mature, conidia are deciduous and separate readily from the parent cell. The parent cell that produces the conidium is usually a specialized, fertile cell, the *conidiogenous cell,* but may sometimes be another conidium. One to several conidiogenous cells may be produced from and supported by a *conidiophore,* a specialized hypha.

Most deuteromycetes rely upon conidia for reproduction. Conidiophores, conidiogenous cells, and conidia are also produced by many ascomycetes to accomplish nonsexual reproduction. Morphological and developmental details of these structures among the ascomycetes are similar to those features described below for the deuteromycetes.

The Conidiophores and Conidiogenous Cells

Conidiophores are usually morphologically distinct from the vegetative hyphae. They may be light or dark, unbranched or branched in a complex manner, and of uniform thickness throughout or with enlarged parts, most often the apices. Conidiophores bear conidiogenous cells that demonstrate various arrangements (e.g., Fig. 6–3). In turn, the conidiogenous cells differ in the manner of conidial

Figure 6–3 Development of the conidial apparatus in *Aspergillus niger:* (a) foot cell originating from hypha and bearing young conidiophore as a vertical branch; (b) developing conidiophore; (c) swelling of the terminal portion of the conidiophore; (d) development of metulae from the conidiophore apex; (e) young sporulating apex showing phialides bearing chains of conidia. (a, b) × 172, (c-e) × 265. [From C. Thom and K. B. Raper, 1945, A manual of the Aspergilli, The Williams & Wilkins Company, Baltimore.]

production, including formation of the conidia singly, in clusters, or in succession. If the conidia are produced in succession, they may occur in chains, either with the oldest conidium located at the apex of the chain and pushed upward by younger ones formed at the base, or conversely with the youngest conidium at the apex.

Some important morphological differences may be distinguished among conidiogenous cells that are related to the way that they produce their conidia (Kendrick and Carmichael, 1973). In some species, the conidiogenous cells may resemble vegetative hyphae and are not specialized for the production of large numbers of conidia. In others, conidia are produced in succession from the tip of determinant, nonelongating conidiogenous cells known as *phialides* (Fig. 6–4). Phialides are usually flask or bottle-shaped fertile cells that terminate in a narrow neck. The neck is sometimes surrounded by a flaring collarette, which is the ruptured outer wall that remains at the phialide apex after differentiation of the first conidium. The *annellidic* conidiogenous cell, unlike that of the phialides, produces a succession of conidia, but the tip of the conidiogenous cell continues to elongate. Growth of the conidiogenous cell tip occurs through a ringlike tear (annellation) remaining after a conidium is shed. A series of these rings is left behind on the tip of the conidiogenous cell (Fig. 6–10). Other variations of conidiogenous cell development also occur in the deuteromycetes.

Figure 6–4 Phialides and their conidia in *Memnoniella echinata*. Left: scanning electron micrograph showing a surface view of a cluster of phialides and their chains of conidia; note that the basal conidia are still incompletely developed, and have not yet formed their rough markings; right: transmission electron micrograph showing a section through the phialide tip and two conidia. [Courtesy R. Campbell, Bristol University; from 1975, Mycologia, 67:760–769].

Spore Types and Morphology

The Conidia: Ontogeny

The actual mode by which a conidiogenous cell forms its conidia is of great importance. Various systems based on light microscopical observations have been proposed to describe developmental types (Hughes, 1953; Subramanian, 1962; Tubaki, 1966). Precise observations of the conidiogenous cells and conidial development are sometimes difficult with light microscopy because of the extremely small size of some conidia and conidiogenous cells and the consequent difficulty in distinguishing critical features such as the nature of wall differentiation and conidial initiation. Conversely, observations with the electron microscope have assisted greatly in resolving some of these finer details of development. Recently, a descriptive system has been proposed that incorporates evidence from ultrastructural studies. According to this system, two major types of development may be distinguished (Cole and Samson, 1979; Kendrick, 1971). These are the *thallic* and *blastic* types.

Thallic development occurs in the formation of the chlamydospores described in the previous section as well as in some conidia (Fig. 6–5). The spore initials are delimited from the remainder of the fertile hypha or conidiogenous cell by septa. The septa are formed only after apical growth of the fertile hypha

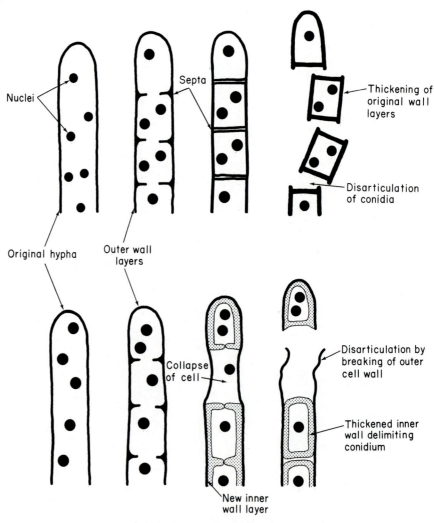

Figure 6–5 Top: holothallic conidium development; bottom: enterothallic conidium development. Successive stages are shown. In each type, a random number of nuclei are included in each conidium.

has ceased. If any enlargement of the spore initial occurs, it takes place only *after* the septa are formed. Spores are released when the parent hypha decays (as in chlamydospores) or when the parent hypha undergoes disarticulation. Conidia that undergo this type of development are often given special names such as *oidia* or *arthrospores*.

In blastic development, conidia are not derived from a preexisting vegetative cell but rather result from the formation of new cells arising from the conidiogenous cell or conidium (Fig. 6–6). Unlike the thallic type, the conidial initial may enlarge considerably before being cut off by a septum. Generally, in this type of development the conidial initial appears to be blown out like a balloon from the parent cell. The portion that is expanding may be small and budlike at first or may originate as a relatively broad extension of the parent cell, differing little from it in width. The conidial initial is surrounded at all stages by a cell wall and continues to increase in size. Both cytoplasm and a nucleus migrate from the parent cell into the conidial initial as it enlarges. The conidial initial undergoes substantial enlargement before it is finally separated from the parent cell by the centripetal growth of a septum. Further changes may occur in the septum, such as the thickening or addition of more layers. Sometimes the septum develops in such a way that a special abscission layer forms where separation later occurs, resulting in conidial release.

In both thallic and blastic development, an important subdivision can be made based on the mode of wall differentiation during conidium formation. All

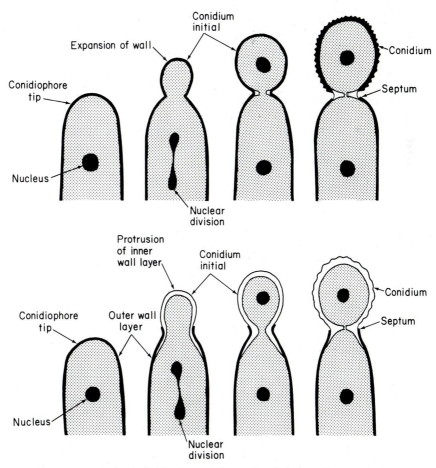

Figure 6–6 Top: holoblastic conidium development; bottom: enteroblastic conidium development. Successive stages are shown in each series.

wall layers of the parent hypha may extend into or be incorporated within all wall layers of the developing conidium, which makes formation of the conidium strictly exogenous, i.e., of the *holo*-type (we designate exogenous thallic development as *holothallic* and exogenous blastic development as *holoblastic*). The vast majority of conidia are apparently either holothallic or holoblastic (Figs. 6–5 and 6–6). Alternatively, some conidia develop endogenously; the outermost wall layer of the parent cell is not included in the conidial cell wall. The new wall of the endogenous conidium may be a modification or continuation of the innermost wall layers of the parent structure or it may be newly synthesized inside the wall of the parent cell. Such development is the *entero*-type, and we may designate the *enterothallic* or *enteroblastic* subdivisions, respectively, for those fungi that form their conidia within the parent structure.

Enterothallic conidia are relatively uncommon, while enteroblastic conidia are much more widely represented. This latter mode of development is best exemplified in the phialides (Figs. 6–4 and 6–7). Phialides may form the first conidium by either the holoblastic (Hanlin, 1976) or enteroblastic (Hammill, 1974) modes. If this first conidium was formed by the enteroblastic method, it ruptures the phialide wall as it emerges. The ruptured wall may then flare outward as a minute collarette. All subsequent conidium development is enteroblastic, each conidium originating when the innermost layer(s) of the phialide wall are blown out through the mouth of the phialide. These wall layers eventually become the outermost layers of the conidium as new wall material is laid down inside the cell. A septum is formed within the phialide itself, separating the exogenously maturing conidium from the endogenous conidium initial. The delimited, maturing conidium is extruded from the phialide by pressure exerted by development of the next conidium (e.g., Hammill, 1972, 1974; Hanlin, 1976; Trinci et al., 1968). Fungi that form ringlike annellations on the conidiogenous cells also produce conidia by enteroblastic development (Hammill, 1971, 1974, 1977).

The Conidia: Morphology

Mature conidia may be smooth or bear ornate markings such as warts. Conidia are often round or ellipsoid but may also be sickle-shaped, curved, coiled, or of

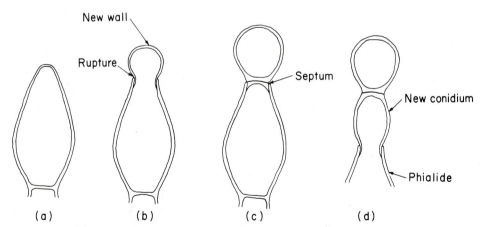

Figure 6–7 Development of conidia from a phialide: (a) slightly elongated phialide showing a single wall; (b) the phialide apex has ruptured, and a new wall is protruding through it; this wall will become the apical wall of the newly developing conidium; (c) the new conidium has been formed, has been delimited by a septum, and has undergone enlargement; (d) the process is being repeated to form a second conidium beneath the older one. [Adapted from G. T. Cole and W. B. Kendrick, 1969. Can. J. Botany 47:779–789.]

other elaborate shapes. Conidia usually consist of a single cell, but they may have two to several cells. They may be either light or dark, and when seen in a mass may occur in a variety of colors such as white, blue, green, yellow, brown, or black. Some of this variation may be noted in Fig. 6–1.

CLASSIFICATION

Because the Form-Subdivision Deuteromycotina is an artificial assemblage of fungi, any further division into taxa is not intended to group closely related organisms together. Each taxon within this form-subdivision is designated as a *form* taxon to indicate that it is based on empirically selected criteria. Classification of the deuteromycetes into form-orders, form-families, and form-genera has been largely looked upon as a convenient way to group these fungi for identification and not as a means to organize related fungi together in such a way that phylogenetic trends may be recognized. These taxa are based primarily on convenient, easily distinguishable morphological characteristics. The criteria that are typically used are the color, shape, and septation of the conidia and whether or not the conidiogenous cells and conidiophores are borne loose or in some organized manner, perhaps in a sporocarp.

The effect of this system has been to lump unrelated fungi together in the same genus and to distribute fungi that are probably related in different genera. For example, many fungi have been placed in the genus *Helminthosporium* because they form dark, ellipsoid conidia with three or more cells on a dark mycelium. All members of *Helminthosporium* are not closely related, however, and when sexual stages have been discovered, they have belonged to the ascomycete genera *Trichometasphaeria* and *Cochliobolus*. To increase this problem, the conidial genus *Curvularia* has also been found to have a telomorph in the ascomycete genus *Cochliobolus*. The assumption may be made that the telomorph may be used as an index of natural relationships. Therefore, this evidence points to the conclusions: (1) that at least two natural but phylogenetically distant or unrelated lines are included in *Helminthosporium,* and (2) that a close phylogenetic relationship exists between some members of *Helminthosporium* and *Curvularia* (Nelson, 1964).

At present, a satisfactory classification system for the deuteromycetes does not exist. However, efforts are being made to devise a classification system for them that will reflect the natural relationships in the group. To do this, it is necessary to find criteria that will indicate such relationships. Attention is being given to the mode of development of conidiogenous cells and conidia, the sequence of conidial maturation in multicellular forms, the type of scars left on the conidiogenous cell after loss of the conidium, the mode of germination of the conidium, and anamorph-telomorph interrelationships. The following artificial taxa are currently utilized:

Form-Subdivision Deuteromycotina
  Form-Class Blastomycetes
  Form-Class Hyphomycetes
  Form-Class Coelomycetes
  Form-Class Mycelia Sterilia

Form-Class Blastomycetes

Members of this form-class include yeastlike fungi. These may be unicellular or form a limited mycelium and reproduce by budding (essentially similar to holoblastic spore formation). Many seem to be similar to the yeasts in the Ascomycotina but fail to produce ascospores. Others appear to have affinities with the Basidiomycotina, as they produce a limited mycelium with clamp connections or produce sterigmata and spores which may be subsequently ejected in a manner similar to that of the basidiospores. These yeasts may be isolated from a variety of habitats, including water, the soil, or plants. Many are pathogenic on animals or humans. Pathogenic nonsexual yeasts include some species of *Cryptococcus* and *Candida,* which we discuss further in Chapter 13.

Sometimes a sexual stage may be discovered for a yeastlike organism that was previously thought to be nonsexual only. These sexual stages are usually found to belong to the Hemiascomycetes or to the basidiomycete orders Tremellales or Ustilaginales (Kreger-Van Rij, 1987).

Form-Class Hyphomycetes

The Hyphomycetes include those deuteromycetes that form a mycelium but lack a sporocarp. These fungi are often commonly called ''molds'' or ''mildews.'' The conidiophores are often scattered singly and at random on the mycelium. A culture of a Hyphomycete can easily be recognized by the powdery or fluffy appearance of the colony. The colony may be uniformly colored owing to the mass of colored conidia overlying the mycelium. The mycelium may be light or dark, and the color of the young colony or nonsporulating edges of the colony is the color of the mycelium and not that of the conidia. Some Hyphomycetes may release pigments into the culture medium, making the reverse side of a culture a different color from the mycelium or spores.

In some species, the conidiophores may be tightly clustered together to form a pulvinate mass, the *sporodochium,* which may be formed upon a stroma. Another type of arrangement is a *synnema,* a cluster of conidiophores that form a fascicle and may adhere to each other for part of their length, forming an elongate, bristlelike structure. Synnemata may be fleshy, brittle, or very hard in consistency. The unattached apical portions of the conidiophores radiate outward, and frequently they produce a mucoid slime in which the conidia are borne (Fig. 6–8).

The Hyphomycetes may be divided in form-families on the basis of whether or not sporodochia or synnemata are formed, as well as on the basis of the morphology of the conidiophores, conidiogenous cells, and conidia. Characteristics of the conidia which are considered are whether or not they are light or dark, their shape and size, and their possible division into more than one cell by septa.

Although the Hyphomycetes are morphologically simple, they are both ecologically and industrially important. They include numerous fungi that are active in soil (Chapter 12) or are plant or human pathogens (Chapter 13), as well as some of the predacious fungi that trap nematodes (Chapter 13). Many industrially important products are derived from genera such as *Penicillium* and *Aspergillus* (Chapter 16).

Form-Class Coelomycetes

In this form-class, the conidiophores are borne on or within fungal structures which resemble ascocarps. The following two form-orders are based on the morphology of these multicellular structures (Sutton, 1973).

FORM-ORDER SPHAEROPSIDALES

The members of the Form-Order Sphaeropsidales are characterized by the production of *pycnidia* (Fig. 6–9). A pycnidium superficially resembles a perithecium as it is a closed ascocarp bearing its conidiophores, conidiogenous cells, and conidia within a cavity. The pycnidia may be discoid, globose, or flasklike.

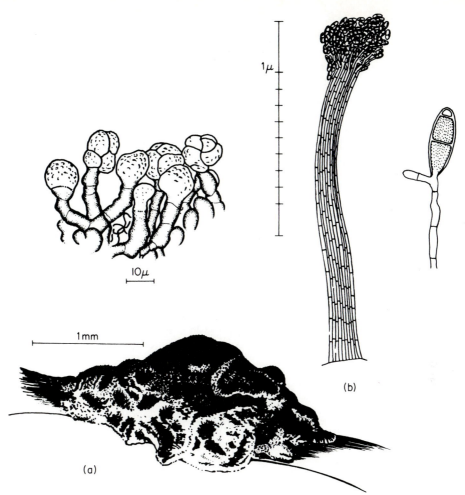

Figure 6–8 (a) Sporodochium formed by *Epicoccum andropogonis* on its host with details of the conidia; (b) synnema of *Calostilbella* and a detail of its conidiogenous cell and conidium. [(a) From drawings by H. H. Swart; (b) from E. F. Morris, 1963, Western Illinois Univ. Ser. Biol. Sci. 3:1–143.]

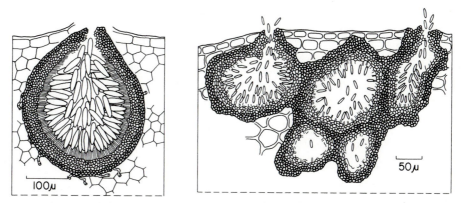

Figure 6–9 Longitudinal sections through pycnidia of two species of *Macrophoma*. [From G. Morgan-Jones, 1971, Can. J. Botany 49:1921–1929.]

Some are originally globose but flatten to become cupulate. Pycnidia may be entirely closed or may open to the outside by an ostiole, slit, or tear. Walls vary from dark and tough, leathery, or carbonaceous to brightly colored and fleshy types. Pycnidia may be immersed within stromata or host tissues or may form on the surface of the substratum. Stromata may enclose one or more pycnidia and may resemble stromata of the Pyrenomycetes or Loculoascomycetes in their variations.

Members of this form-order often occur as saprophytes on decaying plant debris or as parasites on living plants. These fungi are often responsible for a variety of leaf spot diseases in plants; for example, species of *Septoria* may invade celery, azaleas, gladiolus, and wheat.

FORM-ORDER MELANCONIALES

A second type of conidial sporocarp is the *acervulus,* which characterizes the members of the Form-Order Melanconiales (Fig. 6–10). An acervulus is an open mass of closely packed conidiophores and conidiogenous cells which may form a flat discoid cushion of conidia or may become cupulate or slightly pulvinate. Acervuli are usually formed on a plant host and are often erumpent from the epidermis of the host, pushing aside flaps of host tissue as they emerge. Superficially, an acervulus resembles an apothecium. Long, dark bristles may occur among the conidiogenous cells in the acervulus.

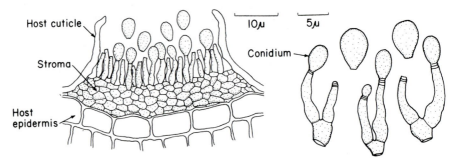

Figure 6–10 Left: longitudinal section through an acervulus of *Cryptocline betularum;* right: details of the conidia and conidiogenous cells. The conidiogenous cells are of the annellidic type and have ringlike markings (annellations). [From G. Morgan-Jones, 1971, Can. J. Botany 49:1921–1929.]

These fungi are primarily plant pathogens, and include many fungi that cause important diseases. An example is *Colletotrichum,* which causes the anthracnose diseases of citrus, banana, cucumber, and other crops.

Form-Class Mycelia Sterilia

Some members of the deuteromycetes lack spores entirely and reproduce by undergoing mycelial fragmentation. Examples include the common genus *Rhizoctonia,* which may enter into a symbiotic relationship with orchids (Chapter 15) or may cause plant diseases or parasitize other fungi (Chapters 13 and 14).

REFERENCES CITED

COLE, G. T., and R. A. SAMSON. 1979. Patterns of development in conidial fungi. Pitman, Belmont, Calif., 190 pp.

HAMMILL, T. M. 1971. Fine structure of annellophores. I. *Scopulariopsis brevicaulis* and *S. koningii.* Am. J. Botany 57:88–97.

——. 1972. Electron microscopy of conidiogenesis in *Chloridium chlamydosporis*. Mycologia 64:1054–1065.

——. 1974. Electron microscopy of phialides and conidiogenesis in *Trichoderma saturnisporum*. Am. J. Botany 61:15–24.

——. 1977. Transmission electron microscopy of annellides and condiogenesis in the synnematal hyphomycete *Trichurus spiralis*. Can. J. Botany 55:233–244.

HANLIN, R. T. 1976. Phialide and condium development in *Aspergillus clavatus*. Am. J. Botany 63:144–155.

HUGHES, S. J. 1953. Conidiophores, conidia, and classification. Can. J. Botany 31:577–659.

KENDRICK, B. 1971. Conclusions and recommendations. *In:* Kendrick, B., Ed. Taxonomy of Fungi Imperfecti. University of Toronto Press, Toronto, pp. 253–262.

——, and J. W. CARMICHAEL. 1973. Hyphomycetes. *In:* Ainsworth, G. C., and A. S. Sussman, Eds. The fungi: an advanced treatise. Academic Press, New York, Vol. 4A, pp. 323–509.

KREGER-VAN RIJ, N. J. W. 1987. Classification of yeasts. *In:* Rose, A. H., and J. S. Harrison, Eds. The yeasts. Academic Press, New York, Vol. 1, pp. 5–61.

NELSON, R. R. 1960. The perfect stage of *Helminthosporium spiciforum*. Mycologia 56:196–201.

SUBRAMANIAN, C. V. 1962. A classification of the Hyphomycetes. Current Sci. (India) 31:409–411.

SUTTON, B. C. 1973. Coelomycetes. *In:* Ainsworth, G. C., F. K. Sparrow, and A. S. Sussman, Eds. The fungi: an advanced treatise. Academic Press, New York, Vol. 4A, pp. 513–582.

TRINCI, A. P. J., A. PEAT, and G. H. BANBURY. 1968. Fine structure of phialide and conidiophore development in *Aspergillus giganteus* "Wehmer." Ann. Botany (London) 32:241–249.

TUBAKI, K. 1966. Sporulatating structures in Fungi Imperfecti. *In:* Ainsworth, G. C., and A. S. Sussman, Eds. The fungi: an advanced treatise. Academic Press, New York, Vol. 2, pp. 113–131.

Additional Recommended References

ABE, S. 1957. The *Penicillia:* atlas of microorganisms. Kanehara Shuppan, Tokyo, 319 pp.

BARNETT, H. L. 1955–1960. Illustrated genera of imperfect fungi, 2nd ed. Burgess, Minneapolis, Minn., 225 pp.

COLE, G. T. 1975. The thallic mode of conidiogenesis in the Fungi Imperfecti. Can. J. Botany 53:2983–3001.

GILMAN, J. C. 1957. A manual of soil fungi, 2nd ed. Iowa State University Press, Ames, Iowa, 450 pp.

HUGHES, S. J. 1971. Annellophores. *In:* Kendrick, B., Ed. Taxonomy of Fungi Imperfecti. University of Toronto Press, Toronto, pp. 132–140.

INGOLD, C. T. 1981. The validity of the concept of conidia as either blastic or thallic. Trans. Brit. Mycol. Soc. 77:194–196.

MINTER, D. W., P. M. KIRK, and B. C. SUTTON. 1982. Holoblastic phialides. Trans. Brit. Mycol. Soc. 79:75–93.

——, ——, and ——. 1983. Thallic phialides. Trans. Brit. Mycol. Soc. 80:39–66.

——, B. C. SUTTON, and B. L. BRADY. 1983. What are phialides anyway? Trans. Brit. Mycol. Soc. 81:109–120.

MORRIS, E. F. 1963. The synnematous genera of the Fungi Imperfecti. Western Illinois Univ. Ser. Biol. Sci. 3:1–143.

PIROZYNSKI, K. A. 1971. Characters of conidiophores as taxonomic criteria. *In:* Kendrick, B., Ed. Taxonomy of Fungi Imperfecti. University of Toronto Press, Toronto, pp. 37–49.

RAPER, K. B., and D. I. FENNEL. 1965. The genus *Aspergillus*. Williams & Wilkins, Baltimore, 686 pp.

SUBRAMANIAN, C. V. 1983. Hyphomycetes: taxonomy and biology. Academic Press, New York, 502 pp.

SUTTON, B. C. 1980. The Coelomycetes: fungi imperfecti with pycnidia acervuli and stromata. Commonwealth Mycological Institute, Kew, London, 696 pp.

7

Variation, Speciation, and Evolution

A particularly striking feature of the fungi is their possession of novel mechanisms for introducing genetic variation in their life cycles, either occurring during sexual reproduction or independently of it. They have mechanisms permitting genetic flexibility that are virtually unsurpassed in other kinds of organisms, and which in turn, have a number of significant implications. The interrelationship of the fungi with their hosts can be affected, making them formidable parasites with a great deal of actual and potential genetic flexibility that enables them to adapt quickly to changing conditions, including new host genotypes. The mechanisms by which fungi form new species is also regulated by their unique reproductive processes.

The mechanisms for genetic variation and speciation are explored in this chapter. In addition, we examine the phylogeny of the fungi.

MECHANISMS FOR VARIATION

The primary source of genetic variation in the fungi is mutation, but further genetic variation can occur as a result of sexual reproduction, as in other organisms. A major attribute of sexual reproduction is that haploid nuclei undergo karyogamy to form a diploid nucleus. Meiosis occurs, resulting in four haploid nuclei. Genetic variability of the progeny results because (1) genetically different cells are brought together to initiate the sexual cycle, and (2) crossing over and random assortment occur during meiosis. In organisms generally, it has been well documented that sexual reproduction is a major source of genetic variability. New gene combinations arise as a result of sexual reproduction and may contribute to the overall fitness of the individual (or, conversely, separate favorable gene combinations leading to a decrease in fitness). Evolution can occur as a result of selection either for or against these new gene combinations.

The vast majority of plants, animals, and other eukaryotic organisms have a comparatively straightforward sexual cycle. Haploid gametes or gametangia unite, forming a diploid zygote. At a later stage, a diploid cell undergoes meiosis, once again restoring the haploid stage. The somatic stage may be either haploid or diploid. Cells are either haploid or diploid, and the genotype is specified by the nucleus in each cell. The Mendelian genetics of such sexually reproducing organisms can be readily analyzed.

Sexual cycles involving haploid and diploid phases also occur in the fungi, but as you are well aware, are usually separated by an intervening dikaryotic $(N + N)$ phase. Nuclear genes involved directly in the sexual cycle conform to the behavior expected in typical Mendelian genetics. Actually, the ascomycetes have proved to be important experimental tools in determining the mechanism for independent segregation and crossing-over.

Somatic Variability

Fungi are unique, however, because they are rarely divided into discrete cells. Instead, they exist as a mycelial system that is incompletely partitioned. The nuclei that exist in the hyphae send genetic signals to the cytoplasm, but the cytoplasm is continuous throughout the entire coenocytic hypha. Further, fungal mycelia undergo somatic fusion and cytoplasm and nuclei from one mycelium can migrate into another. These aspects of the life cycle make it possible for the somatic phase of the fungus to be genetically flexible, expressing phenotypes that are not necessarily part of the sexual cycle.

In most organisms, genetic variability can result only from the sexual cycle, and nonsexual reproduction simply results in the procreation of new individuals that are genetically identical to the parent. Although nonsexual reproduction can be an advantage in a static environment, it does not provide the genetic flexibility required if environmental conditions change. Members of the Deuteromycotina have a distinct advantage over other nonsexually reproducing organisms because of the genetic flexibility that can occur in the absence of sexual reproduction, made possible by *heterokaryosis* and the *parasexual cycle,* described below. These mechanisms potentially provide a significant source of variation for the Deuteromycotina, but they can also occur in a sexually reproducing fungus, providing genetic flexibility in addition to that provided by the sexual cycle. Heterokaryosis and the parasexual cycle apparently occur commonly in the filamentous ascomycetes and basidiomycetes. Genetic variation can also result from *cytoplasmic inheritance,* which can also occur independently of sexual reproduction.

HETEROKARYOSIS

Heterokaryosis is the existence of different nuclear types in the same mycelium (the *heterokaryon*). Heterokaryons may originate from a hypha which has only a single nuclear type (a *monokaryon*). A monokaryon may give rise to a heterokaryon by (1) mutation of a nucleus within the mycelium, thus making that nucleus dissimilar to the other nuclei; (2) fusion of one hypha with another, followed by migration of nuclei and cytoplasm from one hypha to another; and (3) karyogamy of haploid nuclei to form diploid nuclei within the mycelium (*diploidization*) (Fig. 7–1). Both mutation and diploidization may occur spontaneously, and as we have seen from our discussions of sexual cycles, hyphal fusion occurs frequently in the fungi. These events may occur in any combination; for example, a heterokaryon which originated by hyphal fusion may have normal and mutated nuclei from both parents, and these nuclei may exist in the haploid state or form diploid nuclei by karyogamy between similar or dissimilar nuclei.

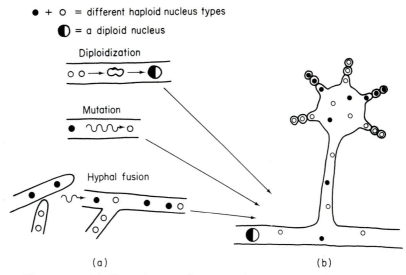

Figure 7–1 (a) Heterokaryon formation by various mechanisms; (b) heterokaryon breakdown occurring at sporulation when the nuclei are individually isolated into spores.

The nuclei in the heterokaryon may divide mitotically for any number of nuclear generations, but these nuclei may finally be separated from each other when uninucleate, nonsexual spores are formed. Upon germination, these spores will give rise to a mycelium with a single nuclear type. Some fungi produce bi- or multinucleate spores that give rise to a heterokaryon if the nuclei in the spore are of different types.

Heterokaryon formation provides for a great deal of genetic variation of the mycelium. Phenotypically, a heterokaryotic mycelium may be quite different from its parental types, as it may contain cytoplasm from different sources as well as genetically different nuclei. In fungi, as in other organisms, the cytoplasm may carry genetic determinants and is sometimes responsible for influencing the phenotype. The phenotype of a heterokaryon is determined by the interaction of all nuclear types with the cytoplasm, and the phenotype is also influenced by the relative numbers of the various nuclei. It is important to realize that a single nuclear type can occur in any frequency, ranging from 0 to 100% of the total nuclei present within a mycelium. Therefore an allele carried by a particular nuclear type may occur in a wide range of dosages (1 to 99%) in a heterokaryon, and if this allele controls a particular phenotypic response, that response may also show a correspondingly flexible range (Fig. 7–2). This variation in relative nuclear numbers allows much more phenotypic diversity than is ordinarily encountered in typical diploid organisms, in which a particular allele is represented in dosages of 100%, 50%, or 0% in a diploid nucleus (consider the relative representation of A in AA, Aa, or aa, for example).

The interaction between nuclei becomes particularly important if the nuclei are biochemically deficient and are unable to direct the synthesis of a needed vitamin, amino acid, or other growth factor. Imagine, for example, that a mycelium which requires exogenous supplies of thiamine but not of the amino acid leucine fuses with a second mycelium which requires exogenous sources of leucine but can manufacture its own thiamine. The resulting heterokaryon, then, has nuclei that can control the biosynthesis of either thiamine or leucine, but not both. *Complementation* takes place, and one nuclear type directs the synthesis of thiamine while the second nucleus controls the synthesis of leucine. Each nuclear type has compensated for the deficiency of the other, and the resulting

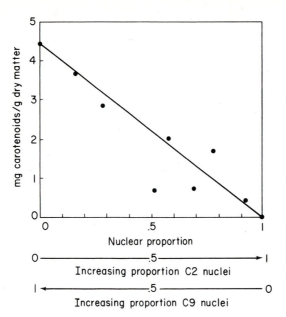

Figure 7–2 Relationship of carotene production to nuclear ratios in heterokaryons of *Phycomyces blakesleeanus*. The white mutant strain C2 is unable to synthesize carotene, while the orange strain C9 produces lycopene, one of the carotenoid pigments. Both mutants have defective enzymes, but when the nuclei are brought together in a heterokaryon, complementation occurs and lycopene may be converted into other carotene pigments (γ-carotene and β-carotene). The wild type has large quantities of β-carotene. [Adapted from M. D. De La Guardia, C. M. G. Aragón, F. J. Murillo, and E. Cerdá-Olmedo, 1971, Proc. Natl. Acad. Sci. USA 68:2012–2015.]

heterokaryon does not need exogenous supplies of either thiamine or leucine. A heterokaryon in which complementation occurs has a wide range of physiological flexibility.

Early investigators of heterokaryosis were particularly impressed by the physiological flexibility provided by complementation. They extrapolated from their laboratory studies to nature and suggested ways in which heterokaryosis could greatly benefit the fungi. One hypothesis was that the fungus may "store" a nuclear type that could not survive alone in the prevailing environment, but with a change of environmental conditions this nuclear type might be necessary for the survival of the heterokaryon. The fungus would be allowed to take advantage of new food sources, for example, by utilizing the metabolic capacities of the stored nucleus. Heterokaryosis would theoretically enhance the chances of survival of a fungus because it would have a large genetic reservoir from which to draw material for adaptations, all accomplished without the delay of passing through sexual cycles. Although laboratory studies generally support these concepts, investigators have had difficulty isolating heterokaryons from nature and they are rarely recovered. Consequently, we have little direct evidence of the actual significance of heterokaryons in nature and the extent to which information derived from laboratory studies can be applied to them (Caten, 1987).

PARASEXUAL CYCLE

The parasexual cycle may take place within a heterokaryon. The parasexual cycle is a sequence involving (1) heterokaryon formation, (2) diploidization of nuclei, and (3) restoration of diploid nuclei to their haploid state (*haploidization*) (Fig. 7–3). Haploidization involves a series of atypical and irregularly occurring mitotic divisions of the diploid nuclei. Frequently, daughter nuclei resulting from mitosis have unequal numbers of chromosomes because sister chromatids failed to separate (nondisjunction) during anaphase. This yields a daughter nucleus with one chromosome too many ($2N + 1$), while the second daughter nucleus is lacking a chromosome ($2N - 1$). Both of these nuclei are now *aneuploids*, as they do not have even chromosome sets such as N or $2N$. The deficient aneuploid nucleus ($2N - 1$) may undergo additional loss of chromosomes in successive mitotic divisions of the same nature until it is reduced to the haploid

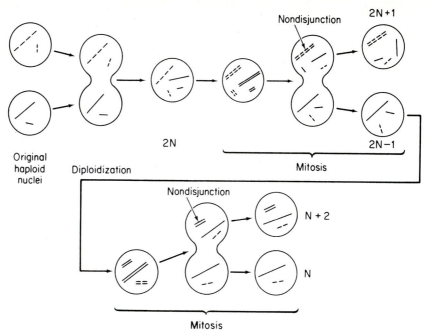

Figure 7–3 Haploidization, beginning with genetically different haploid nuclei in the heterokaryon, and terminating with the formation of a haploid nucleus that is genetically different from the original nuclei.

condition. Pontecorvo (1958) estimates that only 1 out of every 1000 haploid nuclei in a heterokaryon of *Aspergillus niger* has undergone haploidization.

Mitotic crossing-over may accompany the parasexual cycle (Fig. 7–4). In mitotic crossing-over, segments of chromosomes are exchanged for exactly corresponding segments between homologous chromosomes, presumably occurring at the time when the chromosomes replicate. If dominant and recessive genes have been regrouped, subsequent haploidization may produce daughter nuclei which are genetically different from those that would be formed if this exchange had not taken place. For every 10 haploid nuclei derived through the parasexual cycle, perhaps only 1 nucleus is a recombinant derived from mitotic crossing-over (Pontecorvo, 1958).

The parasexual cycle mimics sexual reproduction. Unlike hyphae may fuse

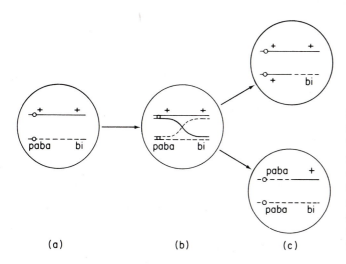

Figure 7–4 Mitotic crossing-over between a single pair of homologous chromosomes. This may occur at some point during haploidization. (a) Diploid nucleus which is heterozygous for the genes *paba* (paminobenzoic acid-requiring) and *bi* (biotin-requiring); (b) crossing-over at chromosome replication (four-strand stage); (c) genetically different nuclei formed after chromosome segregation.

(a) (b) (c)

(plasmogamy) and their (gametic) nuclei occupy the same mycelium. These nuclei may fuse (karyogamy) to form a diploid (zygotic) nucleus. Genetic recombination may occur, and the diploid nucleus is reduced (but without meiosis) to the haploid condition. The recombinant, haploid nuclei are segregated into so-called nonsexual spores, which differ genetically from the parent mycelium.

Although the parasexual cycle is similar to sexual reproduction, it is less efficient in producing recombinant nuclei. Pontecorvo (1956) estimates that the initial frequency of mitotic crossing-over in nuclei is 500 to 1000 times lower than that of meiotic crossing-over. Although this initial number of nuclei produced by mitotic recombination increases through subsequent divisions before spore formation takes place, the possibility of any single nucleus being isolated in a single spore is slight. In sexual reproduction, virtually every spore produced as a result of meiosis contains a recombinant nucleus.

The parasexual cycle is a useful tool for the fungal geneticist. As discussed in Chapter 6, many fungi apparently lack sexual reproduction and therefore genetic studies based on sexual cycles cannot be carried out. Mutants or other genetically distinct strains may be induced to cross and recombinants may be derived from the parasexual cycle. By utilizing appropriate methods of genetic analysis, information may be gained about the location and activity of genes in these fungi. For example, some recent studies are focused on the genetic control of aflatoxin production (aflatoxins are discussed further in Chapter 12) (Papa, 1978).

CYTOPLASMIC INHERITANCE

Up to this point we considered characteristics that were controlled by genes borne in nuclei and which can be analyzed in terms of nuclear behavior. When plasmogamy occurs between gametes or somatic hyphae, there is an intermingling of cytoplasm as well as of nuclei. The cytoplasm can also carry genetic determinants that can contribute to the phenotype of the progeny or of the heterokaryon. This phenomenon is called *cytoplasmic inheritance* to distinguish it from inheritance controlled by nuclear genes.

A characteristic controlled by cytoplasmic inheritance fails to follow the Mendelian pattern of inheritance in sexually reproducing organisms. In nonsexually reproducing fungi that have formed a heterokaryon, nuclear genes can be isolated in spores that give rise to monokaryons, but traits controlled by cytoplasmic inheritance will not segregate with nuclei but instead, behave as though they are borne by particles in the cytoplasm. Traits transmitted by cytoplasmic inheritance are normally inherited from the female parent, which contributes the bulk of the cytoplasm, but not from the male parent.

Cytoplasmic inheritance in fungi can result from the transfer of viral nucleic acids, as is the case for the killer traits and hypovirulence, discussed further in Chapter 14. Transfer of mitochondria can account for some other cases of maternal inheritance, as described below. Both viruses and mitochondria are readily transferred from one cell to another when protoplasmic fusion occurs.

Mitochondria are cytoplasmic organelles that divide autonomously and also contain a large circular molecule of DNA (sometimes called the *mitochondrial chromosome*) which codes for many mitochondrial functions. Alteration of the mitochondrial DNA can result in genotypic changes in the organism that follow the typical behavior for cytoplasmic inheritance (Grossman and Hudspeth, 1985). Traits resulting from alteration of the mitochondrial chromosome include the "petite" variant of the yeast *Saccharomyces cerevisiae*, the "poky" characteristic of *Neurospora crassa*, and "senescence" of *Podospora anserina*. All of these characteristics are due to defective mitochondrial DNA, which in turn results in respiratory deficiencies, although the specific details vary. We will consider poky and senescence further.

Poky is a naturally occurring mutant of *Neurospora* characterized by a slow growth rate. Cytochromes *a* and *b* are greatly reduced in quantity or absent, but there is a greater quantity of cytochrome *c* than in the wild type. Poky results from the deletion of four base pairs in the mitochondrial DNA, in turn causing the production of ribosomal RNA that is missing 38 to 45 nucleotides and resulting in a deficiency of small ribosomal subunits. These defects impair synthesis of proteins in the mitochondrion (Akins and Lambowitz, 1984).

Senescence of *P. anserina* causes vegetatively propagated mycelium to decline in vigor and eventually cease growing, caused by impairments of respiration. As the young mycelium begins to show senescence characteristics, *plasmids* appear in the mitochondria. These plasmids are small circular DNA molecules that replicate autonomously. There are five distinct plasmid types, each having a DNA sequence duplicating a specific but small part of the original mitochondrial chromosome. They contain only 2.6 to about 9.8 kilobase pairs, while the original mitochondrial chromosome has 94 kilobase pairs. The plasmids arise by excision of DNA from the original chromosome, which then becomes circular (Wright et al., 1982). Once formed, the plasmids replicate autonomously to form numerous copies (termed *amplification*) at the expense of the original chromosome. In the final stages of senescence, these plasmids represent virtually all of the DNA present in the mitochondria (Wright and Cummings, 1983).

Sexuality

A fundamental aspect of sexual reproduction involves the union of gametes or gametangia, bringing together their nuclei to form a dikaryon, and, ultimately, a zygote. As you are aware, differentiated sexual organs are sometimes required for this process in the fungi, while in other fungi this process can be accomplished by hyphal fusion alone. In previous chapters (1–6) we explored the range of morphological variation associated with sex in the fungi, as well as the sexual reproductive cycles involving plasmogamy, karyogamy, meiosis, and the formation of haploid spores. There is an additional attribute of sexual reproduction in fungi to be considered, that is, the existence of mechanisms that regulate sexuality, determining whether or not a particular fungus is obliged to mate with a compatible partner (hormonal and environmental controls are considered in Chapter 10). The concepts of *homothallism* and *heterothallism* are used to describe fundamental differences in sexuality in the fungi leading to *inbreeding* and *outbreeding* respectively (Raper, 1960, 1966-a,b).

In organisms generally, the extent of the genetic variability derived from sexual reproduction is often a direct result of the degree to which either inbreeding or outbreeding occurs. Inbreeding involves sexual reproduction between individuals that are more closely related than those of a random sample of the naturally occurring population. The most extreme case of inbreeding involves self-fertilization of a hermaphroditic organism. In contrast, outbreeding occurs when sexual reproduction occurs between individuals that are less closely related than those of a random sample in a naturally occurring population. Inbreeding tends to create a homogenous population, while outbreeding encourages greater genetic heterogeneity, providing the necessary variability for evolution. Having a large number of genotypes within a population or species makes greater adaptation to changing environmental conditions possible than can occur in a population having only a few genotypes, but not all genetic recombinants will have greater adaptive value and may be eliminated as a result of natural selection. In general, the value of sexual reproduction as a vehicle for genetic recombination is enhanced when inbreeding is diminished or made impossible while outbreeding is encouraged or even obligatory in highly evolved organisms that have adapted successfully to their environment.

HOMOTHALLISM

Inbreeding characterizes the homothallic fungi. It is probable that the majority of all fungal species are homothallic. Homothallism predominates among the Mastigomycota and ascomycetes, but also occurs in a small number of basidiomycetes. A homothallic fungus can complete its sexual reproductive cycle within or upon a thallus derived from the germination of a single uninucleate spore, and without the necessity of introducing a second nuclear type. In all cases of homothallism, the single nuclear type contains all genetic requirements for full sexual expression (Raper, 1960, 1966-a). The key to this concept is that karyogamy and meiosis can occur without involving a second mating partner, or, that if a mating with a second partner should occur, there is not a requirement that the nuclei represent different mating types (as outlined below in heterothallism). Examples of reproduction in homothallic fungi follow.

1. A number of fungi form male and female gametangia on the same thallus. This is common among the fungi in the Saprolegniales where male and female gametangia can occur together on the same specialized hyphal branch. A similar situation occurs in the discomycete *Pyronema domesticum,* which forms antheridia and ascogonia borne by a single stalk cell. Self-fertilization can occur in these hermaphroditic fungi if plasmogamy occurs between the neighboring gametangia. The nuclei involved in karyogamy are sister nuclei, derived by mitotic divisions of the nucleus originally present in the spore that gave rise to the thallus.
2. The holobasidiomycetes, which lack sexual organs entirely, can also be homothallic, as, for example, is *Coprinus sterquilinus.* A single basidiospore germinates, giving rise to a mycelium with binucleate cells. This mycelium eventually forms basidiocarps although there is no genetic difference in the nuclei involved.

At first glance, it would seem that genetic variation is not likely to occur in homothallic fungi as karyogamy occurs between sister nuclei (derived mitotically from the original nucleus in the spore). However, many nuclear generations may intervene between the time that the spore germinates and karyogamy occurs. Mutations may occur in one sister nucleus but not the other, eventually leading to karyogamy between nuclei that have some genetic differences. There is also some evidence that fungi derived from different clones (derived from the germination of different spores, and therefore having nonsister nuclei) may come into physical proximity with each other in the natural habitat, resulting in plasmogamy between them and involving karyogamy between genetically dissimilar nuclei. The genetic differences, however, *do not involve mating types.* As a result of these potential mutations and recombination, genetic variation can occur as a result of sexual reproduction in homothallic fungi although it is extremely slow, but there is the advantage that risks are minimized as there will not be great deviations from the parental types (Raper, 1959, 1966-a).

HETEROTHALLISM

Heterothallism involves the obligatory mating between two homokaryotic individuals. These fungi are self-sterile and, unlike the homothallic fungi, do not carry all genetic requirements for sexual development in a single nucleus. Consequently, a thallus developing from the germination of a single uninucleate spore would not contain all genes regulating sexual development, nor would they be present in unicellular organisms representing a single clone. Mating between two compatible cells or homokaryotic thalli is required to produce a zygote having a full complement of the genetic determinants. Heterothallism is most often controlled by genetic incompatibility factors, but can also be controlled by sexual dimorphism in a few fungi (Raper, 1960, 1966-a).

Heterothallism Imposed by Incompatibility Mechanisms. Three distinct types of incompatibility systems are widely distributed among the fungi that produce genetically different mating types. The incompatibility systems differ in the number of genetic loci involved and also in the number of alleles that can occur at each locus.

Two-Allele Heterothallism. A number of heterothallic fungi have two mating types regulated by two alleles at a single genetic locus controlling incompatibility. These mating types are commonly referred to as *plus* (+) and *minus* (−) or *A* and *a*. The mating types segregate out into two spore types as a result of meiosis, and upon germination each produces a clone of uninucleate cells or a thallus with only one nuclear type. In this type of sexuality, a (+) strain cannot mate with another (+) strain but can mate with a (−) strain, as illustrated in Figs. 7–5 and 7–6.

Two-allele heterothallism can occur in fungi that lack morphological sexual differentiation, as for example is the case in the yeasts or smut fungi that have two mating types. However, a number of fungi having this type of heterothallism also have differentiated gametes or gametangia, but the heterothallism is controlled primarily by the mating type differences rather than by morphological

A = 'plus' strain
a = 'minus' strain
+ = compatible cross
− = incompatible cross

Figure 7–5 Mating reactions of possible isolates of a two-allele heterothallic fungus.

Figure 7–6 Mating between plus and minus strains of a heterothallic organism, *Phycomyces blakesleeanus*. A line of zygospores forms where the strains meet. Right: enlarged view of the line of gametangia (light) and zygospores (dark). [Courtesy Plant Pathology Department, Cornell University.]

sexual differences. Such fungi include many members of the Mucorales such as species of *Mucor* and *Rhizopus*, several ascomycetes including some species of *Neurospora* and some discomycetes such as *Ascobolus stercorarius*, and many rust fungi including *Puccinia graminis*. Although these fungi have both male and female reproductive organs on the same thallus, a fertile mating can occur only between female sexual structures and spermatia of opposite mating types, necessarily produced by separate thalli. Mating, however, can be reciprocal between these hermaphroditic fungi.

Multiple-Allele Heterothallism. In multiple-allele heterothallism, there are multiple alleles at each genetic locus controlling incompatibility. Multiple-allele heterothallism includes both the *bipolar* type, which produces two mating types as a result of a single cross, and the *tetrapolar* type, which produces four mating types as a result of a single cross. This type of heterothallism occurs in the basidiomycetes only. In these fungi, the uninucleate basidiospore germinates and produces a monokaryotic hypha. The monokaryotic hypha can grow indefinitely but fusions occur between cells of the same species if they encounter each other during growth. The fusions result in cells containing combined cytoplasm and nuclei of the original cells and occur whether the strains are compatible or incompatible, but subsequent development requires that the mates be compatible. If the strains are compatible, the invading nuclei migrate through the recipient hypha and become associated in pairs in the dikaryon. Dikaryon formation is required for basidiocarp development to occur (Chapter 5).

Bipolar multiple-allele heterothallism. Bipolar multiple-allele heterothallism occurs in most smut fungi, some jelly fungi in the Order Tremellales, and in various members of the Hymenomycetes and Gasteromycetes. It has been estimated that about 25% of the fungi having a genetic incompatibility system have the bipolar multiple-allele type (Raper, 1966-b).

Bipolar multiple-allele heterothallism is characterized by having a single genetic factor, the *A* factor, that controls compatibility. A number of alleles of the *A* factor exist and can be designated as $A_1, A_2, A_3, \ldots, A_n$. A cross is compatible if the mating homokaryotic mycelia carry different alleles. This means, for example, that a mycelium carrying the A_1 allele cannot mate with a mycelium also carrying A_1, but it will mate with a mycelium containing any of the remaining alleles, as illustrated in Fig. 7–7. The dikaryon would then contain both alleles which would then be present in the diploid, and would segregate out at meiosis, once again producing basidiospores which contain either one or the other of the two mating factors.

Tetrapolar multiple-allele heterothallism. Tetrapolar multiple-allele heterothallism occurs in the majority of the Hymenomycetes, Gasteromycetes, and the smut fungus *Ustilago maydis*. It has been estimated that about 65% of the species having an incompatibility mechanism are of this type (Raper, 1966-b).

Two genetic factors, *A* and *B*, control compatibility. These factors are borne on separate chromosomes and segregate independently at meiosis. A large

+ = compatible cross
− = incompatible cross

	A_1	A_2	A_3	A_4
A_1	−	+	+	+
A_2	+	−	+	+
A_3	+	+	−	+
A_4	+	+	+	−

Figure 7–7 Mating reactions of four possible isolates of a bipolar fungus.

Chap. 7 / Variation, Speciation, and Evolution

FL = flat   + = compatible cross
B = barrage   − = incompatible cross

	A_1B_1	A_1B_2	A_2B_1	A_2B_2
A_1B_1	−	FL	B	+
A_1B_2	FL	−	+	B
A_2B_1	B	+	−	FL
A_2B_2	+	B	FL	−

Figure 7–8 Mating reactions of four possible isolates of a tetrapolar fungus.

number of alleles exist for each factor. A compatible mating occurs if both the A and B alleles differ in both mating homokaryons. For example, a mycelium bearing the alleles A_1B_1 will not mate with A_1B_1 but will mate with A_2B_2 because the alleles of both factors are different (Fig. 7–8). A basidiocarp resulting from a compatible cross will bear basidiospores of four mating types after meiosis (A_1B_1, A_1B_2, A_2B_1, and A_2B_2).

In certain fungi (including *Schizophyllum commune*), if crosses are made between hyphae bearing either common A factors or common B factors, dikaryons may be established but basidiocarps are rarely formed. If the dikaryon has common A's but dissimilar B's, the *flat reaction* occurs (Fig. 7–8). The growth of this dikaryon type is sparse, depressed, and with gnarled, irregularly branched hyphae in the older mycelium. The typical growth of a dikaryon formed after mating of hyphae bearing common B factors but dissimilar A factors is also characteristic. The common B dikaryon gives the *barrage* reaction (Fig. 7–8). The hyphae of the mating colonies grow toward each other but near the center growth ceases, leaving a strip (the barrage zone) with little growth separating the two colonies. Common AB dikaryons also have been found to occur, and in this case the colonies grow over each other.

Raper and his co-workers established that in the tetrapolar fungi, the A factor actually consists of two closely linked genes that are expressed together as a functional unit. These genes can be separated by meiotic crossing-over, which occurs at a very low frequency. The genetic loci are designated A-alpha and A-beta, and a number of multiple alleles occur at each locus. A cross between two genetic strains is compatible if they differ in alleles at either the A-alpha or A-beta locus. The alleles at the A-alpha and A-beta loci jointly determine the A factor phenotype. The B factor was more difficult to analyze due to a much lower frequency of crossing-over, but was eventually determined also to consist of two genetic loci, B-alpha and B-beta. In *Schizophyllum commune*, 32 alleles are known to exist at the A-alpha locus, 9 at the A-beta locus, 9 at the B-alpha locus, and 9 at the B-beta locus (Koltin, 1978). A similarly complex situation apparently does not exist in fungi that have bipolar multiple-allele heterothallism—the A factor is known to have only one genetic locus.

The manner in which these genetic factors function in a compatible cross involving unlike A and B factors has been elucidated to some extent in *Schizophyllum commune*. The A factors control specific nuclear pairing, initiation of the clamp, and separation of the clamp by a septum. The B factors control septal breakdown and nuclear migration, and are required to complete clamp formation either alone or in conjunction with the A factor. In addition to the compatibility factors, there are a large number of other genes that are scattered throughout the genome that are expressed during morphogenesis, controlling some specific aspect of reproduction. These are regulated by the incompatibility genes and are considered to be secondary controls of reproduction. Some are expressed as modifiers of the A or B sequence.

Heterothallism Imposed by Sexual Dimorphism.   The majority of the advanced animals and plants have their sexes separated in different individuals that function as males or females. This phenomenon is called *sexual dimorphism*, and it is an effective mechanism promoting outbreeding. Dimorphic fungi include members of the Mastigomycota such as some chytrids, *Blastocladiella*, and some species of *Achlya;* and some members of the Laboulbeniales. Because of the separation of the male and female structures on different thalli, these fungi are automatically self-sterile and require cross-fertilization to bring different nuclear types together. Incompatibility factors are not involved, unlike the previous types of heterothallism, but it is not known if this type of heterothallism has some other type of genetic determinants or is the result of phenotype alone (Raper, 1960, 1966-a).

EVOLUTION OF HETEROTHALLISM FROM HOMOTHALLISM

The majority of the fungi are homothallic, and it may be that homothallism is the primitive condition and that heterothallism regulated by incompatibility factors developed from it. To illustrate the possible way that heterothallism could have evolved, we explore a scheme devised by Whitehouse (1949). The primitive homothallic condition was replaced by a simple form of heterothallism in which the locus controlling incompatibility had only two alleles. Hyphal fusion had previously occurred, but it was limited in significance as only nuclei of a single mating type were involved; but with the evolution of different nuclear types heterokaryons were formed which contained nuclei of different mating types. Sexual organs were no longer required to bring gametic nuclei together and were either partially or wholly lost. With further addition of alleles and another factor, both bipolar and tetrapolar multiple-allelic heterothallism developed and were maintained as a result of natural selection.

Heterothallism has selective advantages over homothallism as it increases outbreeding and the potential for greater variability, simultaneously restricting inbreeding. Consider, for example, the potentials for inbreeding and outbreeding in a fungus having two-allele heterothallism. Because there are only two mating types, any given thallus has the potential of outbreeding with approximately 50% of the existing thalli, which would be of the opposite mating type. In a similar manner, the inbreeding potential is also 50%. However, the situation can be changed dramatically with the addition of multiple alleles of one factor, as occurs in multiple-allelic bipolar fungi. A single incompatibility factor having only ten alleles would increase the outbreeding potential to 90% and can even approach 100% with additional alleles, while the inbreeding potential remains at 50% (only two alleles can be involved in any given mating although multiple alleles of the single factor exist). The presence of a second factor, as occurs in the multiple-allele tetrapolar fungi, does little to increase the outbreeding potential over that of fungi having multiple alleles of a single factor, but further reduces the inbreeding potential by half to 25% (Carlile, 1987; Raper, 1966-b). It has been calculated that *Schizophyllum commune*, which has multiple-allelic heterothallism, has more than 21,000 different mating types (Raper, 1959).

Reproductive Mutants as Models in the Evolution of Heterothallism.   In the preceding section, we considered fairly precise types of fungi exhibiting homothallism or heterothallism. Some fungi are encountered that do not fit neatly into one of these major categories, but can be interpreted as representing intermediate conditions between homothallism and heterothallism. A number of mutants of *Glomerella cingulata* have been studied (Wheeler, 1954; Wheeler and McGahen, 1952), giving some insight into a possible mechanism of evolution of heterothallism from homothallism. Normally *G. cingulata* is homothallic, but a number of mutants were found that cause the strain to be self-sterile because each mutant gene either partially or completely blocks the development of the mature perithecium. The blocks occur at the following stages:

1. *Perithecial initiation.* Genes at the A locus control production of perithecia or conidia, and genes at the B locus control the arrangement (clumped or scattered) of the perithecia. Perithecial initials are not formed if the mutant gene A^1 is present, but their formation is only partially blocked by A^+ with B^+. The latter block is removed if either A^2 or B^1 is substituted at one of these loci (that is, if the genotype is either A^+B^1 or A^2B^+).

2. *Plasmogamy.* Most of the mutant genes block plasmogamy. The nonallelic genes arg^1, bi^1, and th^1, controlling requirements for the growth factors arginine, biotin, and thiamine, respectively, and the genes F^1 and st^1 interrupt plasmogamy.

3. *Karyogamy.* Fungi with the mutant gene B^1 develop normally up to and including the formation of croziers. Nuclear disintegration occurs prior to and during karyogamy. Sterile perithecia result.

4. *Meiosis.* The mutant dw_1 (dwarf) partially blocks meiosis and controls the production of dwarf ascospores. Approximately 70 to 80% of the asci abort as a result of nuclear disintegrations.

The following model demonstrates events leading to the formation of a mature perithecium and also the point at which each mutant exerts its effect:

$$
\begin{array}{c}
& & & & arg^1 & & & & \\
& & & F^1 & bi^1 & & B^2 & & dw^1 \\
& A^+\ B^+ & & & th^1 & & & & \\
\text{mycelium} \longrightarrow & & \text{perithecial} \longrightarrow & & \text{plasmogamy} \longrightarrow & \text{karyogamy} \longrightarrow \\
& A^1 & \text{initials} & st & & & & \\
\end{array}
$$

$$\text{meiosis} \longrightarrow \text{mature perithecium}$$

Matings leading to the production of normal perithecia require that at least one of each homologous pair of genes be of the wild type. After plasmogamy, the wild-type gene will mask the effect of its mutant allele through complementation in the heterokaryotic mycelium until karyogamy occurs. The wild-type genes presumably control synthesis of specific chemical substances which allow development to proceed normally (Wheeler, 1954).

SPECIATION OF FUNGI

Evolution of new species, *speciation*, has been studied widely among plants and animals. This process can begin with a relatively homogeneous population that lives in a particular area and possesses a large gene pool. New gene combinations occur in some individuals, perhaps making it possible for them to adapt to new environmental conditions. Natural selection may then play a role, and the individuals in the group having favorable gene combinations for the prevailing environmental conditions survive and their reproduction is favored. Additional isolation of the subgroups can occur by ecological preferences for a particular habitat, migration to another vicinity, genetic barriers to reproduction, or another *isolating mechanism*. Isolation favors interbreeding within a subgroup and is accompanied by further distinctive genetic change. If the process continues, the subgroups may eventually be incapable of exchanging genes with each other and have evolved into distinct *species* (Fig. 7–9).

First stage
A single population in a homogeneous environment.

Second stage
Differentiation of environment, and migration to new environments produces racial differentiation of races and subspecies (indicated by different kinds of shading).

Third stage
Further differentiation and migration produces geographic isolation of some races and subspecies.

Fourth stage
Some of these isolated subspecies differentiate with respect to genic and chromosomal changes which control reproductive isolating mechanisms.

Fifth stage
Changes in the environment permit geographically isolated populations to exist together again in the same region. They now remain distinct because of the reproductive isolating barriers which separate them, and can be recognized as good sepecies.

Figure 7–9 Sequence of events that leads to the production of different races, subspecies, and species, starting with a homogeneous, similar group of populations. [G. Ledyard Stebbins, *Processes of Organic Evolution*, 3rd Ed., copyright 1977, p. 132. Reprinted by permission of Prentice-Hall, Inc., Englewood Cliffs, New Jersey.]

Speciation in the fungi shares major attributes with the process as it occurs in other organisms, but the distinctive biology of the fungi has contributed to some unique mechanisms. Geographical isolation, which results in *allopatric* speciation, is particularly important in plants and animals, but, in contrast, is comparatively unimportant in fungal evolution. Instead, fungal speciation is especially likely to be *sympatric*, originating from isolating mechanisms arising within an interbreeding population. Unlike the gradual divergence that can occur as outlined above, speciation can occur quite suddenly as a result of *interspecific hybridization,* a process that rarely occurs in the fungi.

Allopatric Speciation

Species can originate as a result of geographical separation of subgroups of the original population, followed by reproductive isolation and genetic divergence. Isolation by virtue of distance can account for divergence of allopatric populations, those that cannot interbreed because they do not occupy the same habitat.

Fungi are distributed worldwide, and it is very common to find that a particular fungal species occurs in North America as well as in Europe, and perhaps even on other continents. An important factor contributing to the wide dispersal of a fungal species is that the exceptionally light spores can be airborne for very long distances. This is well documented for *Puccinia graminis,* as spores of particular races originating in Mexico can be recovered in Canada.

A case history in colonization of a new environment by fungi can be cited. The Krakatau islands are about 40 kilometers from both Java and Sumatra. A volcanic eruption occurring in 1883 essentially destroyed the fungi and vegetation on these islands, which once again became richly covered with vegetation during the next 50 years. A survey was made of the species reappearing during this time, and about 300 species of fungi representing all major taxonomic groups were reported. This number is necessarily low because not all fungi would have been recovered. Boedijn, who conducted this survey, found an occasional tree trunk washing up on the shore and observed that these trunks carried mycelia that could later produce spores. However, he maintained that the majority of the fungi reached the island as airborne spores (Boedijn, 1940).

Although air dispersal of spores can account for long-distance transport of fungi across continents and to new Krakatau, it is unlikely that air dispersal accounts for the transoceanic transport of spores from one continent to another (Bisby, 1943). It is much more likely that parasitic fungi are distributed on their hosts, while spores of saprophytes may be distributed on plant debris. Spores can also be carried by migrating animals, particularly birds, for long distances (see also Chapter 11). Once fungi have been introduced into a new locality, their distribution is affected by other factors, such as plant succession and the prevailing climatic conditions (particularly temperature, humidity, and length of the season).

Geographical isolation alone does not appear to be a particularly important isolating mechanism for evolution of fungal species. Long-distance spore dispersal makes it possible for genes to be exchanged between fungal populations separated by a great distance, and fungi collected on one continent can often mate freely with fungi collected from another. However, the actual amount of gene flow occurring between such geographically isolated populations is unknown.

Sympatric Speciation

Sympatric speciation involves the origin of species as a result of ecological, behavioral, or genetic barriers that arise within a single population. The barriers prevent gene flow between the populations, resulting eventually in the divergence of sympatric species. These species are unable to exchange genes even though they occupy a common habitat. Although this process is relatively uncommon in higher plants or animals, numerous fungal species are known that have apparently arisen by sympatric speciation. Overall, sympatric speciation accounts for the origin of more fungal species than does allopatric speciation. Sympatric speciation in the fungi is particularly likely to occur as a result of genetic barriers or niche selection, a type of ecological barrier (Brasier, 1987; Burnett, 1976, 1983).

GENETIC BARRIERS

In fungi as in other organisms, an interruption of gene flow within an interbreeding population leads to the reproductive isolation of groups within the population, leading to speciation. Genetic barriers can arise in the fungi in a number of ways as diverse as different timing in the release of gametes or hybrid inviability, but mechanisms that interrupt either heterokaryon formation or steps in sporocarp development are particularly significant in the fungi as isolating mechanisms leading to speciation. Outbreeding in sexually reproducing fungi is greatly favored by heterothallism, but even those fungi that are homothallic can exchange genes by somatic mechanisms such as heterokaryosis. Lacking a sexual cycle does not prevent gene flow, which can be accomplished by heterokaryosis. Reproductive isolation often arises from the ability of hyphae to form heterokaryons or to complete either somatic or sexual plasmogamy. Two lines of evidence support the manner in which genetic differences can lead to reproductive isolation of fungi (Burnett, 1983):

1. *Genetic control of heterokaryon formation.* Some fungi possess mutant genes that prevent vegetative cells from forming heterokaryons. This

phenomenon has been extensively studied in *Aspergillus nidulans*, which has at least 8 specific *het* genes that control vegetative incompatibility. Heterokaryon formation can occur only if the *het* alleles are identical in the paired hyphae, and consequently a difference between alleles at any *het* locus can prevent heterokaryon formation. As a result, there are a number of genetically distinct strains that differ in their ability to form heterokaryons with each other (Croft, 1987). Heterokaryon incompatibility can lead to speciation in a fungus that reproduces nonsexually only.

2. *Genetic control of postfusion events.* Sexual reproduction can be initiated by fusion between gametangia, gametes, or hyphae, but sometimes plasmogamy is not completed because of unfavorable reactions between the nuclei and cytoplasm of the mating partners. In some basidiomycetes, the cells that fuse may become darkened, vacuolated, and then disintegrate (Burnett, 1983). Similar lethal reactions occur in the Pyrenomycete *Nectria haematococca*. Spermatia of a particular mutant strain can undergo plasmogamy normally with a trichogyne, but adverse nucleocytoplasmic interactions occur, preventing perithecial formation (Bistis and Georgopoulos, 1979). Numerous postfusion blocks occur in *Glomerella cingulata* (page 257).

Burnett (1983) maintains that mutations result in the inability of hyphae or gametes to fuse with each other, and this in turn leads to the development of isolated groups within a single population. Isolation of reproductive groups can occur quickly, but once this occurs, further adaptation to different niches or differentiation through genetic drift can follow. These fungal groups would then become sympatric species when they are incapable of mating with each other.

Sibling Species. In some cases, sympatric speciation results in the evolution of *sibling species*, which are reproductively isolated although morphologically indistinguishable. The evolution of sibling species appears to be particularly common among the fungi. Exceptionally genetically variable isolates of some fungi, previously considered as isolates of a single species, are in some cases found to be a complex of sibling species (Burnett, 1983). This is the case for the honey fungus *Armillaria mellea*. Although presently classified as a single species, mating experiments with isolates from North America and Europe revealed that a group of morphologically similar but biologically distinct species comprise *A. mellea*, which has tetrapolar multiple-allelic heterothallism. At least 10 sibling species were identified from North America and another five from Europe. The North American species failed to mate at all when monokaryons were paired, as did the European species when paired with other European species. However, mating could occur at a low frequency between some of the North American and European species (Fig. 7–10). These results indicate that

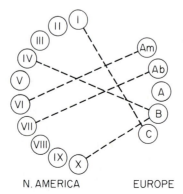

N. AMERICA   EUROPE

Figure 7–10 Relationships between North American and European biological species of *Armillaria* as revealed by mating experiments. Dashed lines indicate partial interfertility. All the other combinations between different groups are intersterile. [From J. B. Anderson, K. Korhonen, and R. C. Ullrich, 1980, Exptl. Mycol. 4:87–95.]

these sibling species were isolated by reproductive mechanisms. Complete isolation among fungal groups occupying a common habitat probably occurred because of greater genetic selection pressures operating on genetically divergent populations occupying the same geographical habitat. Similarly, intense selection pressures could not operate on these genetic mechanisms when the fungi were geographically separated, and those strains displaying some fertility with each other are in the process of speciation (Anderson et al., 1980).

ECOLOGICAL BARRIERS

It is well known that the fungi generally are adapted to specific habitats, and consequently certain species can be found only in certain habitats such as open fields, woodlands, or bogs. Further, the fungi occupy special ecological niches, often determined primarily by nutritional relationships. Saprophytic fungi may degrade a specific type of substratum such as leaf litter or dung (perhaps even the dung of a specific kind of animal). The parasitic fungi are often restricted to a particular host species and sometimes even to a particular genetic strain. Other unique relationships of the fungi to their food source include the Trichomycetes, which are commensal, and the symbiotic fungi such as those occurring in lichens and mycorrhizae.

In their natural habitats, fungi are constantly subjected to a number of ecological factors. These include the availability of suitable substrata that can be decayed or hosts that can be parasitized to provide nutrients; interaction with competing organisms, predators, and parasites; and abiotic factors such as moisture and temperature. Agricultural practices by humans constitute another type of ecological factor—massive plantings of crops of a particular genotype may be made, for example. The particular combination of these factors characterizes a particular ecological niche, and they can operate in the natural selection of one particular fungal strain over another.

There is evidence that many species acquired their ecological specialization as a result of sympatric speciation (Brasier, 1987). Sometimes fungal species that closely resemble each other can be found in the same habitat but occupying different niches and not interbreeding. An example is that of *Lophodermium*, which has three species occurring on pine. One species primarily invades the young green needles, another invades the killed needles on the tree, and the third occurs on the needle litter. These appear to be sympatric species that became specialized to particular ecological niches.

Although niche specialization is extremely important among the fungi, there is some uncertainty as to the exact manner in which it occurs. Burnett (1983) maintains that genetic isolation occurs and is followed by further divergence caused by adaptation to the ecological niches, as described below. According to this concept, genetic isolation would be the primary isolating mechanism reinforced by ecological divergence. An alternative view is that ecological factors initiate niche selection, resulting in genetic differences that follow (Brasier, 1987). According to this concept, two or more available ecological niches are each occupied by a subunit of the original population that subsequently acquires a phenotype by natural selection that is optimal for each niche. This process may result in the divergence of the original population into subgroups that are specialized for different niches, eventually becoming different species. Although the precise nature of the original isolating mechanism involved in niche specialization remains to be resolved, genetic and ecological factors are very much intertwined.

The effects of niche selection together with genetic isolation are readily apparent in the evolution of a plant pathogenic species, *Phytophthora megasperma*. Two distinct evolutionary lines apparently diverged from an ancestral *P. megasperma* (Fig. 7–11). Fungi in one line parasitized legumes while fungi in

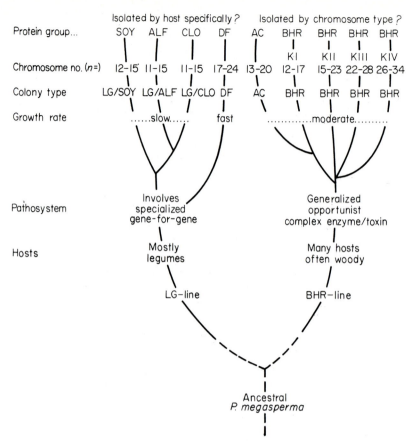

Protein group...	Isolated by host specifically?					Isolated by chromosome type?			
	SOY	ALF	CLO	DF	AC	BHR	BHR	BHR	BHR
							KI	KII	KIII	KIV
Chromosome no. (*n=*)	12–15	11–15	11–15	17–24	13–20	12–17	15–23	22–28	26–34
Colony type	LG/SOY	LG/ALF	LG/CLO	DF	AC	BHR	BHR	BHR	BHR
Growth rate	......slow......			fast		............moderate..........			
Pathosystem	Involves specialized gene-for-gene					Generalized opportunist complex enzyme/toxin			
Hosts	Mostly legumes					Many hosts often woody			
	LG–line					BHR–line			
				Ancestral *P. megasperma*					

Figure 7–11 Hypothetical interpretation of the phylogenetic structure of *P. megasperma*. The branch ends represent minor subgroups or emerging biological species, reproductively isolated from one another. [From E. M. Hansen, C. M. Brasier, D. S. Shaw, and P. B. Hamm, 1986, Trans. Brit. Mycol. Soc. 87:557–573.]

the second line occurred on a variety of hosts, including woody plants (designated the LG line and the BHR, or broad host line, respectively). The legume-inhabiting fungi acquired the highly specialized gene-for-gene mechanism that selects for host genotype (see Chapter 13) while the fungi in the BHR line did not acquire this trait. Polyploidy may have originated in the BHR line. Each line apparently gave rise to at least four to five subgroups that presently differ in terms of their growth rate, colony morphology, chromosome number, and protein type (Fig. 7–12). The taxonomic status of these subgroups is unclear, but they appear to be in the process of becoming different species. Speciation may be occurring as a result of isolation by host in the legume line and by karyotype in the other line. Isolates belonging to the same subgroup can be obtained from widely separated geographical locations (e.g., the United States and Japan), indicating that the fungus is probably transported widely in infected root stocks or infested soil as a result of agricultural activities (Hansen et al., 1986).

Many fungal parasites have undergone *coevolution* with their hosts. As the characteristics of the host changed through time, the characteristics of the fungi occupying them were also forced to change. There are a few striking examples where the fungal parasites have coevolved not only with their plant hosts but also with their insect vectors. *Monilinia vaccinii-corymbosi*, a Discomycete in the Helotiales, forms conidia on the leaves of its blueberry host in the spring. Insects that normally pollinate the flowers of these hosts also visit the young diseased

Figure 7–12 Major protein bands in a portion of a total protein separation on polyacrylimide gels. Four isolates of the BHR group are compared with isolates from the groups infecting soybean (SOY) and alfalfa (ALF). The BHR isolates represent an apparent polyploid series, with chromosome numbers of $n = 11$–15, 17–24, 22–28, and 26–34, respectively. [From E. M. Hansen, C. M. Brasier, D. S. Shaw, and P. B. Hamm, 1986, Trans. Brit. Mycol. Soc. 87:557–573.]

leaves and exhibit behavior normally associated with their visits to the flowers, such as licking the leaf and making distinctive movements. The insects come into contact with conidia which are transferred to the flowers that the insects subsequently visit. The fungus infects the ovaries of the flower, which then develop into the stromatized fruits ("mummies") that characterize this disease. Completion of the fungal life cycle depends on the insect transmission of the conidia, and as a result of coevolution, the diseased leaves acquired some insect-attracting characteristics that mimic those normally found in the flower. These are the exudation of a sugary mixture resembling nectar, a distinctive fermentative odor, and ultraviolet patterns similar to nectar guides (Batra and Batra, 1985). A second example may be found in the smuts (page 184).

Niche selection may actually be responsible for some apparent cases of geographical isolation. This may be illustrated by *Auricularia auriculata*, which appears to be undergoing allopatric speciation (Duncan and MacDonald, 1967). This jelly fungus occurs in one European subgroup that inhabits deciduous substrates and two North American subgroups, one occurring on deciduous substrates while the other occurs on coniferous substrates. These populations also have some morphological differences. In experimental crosses of monokaryons, the European isolates were virtually incapable of mating with either North American group, but the North American "deciduous" group did mate at a very low frequency (about 26%) with North American "coniferous" isolates. Mating behavior could not be correlated with either particular incompatibility factors (this fungus has bipolar heterothallism) or host preferences. Duncan and MacDonald (1967) suggest that these fungi actually represent two species (a European and a North American species) that originated primarily as a result of ecological isolation. The European species is particularly adapted to growth on *Sambucus nigra*, a plant that does not occur in North America.

Speciation by Hybridization

It is possible for new fungal species to arise by hybridization of existing species, although there are comparatively few known instances in which this has occurred. One of the best documented examples of speciation by hybridization occurs in the genus *Allomyces*, a member of the Blastocladiales (Emerson and Wilson, 1954). Two of the species are naturally occurring polyploids. Isolates of *A. macrogynus* have chromosomes in multiples of 14 (14, 28, and probably 56), while isolates of *A. arbuscula* similarly represent a polyploid series with a multiple of 8 chromosomes (8, 16, 24, and 32). The antheridia are borne above the female gametangia in *A. macrogynus* but below in *A. arbuscula*. In addition, there is a third species, *A. javanicus*, that arose naturally from the hybridization of these species and can be resynthesized experimentally by mating the two parental species. *A. javanicus* has chromosome numbers ranging from 13 to 21 and can bear its antheridia either above or beneath the female gametangia.

PHYLOGENETIC EVIDENCE

Understanding the relationships of fungi to each other is essential in the study and practice of mycology. Phylogenetic schemes are based on presumed relationships, as are taxonomic schemes. Whether or not fungi are grouped together in a particular taxon is essentially a statement about their relationships—fungi having similar characteristics are classified together in a genus, family, order, or higher taxon. Their possession of similar characteristics is based on the assumption that they had a common ancestry, but an ancestry different from that producing the fungi assigned to other taxa. The closeness of the relationship between species must be accurately evaluated for the construction of valid taxonomic schemes. Closeness and distance in phylogeny refer to the relative time in the past when species diverged from a common parental line, which in turn is reflected by the number of characters that the species continue to share.

In the past, mycologists have relied strongly upon morphological evidence. Simply expressed, the more features two species have in common, the more likely they are to be closely related. This reasoning has resulted in the classification scheme that is used in this book, in which the taxa are based primarily upon morphology and life cycles. Much of the morphology is based on macroscopic features, but as you may have noted, a number of microscopical features (e.g., operculation of sporangia and asci, spore characteristics, cuticle structure in the agarics) are utilized. Mycologists continue to refine the collection and evaluation of morphological evidence, now using ultrastructural evidence as well.

Other lines of evidence, based on comparative physiology, biochemistry, and molecular biology, are being used increasingly. Such techniques include comparison of the genomes represented by base composition and sequence within DNA, determination of the serological response elicited by fungal proteins when used as antigens, comparison of electrophoretic properties of enzymes and other proteins, and comparison of the enzymatic ability to utilize certain carbohydrates or other substrates.

In particular, some emphasis has been placed on an analysis of the relative proportion of the nucleic acid bases guanine and cytosine compared with the other two bases, adenine and thymine, in the DNA molecule. GC values in the fungi range from about 34 to 63% but are similar among closely related organisms. From a survey of about 300 fungal species it was determined that the ones generally considered to be more primitive based on morphological evidence have a lower GC content than those considered to be more advanced, and it appears that the evolution of fungi was accompanied by a progressive rise in the GC content (Storck and Alexopoulos, 1970). Having close GC values gives some indication of a close phylogenetic relationship, but distantly related species can also have similar GC values.

The extent of hybridization that can occur between single DNA strands isolated from different fungi is also an important technique used to determine closeness of relationship. Hybridization can occur if the base sequence is complementary between the two DNA strands. If hybridization is at least 40 to 60% between two fungi, they are considered to be distantly related, while a value greater than 95% indicates that they belong to the same species.

Investigations of mitochondrial rather than nuclear DNA may contribute to our understanding of fungal evolution in the future. The mitochondrial DNA is a favorable tool for such studies because it is relatively small, making it possible to analyze the entire molecule, and its composition is not complicated by the sexual recombination that occurs in nuclei as a result of sexual reproduction (Taylor, 1986).

PHYLOGENY OF THE FUNGI

The precise phylogenetic history of the fungi can be open to various interpretations, in part because of their poor fossil record and comparatively simple morphology. A particularly long-standing discussion has centered around whether or not the ascomycetes were derived from the red algae (discussed further on page 268). A widely held view of fungal phylogeny assumes that there is a main evolutionary lineage extending from the chytrids through the Zygomycetes, ascomycetes, and finally the basidiomycetes (Luttrell, 1955; Savile, 1968) (Fig. 7–13). This view is supported by the DNA hybridization studies of Kwok et al. (1986), which provides evidence that the Basidiomycotina have affinities with the chytrids, but not the red algae. This is important evidence because it supports the view that the Basidiomycotina and chytrids probably evolved from a common ancestor, the approach that is adopted here.

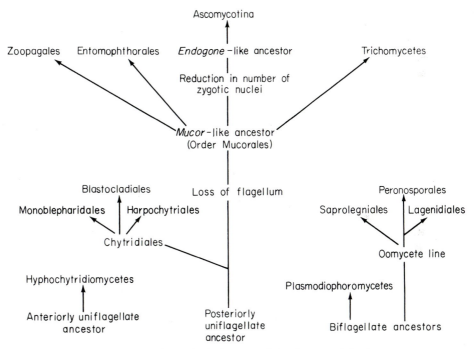

Figure 7–13 Hypothetical phylogeny of the Division Mastigomycota.

Evolution within the Division Mastigomycota

The position and number of the flagella have been considered to be of great importance in the classification of the Division Mastigomycota and the different classes were probably derived from different ancestral types (Fig. 7–13). A posteriorly uniflagellate protozoan probably gave rise to the present-day Chytridiomycetes, which are also characterized by posterior uniflagellate zoospores (Sparrow, 1958, 1960). Utilizing evidence based on zoospore ultrastructure, Barr (1978) proposed that the most primitive type of chytrid was one having simple zoospores without specialized organelles (e.g., Golgi bodies or rumposomes) or a complex internal organization of organelles. The ancestral chytrid type was monocentric and the zoosporangial walls dissolved to release the zoospores. An ancestral *Rhizophydium* was produced which perhaps had multiple papilla. From this ancestor two lines diverged, each having monocentric and polycentric chytrids. One line gave rise to chytrids with a reduced number of papilla, while the second line retained the multiple papilla and eventually culminated in the evolution of the Blastocladiales. The zoospore in the Blastocladiales is of the advanced type with a nuclear cap, side-body complex consisting of lipid globules and a microbody, and microtubles that radiate from the kinetosome. The Harpochytriales and Monoblepharidales may have been derived from polycentric chytrids.

The remaining members of the Division Mastigomycota are distantly related, if at all, to the Chytridiomycetes (Cantino, 1966; Sparrow, 1958). Following Sparrow's (1958, 1960, 1976) scheme, an anteriorly uniflagellate ancestor probably gave rise to the Hyphochytridiomycetes. The remaining two groups, the Plasmodiophoromycetes and the Oomycetes, have biflagellate zoospores. The Plasmodiophoromycetes have a number of characteristics unlike any other fungal group. These include the plasmodial stages, production of the resting spores in clusters, and unique mitotic divisions. They probably arose from ancestors other than those giving rise to the Oomycetes, a view supported by a comparative study of ribosomal RNA (Fraser and Buczacki, 1983). The Oomycetes share a number of features that differ from other fungi. These include cellulose cell walls and a different biochemical pathway for lysine synthesis (Kwok et al., 1986), as well as a number of ultrastructural features (Beakes, 1987). The Oomycetes are morphologically similar to the siphonaceous algae (such as *Vaucheria*), and Beakes (1987) and Sparrow (1958) suggest that at least some of the Oomycetes may have been derived from algae by the loss of chlorophyll. Although there has been no resolution of whether or not the Oomycetes were derived from the algae, Savile (1968) maintains that their similarity may be due to convergent evolution.

The Oomycetes themselves have apparently had a complex phylogeny. There is no clear indication of how the major orders are related to each other (Beakes, 1987). Sparrow (1976) recognizes two groups within the Oomycetes. These are (1) the Saprolegnian Galaxy, which has dimorphic zoospores, one to several eggs without periplasm, and oospores with a peripheral reserve globule; and (2) the Peronosporacean Galaxy, with monomorphic zoospores, periplasm surrounding a single egg, and an oospore with a single, central reserve body.

Evolution of the Zygomycotina

The primitive fungi were aquatic and became increasingly adapted to terrestrial habitats. Loss of the flagella occurred as well as evolution of zygospore formation, giving rise to the Division Zygomycotina (Fig. 7–13). They appear to constitute a natural group of closely related organisms, sharing a number of important characteristics. These are their nonmotile spores, sexual reproduction as a result of gametangial fusion, and walls containing chitosan and chitin. The most primitive member of the Mucorales was similar to *Mucor*, which formed large multispored sporangia, each having a columella. The multispored sporangium is considered to be primitive, with evolution proceeding from that type to the few-spored sporangioles, to the monosporous sporangiole with a dehiscent outer wall, and finally to those monosporous sporangioles with a double, fused wall (Benjamin, 1959; Hesseltine and Ellis, 1973). Hesseltine and Ellis also recognize two additional evolutionary trends. One was a major evolutionary trend from the primitive saprophytic mode of nutrition to the more advanced nutritional modes. The steps along this pathway proceeded from those fungi, such as many members of the Mucorales, that live as saprophytes with no special requirements; to some, such as *Pilobolus*, which are saprophytic but require a specialized growth factor (ferrichrome) found in dung; to the Entomophthorales, which can be grown in culture but have complex growth requirements; and finally, to those fungi in the Endogonaceae that cannot be cultured. A third major evolutionary line was from passive spore discharge in most members of the Mucorales to the more elaborate forcible spore discharge mechanisms occurring in *Pilobolus* and the Entomophthorales. The Entomophthorales may be more advanced than *Pilobolus* in this respect, as the spore is capable of repeated discharge until it alights on a favorable substratum.

The Trichomycetes present an evolutionary enigma (Moss and Young, 1978; Lichtwardt, 1986). The Trichomycetes in the Order Harpellales or closely related fungi appear to have close relationships with the Family Kickxellaceae, the most advanced family in the Mucorales, and are perhaps derived from them. These fungi all produce elongated, monosporous sporangioles which are borne in a similar fashion on the thallus; the supporting structures associated with the sporangioles resemble each other and may be homologous; and a flared septum with a lenticular plug occurs in the Kickxellaceae as well as in the Harpellales of the Trichomycetes. The Eccrinales and Asellariales seem to have fewer direct links with the Kickxellaceae and were probably derived from the Harpellales. The Amoebidales form peculiar amoeboid sporangiophores and lack chitin or cellulose in their cell walls, unlike the remainder of the Trichomycetes. They may not be related to the other three Trichomycete orders, but resemble them because of convergent evolution from a protozoan ancestor.

Evolution within the Ascomycotina

A number of similarities can be found between *Endogone* in the Zygomycotina and *Dipodascopsis* in the Hemiascomycetes, and it is possible that the connection between these two groups occurred between fungi resembling these organisms (Gäumann and Dodge, 1928). The gametangia in *Endogone* resemble those formed by *Dipodascopsis*, and in each fungus, only one pair of nuclei undergoes karyogamy (unlike the more primitive *Mucor*, where several nuclei undergo karyogamy). In *Dipodascopsis*, the product of the sexual union is the saclike ascus, which contains numerous spores derived from the zygote and cleaved from the cytoplasm by free cell formation, excluding those nuclei that had not undergone karyogamy. Although free cell formation occurs in the ascus, it has a superficial resemblance to a multispored sporangium.

The remaining members of the Hemiascomycetes were derived from the *Dipodascopsis*-like ancestor (Figs. 7–13 and 7–14). The unicellular yeasts may have been derived by reduction. Early Hemiascomycetes also evolved into more advanced fungi having an extended dikaryophase (Atkinson, 1915). The dikaryotic line gave rise to the Taphrinales, also members of the Hemiascomycetes, as well as to those lines extending to the more advanced Ascomycotina and Basidiomycotina.

A main evolutionary line derived from the Hemiascomycetes probably gave rise to the remaining members of the Ascomycotina. The fungi in this line had ascogenous (dikaryotic) hyphae, produced multicellular ascocarps, and also had unitunicate asci. Evolution of the bitunicate ascus occurred, producing a divergent line that terminated in the Loculoascomycetes. The bitunicate ascus is correlated with the ascostromatic structure of the ascocarp, although this type of structure can be found elsewhere (Luttrell, 1955). Some lichenized Discomycetes, which have a bitunicate ascus, were derived from this line (Korf, 1958).

In a continuation of the unitunicate line, evolution proceeded through fungi similar to the members of the Class Plectomycetes, which contain randomly organized asci within a cleistothecium, to the Class Pyrenomycetes, which contain asci in fascicles or hymenia in a cleistothecium or perithecium. This Pyrenomycete line may have diverged at various points to produce the

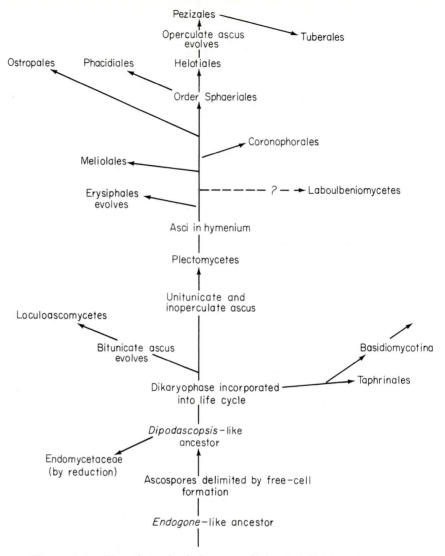

Figure 7–14 Hypothetical phylogeny of the Subdivision Ascomycotina.

various orders in the Class Pyrenomycetes, and culminated in the evolution of the family Xylariaceae in the Order Sphaeriales, a pivotal group of fungi (Luttrell, 1955).

Evolution of the apothecium that characterizes the Discomycetes occurred in a number of different lines (Korf, 1958). Some lichenized Discomycetes arose from the bitunicate line, while the remaining Discomycetes have the unitunicate ascus. The Order Sphaeriales gave rise to two inoperculate Discomycete lines, one leading to the Order Phacidiales and the second to the Order Helotiales. The operculate ascus developed in a line derived from the Helotiales, producing in turn the Pezizales and Tuberales. The Ostropales were derived from a separate evolutionary line derived from Pyrenomycetes having elongated multicelled spores (in the Clavicepitaceae).

A RED ALGAL ANCESTOR FOR THE LABOULBENIALES?

The question of whether or not the red algae were ancestral to the ascomycetes has been a lively topic for discussion among mycologists since Dodge (1914) proposed this connection. Dodge maintained that the ascomycetes originated from the red algae through the loss of chlorophyll. He noted numerous morphological similarities between the archicarps, trichogynes, and fertilization by spermatia that occur in both groups. According to Dodge, the greatest obstacle to accepting this concept is that nothing resembling cleavage of ascospores from the cytoplasm—or anything morphologically similar to the ascus—is found among the red algae. Recent proponents of the red algal origin include Denison and Carrol (1966) and Kohlmeyer (1973, 1975), who note additional similarities, including lack of motility, presence of a septal pore, thallus organization, and the resemblance of the red algal pericarp to the ascomycetous perithecium.

The relationships of the Laboulbeniales have been particularly difficult to establish. They appear to be more closely related to the Pyrenomycetes than to other ascomycetes, although this apparent relationship is questionable (Luttrell, 1955). Kohlmeyer (1973, 1975) described some fungi in the genus *Spathulospora* which are parasitic on red algae and are intermediate in their characteristics between the Laboulbeniales and the Pyrenomycetes. These fungi resemble the Laboulbeniales in that they form a trichogyne and spermatia, have deliquescent asci, and lack mycelium or an ascogenous system in the form of hyphae. They differ from the Laboulbeniales but resemble the Pyrenomycetes by having a perithecium with an ostiole and thick wall typical for the fungi in the Sphaeriales and also by forming a stroma between the cells of the host or a crustlike pad. The perithecia are not immersed in the stroma. Kohlmeyer suggests that the ancestral ascomycete was similar to *Spathulospora* and was derived from red algae that had become reduced in form and lived as symbionts or parasites on other red algae. According to Kohlmeyer, this ancestral type gave rise (1) to the Laboulbeniales, first occurring on marine arthropods and later moving onto land along with their hosts; (2) to parasites on plants similar to the present-day *Spathulospora* group; and (3) then to other Pyrenomycetes.

An alternative argument is that morphological similarities of some ascomycetes and red algae resulted from convergent evolution. As a result of an ultrastructural study on a member of the Laboulbeniales, Hill (1977) noted that a number of unique cellular features are shared with the more typical ascomycetes. These are a Woronin body in the septal pore, lack of a Golgi apparatus, and delimition of their ascospores by free cell formation involving an ascus vesicle. These features are quite different from those of the red algae. He concludes that cellular structure is more conservative than thallus morphology and that the Laboulbeniales are not necessarily closely related to the red algae.

Evolution in the Basidiomycotina

Numerous theories of the origin and evolution of the Subdivision Basidiomycotina have been proposed, and there is no universally accepted scheme. A tenable theory has been presented by Savile (1955) and is considered here.

The phylogenetic age of the host plant may be used as an index of the approximate age of the parasite (Fig. 7–15). The most primitive existent basidiomycete genus is assumed to be *Uredinopsis*, a rust genus with species parasitic on ferns in the genus *Osmunda*, the phylogenetically oldest genus with members that are parasitized by rust fungi. Presumably, the hypothetical ancestral Protobasidiomycete was a rust that shared some resemblance to *Uredinopsis*. This rust had no pycnia, aecia, or uredinia, and it had simple teliospores that may have merely been rounded hyphal cells. The mycelium inhabited fern fronds and was self-sterile, clampless, and dikaryotic. The teliospores each produced a basidium bearing four spherical basidiospores upon sterigmata. Presumably the spores germinated upon the fern frond, and the hyphae from two compatible spores fused either immediately upon or after infection. This hypothetical rust ancestor is remarkably similar to the ascomycete genus *Taphrina*, which also forms a dikaryotic mycelium in its host and has rounded cells from which the ascus forms. Under abnormal conditions, *Taphrina* may bud externally, producing subspheri-

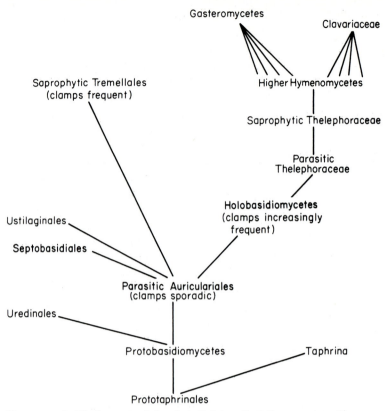

Figure 7–15 Phylogeny of the Subdivision Basidiomycotina. [Reproduced by permission of the National Research Council from D. B. O. Savile, 1955, Can. J. Botany 33:60–104.]

cal spores upon sterigmalike projections. Both the rusts and members of *Taphrina* parasitize plants of ancient lineage, the ferns and woody dicots, indicating that they existed during the same geological age span; of greatest importance is the fact that a fungus similar to *Taphrina* would be unspecialized and plastic enough to be ancestral to phylogenetic lines that must adapt to different environments. A *Taphrina*-like fungus, or a Prototaphrina ancestor, is assumed to have given rise to both the present-day *Taphrina* and the basidiomycetes.

Both the Prototaphrina and Protobasidiomycete ancestors are extinct. The Protobasidiomycetes gave rise to the Uredinales, beginning with the *Uredinopsis*-like forms. Life cycles of the rust fungi lengthened with the introduction of more spore forms. The first heteroecious rusts originated when part of the life cycle transferred from the telial to the aecial host.

The Protobasidiomycete ancestor also gave rise to a primitive parasitic fungus in the Phragmobasidiomycetidae. The unstable and highly variable heterobasidium is assumed to be primitive and not likely to be derived from the undivided basidium that has become a stable feature in the Holobasidiomycetidae. This primitive fungus was of the type found in the family Auriculariaceae in the Auriculariales, which has a transversely septate basidium (Fig. 5–20). This basidium type is considered to be primitive because it resembles a mycelium and because the fungi in which it appears are morphologically simple parasitic forms on phylogenetically primitive hosts and, most importantly, because clamps appear sporadically in this group. This primitive, *Auricularia*-type ancestor gave rise to (1) the Septobasidiales (parasitic and with no clamps), (2) the Ustilaginales (parasitic with clamps), (3) the Tremellales (saprophytic with clamps), and (4) the Holobasidiomycetidae.

Chap. 7 / Variation, Speciation, and Evolution

The early members of the Holobasidiomycetidae line were parasitic fungi like some Aphyllophorales (the family Thelephoraceae) that form a flat hymenium on a resupinate basidiocarp, either with or without clamps. Through evolutionary pressure, the hymenophore developed a greater spore-producing surface by forming ridges, teeth, gills, and pores. In addition, there was a tendency for the hymenophore to project downward, as this increased the efficiency of spore discharge. This evolutionary line gave rise to the more advanced Hymenomycetes. The Gasteromycetes represent a polyphyletic group in which convergent evolution under the pressures of the same habitat produced superficial similarities. Both the Gasteromycetes and the Clavariaceae (containing *Clavaria*) represent polyphyletic groups originating from the Hymenomycetes.

REFERENCES CITED

AKINS, R. A., and A. M. LAMBOWITZ. 1984. The (*poky*) mutant of *Neurospora* contains a 4-base-pair deletion at the 5′ end of the mitochondrial small RNA. Proc. Natl. Acad. Sci. USA 81:3791–3795.

ANDERSON, J. B., K. KORHONEN, and R. C. ULLRICH. 1980. Relationships between Europe and North American biological species of *Armillaria mellea*. Exptl. Mycol. 4:87–95.

ATKINSON, G. F. 1915. Phylogeny and relationships in the ascomycetes. Ann. Missouri Botan. Garden 2:315–376.

BARR, D. J. S. 1978. Taxonomy and phylogeny of chytrids. Biosystems 10:153–165.

BATRA, L. R., and S. W. T. BATRA. 1985. Floral mimicry induced by mummy-berry fungus exploits host's pollinators as vectors. Science 228:1011–1012.

BEAKES, G. W. 1987. Oomycete phylogeny: ultrastructural perspectives. *In:* Rayner, A. D. M., C. M. BRASIER, and D. MOORE, Eds. Evolutionary biology of the fungi. Cambridge University Press, Cambridge, pp. 405–421.

BENJAMIN, R. K. 1959. The merosporangiferous Mucorales. Aliso 4:149–169.

BISBY, G. R. 1943. Geographical distribution of fungi. Botan. Rev. 9:466–482.

BISTIS, G. N., and S. G. Georgopoulos. 1979. Some aspects of sexual reproduction in *Nectria haematococa* var. *cucurbitae*. Mycologia 71:127–143.

BOEDIJN, K. B. 1940. The mycetozoa, fungi and lichenes of the Krakatau group. Bull. Jard. Botan. Buitenzorg III 16:358–429.

BRASIER, C. M. 1987. The dynamics of fungal speciation. *In:* Rayner, A. D. M., C. M. Brasier, and D. Moore, Eds. Evolutionary biology of the fungi. Cambridge University Press, Cambridge, pp. 231–260.

BURNETT, J. H. 1976. Fundamentals of mycology. Second ed. Edward Arnold, London, 673 pp.

——. 1983. Speciation in fungi. Trans. Brit. Mycol. Soc. 81:1–14.

CANTINO, E. C. 1966. Morphogenesis in aquatic fungi. *In:* Ainsworth, G. C., and A. S. Sussman, Eds. The fungi: an advanced treatise. Academic Press, New York, Vol. 2, pp. 283–337.

CARLILE, M. J. 1987. Genetic exchange and gene flow: their promotion and prevention. *In:* Rayner, A. D. M., C. M. Brasier, and D. Moore, Eds. Evolutionary biology of the fungi. Cambridge University Press, Cambridge, pp. 203–214.

CATEN, C. E. 1987. The genetic integration of fungal life styles. *In:* Rayner, A. D. M., C. M. Brasier, and D. Moore, Eds. Evolutionary biology of the fungi. Cambridge University Press, Cambridge, pp. 215–229.

CROFT, J. H. 1987. Genetic variation and evolution in *Aspergillus*. *In:* Rayner, A. D. M., C. M. Brasier, and D. Moore, Eds. Evolutionary biology of the fungi. Cambridge University Press, Cambridge, pp. 311–323.

DENISON, W. C., and G. C. CARROLL. 1966. The primitive ascomycete: a new look at an old problem. Mycologia 58:249–269.

DODGE, B. O. 1914. The morphological relationships of the Florideae and the ascomycetes. Bull. Torrey Botan. Club 41:157–202.

DUNCAN, E. G., and J. A. MACDONALD. 1967. Micro-evolution in *Auricularia auricula*. Mycologia 59:803–818.

EMERSON, R., and C. M. WILSON. 1954. Interspecific hybrids and the cytogenetics and cytotaxonomy of *Euallomyces*. Mycologia 46:393–434.

FRASER, R. S. S., and S. T. BUCZACKI. 1983. Ribosomal RNA molecular weights and the affinities of the Plasmodiophorales. Trans. Brit. Mycol. Soc. 80:107–112.

GÄUMANN, E. A., and C. W. DODGE. 1928. Comparative morphology of fungi. McGraw-Hill, New York, 701 pp.

GROSSMAN, L. I., and M. E. S. HUDSPETH. 1985. Fungal mitochondrial genomes. *In:* Bennet, J. W., and L. L. Lasure, Eds. Gene manipulations in fungi. Academic Press, New York, pp. 65–103.

HANSEN, E. M., C. M. BRASIER, D. S. SHAW, and P. B. HAMM. 1986. The taxonomic structure of *Phytophthora megasperma:* evidence for emerging biological species groups. Trans. Brit. Mycol. Soc. 87:557–573.

HESSELTINE, C. W., and J. J. ELLIS. 1973. Mucorales. *In:* Ainsworth, G. C., F. S. Sparrow, and A. S. Sussman, Eds. The fungi: an advanced treatise. Academic Press, New York, Vol. 4B, pp. 237–243.

HILL, T. W. 1977. Ultrastructure of *Herpomyces* sp. (Laboulbeniales) and its phylogenetic implications. Can. J. Botany 55:2015–2032.

KOHLMEYER, J. 1973. Spathulosporales, a new order and possible missing link between Laboulbeniales and Pyrenomycetes. Mycologia 65:614–647.

——. 1975. New clues to the possible origin of ascomycetes. Bioscience 25:86–93.

KOLTIN, Y. 1978. Genetic structure of incompatibility factors: the ABC of sex. *In:* Schwalb, M. N., and P. G. Miles, Eds. Genetics and morphogenesis in the basidiomycetes. Academic Press, New York, pp. 31–54.

KORF, R. P. 1958. Japanese Discomycete Notes I–VIII. Sci. Rept. Yokohama Natl. Univ. Sect. II (7):7–35.

KWOK, S., T. J. WHITE, and J. W. TAYLOR. 1986. Evolutionary relationships between fungi, red algae, and other simple eucaryotes inferred from total DNA hybridizations to a cloned basidiomycete DNA. Exptl. Mycol. 10:196–204.

LICHTWARDT, R. W. 1986. The Trichomycetes: fungal associates of arthropods. Springer-Verlag, New York, 343 pp.

LUTTRELL, E. S. 1955. The ascostromatic ascomycetes. Mycologia 47:511–532.

MOSS, S. T., and T. W. K. YOUNG. 1978. Phyletic considerations of the Harpellales and Asellariales (Trichomycetes, Zygomycotina) and the Kickxellales (Zygomycetes, Zygomycotina). Mycologia 70:944–963.

PAPA, K. E. 1978. The parasexual cycle in *Aspergillus parasiticus*. Mycologia 70:766–773.

PONTECORVO, G. 1956. The parasexual cycle in fungi. Ann. Rev. Microbiol. 10:393–400.

——. 1958. Trends in genetic analysis. Columbia University Press, New York, 145 pp.

RAPER, J. R. 1959. Sexual versatility and evolutionary processes in fungi. Mycologia 51:107–124.

——. 1960. The control of sex in fungi. Am. J. Botany 47:794–808.

——. 1966-a. Life cycles, basic patterns of sexuality, and sexual mechanisms. *In:* Ainsworth, G. C., and A. S. Sussman, Eds. The fungi: an advanced treatise. Academic Press, New York, Vol. 2, pp. 473–511.

——. 1966-b. Genetics of sexuality in higher fungi. Ronald Press, New York, 283 pp.

SAVILE, D. B. O. 1955. A phylogeny of the basidiomycetes. Can. J. Botany 33:60–104.

——. 1968. Possible interrelationships between fungal groups. *In:* Ainsworth, G. C., and A. S. Sussman, Eds. The fungi: an advanced treatise. Academic Press, New York, Vol. 3, pp. 60–104.

SPARROW, F. K. 1958. Interrelationships and phylogeny of the aquatic Phycomycetes. Mycologia 50:797–813.

——. 1960. Aquatic Phycomycetes, second ed. University of Michigan Press, Ann Arbor, Mich., 1187 pp.

——. 1976. The present status of classification in biflagellate fungi. *In:* Jones, E. B. G., Ed. Recent advances in aquatic mycology. Wiley, New York, pp. 213–222.

STORCK, R., and C. J. ALEXOPOULOS. 1970. Deoxyribonucleic acid of fungi. Bacteriol. Rev. 34:26–154.

TAYLOR, J. W. 1986. Fungal evolutionary biology and mitochondrial DNA. Exptl. Mycol. 10:259–269.

WHEELER, H. E. 1954. Genetics and evolution of heterothallism in *Glomerella*. Phytopathology 44:342–345.

——, and J. W. McGAHEN. 1952. Genetics of *Glomerella*. X. Genes affecting sexual reproduction. Am. J. Botany 39:110–119.

WHITEHOUSE, H. L. K. 1949. Heterothallism and sex in the fungi. Cambridge Phil. Biol. Rev. Cambridge Phil. Soc. 24:411–47.

WRIGHT, R. M., and D. J. Cummings. 1983. Integration of mitochondrial gene sequences within the nuclear genome during senescence in a fungus. Nature 302:86–88.

——, M. A. HORRUM, and D. J. CUMMINGS. 1982. Are mitochondrial structural genes selectively amplified during senescence in *Podospora anserina*? Cell 29:505–515.

Additional Recommended References

BARRETT, J. A. 1987. Molecular variation and evolution. *In:* Rayner, A. D. M., C. M. Brasier, and D. Moore, Eds. Evolutionary biology of the fungi. Cambridge University Press, Cambridge, pp. 83–95.

BRASIER, C. M. 1984. Inter-mycelial recognition systems in *Ceratocystis ulmi:* their physiological properties and ecological importance. *In:* Jennings, D. H., and A. D. M. Rayner, Eds. The ecology and physiology of the fungal mycelium. Cambridge University Press, Cambridge, pp. 451–497.

BURNETT, J. H. 1956. The mating systems of fungi. I. New Phytologist 50:50–90.

——. 1975. Mycogenetics: an introduction to the general genetics of fungi. Wiley, New York, 375 pp.

——, and M. E. BOULTER. 1963. The mating systems of fungi. II. Mating systems of the Gasteromycetes *Mycocalia denudata* and *M. Duriaerna.* New Phytologist 62:217–236.

BUXTON, E. W. 1960. Heterokaryosis, saltation, and adaptation. III. Heterokaryosis as a factor in variability. *In:* Horsfall, J. G., and A. E. Dimond, Eds. Plant Pathology: an advanced treatise. Academic Press, New York, Vol. 2, pp. 365–380.

CAIN, R. F. 1972. Evolution of the fungi. Mycologia 64:1–14.

CHESTERS, C. G. C. 1968. Morphology as a taxonomic criterion. *In:* Ainsworth, G. C., and A. S. Sussman, Eds. The fungi: an advanced treatise. Academic Press, New York, Vol. 3, pp. 517–542.

CROFT, J. H., and R. B. G. DALES. 1984. Mycelial interactions and mitochondrial inheritance in *Aspergillus*. *In:* Jennings, D. H., and A. D. M. Rayner, Eds. The ecology and physiology of the fungal mycelium. Cambridge University Press, Cambridge, pp. 433–450.

DAVIS, R. H. 1966. Mechanisms of inheritance. 2. Heterokaryosis. *In:* Ainsworth, G. C., and A. S. Sussman, Eds. The fungi: an advanced treatise. Academic Press, New York, Vol. 2, pp. 567–588.

ESSER, K. 1966. Incompatibility. *In:* Ainsworth, G. C., and A. S. Sussman, Eds. The fungi: an advanced treatise. Academic Press, New York, Vol. 2, pp. 661–676.

FRAVEL, D. R., and P. B. ADAMS. 1986. Estimation of United States and world distribution of *Talaromyces flavus*. Mycologia 78:684–686.

FRIES, N. 1948. Heterothallism in some Gasteromycetes and Hymenomycetes. Svensk Botan. Tidskr. 42:158–168.

HALISKY, P. M. 1965. Physiologic specialization and genetics of the smut fungi. III. Botan. Rev. 31:114–150.

HERSKOWITZ, I. 1988. Life cycle of the budding yeast *Saccharomyces cerevisiae*. Microbiol. Rev. 52:536–553.

JAHNKE, K.-D., and G. BAHNWEG. 1986. Assessing natural relationships in the basidiomycetes by DNA analysis. Trans. Brit. Mycol. Soc. 87:175–191.

JOHNSON, T. 1968. Host specialization as a taxonomic criterion. *In:* Ainsworth, G. C., and

A. S. Sussman, Eds. The fungi: an advanced treatise. Academic Press, New York, Vol. 3, pp. 543–556.

KURTZMAN, C. P. 1985. Molecular taxonomy of the fungi. *In:* Bennett, J. W., and L. L. Lasure, Eds. Gene manipulations in fungi. Academic Press, New York, pp. 35–63.

LEPPIK, E. E. 1965. Some viewpoints on the phylogeny of rust fungi. V. Evolution of biological specialization. Mycologia 57:6–22.

MÜLLER, E. 1974. Taxonomy and phylogeny of fungi. Prog. Botan. 36:247–262.

OLIVE, L. S. 1958. On the evolution of heterothallism in fungi. Am. Naturalist 92:233–251.

PERKINS, D. D., and B. C. TURNER. 1988. *Neurospora* from natural populations: toward the population biology of a haploid eukaryote. Exptl. Mycol. 12:91–131.

PIROZYSKI, K. A. 1968. Geographical distribution of fungi. *In:* Ainsworth, G. C., and A. S. Sussman, Eds. The fungi: an advanced treatise. Academic Press, New York, Vol. 3, pp. 487–504.

——. 1976. Fossil fungi. Ann. Rev. Phytopathol. 14:237–246.

RAPER, C. 1978. Control of development by the incompatibility system in basidiomycetes. *In:* M. N. Schwalb and P. G. Miles, Eds. Genetics and morphogenesis in the basidiomycetes. Academic Press, New York, pp. 3–29.

RAPER, J. R. 1953. Tetrapolar sexuality. Quart. Rev. Biol. 28:233–259.

ROPER, J. A. 1966. Mechanisms of inheritance. 3. The parasexual cycle. *In:* Ainsworth, G. C., and A. S. Sussman, Eds. The fungi: an advanced treatise. Academic Press, New York, Vol. 2, pp. 589–617.

SAVILE, D. B. O. 1971. Coevolution of the rust fungi and their hosts. Quart. Rev. Biol. 46:211–218.

——. 1971. Co-ordinated studies of parasitic fungi and flowering plants. Nat. Can. 98:535–552.

——. 1976. Evolution of the rust fungi (Uredinales) as reflected by their ecological problems. Evol. Biol. 9:137–207.

TINLINE, R. D., and B. H. MACNEILL. 1969. Parasexuality in plant pathogenic fungi. Ann. Rev. Phytopathol. 7:147–170.

TYRELL, D. 1969. Biochemical systematics and fungi. Botan. Rev. 35:305–316.

ULLRICH, R. C., and J. R. RAPER. 1977. Evolution of genetic mechanisms in fungi. Taxon 26:169–179.

WESSELS, J. G. H. 1978. Incompatibility factors and the control of biochemical processes. *In:* Schwalb, M. N., and P. G. Miles, Eds. Genetics and morphogenesis in the basidiomycetes. Academic Press, New York, pp. 81–104.

PART II PHYSIOLOGY AND REPRODUCTION

8

Growth

All organisms have the potential to increase in mass by cell division, cell enlargement, or both. Such an increase in mass is termed *growth*. These processes play roles of varying importance in different organisms. In comparatively simple, unicellular organisms (including yeasts), growth occurs predominantly by cell division, which increases the population and is therefore a reproductive process. Cell enlargement, accompanied by nuclear division and synthesis of cytoplasm, is primarily responsible for growth of fungi with coenocytic, nonmycelial thalli such as members of the Blastocladiales. Mycelial fungi, like other complex multicellular organisms, grow through a combination of cell division and enlargement.

GROWTH IN MULTICELLULAR FUNGI

A fungal hypha consists primarily of a rigid cell wall surrounding a vacuolated cytoplasm and a hyphal tip (with a plastic wall, a disproportionately large number of nuclei, and a nonvacuolated cytoplasm). This hyphal tip constitutes approximately the last 50 to 100 micrometers of the hypha. As described in Chapter 1, growth in mycelial fungi occurs by extension of the hyphal tip, while the older portions of the hyphae are incapable of growth. Although the older hypha is not capable of growth, it has an important role in supporting growth of the tip as new protoplasm is formed throughout the hypha and transported to the tip by active cytoplasmic streaming. (The use of radioactive precursors indicates that the rates of protein and ribonucleic acid synthesis do not decrease substantially in hyphal regions quite distant from the hyphal tip.) It is this active translocation of synthesized components that makes possible the rapid growth rate of the hyphal tip (Zalokar, 1959).

The developing hypha usually forms branches in acropetal succession (proceeding towards the tip) behind the hyphal tip. Branch formation is preceded by a localized softening of the previously hardened hyphal wall, and a branch develops from this site. This primary branch in turn usually gives rise to secondary and tertiary branching systems. When a branch is initiated, a thickened cell wall must be differentiated and a new growing apex established. This process is extremely rapid, requiring only about 40 to 60 seconds (Burnett, 1976). Apical dominance is typically apparent; that is, a hypha that is producing branches continues to grow at a more rapid rate and is usually longer than the branches it is producing. The typical growth pattern, therefore, is that of a long central axis from which shorter axes branch; each central axis is longer than the branches that it bears, the whole strongly resembling a two-dimensional replica of a pine tree. This growth pattern may break down in culture, where the mycelium shows a strong tendency to occupy all medium within the circumference of the developing colony. In this case, apical dominance of the lead hyphae may be lost and branches from the lead hyphae may grow vigorously into unoccupied space. The effect of this type of growth is to form a circular colony on a solid medium on which there are no obstructions or to form a spherical colony within a liquid medium. Mechanisms controlling branching and apical dominance are not known.

Complex sporocarps and sclerotia are composed of mycelium that is modified to form distinctive cells and tissues. Sporocarps are characterized by zones of tissues which elongate, expand, and change during development to give the sporocarp its final mature form (Fig. 8–1). Some of the developing tissue zones

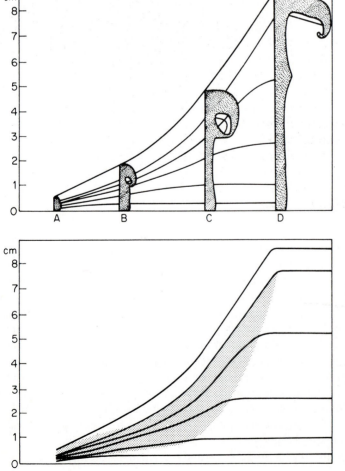

Figure 8–1 Above: A diagrammatic growth curve of *Agaricus campestris* showing four different stages. The lines are drawn through homologous points. The horizontal axis is only a very rough approximation of time. Below: This is the same graph as above, but those regions of the stalk that are in the process of elongation (i.e., the growth zone) lie within the shaded area. In this way it is possible at each stage of development to see the extent of the growth zone. [Figures and legend from J. T. Bonner et al., 1956, Mycologia 48:13–19.]

Chap. 8 / Growth

(usually consisting of tightly packed parallel hyphae) appear to be actively growing from their apices, which have the dense cytoplasm and numerous nuclei characteristic of a growing hyphal tip. Unlike hyphae, the cells in an ascocarp or basidiocarp may undergo intercalary enlargement, elongation, or septation after their formation (Fig. 8–2). In an ascocarp, additional hyphae may push their way up through partially developed tissue, thereby increasing the mass of the sterile tissue of the ascocarp and the hymenium. Sclerotia which lack the somewhat precise growing points found in ascocarps and basidiocarps achieve their final size by prolific hyphal elongation and branching, followed by a marked enlargement of the cells.

Figure 8–2 Growth of a sporocarp is determined by the growth rate and number of all individual cells, as illustrated in the agaric *Flammulina velutipes*. [Adapted from W. M. Wong and H. E. Gruen, 1977, Mycologia 69:899–913.]

MEASUREMENT AND KINETICS OF GROWTH

The simplest method of assessing fungal growth is by a linear measurement. The change in the radius of a developing colony on agar may be observed over a period of time. Alternatively, a horizontal growth tube of about 40 centimeters in length containing a layer of agar may be employed by measuring from the point of inoculation to the advancing edge of the mycelium at intervals or by the microscopic measurement of individual hyphae. These methods have the virtue of being extremely simple, are of value in making rough estimates of growth, and are nondestructive; this allows repeated observations of the same mycelium. They fail to account for differences in aerial mycelium or mycelium submerged in the agar and also fail to distinguish between mycelial spread over the medium and the actual total production of mycelium.

A more accurate method of assaying growth of a fungus is assessment of dry-weight gain. A liquid medium is inoculated with the fungus and, after the growth period, the resulting mycelium is removed by filtration or centrifugation, is ashed, and then is weighed. Unlike the linear techniques, this method is a destructive one and large numbers of cultures must be used successively in order to obtain data about growth rates over a period of time, and it may measure accumulated storage products rather than new protoplasm.

Growth of single-celled fungi, such as the yeasts, may be assessed by bacteriological methods, such as changes in the light absorption or light scattering properties of the liquid culture medium. These techniques are simple to carry out and allow repeated observations on a single culture.

If the growth rate of single-celled organisms such as bacteria or yeasts is followed in a liquid culture from the time of inoculation until decline, a typical pattern evolves which may be divided into stages (Fig. 8–3). Stage I (the *lag* phase) is a period occurring after inoculation and before any cell division takes place. Stage II (the *acceleration* phase) is initiated when cell division begins and marks the transition between dormancy and active growth. Stage III (the *exponential* phase) is attained when cell division reaches and maintains a uniform rate. It is during stage III that the greatest increase in cell numbers takes place; this is an exponential increase in numbers. Stage IV (the *deceleration* phase) is characterized by a decrease in cell division until stage V (the *stationary* phase) is reached, where either there is no additional growth or production of new cells is exactly balanced by death of old cells. Senescence occurs eventually, and is marked by a decrease in the number of viable cells (stage VI, the *decline* phase). The actual time required for each phase varies with the particular organism and with the specific conditions under which it is cultured.

Although stages can be rather precisely defined in the case of unicellular organisms in a liquid medium, mycelial fungi do not lend themselves to such a precise quantitative analysis. Division of each cell into two cells resulting in exponential increase in unicellular organisms is lacking and there is differential activity of the cells. Growth is restricted to the hyphal tip only, which grows more or less at a constant linear rate and therefore extends the margin of the colony at a linear rate. Although each hypha extends at a linear rate, growth of *Chaetomium globosum* occurs exponentially during the early stages of colony development if the entire branching mycelium is considered. During the later stages of colony development, however, growth in the center of the colony de-

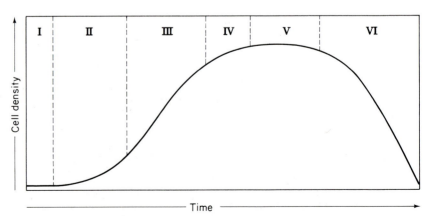

Figure 8–3 Idealized growth curve for a single-celled organism such as a yeast. See the text for a discussion of the stages indicated.

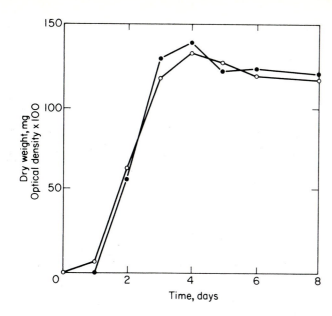

Figure 8–4 Growth of the mycelium of *Fusarium solani* in an aerated medium. Open circles indicate dry weight; closed circles indicate optical density. [From V. W. Cochrane, 1958, Physiology of fungi, John Wiley & Sons, Inc., New York.]

clines while the marginal hyphae maintain their linear rate of growth (Plomely, 1959). In general, mycelial growth may be qualitatively divided into the following: stage I (lag phase), with no apparent growth; stage II (*linear* phase), with rapid and approximately linear growth; and stage III (decline phase), during which there is no growth or there is a decline in dry weight owing to autolysis (Fig. 8–4).

NUTRITION

Digestion

Fungi are in direct contact with their nutrients in the environment. Smaller molecules (such as simple sugars and amino acids) in solution in the watery film surrounding the hyphae can be directly absorbed by the hyphae. Larger insoluble polymers such as cellulose, starch, and proteins must undergo a preliminary digestion before they can be used.

Molecules that are too large to be absorbed by the fungus are attacked by extracellular enzymes. Like all digestive enzymes, the digestive enzymes of fungi control hydrolysis reactions which cleave the large molecules into simpler components. The digestive enzymes are highly specific and are able to control hydrolysis of particular molecules only. Complete digestion of a large polymer is a stepwise process involving different enzymes, until finally a simple, soluble molecule is released. It is this simple molecule which is taken up by the fungus. Once it is absorbed into the cell, this small molecule is further acted upon by intracellular enzymes (Fig. 8–5).

The ability to utilize large molecules ultimately depends on the ability of the fungus to digest them, which in turn depends on the enzymes with which the fungus is equipped. Fungi typically have a large number of enzymes but, for the most part, many of them lie idle until the fungus comes into contact with a substrate on which particular enzymes can act. Growth of the fungus occurs equally well on a medium containing either complex or simple nutrients. This is not always the case, however, and the necessary enzymes may be entirely lacking and the fungus may be unable to grow on a medium that contains an undigestible substrate.

Figure 8–5 Digestion and uptake by a fungus.

Uptake

All ions and molecules entering the fungal cell must pass through both the cell wall and the plasmalemma. The wall itself is somewhat porous, allowing ions and molecules to pass through it, perhaps through minute pores or channels (Burnett, 1976). The plasmalemma is a semipermeable membrane that can regulate the movement of solutes into the cell. A nonelectrolyte in solution may pass through the plasmalemma in response to a concentration gradient, moving from a greater to a lesser solute concentration. Ions in solution will similarly pass across the membrane from a greater to a lesser solute concentration, but there the driving force is a difference in electrochemical potential rather than the difference in concentration. In many cases, nonelectrolytes or ions are accumulated by the cell, but this accumulation cannot be accounted for by the *passive* mechanisms just described. Instead, accumulation may occur in the direction opposite to that predicted by the concentration gradient. Solutes may be moved into a cell by *active transport* (or alternatively, moved outward). This mechanism involves the binding of the solute to a *carrier* molecule that exists within the plasmalemma. The carrier transports the solute across the plasmalemma and releases it inside the cell. Operation of the carrier system requires metabolic energy provided by ATP. *Facilitated diffusion* also involves a carrier system, but unlike active transport, the solute is not moved against a gradient. Instead, the solute is transported across the plasmalemma from its greater to its lesser concentration. Much of our knowledge concerning ion uptake has been derived from extensive studies on yeasts, and to a lesser extent, from studies on filamentous fungi which support many of the observations made from studies of yeasts.

Almost all of the important ions are taken up by active transport. These include cations such as those of potassium, magnesium, manganese, iron, and the ammonium ion. Important anions taken up by active transport include the phosphate and sulfate ions. Uptake of the potassium cation (K^+) is remarkably efficient and can occur against a concentration gradient as high as 5000:1 as long as an exogenous substrate is available to provide energy from respiration. The uptake of K^+ is normally matched by the outward transport of H^+. In some instances, uptake of one ion is affected by the uptake of another ion. Transport of Mg^{2+}, Ca^{2+}, and Mn^{2+} occurs only when phosphate uptake is occurring, and potassium stimulates this uptake as well as the total uptake of phosphate in yeast (Griffin, 1981). There has been some disagreement about whether the nitrate ion enters by an active or a passive mechanism.

The nonelectrolytes also enter almost exclusively by specialized transport systems. Uptake of the various sugars has been reported to occur by either active transport or facilitated diffusion in various fungi. Uptake of amino acids occurs by active transport, while the uptake of other organic acids appears to occur both by specialized transport and by passive diffusion. Specific carrier molecules (permeases) are typically produced for each nutrient transported. Many factors, such as the pH, temperature, particular combination of nutrients, or physiological condition of the fungus may affect uptake of a given nutrient.

Enzyme Induction and Repression

Uptake of nutrients depends to a great extent upon the activity of enzymes. Digestive enzymes are responsible for breaking down the large polymers in the medium, while the carrier molecules that occur in the membranes are also enzymes. Once a nutrient has been taken up into the cell, its fate is determined in part by enzymes associated with specific metabolic pathways.

An enzyme for a particular function may not be present in the cell at all times. However, the synthesis of this enzyme may be *induced* by the availability of a particular nutrient. A fungus may not have the enzymes necessary to digest starch, but if it is transferred to a medium containing starch, it may eventually produce the necessary enzymes and utilize starch. Induction of enzymes required to take up and utilize nitrate and nitrite as nutrient sources also can occur when a fungus is exposed to these nutrients.

In many cases, the uptake of a particular nutrient will be affected by the presence of another nutrient which can meet the same requirement. Given a mixture of sugars, for example, a fungus may utilize glucose preferentially while not taking up the available fructose until the glucose is exhausted. This phenomenon is known as *catabolite repression*. Other cases of catabolite repression occur. Phosphates can repress the uptake of other phosphorus-containing compounds; amino acids containing sulfur can repress the uptake of sulfates; and finally, the presence of ammonia or glutamine can repress the uptake of nitrates. Repression, which is under genetic control, may be expressed at two levels. Either the carrier molecule responsible for transporting the nonfavored metabolite across the membrane is not manufactured properly, or enzymes in the pathway that would use this metabolite are altered.

Both enzyme induction and repression are under genetic control, involving both structural and regulatory genes. A structural gene is one that codes directly for the enzyme, while a regulatory gene modifies its expression. Regulatory genes may be *positive-acting*; that is, their action is required for the expression of the structural genes, or *negative-acting*, with their action resulting in the repression of the expression of the structural genes. An example of the complex interplay of structural and regulatory genes is given in conjunction with nitrogen uptake on page 289, although similar data exist for other pathways.

Translocation

Once the nutrients and water have gained entrance to the cell, they are often translocated from one site to another. Translocation is especially important in fungal hyphae, where growth at the tips must be supported, and when macroscopic structures such as sporocarps or slerotia are formed. Translocation of ions and sugars in hyphae can be studied by tracing the movement of radioactive substrates. Radioactive glucose, amino acids, phosphorus, and magnesium were made available to a mycelium of *Serpula lacrimans* growing on pine blocks.

The mycelium was forced to grow across a nonnutritive surface in a special cabinet, and the translocation of the radioactive substances from the food source to the hyphal tips was traced by autoradiography. The translocation rate of each substance occurred at approximately the same rate, 20 to 30 centimeters an hour. Also in *S. lacrimans*, addition of a respiratory inhibitor or a concentrated glucose solution at the food source inhibited translocation along the full length of the mycelium. These results indicate that energy is required at the source, and also that the flow is generated by osmotic forces (Brownlee and Jennings, 1982). A bulk flow model can explain the movement of water and its dissolved substances from the food source to the hyphal tips. According to this model, bulk flow of the solution is driven by hydrostatic pressure through hyphae in which an osmotic gradient exists. The hydrostatic pressure is generated by the hyphae that are in contact with the food source, which take up nutrients such as sugars, increasing the osmotic potential. Water moves by osmosis from the surrounding hypotonic medium into the hyphae, and the resulting hydrostatic pressure drives the fluid toward the hyphal tips. The flow can be maintained because water is lost from the hyphal tips as droplets that are extruded through the highly permeable walls by turgor pressure. Water loss from the hyphal tips helps maintain a high osmotic potential there (Jennings, 1984; Thain and Girvin 1987). It is also well documented that evaporation occurs at a great rate from basidiocarps (Plunkett, 1958), which makes this model applicable to water and solute transport in them as well (Jennings, 1984). The extent to which the bulk flow may be responsible for cytoplasmic streaming itself is unknown but may be an even more important factor than cytoplasmic contractions (see also Chapter 1) (Brownlee and Jennings, 1982; Jennings, 1984).

Nutritional Requirements

Fungi grow normally as saprophytes or parasites on naturally occurring animal or plant products; these are extremely complex materials that provide a number of unknown nutrients for growth of the fungus. It is usually impossible or at least extremely inconvenient for mycologists to study physiology of fungi in their natural habitats, and it is desirable to grow fungi in the laboratory under conditions where nutritional and environmental factors can be controlled and duplicated. Aside from physiology, many other aspects of mycology require that fungi be grown and readily available in the laboratory (such as routine identification of fungi isolated from natural substrata, genetic or developmental studies, and teaching). It becomes necessary to create in the laboratory nutritional and physical conditions that will allow fungi to grow and develop as "normally" as possible.

Selection of a suitable substratum is most important for cultivation of fungi. The simplest of these are the materials on which the fungus would ordinarily occur in nature (dung, grains, wood, etc.). These are autoclaved and used directly, or fungi may ai o be grown on water in which these materials were soaked and heated. Media of this type are extremely cheap and easy to prepare and will often support fungi that will not grow on other media. A disadvantage of natural media is that they can never be precisely duplicated for physiological studies as they are of an unknown composition. Alternatively, a completely synthetic medium consisting of chemicals (such as glucose, asparagine, magnesium sulfate, or biotin) which are of both known composition and concentration may be used. A synthetic medium is the only type of medium which may be precisely duplicated and is suitable for meticulous physiological studies, but unfortunately some fungi may grow poorly or not at all on such media. Such fungi may grow quite well on a semisynthetic medium, one which is basically similar to a synthetic medium but to which some natural products such as malt extract or

Basal semisynthetic medium

Glucose_____ 10 g.

Asparagine_____ 2 g.

KH_2PO_4 _____ 1 g.

$MgSO_4 \cdot 7H_2O$_____ 0.5 g.

Fe^{+++}_____ 0.2 mg.

Zn^{++}_____ 0.2 mg.

Mn^{++}_____ 0.1 mg.

Biotin _____ 5 μg

Thiamine_____ 100 μg

Distilled water to make_____ 1 liter

Agar (for solid media)_____ 20 g.

Figure 8–6 Semisynthetic medium that supports growth of a wide variety of fungi. [From V. G. Lilly and H. L. Barnett, 1951, Physiology of the fungi. Used with permission of McGraw-Hill Book Company, New York.]

yeast hydrolysate have been added (Fig. 8–6). The natural materials provide unknown nutrients which stimulate growth of the fungus.

A semisynthetic or synthetic growth medium may be used as a liquid or as a liquid solidified with agar. Agar is a gelatinous polysaccharide prepared from red algae. Agar is mixed with the nutrient liquid, heated to melt the agar, and then allowed to cool. When the preparation reaches about 45°C, the agar solidifies to a gel in which the nutritional fluid is entrapped. The preparation may be placed in flasks or tubes and then sterilized by autoclaving; this kills contaminating microorganisms. The medium can then be inoculated with spores or mycelium of the fungus.

Some elements are essential for the growth of fungi, and if an essential element is lacking from the medium or substratum, the fungus will not survive no matter how abundant other elements are. Studies to determine which elements are essential may be of two basic types. First, mycelia and spores may be analyzed to determine what elements are present. If certain elements (such as carbon, nitrogen, or potassium) are always present regardless of the composition of the medium on which the fungus was grown, one may assume that these elements are essential for growth of the fungus. The second, and more commonly used method, is that of conducting nutritional studies using synthetic media from which the elements in question are eliminated one at a time. If a fungus is unable to grow when sulfur, for example, is eliminated, it is assumed that sulfur is essential. Nutritional studies require that sources of chemical contamination be eliminated as nearly as possible; for this reason, chemicals must be of the greatest purity, water must be glass-distilled, glassware must be chemically clean and preferably of quartz, and agar either must not be used or must be highly purified (agar normally contains several trace elements and vitamins that exert a favorable effect on the growth of fungi).

ESSENTIAL MACROELEMENTS

Almost the entire fungus mycelium is composed of the nonmetallic elements carbon, nitrogen, hydrogen, oxygen, sulfur, and phosphorus and the metallic elements potassium and magnesium. These elements are the *macroelements* which are required in comparatively large quantities by fungi. Carbon, nitrogen, hydrogen, and oxygen are used in the formation of fungal walls, and all these elements take a functional role in the constantly occurring metabolic events in the protoplasm. Hydrogen and oxygen are obtained from water or when organic compounds are metabolized. We shall concentrate our attention on carbon, nitrogen, and the remaining macroelements.

Carbon. About half of the dry weight of fungus cells consists of carbon, which gives an indication of the important role of carbon compounds within the cell. Organic compounds are used as structural materials and also provide energy to the cell upon oxidation. Carbon is required in greater quantities than any other essential element by the fungus, and carbon nutrition is of paramount importance to the fungus.

Fungi may use a wide variety of organic compounds or carbon dioxide as a source of carbon. Even though atmospheric carbon dioxide can be utilized by some fungi, it cannot supply enough carbon to serve as the only source. Organic compounds which may be used by the fungi include carbohydrates (mono-, di-, oligo-, and polysaccharides) and organic acids. Carbohydrates constitute the most important carbon sources. Fungi differ widely in their abilities to use different carbon sources, and their ability to utilize a particular carbon source may be altered by the combination of nutrients present or by other cultural conditions, such as pH. If a fungus is provided with a mixture of carbon sources, it will sometimes use one carbon source preferentially over the others present. Also, a mixture of carbon sources may support more growth than a single carbon source (e.g., galactose and glucose in combination may give much more growth than galactose or glucose alone) (Burnett, 1976).

Carbohydrates.

Monosaccharides and their derivatives. Monosaccharides are simple sugars which have either five carbon atoms (pentoses) or six carbon atoms (hexoses). Each sugar molecule has an aldehyde (–CHO) and ketone (=CO) group. These groups may be reduced to form an alcohol derivative of the sugar or oxidized to form an acid derivative of the sugar. The sugars may differ in the spatial configuration of the molecule; one series consists of the D and L forms which differ in the position of the hydroxyl group nearest the aldehyde grouping. Only some of the D and L forms occur naturally.

The sugar supporting growth of almost all fungi is D-glucose, a naturally occurring hexose. Many fungi can make equally as good growth on D-fructose and D-mannose as on glucose (exceptions to this general rule are found among the Chytridiales, Blastocladiales, and Saprolegniales). D-Galactose is used by the majority of the fungi, but few of them grow as well on D-galactose as on D-glucose. Fungi which at first may make no growth on D-fructose, D-mannose, or D-galactose may initiate growth after a period of time due to adaptive enzyme formation. Although a sugar other than glucose may give maximum growth of a fungus, it appears that the more closely the configuration of a sugar resembles glucose, the more fungi use it. This is perhaps because the ability of a fungus to use a particular sugar depends on how easily it can be converted to a phosphorylated derivative of glucose which can enter the respiratory pathways. Glucose very often causes catabolite repression.

Of the naturally occurring pentoses, D-xylose supports growth of the largest number of fungi. Growth of a particular fungus on D-xylose may be equal to or better than that on D-glucose, although some fungi may not grow at all on D-xylose. L-Arabinose may support satisfactory growth of some fungi but the amount of growth is usually less than on either D-glucose or D-xylose. The remaining pentoses may be utilized by some fungi but in general appear to be poor sources of carbon.

Many sugar alcohols, such as sorbitol, glycerol, and mannitol, occur in nature. These sugar alcohols may be utilized by fungi as a carbon source, but usually somewhat less satisfactorily than the corresponding simple sugar. An exception is mannitol, derived from the reduction of D-fructose or D-mannose, which may give growth equivalent to that obtained on glucose for some fungi. However, many fungi which have wide substrate ranges will not grow at all on mannitol. Acids derived from simple sugars and glycosides may serve as a carbon source for certain fungi, but data are rare and it is not possible to make generalizations about the utilization of these compounds by fungi.

Disaccharides and polysaccharides. Simple sugars or their derivatives may combine to form complex, chainlike polymers. Either like or unlike sugar units (monomers) may combine. Repeating units of the same sugar may form two types of polymers which differ only in the spatial configuration (alpha or beta) of the glycoside linkage. If the polymer consists of only two monomers, it is a disaccharide; longer polymers are polysaccharides.

Both disaccharides and polysaccharides are important sources of carbon in nature. High-molecular-weight polysaccharides constitute the bulk of the carbohydrates produced by animals and plants. Disaccharides are abundantly produced and also may be freed from polysaccharides upon their breakdown. In order to utilize either di- or polysaccharides, the fungi must produce extracellular digestive enzymes which will cleave the glycoside linkages between the monomers. After the sugars or their derivatives are freed by digestion, the fungus can usually absorb and utilize the simple sugar in the same fashion that it would utilize a freely occurring sugar. It is evident that the ability of the fungus to utilize these large compounds depends on its ability (1) to digest them and (2) to absorb the component simple sugars. Usually, any inability to utilize a complex carbohydrate is due to the failure to hydrolyze it and not to the inability to absorb the simple sugars. A fungus capable of hydrolyzing a given polymer is usually able to grow on any of the component sugars if they are provided in their simple form (any fungus that can grow on a glucose polymer can grow on free glucose).

The most commonly occurring disaccharides are maltose, cellobiose, lactose, and sucrose. Maltose, a hydrolysis product of starch, consists of glucose molecules joined by an *a*-glycoside linkage. Cellobiose, like maltose, also consists of glucose molecules but differs by having a *β*-glycoside linkage. Cellobiose is a hydrolysis product of cellulose. Lactose, a component of milk, is composed of glucose and a molecule of galactose. Sucrose, which occurs in plants, contains one molecule of glucose and one of fructose. These disaccharides are digested to their component molecules by the enzymes maltase, cellobiase, lactase, and sucrase, respectively. Maltose and cellobiose are widely available to fungi; fewer fungi use sucrose; and even fewer use lactose.

Polysaccharides widely available in nature include the pentosans (polymers of pentose); glycogen, starch, and cellulose (all polymers of D-glucose but differing in degree of branching or configuration of glycoside linkage); pectins (polymers of D-galacturonic acid, an acid derivative of D-galactose); and hemicelluloses, lignins, and gums (all of a complex and poorly understood composition). All of these polysaccharides are utilized by at least some fungi. Starch and cellulose are widely utilized as carbon sources by fungi. The breakdown of starch, a storage product of plants, is diagrammed in Fig. 8–7. Digestion of cellulose, hemicelluloses, and lignins and the role of these in nature is discussed more fully in Chapter 12.

Organic acids. An organic acid possesses one or more carboxyl groups. Organic acids include the monocarboxylic acids (fatty acids that form a fat when esterified with glycerol), dicarboxylic acids (the four-carbon acids of the tricarboxylic acid cycle), and both monocarboxylic and dicarboxylic amino acids that may combine to form a protein. Both fatty acids and amino acids may be derived from these larger molecules by digestion. Most fungi apparently produce the enzyme lipase, which hydrolyzes fats to release fatty acids. Similarly, peptones may be hydrolyzed to release their component amino acids.

The organic acids may serve as a sole carbon source for some fungi but, in general, fungi grow poorly or not at all on them. Aside from the intrinsic incapability of many cells to metabolize organic acids, the general unavailability of

Figure 8–7 Digestion of starch.

these acids may be partially a result of impermeability of the cell to them and of their tendency to interfere with growth by causing an increase in pH of the culture medium if used in the neutralized form (Cochrane, 1958).

Nitrogen. Nitrogen is required by all organisms to synthesize amino acids and, from these, proteins, which are required to build protoplasm. Without protein, growth could not occur. Nitrogen is also a component of the nucleic acids and some vitamins.

Fungi may use inorganic nitrogen in the form of nitrates, nitrites, or ammonia or organic nitrogen in the form of amino acids. It is questionable whether fixation of atmospheric nitrogen, which is so well known among the bacteria, occurs in the fungi. Not all fungi use nitrogen sources with equal facility, and a fungus may have a requirement for nitrogen in a specific form.

Nitrates. Numerous fungi utilize nitrates as a form of nitrogen, but inabilities to use nitrates are common among the Blastocladiales, Saprolegniaceae, yeasts, and the higher basidiomycetes. The nitrate ion may be incorporated into the medium as ammonium nitrate, potassium nitrate, sodium nitrate, and calcium nitrate (Cochrane, 1958).

The nitrate ion must be reduced to the oxidation level of ammonia (NH_3) before the nitrogen can be assimilated into organic compounds. Overall, this stepwise reduction may be summarized as follows:

$$\text{nitrate} \longrightarrow \text{nitrite} \longrightarrow \underset{\text{intermediate}}{\text{unknown}} \longrightarrow \text{hydroxylamine} \longrightarrow \text{ammonia}$$

Oxidation states:
$$+5 \qquad +3 \qquad +1 \qquad -1 \qquad -3$$

The first step in the reduction sequence involves the reduction of the nitrate (NO_3-) ion to the nitrite (NO_2-) ion: this is mediated by the enzyme nitrate reductase. Nitrate reductase obtained from *Neurospora* has been studied extensively (Garrett and Amy, 1978). It is a complex enzyme containing a heme moiety (cytochrome b_{557}), sulfhydryl group(s), FAD (flavin adenine dinucleotide), and molybdenum (Garrett and Nason, 1969). When the reduction of nitrate occurs, the enzyme moieties participate in the sequential transfer of electrons. This sequence begins when NADPH (reduced nicotinamide adenine dinucleotide phosphate) serves as the electron donor and the FAD, heme (cytochrome b_{557}), and molybdenum of the enzyme function as sequential carriers. Ultimately, the nitrate ion receives electrons and is reduced to nitrite. This pathway operates in close association with cytochrome c, which may receive part of the electrons. We may summarize the flow of electrons in *Neurospora* as follows (Garrett and Nason, 1969):

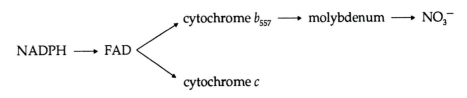

$$\text{NADPH} \longrightarrow \text{FAD} \left\langle \begin{array}{l} \text{cytochrome } b_{557} \longrightarrow \text{molybdenum} \longrightarrow \text{NO}_3^- \\ \\ \text{cytochrome } c \end{array} \right.$$

The final portion of the scheme above, which culminates in nitrate reduction is

$$\text{Mo}^{5+} + 2e \; \big\rangle\!\big\langle \; \begin{array}{l} \text{NO}_3^- \\ \text{NO}_2^- + \text{H}_2\text{O} \end{array}$$
$$\text{Mo}^{6+}$$

Subsequently, reduction of the nitrite ion (NO_2-) to ammonia occurs with two electron changes made for each step in the following sequence (Nason, 1962):

$$\text{NO}_2^- \xrightarrow[\text{reductase}]{\text{nitrite}} \begin{array}{l} \text{?(HNO) nitroxyl group} \\ \text{or} \\ \text{?NO}_2\text{NH}_2 \text{ nitramide} \\ \text{or} \\ \text{?H}_2\text{N}_2\text{O}_2 \text{ hyponitrous acid} \end{array} \longrightarrow \text{NH}_2\text{OH} \xrightarrow[\text{reductase}]{\text{hydroxylamine}} \text{NH}_3$$

The identity of the $+1$ oxidation state intermediate is unknown. Nitrite reductase, which reduces nitrite, and hydroxylamine reductase, which reduces hydroxylamine, have been detected in extracts from filamentous fungi. Both are pyridine-nucleotide enzymes conjugated with an unidentified metal. Like nitrate reductase, nitrite reductase has a heme moiety (Garrett and Amy, 1978). Nitrate utilization is under genetic control, as illustrated in Fig. 8–8.

In general, any fungus that can reduce and assimilate the nitrate ion can utilize any form of nitrogen with an oxidation level equivalent to that of ammonia (such as the ammonium ion or organic nitrogen). There are, however, some fungi that utilize nitrate exclusively.

Nitrites. Nitrites may serve as the sole source of nitrogen for some fungi, and some nitrate fungi may be able to utilize nitrites. Nitrites tend to be toxic to most species of fungi, however, especially if they accumulate in the medium and cannot be used to satisfy the nitrogen requirement. Nitrite can exert its toxic effects by deaminating amino acids or by interfering with sulfur metabolism. Nitrites are not routinely used in the preparation of laboratory media.

Ammonium. Numerous fungi are unable to utilize nitrates as a source of nitrogen but require a more reduced form of nitrogen, presumably because they cannot reduce the nitrate ion. These fungi may use nitrogen in the form of the ammonium ion or in the form of organic nitrogen, which has the same oxidation level as the ammonium ion. The ammonium ion may be supplied as an ammonium salt. Ammonium ions of different salts are equivalent physiologically, but

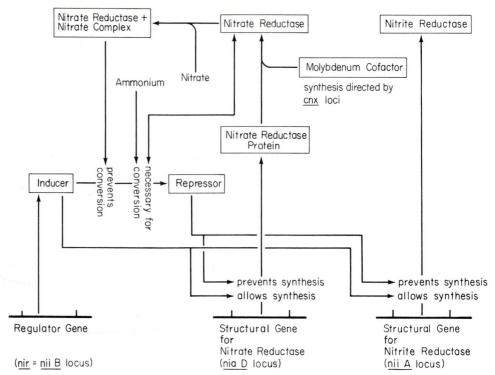

Figure 8–8 Model illustrating the possible mechanisms of the control of nitrate and nitrite production in *Aspergillus nidulans*. The *nir* gene produces a substance that is necessary for expression of the structural genes. This substance is converted to a repressor, which prevents the expression of the structural genes, either by ammonium or by nitrate reductase when it is not complexed with nitrate. Complexed with nitrate, nitrate reductase interferes with the conversion of the substance to a repressor. [From D. J. Cove and J. A. Pateman, 1969, J. Bacteriol. 97:1374.]

the anions are not. Ammonium ions of organic salts do not cause a drop of the pH of the medium due to the buffering capacity of the organic ion. If ammonium nitrate is added to the medium, preferential utilization of the ammonium ion causes a decrease in the pH of the medium, perhaps to the point where it limits growth. When ammonium nitrate is added to the medium, preferential utilization of the ammonium ion is the rule, although the nitrate ion is present (Fig. 8–8). Often the nitrate ion will not be utilized until the ammonium ion is exhausted. This preferential utilization is probably the result of a feedback repression, whereby the end product represses the activity of an enzyme(s) producing it. Ammonia will repress nitrate reductase activity, thereby repressing conversion of nitrate to nitrite. It is possible that preferential utilization of ammonia over nitrate is of some ecological value for soil fungi. In the soil, both nitrates and ammonia usually occur together, but it requires less energy expenditure by the fungus to use the reduced form of nitrogen.

Organic Nitrogen. Essentially all fungi utilize nitrogen supplied to them in the organic form, although only a comparatively small number must obtain their nitrogen in this form. The amino acid-requiring fungi include some Oomycetes (*Leptomitus lacteus* and *Apodachlya brachynema*) and species of *Blastocladia*. Fungi as a group actively decompose proteins in nature to their component amino acids, which can be assimilated. In culture, organic nitrogen may be supplied as amino acids, peptides, or peptones, the last two yielding amino acids upon hydrolysis. The fungus species vary greatly in this response to the different amino acids and use some of them much more readily than others. Although the response of different species to any single amino acid varies widely, asparagine most often gives good growth. Good growth is generally obtained also with glycine, glutamic acid, and aspartic acid. That amino acid most often giving poor growth is leucine. Growth is usually better with a mixture of amino acids than with a single acid.

Some Regulatory Aspects of Nitrogen Utilization. There are many facets of nitrogen nutrition which are under complex genetic control. These include urea, purine, nitrate, amino acid, and ammonia uptake and metabolism. To explore this topic completely is beyond the scope of this book, but to give an example of the type of regulation that exists, we examine that involved with nitrate uptake in *Aspergillus nidulans.*

In *A. nidulans,* the enzymes involved with nitrate reduction can be induced by the presence of nitrate or nitrite, but are repressed by presence of ammonia. Both the induction and repression directly affect the synthesis of these key enzymes (Fig. 8–8). A number of genes are known to be involved in nitrate reduction. Two are structural genes directly coding for the enzymes: *niaD* is the structural gene for nitrate reductase and *niiA* is the structural gene for nitrite and hydroxylamine reductase. The normal functioning of these genes is required for the production of the enzyme molecule, and a recessive mutation in either can result in the inability to utilize nitrate or nitrite, respectively. Five *cnx* genes code for cofactors, and mutation of these genes affects electron transfer. Mutations in the cofactor genes and the structural gene for nitrate reductase can result in (1) the loss of nitrate reductase activity, (2) constitutive synthesis of all enzymes in the nitrate-reduction pathway in the absence of ammonia, and (3) altered response to nitrate induction in the presence of ammonia. Another gene, *nirA,* regulates *nirD* and *niiA,* and its activity is necessary for the induction of nitrate and nitrite reductase. Mutations in this positive-acting regulator gene result in the loss of enzyme activity (Pateman and Kinghorn, 1976).

A second regulatory gene *areA* affects virtually all enzymes whose synthesis is subject to nitrogen metabolite repression by glutamine or ammonia, which are the preferred nitrogen sources when they are present. *AreA* is a positive-acting regulatory gene and its normal expression favors the production of nitrate reductase and other enzymes. Mutants of *areA* have a loss of function, resulting in the repressed levels of enzymes and permeases involved in nitrogen utilization. As a result, the fungus is able to utilize ammonia or glutamine only, resulting in catabolite repression (Arst and Scazzocchio, 1985). There is indirect evidence that nitrogen repression operates by interfering with the normal mRNA transcription (Waime et al., 1985).

Another gene, *crnA,* also is involved in nitrogen metabolism. This negative-acting regulatory gene primarily regulates carbon catabolite repression, but under certain conditions mutants of this gene can reverse the repression caused by *areA* mutants. Three of the genes affecting nitrate metabolism occur clustered in linkage group VIII, where they are aligned in the order *crnA, niiA, niiD.* This arrangement of genes affecting nitrate uptake has some resemblance to the linear arrangement of genes that occurs in the operons of prokaryotes. However, this is not considered to be an operon because the messenger RNA produced by the fungi in general transcribe only one gene. This is unlike the prokaryote operon, which results in the transcription of a number of genes sequentially on a single mRNA molecule (Arst, 1981).

Sulfur. Fungi have a requirement for sulfur, which is generally satisfied in culture by the incorporation of the sulfate ion, SO_4^{2-} (usually as magnesium sulfate), in the medium. Fungi reduce the sulfate ion by an unknown mechanism and metabolize needed sulfur compounds from the reduced sulfur. Sulfur compounds detected in fungi include amino acids (cysteine, cystine, and methionine), the tripeptide glutathione, vitamins (thiamine and biotin), antibiotics (penicillin and gliotoxin), and other miscellaneous compounds.

Some fungi are unable to reduce the sulfate ion and have a requirement for reduced sulfur. These fungi include several members of the Saprolegniales, all members of the Blastocladiales, and some induced mutants of higher fungi. Some fungi may be able to utilize organic sulfur although they do not require it. Reduced sulfur may be supplied as cysteine, cystine, glutathione, and methionine, although the fungus may utilize one form better than another.

Phosphorus, Potassium, Magnesium. Approximately 0.001 to 0.004 M concentrations of phosphorus, potassium and magnesium are sufficient to support normal fungus growth. The inorganic form is utilized. Phosphorus metabolism is closely interrelated with respiration and carbohydrate metabolism, and a high concentration of phosphorus in the medium increases utilization of carbohydrates. Magnesium is involved with activation of enzyme systems, particularly those utilized in aerobic and anaerobic respiration. Although fungi will not grow in the absence of potassium, its role in the cell is not known.

ESSENTIAL MICROELEMENTS

Metallic elements required in trace quantities by fungi are iron, zinc, copper, manganese, molybdenum, and either calcium or strontium. Some fungi may have a requirement for a microelement which is not generally shared by other fungi. *Aspergillus niger,* for example, requires gallium (Steinberg, 1938) and scandium (if glycerol is used as a carbon source) (Steinberg, 1939), while other fungi may require cobalt. Boron, an essential microelement for green plants, is needed in only trace quantities, if at all, by fungi.

Microelements are usually required in concentrations ranging from 0.0001 to 0.5 ppm (parts per million). Molybdenum is needed in such small quantities that it is difficult to establish the actual amount needed, and estimates range between 0.1 and 10 ppb (parts per billion). Quantities of microelements greatly in excess of the required amount are generally toxic. When media are prepared, these microelements are not usually directly added, since chance impurities present on glassware, in the distilled water, or from chemicals usually provide them in sufficient amounts. Some of them may even be introduced on dust particles that fall into the medium during preparation. All the essential microelements, in addition to 15 other minor elements, occur in yeast extract, often used in media (Grant and Pramer, 1962).

Microelements play diverse roles in the cell but are chiefly associated with enzymes. An enzyme may be activated by the microelement or may contain the microelement as part of its structure. Molybdenum, for example, participates as an electron carrier in the enzymatic reduction of nitrates in some fungi (see also page 287). The microelements may also be structural components of vitamins and other metabolites and are therefore required for their synthesis. Iron, for example, is contained within the enzyme catalase, the cytochromes involved in electron transport, and other metabolites, including growth factors and pigments. Microelements are required for normal growth and sporulation, and deficiencies may have various effects on the fungus (Fig. 8–9). The effect of a deficiency for a particular microelement is largely determined by the function of that microelement; for example, a deficiency of zinc or manganese, which normally activate enzymes of the tricarboxylic acid cycle, may impair the normal operation of this cycle. Manganese may also decrease the rate of sporulation generally, while a deficiency of copper may result in the reduction of pigmentation in the spores of some fungi (Cochrane, 1958).

Fungi may also take up metallic elements that are not known to be essential. These include mercury, nickel, lead, uranium, and possibly others. High concentrations of any of these elements may be toxic, and as mentioned above,

(a)

(b)

(c)

Figure 8–9 (a, b) Severe manganese deficiency in the Pyrenomycete *Chaetomium globosum*; symptoms are a decrease in elongation of the hypha, increased branching, and formation of globose cells; (c) normal hyphae growing in medium with 0.5 ppm manganese. All × 750. [From H. L. Barnett and V. G. Lilly, 1966, Mycologia 58:585–591.]

excessive quantities of the essential microelements may also be toxic. Mercury and the essential microelement copper are widely used as major constituents of fungicides. In toxic concentrations, much of the copper binds to the plasmalemma and interferes with transport of other materials into the cell. Copper which has gained entrance to the cell may cause genetic mutations (Ross, 1975).

VITAMINS

Vitamins are organic compounds that function as coenzymes or as constituent parts of coenzymes which catalyze specific reactions. They are effective in very small quantities and are not used in the manufacture of structural parts of the cell. All organisms require at least some vitamins, but the synthetic capacities vary. Green plants manufacture their own vitamins, but animals must obtain their vitamins from an outside source.

Some fungi, if supplied with a sugar, nitrogen source, and minerals, are able to synthesize all the vitamins that they require. They may even synthesize vitamins in such quantity that they are excreted into the medium. Other fungi may have a partial vitamin deficiency and make limited growth on a vitamin-free medium but much better growth if an exogenous vitamin source is provided (Fig. 8–10). Other fungi have a complete deficiency for a given vitamin and must rely totally on an exogenous supply. Fungi may have only a single vitamin deficiency, or they may have multiple deficiencies and require more than one vitamin. The vitamin deficiency may be absolute and unvarying, or it may be a conditioned deficiency which disappears as the mycelium ages or when external conditions are altered. Vitamin deficiencies that can be conditioned by external factors are partial deficiencies; complete deficiencies are virtually unalterable. A partial requirement becomes more evident as the upper temperature limits allowing growth are reached and sometimes when the composition of the medium is changed. As an example of the latter, *Pythium butleri* becomes thiamine-requiring when the mineral salt concentration exceeds a certain level (Robbins

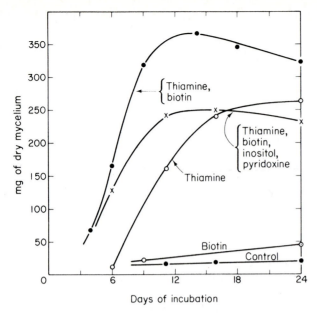

Figure 8–10 Growth of *Lambertella pruni* in medium containing various vitamins. Partial deficiencies for both thiamine and biotin are evident, being greater for thiamine. Note that the addition of inositol and pyridoxine to media containing thiamine and biotin depressed growth. [From V. G. Lilly and H. L. Barnett, 1951, Physiology of the fungi. Used with permission of McGraw-Hill Book Company, New York.]

and Kavanagh, 1938). *Pellicularia koleroga* requires both thiamine and biotin when growing on sucrose; when growing on glucose, it requires thiamine alone (Mathew, 1952).

Fungi apparently do not have a need for vitamins A, D, and E, which are required by animals, and synthesis of these vitamins has not been detected in fungi. Fungi do synthesize and require the water-soluble B vitamins and vitamin H.

Fungi often respond with increased growth to additions of natural materials such as tomato or coconut juice, dung or wood extracts, extracts of the host plant, or ash. In many cases, an equivalent growth cannot be obtained by the addition of specific purified vitamins or minerals, and the increase in growth is attributed to unknown *growth factors* present in the natural materials. Those materials giving increased growth may eventually prove to be vitamins, specific minerals, amino acids, or other nutrients. In many cases the growth factors prove to be a chemical which does not ordinarily serve as a nutrient; for example, choline, hemin, or sterols may serve as growth factors for some fungi. Many growth factors have not been identified.

Vitamins manufactured by fungi are thiamine (vitamin B_1), biotin (vitamin H), pyridoxine (vitamin B_6), nicotinic acid, pantothenic acid, riboflavin (vitamin B_2), inositol, and *p*-aminobenzoic acid (Fig. 8–11).

Thiamine. The greatest number of deficiencies are for thiamine, and any filamentous fungus which has a single vitamin deficiency is most likely to be thiamine-deficient (Fig. 8–12). All species of *Phytophthora* are thiamine-deficient, but only a few basidiomycetes lack this vitamin. Next to biotin, yeasts are most often deficient for thiamine. The thiamine molecule contains two ring structures, a substituted pyrimidine and a substituted thiazole:

thiamine pyrimidine

thiamine thiazole

Thiamine

Biotin

Pyridoxine

Nicotinic acid

Pantothenic acid

Inositol

Para-aminobenzoic acid

Riboflavin

Figure 8–11 Structural formulas of some vitamins important to fungi.

Figure 8–12 Effect of thiamine on growth of *Phycomyces blakesleeanus*. Growth is absent in the thiamine-deficient medium on the left, and is abundant in the thiamine-supplemented medium on the right. [Courtesy Plant Pathology Department, Cornell University.]

Thiamine-deficient fungi vary in their capacity to utilize either the pyrimidine or thiazole moiety or the entire thiamine molecule. Many thiamine-deficient fungi must be supplied with the entire preformed thiamine molecule and are apparently unable to synthesize any part of it. Numerous fungi can make the thiazole part of the molecule and will synthesize the thiamine molecule if provided with the pyrimidine moiety. Some fungi have the reverse synthetic capacity and can manufacture the pyrimidine ring only. *Phycomyces* can combine the pyrimidine and thiazole rings to make thiamine, but it is unable to synthesize either intermediate.

Thiamine has a role in the regulation of carbohydrate metabolism. The active form of thiamine is thiamine pyrophosphate (a pyrophosphoric ester). This is also known as cocarboxylase, the coenzyme of carboxylase. Pyruvic acid, a key intermediate in carbohydrate metabolism, is transformed into acetaldehyde and carbon dioxide by the enzyme carboxylase. Pyruvic acid tends to accumulate in cultures of thiamine-deficient fungi.

Biotin. Numerous fungi are deficient for biotin, although this deficiency is not nearly so common as that for thiamine. Most yeasts are biotin-deficient. The biotin molecule is synthesized according to the following scheme:

$$\text{pimelic acid} \longrightarrow \text{desthiobiotin} \longrightarrow \text{biotin}$$

Some biotin-deficient fungi can grow if provided with the precursor desthiobiotin and are apparently unable to convert pimelic acid to desthiobiotin. Such fungi include the Pyrenomycetes *Ceratostomella ips, Neurospora crassa,* and some yeasts. Other fungi cannot grow if provided with desthiobiotin but require biotin to satisfy their requirements. These fungi, which include the Pyrenomycetes *Ceratostomella pini* and *Sordaria fimicola,* have a metabolic block which makes it impossible for them to convert desthiobiotin to biotin.

The role of biotin is only vaguely understood. It is apparently involved in the synthesis of aspartic acid, which may partially replace a biotin requirement; in the synthesis of fatty acids such as oleic acid, it may partially replace biotin for a *Neurospora crassa* mutant and for *Ceratostomella pini* (Cochrane, 1958).

The Remaining Vitamins. Pyridoxine (vitamin B_6) is required by fewer fungi than either thiamine or biotin. Partial or complete deficiencies for pyridoxine are known among some yeasts and are especially common in species of *Ceratostomella.* The coenzyme form of pyridoxine serves as a coenzyme for various enzymes involved in amino acid metabolism.

A deficiency for pantothenic acid is fairly common among the yeasts, especially among species of *Saccharomyces,* but it is virtually unknown among the filamentous fungi. Pantothenic acid is a constituent of coenzyme A which mediates in acyl transfers and related reactions. Pantothenic acid favors the accumulation of glycogen and increases respiratory activity in yeasts.

Deficiencies for inositol, riboflavin, and *p*-aminobenzoic acid are rare. Riboflavin provides the prosthetic group for the flavin enzymes, required in respiration, and for various enzymes involved in nitrogen metabolism. The roles of inositol and *p*-aminobenzoic acid are unknown.

BIOASSAY

Bioassays can be used to determine amounts of a substance within a complex material such as food or soil with the aid of a microorganism. Some bioassays are carried out by utilizing fungi deficient for vitamins or minerals. These include *Phycomyces blakesleeanus* for thiamine (Fig. 8–12), *Saccharomyces carlsbergensis* for pyridoxine and panthothenic acid, mutants of *Neurospora crassa* for *p*-aminobenzoic acid and inositol, and *Saccharomyces cerevisiae* for biotin. *Asper-*

gillus niger can be used to determine available copper, magnesium, potassium, and molybdenum in the soil. Few fungi isolated from nature are deficient for amino acids, so that the majority of amino acid bioassays are done with bacteria.

Bioassay requires the selection of an appropriate microbe deficient for the test substance, and which is sensitive enough to its presence to show a growth response to dosage increases. The growth/dosage response is established by using a synthetic medium that lacks the test substance but contains all other necessary nutrients. Chemically pure samples of the vitamin, amino acid, or other test substance are added to the medium in a series of increasing concentrations. Using carefully standardized inoculum as well as rigidly controlled cultural conditions (e.g., light, temperature, etc.), the growth of the deficient fungus on these media is evaluated and a standard growth curve is derived. Then, using similar techniques, the growth of the fungus on the test material can be evaluated and compared with the standard growth curve to determine the concentration of the test substance.

Bioassay sometimes offers advantages over chemical analysis. It is not necessary to obtain the substance in chemically pure form and minute quantities of the substance may suffice for bioassay. In some cases, bioassays may be more accurate than a chemical analysis—bioassay of the mineral content of the soil measures the minerals that are available to plants, not the absolute amounts present.

PHYSICAL REQUIREMENTS FOR GROWTH

In addition to having nutrients, fungi must be exposed to favorable temperature, moisture, pH, and light conditions for growth to take place. For each of these physical environmental factors, there is a range in which growth will occur, delimited by a minimum point below which no growth will occur and a maximum point above which no growth will occur. There is usually an optimum over some small part of the range, indicating that the greatest growth rate occurs there (Fig. 8–13). Growth curves that are drawn as a function of temperature or other physical factors are characterized by the minimum, optimum, and maximum points. These cardinal points are a useful index to the behavior of a given fungus and as a point of comparison between fungi, but these are not absolute values because they may change with an alteration in cultural conditions, the age of the mycelium, or the particular genetic strain of the fungus.

Virtually all reactions within the cell are affected by physical factors, especially temperature and pH, and each reaction tends to have its own minimum, optimum, and maximum. Cell processes have widely different ranges in which they can operate; a given temperature, for example, favors some reactions and not others. The optimum value for growth, then, is that point at which a given

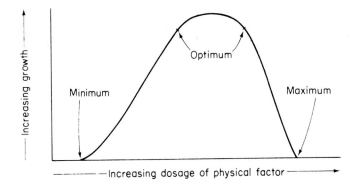

Figure 8–13 Theoretical growth curve of a fungus in response to a physical factor, showing minimum, optimum, and maximum points.

physical factor is optimum, or at least favorable, for the greatest number of metabolic processes.

A change in the physical environment may alter the response of the fungus to other factors influencing growth. Conditional vitamin requirers, for example, may require an exogenous source of a vitamin at certain temperatures and pH values but not at others (Fig. 8–14).

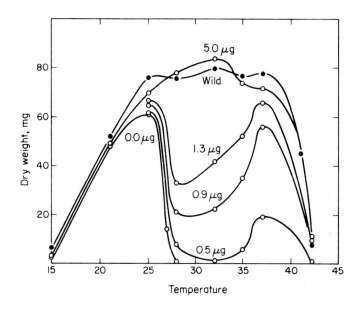

Figure 8–14 Growth of a wild-type *Neurospora* and a temperature-sensitive riboflavin-deficient mutant in response to temperature and to different concentrations of riboflavin (indicated on curves in micrograms per 20 milliliters of medium). Below 25°C, no riboflavin is required for growth, while no growth occurs above 28°C without an exogenous source riboflavin. [From H. K. Mitchell and M. B. Houlahan, 1946, Am. J. Botany 33:31–35.]

Temperature

Temperature is extremely important in determining the amount and rate of growth of an organism. Increasing temperature has the general effect of increasing enzyme activity and chemical activity. Many chemical reactions increase tenfold for each 10°C rise in temperature, but enzyme activity usually increases only twofold for each 10°C increase. Enzymes eventually become inactivated at high temperature levels; some enzymes are inactivated at temperatures as low as 30°C. In addition to having a general effect on the metabolic rate, temperature may directly affect the synthesis of vitamins, amino acids, or other metabolites—failure of some fungi to grow at high temperatures is directly due to an inability to synthesize such a needed component.

Beginning with the minimum temperature permitting growth, the amount of growth of the fungus increases with the increase in temperature until the optimum is reached. Throughout the optimum range, the growth rate becomes more or less steady. When the temperature increases above the optimum, there is a decline in growth rate until the maximum point permitting growth is reached (Fig. 8–15). If several cultures are grown at different temperatures and the amount of growth plotted, one can derive a typical growth curve illustrating this response.

The vast majority of the fungi in the temperate regions are *mesophiles,* thriving at moderate temperatures between 10 and 40°C (Fig. 8–16). A typical mesophilic fungus has a minimum temperature of 0 to 5°C. In most cases, the optimum temperature range lies somewhere between 15 and 30°C, causing the temperature growth curve to be skewed to the right. Few mesophilic fungi are able to grow above the maximum temperatures of 35 to 40°C. Some mesophilic fungi are able to survive and able to grow very slowly at either higher or lower temperature extremes. Many can cause spoilage of refrigerated food because of this ability to tolerate low temperatures. The point at which high temperatures

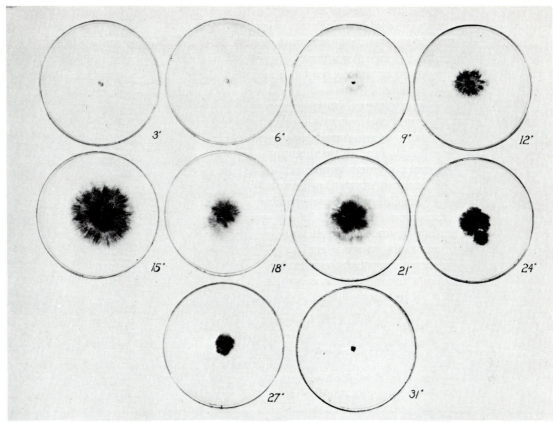

Figure 8–15 Differential effect of temperature on the growth of *Phoma apiicola* in culture. [Courtesy Plant Pathology Department, Cornell University.]

can cause death may be much higher than that permitting growth. The thermal death point for many fungi lies approximately between 50 and 60°C in moist heat, but temperatures as high as 105°C may be required for 12 hours to kill all the wood-rotting fungi in dry blocks of wood (Cartwright and Findlay, 1934).

The *psychrophiles* grow at cooler temperatures in environments where the temperatures rarely rise above 10°C, and have a growth optimum of 10°C or below (Deverall, 1968). These fungi occur abundantly in environments characterized by extreme cold where ice and snow cover the ground much of the year. Such environments include the arctic, where air temperatures climb to 10°C during only one month of the year, and near the tops of high mountains such as those in the Alps or western United States. Some mushrooms have been recovered adjacent to the melting edges of the snow at high altitudes (Deverall, 1968). Some plant pathogens are most dangerous when temperatures are low and the plants are covered by deep snow on unfrozen ground and cause diseases generally known as snowmolds. The temperature underneath the snow can be at 0°C or slightly below. Hess et al. (1985) investigated ultrastructural changes in a snowmold associated with growth at different temperatures. This fungus can grow at −4°C and causes disease of its host at about 1°C, although its optimum temperature for growth is between 8 and 12°C. Large reserves of glycogen accumulate when it is grown at the lower temperatures, and may be correlated with the ability to grow at the low temperatures.

Only a few dozen *thermophilic* fungi are known. These thrive at extremely high temperatures, having an optimum temperature for growth at 40°C or above. The minimum temperature for the growth of thermophiles is 20°C or above, and their maximum may be more than 50°C. Thermophilic fungi occupy

Figure 8–16 Temperature–growth curves of eight fungi which cause potato-rot and wilt diseases. [From H. A. Edson and M. Shapovalov, 1920, J. Agr. Res. 18:511–524.]

specialized ecological niches such as dung and compost, where heat rises naturally due to microbial activity. They are also an important part of the microbial flora in soils thermally heated in volcanic regions. The high temperatures are actually favorable to the heat-loving fungi because their mesophilic competitors are eliminated (Emerson, 1968).

Hydrogen Ion Concentration

All fungi are in contact with aqueous solutions in nature or in the laboratory, and the hydrogen ion concentration (pH) of these solutions exerts control over fungal growth. Effects of pH are exerted in various ways.

One effect of pH is on the availability of certain metallic ions. Metallic ions may form complexes that become insoluble at certain pH ranges. Magnesium and phosphate ions may coexist in their free form at a low pH, but at a higher pH they form an insoluble complex, reducing the availability of these ions to the fungus. Iron deficiency may result in an alkaline medium as the ferric ion complex becomes insoluble at a high pH. The effect is similar with the calcium and zinc ions.

Cell permeability is altered with different degrees of acidity or alkalinity. The effect is particularly noticeable on compounds that ionize. A possible explanation is that at a lower pH the plasmalemma becomes saturated with hydrogen ions so that passage of essential cations is limited, while at a higher pH this membrane becomes saturated with hydroxyl ions and thereby limits the entrance of essential anions. *p*-Aminobenzoic acid exists predominantly as a free acid at a very low pH, and uptake of this vitamin is greatest at a low pH. About eight times as much *p*-aminobenzoic acid is required at pH 6.0 as at pH 4.0 to support the same amount of growth.

External hydrogen ion concentration also affects pH within the cell, which in turn affects enzyme activity. Enzymes are inactivated at either pH extreme, but they have different optimum pH levels for activity. Some are more active in a weakly acidic solution, while others are more active in a weakly alkalinic solution. Optima for most enzymes lie between pH 4 and 8 (Fig. 8–17). An unfavorable pH may alter normal synthetic ability of the cell. *Sordaria fimicola*, for example, is usually able to synthesize its own thiamine; but when it is cultured on a medium with an initial pH of 3.6 to 3.8, it is unable to grow unless exogenous thiamine is provided (Lilly and Barnett, 1947).

Fungi have a comparatively broad pH range over which they can grow and the optimum pH for most fungi is on the acidic side of the scale, under pH 7. Any pH-growth curve is a summation of all the effects of pH on the numerous factors which control growth and does not represent a unitary effect. Growth at low pH may be the direct result of greater iron availability, while a growth at higher pH may be due to the increased activity of enzymes with high pH op-

Figure 8–17 Differential effect of pH on the growth of *Ceratocystics ulmi* in culture. [Courtesy Plant Pathology Department, Cornell University.]

tima. It is not surprising that many pH-growth curves show two optimal hydrogen ion concentrations. Such a case are species of *Coprinus*, for which the minima represent pH-dependent unavailability of different ions. If iron, zinc, and calcium are provided in available form, the apparent double optima merge into a single, broad optimum range (Fries, 1956).

pH-Growth curves are not constant but may be modified by an alteration of any number of environmental factors, including temperature, age of the mycelium at the time of harvest, calcium and magnesium levels, and nitrogen source. The ease with which minima, optima, and maxima in pH-growth relationships may be shifted is further testimony to the intricate involvement of hydrogen ion concentration with a large number of growth processes.

HYDROGEN ION CHANGES CAUSED BY THE FUNGUS

Fungi almost invariably alter the pH of the medium in which they grow. An uptake of either anions or cations by the fungus can drive the pH in the opposite direction (Fig. 8–18). A good example of this may be found with different nitrogen sources. If nitrogen is provided as an ammonium salt, utilization of the ammonium ion will make the medium become more acidic. Alternatively, if nitrogen is provided as sodium nitrate ($NaNO_3$), the medium becomes more alkaline as the nitrate ion is removed.

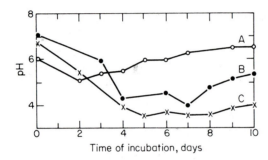

Figure 8–18 Changes in pH of *Penicillium notatum* cultures using glucose and varying sources of nitrogen: (a) $NaNO_3$; (b) mixture of $NaNO_3$ and the amino acids tryptophan, asparagine, and cystine; (c) tryptophan, asparagine, and cystine. [From A. E. Dimond and G. L. Peltier, 1945, Am. J. Botany 32:46–50.]

Conversion of a neutral component to an acidic or alkaline metabolite may also be responsible for a shift in pH of the medium. A common cause of a decrease in pH is the accumulation of organic acids (especially gluconic, pyruvic, citric, and succinic acids) formed from the metabolism of sugars. Carbon dioxide, a byproduct of carbohydrate metabolism, combines with water to form carbonic acid. Carbonic acid may drive the pH down, but it is dissociated in the presence of stronger acids and forms bicarbonates under alkaline conditions. In a similar fashion, the release of ammonium ions from the deamination of amino acids and protein may cause the pH to rise. The ultimate pH of the culture medium depends on the relative rate and occurrence of these pH-changing processes and the extent to which the medium is buffered.

Moisture

Fungi require relatively high moisture levels, although many of the higher fungi are able to grow in the absence of free water. A high relative humidity is also required. Maximum growth for most fungi occurs at a relative humidity of 95 to 100%, and growth declines or ceases in humidity of 80 to 85%. A few fungi will grow at a relative humidity as low as 65%.

Conditions for adequate moisture are satisfied in routine cultural techniques. Agar media contain water bound within a gel, and the atmosphere within a petri dish or flask has a relative humidity near 100%. In nature, the ability to grow on a given substrate is partially determined by the moisture content of the substrate. Wood, for example, is not decayed by fungi if the moisture content is below 20%. Soil fungi grow better in moderate than in high moistures, but this is because soil aeration (and therefore oxygen supply) is limited when the moisture content is high.

Fungi grow either within or in contact with watery solutions normally having at least some dissolved sugars or salts. As you probably know, the concentration of solutes may increase the osmotic concentration of the solvent in which they are dissolved. Most fungi grow best with a relatively low osmotic concentration (0.5 M or less). Many can tolerate higher osmotic concentrations, especially if they are allowed to adapt gradually to increasing levels of higher osmotic concentrations. The degree of tolerance differs from species to species. Most fungi cease growth or are inhibited when the concentration of soluble sugar exceeds about 2.0 M (Burnett, 1976). Fungi tolerant of the higher osmotic concentrations are often found growing on concentrated sugary materials such as jams, nectar, honey, and fruit juices. Such fungi may even require this high osmotic concentration; an example is the yeast ally *Eremascus albus* (Paugh and Gray, 1969). *E. albus* will not grow in a medium with a low sucrose concentration (less than 5%) but will grow directly on sucrose crystals. Optimum growth occurs on a medium with a sucrose concentration of approximately 1.4 M, but it will still grow well on 2.33 M sucrose. Optimum growth of *E. albus* is not dependent upon the availability of sucrose as a nutrient (an inert material producing a high osmotic concentration will give similar results). Other fungi are adapted to the high concentration of salts that may be found in brine or in seawater. Marine fungi typically yield optimum growth when provided with a medium consisting of approximately 50% to 100% seawater, but similar salt concentrations inhibit growth or reproduction of terrestrial fungi. It has been suggested that this difference in degree of tolerance to salinity may be the principal factor maintaining the marine fungi as an ecological group distinct from terrestrial fungi (Jones and Byrne, 1976). Salt-tolerant fungi may be isolated from salted fish, and one of these, *Aspergillus penicilloides*, can actually grow on a medium saturated with sodium chloride. The growth of a few others in culture is favored by salt (Andrews and Pitt, 1987).

Light

Light is a natural component of any environment. The vast majority of fungi are exposed to alternating cycles of daylight and darkness, while other fungi occurring deep within the soil or host tissues are likely to be in total darkness. Light intensities vary in nature: a fungus growing beneath a leaf in the forest is not exposed to as great a light intensity as a fungus growing above the ground or on the surface of a plant in an unshaded clearing.

Although the growth rate and/or the synthetic capacity of some fungi may be controlled by light, the growth of most fungi is apparently not sensitive to light. The most widespread effect of light on fungal growth is that of inhibition in strong light. This inhibition may sometimes be overcome by adding natural materials to the medium, which suggests that strong light may destroy required vitamins (e.g., Barnett, 1968; Robbins and Hervey, 1960). However, the vegetative growth of some aquatic fungi may be stimulated by light. Stimulation of growth has been observed in the marine ascomycete *Buergenerula spartinae* (Gessner, 1976) and in *Blastocladiella emersonii* (a member of the Blastocladiales). The dry weight of *B. emersonii* grown in light may be as much as 141% of the dry weight of cultures grown in the dark (Cantino and Horenstein, 1956). Light may exert both stimulatory and inhibitory effects on the vegetative structures of the same fungus. An example is *Candida albicans*, which has growth inhibition when exposed to light but also has enhanced carbohydrate synthesis (Saltarelli and Coppola, 1979).

Blue light depresses synthesis of the pigment melanin and of tyrosinase in a mutant of *Neurospora crassa,* while it has an opposite effect in *Aureobasidium pullulans* and *Cladosporium mansonii* (Page, 1965).

Light may also affect the formation of reproductive structures or may control oriented phototropic movements of reproductive structures. Generally, many more instances are known of light affecting reproduction than vegetative growth. Additional information about the effects of light and the mode of its interaction with fungi is presented in Chapter 10.

Aeration

Two components of the air are of special importance to the fungi: these are oxygen and carbon dioxide. Oxygen is vital to cellular respiration, in which an energy source is oxidized to carbon dioxide and water and energy is made available to the cell. Carbon dioxide may accumulate as a result of cellular respiration by either the fungi or other organisms. Respiration is discussed further in Chapter 9. In addition, as mentioned on page 284, some fungi are able to fix atmospheric carbon dioxide and use this as a source of carbon.

It has been widely recognized that most fungi are *obligate aerobes* and require at least some free molecular oxygen in the atmosphere. Fungi are generally found growing on or near the surface of their substrata (e.g., foods such as meats or cheese, wood of living trees, or soil) but are occasionally isolated from substrata far removed from the air. Many fungi will fail to grow in cultures if molecular oxygen is not freely available. Oxygen pressure in the atmosphere is normally measured as 160 millimeters Hg, but some fungi grow well in extremely small amounts of oxygen (as low as 10 to 40 millimeters Hg). This apparently widespread ability to adjust to wide ranges of available oxygen probably indicates that oxygen variation in nature is ordinarily not critical for fungal growth (Tabak and Cooke, 1968-a). Some fungi are *facultative anaerobes* and not only can survive in the absence of measurable oxygen but also can grow and sporulate under such conditions. Facultative anaerobes can use combined oxygen in addition to free molecular oxygen in the atmosphere. Fungi which are facultative anaerobes are able to survive in many environments having extremely limited oxygen levels; these include sewage sludge and polluted streams (Tabak and Cooke, 1968-b). Although many bacteria are *obligate anaerobes* which cannot grow in the presence of free molecular oxygen, we do not know of any fungi which are obligate anaerobes (Tabak and Cooke, 1968-a).

Effective utilization of carbon or nitrogen compounds as sources of nutrients may be affected by the amount of oxygen available in the atmosphere. For example, in an aerobic atmosphere, *Mucor rouxii* is able to utilize a large variety of carbon sources and amino acids for growth, but under anaerobic conditions it can utilize only hexoses as a carbon source and poorly utilizes amino acids as a nitrogen source (nitrate or ammonium ions serve as a better source of nitrogen under anaerobic conditions). Insufficient oxygen also increases nutritional demands in *M. rouxii* and decreases growth (Bartnicki-Garcia and Nickerson, 1962). Similarly, the growth of many fungi under anaerobic conditions may be improved by the addition of biotin and thiamine (Tabak and Cooke, 1968-b) or yeast extract to the medium (Deploey and Fergus, 1975).

Air usually contains about 0.3% carbon dioxide, but high concentrations of carbon dioxide resulting from cellular respiration may accumulate in tightly closed culture containers or in nature. Natural environments where high carbon dioxide levels occur often result from cellular respiration by various organisms in combination with poor aeration. These environments include stagnant waters and water-logged soils where bacteria may thrive and the woody tissues of a host plant where carbon dioxide produced by the host plant may accumulate. Fungi are usually inhibited by concentrations of carbon dioxide greater than 10 to 15%, but the amount of carbon dioxide tolerated varies among the fungi. *Fusarium oxysporum* and *F. eumartii*, which grow under normal atmospheric conditions, can tolerate a carbon dioxide concentration up to 75.3% (Hollis, 1948). Some fungi that inhabit stagnant water can tolerate a higher carbon dioxide concentration. An example of such a fungus is *Aqualinderella fermentans*, a member of the Oomycetes. *A. fermentans* grows poorly under normal atmospheric conditions, but its growth is actually improved as the carbon dioxide concentration increases to 20%, and it can tolerate a concentration up to 99% (Emerson and Held, 1969). *A. fermentans* is unique because it actually requires carbon dioxide for optimum growth. In addition, *A. fermentans* can grow in the absence of aerobic oxygen. Growth of fungi may sometimes be inhibited by the total absence of carbon dioxide or by concentrations of oxygen much greater than those that normally occur in the atmosphere (Tabak and Cooke, 1968-a).

Another property of the air is that it may contain volatile organic compounds. Fungi are known to produce a wide variety of volatile metabolites, including alcohols, aldehydes, and other organic compounds such as coumarin. Volatile metabolites are produced in nature by higher plants, bacteria, and actinomycetes. Volatiles may accumulate either in closed culture containers or in poorly aerated habitats. Volatiles may exert either inhibitory or stimulatory effects on the growth and reproduction of the fungi: the differential effects depend upon the type and concentration of the volatiles as well as upon the fungal species. Deformation of the hyphae and abnormally thick walls have been observed in some fungi exposed to volatiles produced by bacteria (Moore-Landecker and Stotzky, 1973). Probably most of these volatiles exert their effects upon fungi by regulating internal metabolism and not by serving as a nutrient (Fries, 1973).

REFERENCES CITED

ANDREWS, S., and J. I. PITT. 1987. Further studies on the water relations of xerophilic fungi, including some halophiles. J. Gen. Microbiol. 133:233–238.

ARST, H. N., Jr. 1981. Aspects of the control of gene expression in fungi. *In:* Glover, S. W., and D. A. Hopwood, Eds. Genetics as a tool in microbiology. Cambridge University Press, Cambridge, pp. 131–160.

——, and C. SCAZZOCCHIO. 1985. Formal genetics and molecular biology of the control of gene expression in *Aspergillus nidulans. In:* Bennett, J. W., and L. L. Lasure, Eds. Gene manipulations in fungi. Academic Press, New York, pp. 309–343.

BARNETT, H. L. 1968. The effects of light, pyridoxine, and biotin on the development of the mycoparasite *Gonatobotyrum fuscum.* Mycologia 60:244–251.

BARTNICKI-GARCIA, S., and W. J. NICKERSON. 1962. Nutrition, growth, and morphogenesis of *Mucor rouxii.* J. Bacteriol. 84:841–858.

BROWNLEE, C., and D. H. JENNINGS. 1982. Long distance translocation in *Serpula lacrimans:* velocity estimates and the continuous monitoring of induced perturbations. Trans. Brit. Mycol. Soc. 79:143–148.

BURNETT, J. H. 1976. Fundamentals of mycology, 2nd ed. Crane Russak, New York, 673 pp.

CANTINO, E. C., and E. A. HORENSTEIN. 1956. The stimulatory effect of light upon growth and CO_2 fixation in *Blastocladiella.* I. The S.K.I. cycle. Mycologia 48:777–799.

CARTWRIGHT, K. St. G., and W. P. K. FINDLAY. 1934. Studies in the physiology of wood-destroying fungi. II. Temperature and rate of growth. Ann. Botany (London) 48:481–495.

COCHRANE, V. W. 1958. Physiology of fungi. Wiley, New York, 524 pp.

DEPLOEY, J. J., and C. L. FERGUS. 1975. Growth and sporulation of thermophilic fungi and actinomycetes in O_2–N_2 atmospheres. Mycologia 67:780–797.

DEVERALL, B. J. 1968. Psychrophiles. *In:* Ainsworth, G. C., and A. S. Sussman, Eds. The fungi: an advanced treatise. Academic Press, New York, Vol. 3, pp. 129–135.

EMERSON, R. 1968. Themophiles. *In:* Ainsworth, G. C., and A. S. Sussman, Eds. The fungi: an advanced treatise. Academic Press, New York, Vol. 3, pp. 105–128.

——, and A. A. HELD. 1969. *Aqualindrella fermentans* gen. et sp. n., a phycomycete adapted to stagnant waters. II. Isolation, cultural characteristics, and gas relations. Am. J. Botany 56:1103–1120.

FRIES, N. 1956. Studies in the physiology of *Coprinus.* II. Influence of pH, metal factors, and temperature. Svensk Botan. Tidskr. 50:47–96.

——. 1973. Effects of volatile organic compounds on the growth and development of fungi. Trans. Brit. Mycol. Soc. 60:1–21.

GARRETT, R. H., and N. K. AMY. 1978. Nitrate assimilation in fungi. Advan. Microb. Physiol. 18:1–65.

——, and A. NASON. 1969. Further purification and properties of *Neurospora* nitrate reductase. J. Biol. Chem. 244:2870–2882.

GESSNER, V. 1976. In vitro growth and nutrition of *Buergenerula spartinae,* a fungus associated with *Spartina alterniflora.* Mycologia 68:583–599.

GRANT, C. L., and D. PRAMER. 1962. Minor element composition of yeast extract. J. Bacteriol. 84:869–870.

GRIFFIN, D. H. 1981. Fungal physiology. Wiley, New York, 383 pp.

HESS, W. M., D. L. NELSON, and D. L. STURGES. 1985. Morphology and ultrastructure of a snowmold fungus on sagebrush *(Artemisia tridentata).* Mycologia 77:637–645.

HOLLIS, J. P. 1948. Oxygen and carbon dioxide relations of *Fusarium oxysporum* Schlecht. and *Fusarium eumartii* Carp. Phytopathology 38:761–775.

JENNINGS, D. H. 1984. Water flow through mycelia. *In:* Jennings, D. H., and A. D. M. Rayner, Eds. The ecology and physiology of the fungal mycelium. Cambridge University Press, Cambridge, pp. 143–164.

JONES, E. B., and P. J. BYRNE. 1976. Physiology of the higher marine fungi. *In:* Jones, E. B. H., Ed. Recent advances in aquatic mycology. Wiley, New York, pp. 135–175.

LILLY, V. G., and H. L. BARNETT. 1947. The influence of pH and certain growth factors on mycelial growth and perithecial formation by *Sordaria fimicola.* Am. J. Botany 34:131–138.

MATHEW, K. T. 1952. Growth-factor requirements of *Pellicularia koleroga* Cooke in pure culture. Nature 170:889–890.

MOORE-LANDECKER, E., and G. STOTZKY. 1973. Morphological abnormalities of fungi induced by volatile microbial metabolites. Mycologia 65:519–530.

NASON, A. 1962. Symposium on metabolism of inorganic compounds. II. Enzymatic pathways of nitrate, nitrite, and hydroxylamine metabolism. Bacteriol. Rev. 26:16–41.

PAGE, R. M. 1965. The physical environment for fungal growth. 3. Light. *In:* Ainsworth, G. C., and A. S. Sussman, Eds. The fungi: an advanced treatise. Academic Press, New York, Vol. 1, pp. 559–574.

PATEMAN, J. A., and J. R. KINGHORN. 1976. Nitrogen metabolism. *In:* Smith, J. E., and D. R. Berry, Eds. The filamentous fungi. Edward Arnold, London, Vol. 2, pp. 159–237.

PAUGH, R. L., and W. D. GRAY. 1969. Studies on the growth of the osmiophilic fungus *Eremascus albus.* Mycologia 61:281–288.

PLOMELY, N. J. B. 1959. Formation of the colony in the fungus *Chaetomium.* Australian J. Biol. Sci. 12:53–64.

PLUNKETT, B. E. 1958. Translocation and pileus formation in *Polyporus brumalis.* Ann. Botany N.S. (London) 22:237–249.

ROBBINS, W. J., and A. HERVEY. 1960. Light and the development of *Poria ambigua.* Mycologia 52:231–247.

——, and F. KAVANAGH. 1938. Thiamine and growth of *Pythium butleri.* Bull. Torrey Botan. Club 65:453–461.

ROSS, I. S. 1975. Some effects of heavy metals on fungal cells. Trans. Brit. Mycol. Soc. 64:175–193.

SALTARELLI, C. G., and C. P. COPPOLA. 1979. Effect of light on growth and metabolite synthesis in *Candida albicans.* Mycologia 71:773–785.

STEINBERG, R. A. 1938. The essentiality of gallium to growth and reproduction of *Aspergillus niger*. J. Agr. Res. 57:569–574.

———. 1939. Relation of carbon nutrition to trace-element and accessory requirements of *Aspergillus niger*. J. Agr. Res. 59:749–763.

TABAK, H. H., and W. B. COOKE. 1968-a. The effects of gaseous environments on the growth and metabolism of fungi. Botan. Rev. 34:126–252.

———, and ———. 1968-b. Growth and metabolism of fungi in an atmosphere of nitrogen. Mycologia 60:115–140.

THAIN, J. F., and D. GIRVIN. 1987. Translocation through established mycelium of *Neurospora crassa* on a nutrient-free substrate. Trans. Brit. Mycol. Soc. 89:45–49.

WAIME, J.-M., M. GRENSON, and H. N. ARST, JR. 1985. Nitrogen catabolite repression in yeasts and filamentous fungi. Advan. Microb. Physiol. 26:1–88.

ZALOKAR, M. 1959. Growth and differentiation of *Neurospora* hyphae. Am. J. Botany 46:602–610.

Additional Recommended References

AGARWALA, S. C., N. NAUTIYAL, and C. CHATTERJEE. 1986. Iron nutrition of four *Aspergillus* species. Trans. Brit. Mycol. Soc. 86:561–569.

BARNETT, J. A., and A. P. SIMS. 1982. The requirement of oxygen for the active transport of sugars into yeasts. J. Gen. Microbiol. 128:2303–2312.

COOPER, T. G. 1982. Transport in *Saccharomyces cerevisiae*. *In:* Strathern, J. N., E. W. Jones, and J. R. Broach, Eds. The molecular biology of the yeast *Saccharomyces:* metabolism and gene expression. Cold Spring Harbor Laboratory, Cold Spring Harbor, N.Y., Vol. 2, pp. 399–461.

FRIES, N. 1965. The chemical environment for fungal growth. 3. Vitamins and other organic growth factors. *In:* Ainsworth, G. C., and A. S. Sussman, Eds. The fungi: an advanced treatise. Academic Press, New York, Vol. 1, pp. 491–523.

JENNINGS, D. H. 1974. Sugar transport into fungi: an essay. Trans. Brit. Mycol. Soc. 62:1–24.

———1976. Transport and translocation in filamentous fungi. *In:* Smith, J. E., and D. R. Berry, Eds. The filamentous fungi. Edward Arnold, London. Vol. 2, pp. 34–64.

KURTZ, M. B. 1980. Regulation of fructose transport during growth of *Aspergillus nidulans*. J. Gen. Microbiol. 118:389–396.

LILLY, V. G. 1965. The chemical environment for fungal growth. I. Media, macro- and micronutrients. *In:* Ainsworth, G. C., and A. S. Sussman, Eds. The fungi: an advanced treatise. Academic Press, New York, Vol. 1, pp. 465–478.

MANDELS, G. R. 1965. Kinetics of fungal growth. *In:* Ainsworth, G. C., and A. S. Sussman, Eds. The fungi: an advanced treatise. Academic Press, New York, Vol. 1, pp. 599–612.

NICHOLAS, D. J. D. 1965. Utilization of inorganic nitrogen compounds and amino acids by fungi. *In:* Ainsworth, G. C., and A. S. Sussman, Eds. The fungi: an advanced treatise. Academic Press, New York, Vol. 1, pp. 349–376.

PERLMAN, D. 1965. The chemical environment for fungal growth. 2. Carbon sources. *In:* Ainsworth, G. C., and A. S. Sussman, Eds. The fungi: an advanced treatise. Academic Press, New York, Vol. 1, pp. 479–489.

RAO, S., and G. VENKATESWERLU. 1986. Gutamine metabolism in *Neurospora crassa* under conditions of copper toxicity. J. Exptl. Botany 37:947–955.

ROBERTSON, N. F. 1965. The mechanism of cellular extension and branching. *In:* Ainsworth, G. C., and A. S. Sussman, Eds. The fungi: an advanced treatise. Academic Press, New York, Vol. 1, pp. 613–623.

SNELL, E. E. 1948. Use of microorganisms for assay of vitamins. Physiol. Rev. 28:255–282.

STEINBERG, R. A. 1950. Growth of fungi in synthetic nutrient solutions. II. Botan. Rev. 16:208–228.

SUOMALAINEN, H., and E. OURA. 1971. Yeast nutrition and solute uptake. *In:* Rose,

A. R., and J. S. Harrison, Eds. The yeasts. Academic Press, New York, Vol. 2, pp. 3–74.

TRINCI, A. P. J. 1984. Regulation of hyphal branching and hyphal orientation. *In:* Jennings, D. H., and A. D. M. Rayner, Eds. The ecology and physiology of the fungal mycelium. Cambridge University Press, Cambridge, pp. 23–52.

9

Metabolism

In Chapter 8 we considered nutrient uptake and those nutritional and environmental conditions that promote growth. The fate of nutrients is not ended with their preliminary digestion and entrance into the cell, because they will become involved in the metabolism of the cell. Metabolic pathways are of two general types: (1) catabolic pathways that degrade a substance to simpler forms (digestion, considered in Chapter 8, comprises the first catabolic reactions) and (2) anabolic pathways through which syntheses of cell components take place. Synthetic processes include formation of carbohydrates, proteins, lipids, and nucleoproteins which are common to all forms of life and of some specialized fungal metabolites.

The fungi have many metabolic pathways in common with other organisms, as well as some specialized ones. In this chapter we consider some aspects of carbon, nitrogen, and lipid metabolism.

CARBON METABOLISM

Fermentation and Respiration

Organisms derive useful energy and intermediates for syntheses from oxidation of compounds as a result of *fermentation* and *respiration*. The substrate most commonly oxidized by fungi, plants, and animals is glucose. Since glucose is broken down as a result of its oxidation, it is also said to be *dissimilated*.

Oxidation is the removal or loss of electrons from a substrate (substrate *A*, for example). These electrons cannot remain free but must be gained by a second substrate (substrate *B*), which is reduced. An oxidation reaction must always be accompanied by a reduction reaction. Most organic compounds lose hydrogen ions upon the removal of electrons, and in these cases, oxidation is equivalent to dehydrogenation. Therefore, substrate *A*, which is oxidized, is a hydrogen donor and the reduced substrate *B* is a hydrogen acceptor. These two reactions would be coupled in the following manner:

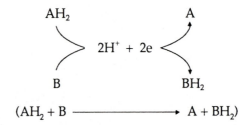

A principal role of fermentation and respiration is to release chemical energy from the substrate being oxidized and to make the energy available to drive other synthetic reactions within the cell. Oxidation is an *exergonic* chemical reaction, meaning that energy is released. The energy is released in a chemical form that may drive *endergonic* reactions (those that require energy). Much of the energy derived from the substrate is in the form of hydrogen ions and their electrons, which are bound in a reduced compound such as NADH (reduced nicotinamide adenine dinucleotide), and in turn can drive an endergonic reaction.

Finally, the oxidation-reduction sequence ends when the *terminal electron acceptor* is reduced by accepting the electrons. The terminal electron acceptor is an organic compound in fermentation, while it is oxygen in aerobic respiration. Fungi must have molecular oxygen from the atmosphere to carry out aerobic respiration, whereas fermentation can occur in the presence or absence of oxygen. Harvest of the energy is considered further on page 317.

GLYCOLYSIS

The initial steps of fermentation and respiration are the same, and occur under either anaerobic or aerobic conditions. The process begins with *glycolysis*, the stepwise conversion of glucose to pyruvic acid. After glycolysis is completed, fermentation is favored if oxygen is lacking or in short supply. Alternatively, if aerobic pathways are favored by an abundant supply of oxygen, the pyruvic acid will be utilized in aerobic respiration.

The principal glycolytic pathways are the Embden–Meyerhof (EM) and the hexosemonophosphate (HMP) pathways, occurring in animals, plants, bacteria, and fungi. A third glycolytic pathway is the Entner–Doudoroff (ED) pathway, known to occur primarily in bacteria, and possibly in a few fungi. The details of the first two pathways are presented in Figs. 9–1 to 9–3. Biochemical information about the glycolytic pathways in fungi can be gained from determining the balance of products, study of enzymatic steps, and tracing the distribution of radioactive isotopes (Blumenthal, 1965; Cochrane, 1976).

The EM pathway is the principal glycolytic pathway in fungi, as in most other organisms, and usually accounts for most of the glucose dissimilated. The HMP pathway also occurs widely in the fungi. In a given fungus, glycolysis can occur via both pathways or almost exclusively by one or the other. Commonly, the EM pathway accounts for more than half of the glucose dissimilated, with the HMP pathway dissimilating proportionately less of the glucose. The HMP pathway may be the major route of glycolysis in a small number of fungi, such as the yeast *Rhodotorula gracilis*, which metabolizes 60 to 80% of its glucose via the HMP cycle (Suomalainen and Oura, 1971). It has been suggested that some species of *Rhodotorula* do not have an EM pathway and dissimilate all of their glucose via the HMP pathway (Cochrane, 1976).

The EM pathway is of great importance because most glucose dissimilation to pyruvate occurs as a result of its reactions. The HMP pathway, however, is

HC=O
|
HCOH
|
HOCH
|
HCOH —— Glucose
|
HCOH
|
H_2COH

HC=O
|
HCOH
|
HOCH
|
HCOH
|
HCOH
|
H_2C—O—P

ATP
ADP

Glucose – 6 – phosphate —— HCOH

H_2COH
|
C=O
|
HOCH —— Fructose – 6 – phosphate
|
HCOH
|
HCOH
|
H_2C—O—P

H_2C—O—P
|
C=O
|
HOCH
|
HCOH
|
HCOH
|
H_2C—O—P

ATP
ADP

Fructose–1, 6–diphosphate —— HCOH

HC=O
|
HCOH —— Glyceraldehyde –3– phosphate
|
H_2C—O—P +H_3PO_4

H_2—C—OH
|
Dihydroxyacetone —— C=O
phosphate H_2C—O—P

NAD$^+$
NADH

P~O—C=O
|
HCOH ——1, 3, diphosphoglyceric acid
|
H_2C—O—P

ADP
ATP

HO—C=O
|
HCOH ——3 – phosphoglyceric acid
|
H_2C—O—P

HO—C=O
|
HC—O—P ——2 – phosphoglyceric acid
|
H_2C—OH

HO—C=O
|
C—O~P ——Phosphoenolpyruvic acid
||
CH_2

$2H_2O$

ADP
ATP

HO—C=O
|
C=O ——— Pyruvic acid
|
CH_3

Figure 9–1 The Embden–Meyerhof (EM) pathway begins with the phosphorylation of glucose which is accomplished by the transfer of a phosphate group from ATP to glucose. The glucose-6-phosphate which is formed by this reaction undergoes a slight rearrangement to form fructose-6-phosphate, and a second phosphate group is added to form fructose-1,6-diphosphate. Fructose-1,6-diphosphate then splits to give a mixture of an aldehyde (glyceraldehyde-3-phosphate) and a ketone (di-hydroxyacetone-phosphate), both three-carbon molecules. The ketone may be transformed into the aldehyde, thereby giving two molecules of glyceraldehyde-3-phosphate from one fructose-1,6-diphosphate mole-cule. The glyceraldehyde-3-phosphate molecules are oxidized, and two hydrogen ions and their electrons are removed by NAD$^+$, which is re-duced to NADH. Accompanying this reaction is the addition of a second phosphate group from phosphoric acid (H_3PO_4). The compound formed is 1,3-diphosphoglyceric acid. An energy-rich group is lost from 1,3-disphosphoglyceric acid and a molecule each of ATP and 3-phosphogly-ceric acid are formed. 3-phosphoglyceric acid is converted into 2-phospho-glyceric acid, and then dehydrated to form phosphoenolpyruvic acid. The phosphate group of phosphoenolpyruvic acid is then transferred to ADP to form ATP and a molecule of pyruvic acid, the end product of this sequence. Each of the above steps is controlled by specific enzymes. The net reaction is

$$1 \text{ glucose } + 2\text{ADP} + 2\text{NAD}^+ + 2H_3PO_4 \longrightarrow$$
$$2 \text{ pyruvic acid } + 2\text{ATP} + 2\text{NADH} + 2H^+ + 2H_2O$$

HC=O   COOH H_2COH
|
HCOH   NADP$^+$ NADPH+H$^+$ HCOH   NADP$^+$ NADPH+H$^+$ C=O
|
HOCH   HOCH   HCOH
|
HCOH   Glucose-6-dehydrogenase   HCOH   Phosphogluconate dehydrogenase   HCOH
|
HCOH   HCOH CO_2 H_2C-O-P
|
H_2C-O-P H_2C-O-P

Glucose-6-phosphate   6-phosphogluconic acid   Ribulose-5-phosphate

Figure 9–2 First set of reactions of the HMP pathway, which are oxidative. Like the EM pathway, the HMP pathway begins with glucose and the phosphorylation of glucose to form glucose-6-phosphate. Glucose-6-phosphate is usually oxidized to 6-phosphogluconic acid by the removal of electrons which are accepted by NADP (nicotinamide adenine dinucleotide phosphate). 6-Phosphogluconic acid is oxidized by a second NADP molecule, and the first carbon atom (in the C_1 position) is removed as carbon dioxide, leaving a five-carbon intermediate, ribulose-5-phosphate. The sequence of reactions above completes the *oxidative* phase in which NADPH is generated and made available as a hydrogen donor. The second phase of the HMP pathway is nonoxidative, consisting of the complex series of rearrangements shown in Fig. 9–3.

also important and is not simply an alternative glycolytic route. The HMP pathway is the biosynthetic source of pentoses, required for the synthesis of nucleic acids. Unlike NADH, which is generated from the EM pathway, the NADPH, formed from the HMP pathway, apparently does not generate ATP through the respiratory chain (page 316). Therefore, energy release through aerobic glycolysis occurs only from the EM cycle. The primary role of NADPH is to function as a hydrogen donor, making possible the reduction of other compounds. These reactions include fatty acid, glycogen, and glutamic acid synthesis. In view of the role of the NADPH in syntheses, it is not surprising that the amount of glucose dissimilated by the HMP pathway is correlated with growth phases. Typically the HMP cycle is more active in growing cells than in resting cells, as can be seen in *Candida utilis*, in which the rate of glucose dissimilation in the HMP cycle rises from 40 to 50% in actively growing cells. Similar increases have been noted for the Pyrenomycetes *Claviceps purpurea* and *Neurospora crassa* (Blumenthal, 1965). A rise in the activity of the HMP cycle can sometimes be detected just prior to or during sporulation (Cochrane, 1976).

The relative balance between the EM and HMP pathways can be affected by whether or not oxygen is available. Both pathways can occur under aerobic and anaerobic conditions, but the HMP pathway becomes much less active in the absence of free oxygen. Under anaerobic conditions, the EM pathway becomes increasingly important due to impairment of the HMP pathway, and may increase sufficiently to account for all of the glucose dissimilated. Because the HMP pathway is impaired at low oxygen levels, the EM pathway is probably chiefly responsible for meeting the energy requirements of fungi in these circumstances. Examples include the yeast *Candida utilis*, which dissimilates 41% of the glucose via the HMP pathway under aerobic conditions but only 4% under anaerobic conditions (Blumenthal, 1965).

These two glycolytic pathways share a number of biochemical features. Most of the enzymes required for the EM pathway also operate in the HMP pathway. One enzyme, phosphofructokinase, is unique for the EM pathway. Phosphofructokinase, as well as some other respiratory enzymes, requires zinc for its synthesis. A deficiency of zinc impairs the function of the EM pathway while favoring the HMP pathway. Both the EM and HMP pathways produce

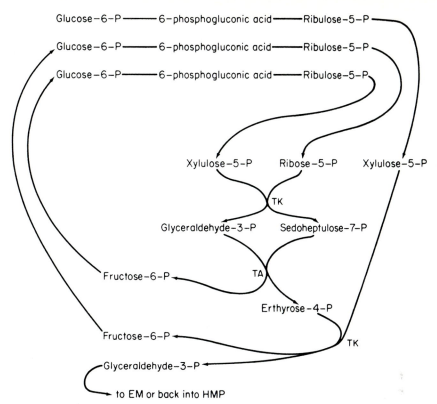

Figure 9–3 HMP pathway. The conversions of glucose-6-phosphate to ribulose-5-phosphate are oxidative (see the text). The nonoxidative stage begins with the enzymatic conversion of ribulose-5-phosphate to ribose-5-phosphate and xylulose-5-phosphate. An enzyme, transketolase, transfers an active glycoaldehyde ($CH_2OH-\overset{\overset{\displaystyle O}{\displaystyle \|}}{C}-$) group from xylulose-5-phosphate to ribose-5-phosphate, thereby producing a 7-carbon intermediate (sedoheptulose-7-phophate) and the three-carbon glyceraldehyde-3-phosphate. In a similar reaction, a transaldolase enzyme cleaves an active dihydroxyacetone group ($H_2COH-CO-H_2COP$) from the sedoheptulose-7-phosphate (leaving behind a hexose, erythrose-4-phosphate) and transferring it to glyceraldehyde-3-phosphate, producing a molecule of fructose-6-phosphate (a hexose). A second transketolase reaction converts xylulose-5-phosphate into glyceraldehyde-3-phosphate and erythrose-4-phosphate into fructose-6-phosphate. The fructose-6-phosphate molecules may be directly converted into glucose-6-phosphate and enter the cycle again, while the glyceraldehyde-3-phosphate may be diverted into the EM pathway or two glyceraldehyde-3-phosphate molecules may enter a transaldolase reaction to give another molecule of fructose-6-phosphate. The overall reaction is

$$3 \text{ glucose-6-P} + 3NADP^+ \longrightarrow$$
$$3CO_2 + 3NADPH + 3H^+ + 3 \text{ glyceraldehyde-3-P}$$

glyceraldehyde-3-phosphate and fructose-6-phosphate. Either of these intermediates that are common to both pathways may be diverted into one or the other of the pathways, regardless of which one originally produced them.

FERMENTATION

In fermentation, an intermediate of glycolysis or the end product of glycolysis may serve as the terminal electron acceptor, ending with the production of organic compounds. As outlined on page 317, the energy yield from fermentation is comparatively low, as the glucose has not been completely dissimilated. Fermentation can occur under aerobic conditions, occurring concurrently with pyruvic acid oxidation via the aerobic tricarboxylic acid cycle, but it is usually favored by anaerobic conditions in which it does not compete with the aerobic tricarboxylic acid cycle for pyruvic acid. Because the EM pathway is the major glycolytic pathway in most fungi, it accounts for the greatest quantity of fermentative end products formed.

Fermentation to Alcohols. Ethyl alcohol is the most important fermentation product commercially. Fermentation to alcohol was originally elucidated in the yeast *Saccharomyces cerevisiae*, and this organism continues to be the most important source of industrial alcohol (see Chapter 16). Several fungi produce ethyl alcohol, including the zygomycetes *Mucor* and *Rhizopus;* the ascomycetes *Ashbya gossypii* and *Neurospora crassa;* and the deuteromycetes *Aspergillus, Fusarium, Penicillium,* and others (Cochrane, 1958).

Ethyl alcohol formation is initiated by the union of pyruvic acid with a coenzyme, thiamine diphosphate (TPP), to form an *active pyruvate* complex on the surface of the enzyme pyruvate decarboxylase. This complex is decarboxylated to yield an active acetaldehyde complex, which breaks down to give acetaldehyde and TPP. The reactions may be summarized as follows:

$$TPP + CH_3CO \text{---} COOH \longrightarrow$$
$$\text{pyruvic acid}$$

$$CH_3CO \text{---} COOH \text{---} TPP \xrightarrow[\text{decarboxylase}]{\text{pyruvate}} \begin{array}{c} CO_2 \\ CH_3COH \text{---} TPP \end{array} \longrightarrow \begin{array}{c} TPP \\ CH_3CHO \\ \text{acetaldehyde} \end{array}$$

The acetaldehyde is reduced to ethyl alcohol, accepting hydrogen ions from NADH, by the enzyme alcohol dehydrogenase:

$$H^+ + CH_3CHO + NADH \xrightarrow[\text{dehydrogenase}]{\text{alcohol}} CH_3CH_2OH + NAD^+$$
$$\text{ethyl alcohol}$$

If the EM pathway was utilized to provide pyruvic acid, the NADH required above may be provided by that pathway. An overall summary of these reactions is:

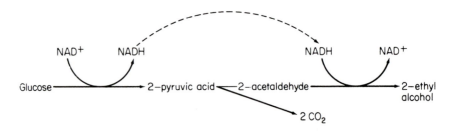

Fermentation to Lactic Acid. Some fungi in the Chytridiomycetes and Oomycetes are known to form lactic acid as the only fermentation product under anaerobic conditions. Production of lactic acid as the sole end product is similar to that of ethyl alcohol. Pyruvic acid is reduced by NADH or NADPH and is catalyzed by lactic dehydrogenase as follows:

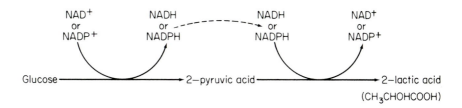

Lactic acid may also appear in combination with other end products. Species of *Rhizopus* produce lactic acid, carbon dioxide, and alcohol. Lactic acid may also appear with other compounds, such as acetaldehyde or succinic acid.

Fermentation to Glycerol and Other Polyhydric Alcohols. Under alkaline conditions, *Saccharomyces cerevisiae* forms glycerol. Part of the dihydroxyacetone phosphate which is produced in the EM pathway is reduced by NADH to glycerol-3-phosphate, which is dephosphorylated to glycerol. This reduction results in a deficiency of NADH available for the reduction of acetaldehyde. Rather than accumulating, acetaldehyde undergoes a conversion to a mixture of ethyl alcohol and acetic acid (Fig. 9–4). A number of yeasts which are able to grow in the presence of high concentrations of carbohydrate carry out similar conversions in the absence of alkaline conditions. These yeasts form polyhydric alcohols such as arabitol, erythritol, and mannitol.

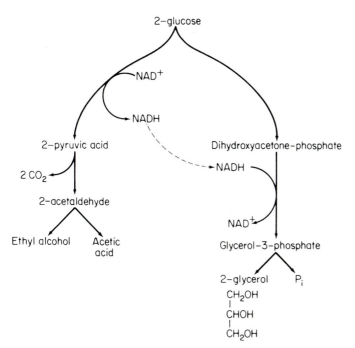

Figure 9–4 Fermentation to glycerol.

AEROBIC RESPIRATION

In aerobic respiration the pyruvic acid is converted to acetyl-CoA, which enters the tricarboxylic acid cycle (TCA) by combining with citric acid. The subsequent reactions result in its stepwise oxidation (Fig. 9–5). This cycle is an aerobic pathway and requires the presence of molecular oxygen. The TCA cycle oc-

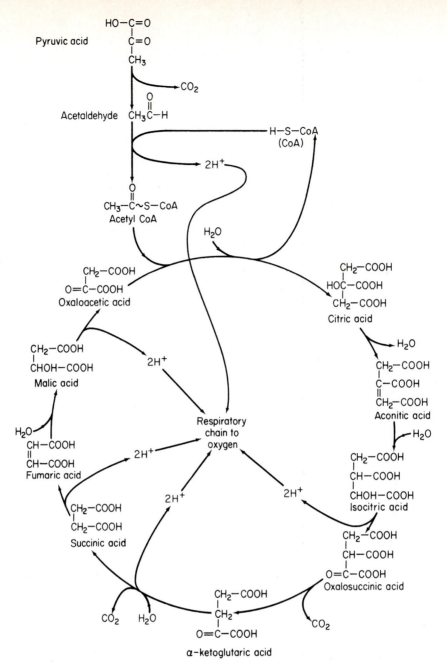

Figure 9–5 Conversion of pyruvic acid to acetyl-CoA and the tricarboxylic acid (TCA) cycle. Pyruvic acid is decarboxylated to leave an active two-carbon acetaldehyde, which combines with coenzyme A to form acetyl-CoA. During the formation of acetyl-CoA, two hydrogen ions and two electrons are transferred first to NAD^+ and then into the cytochrome system (the respiratory chain to oxygen). The acetate portion (two-carbon unit) of acetyl-CoA then joins with oxaloacetic acid (a four-carbon acid) to form citric acid (a six-carbon acid), the first compound in the TCA cycle. Water is both removed from and added to citric acid, resulting in a rearrangement which produces isocitric acid. Isocitric acid is oxidized to form oxalosuccinic acid, which is decarboxylated to form α-ketoglutaric acid. A molecule of carbon dioxide and two hydrogen ions and two electrons are removed from α-ketoglutaric acid. The hydrogen ions and electrons are accepted by NAD^+. A molecule of water enters at this stage, and succinic acid is formed with the concurrent generation of one ATP molecule. Succinic acid is oxidized to fumaric acid by the loss of two additional hydrogen ions and two electrons. The oxidizing agent in this last reaction is flavin. Fumaric acid combines with a molecule of water to form malic acid. Malic acid is then oxidized, donating two hydrogen ions and two electrons to NAD^+ and forming oxaloacetic acid. This oxaloacetic acid is then ready to combine with another active acetate group and to initiate the cycle again. If no intermediates are drained from the TCA cycle, the reactions may be summarized as follows:

$$CH_3-CO-COOH + CoA-SH \longrightarrow CH_3-CO-SCoA + H_2 + CO_2$$

pyruvic acid   CoA   acetyl-CoA

$$CH_3-CO-SCoA + 3H_2O \longrightarrow CoA-SH + 4H_2 + 2CO_2$$

acetyl-CoA   CoA

curs widely among the fungi, and in its biochemical details is similar to that occurring in other organisms (Kubicek, 1988).

A number of bacteria and fungi can utilize 2-carbon compounds (such as acetate and ethyl alcohol) as the sole source of carbon. Acetate may enter the TCA cycle via the *glyoxylic bypass* (Fig. 9–6). Isocitric acid, which is formed in the TCA cycle, is split by isocitrate lyase to succinic and glyoxylic acids. The succinic acid reenters the TCA cycle, while the 2-carbon glyoxylic acid combines with the 2-carbon acetyl-CoA, derived from the acetate, to form malic acid. This last condensation reaction is mediated by the enzyme malate synthetase. Malic acid then enters the TCA cycle. It is possible that other 2-carbon compounds may also enter the TCA cycle in this manner (Collins and Kornberg, 1960; Niederpruem, 1965). Additional enzymes, which have not been completely characterized, are required for the glyoxylic bypass (Casselton, 1976). As you may recall from Chapter 1, the enzymes required for the glyoxylic bypass are contained within specialized microbodies (glyoxysomes). The cellular location where the glyoxylic bypass occurs is unknown, although the TCA cycle is associated with mitochondria.

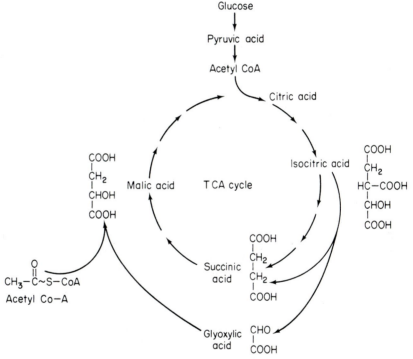

Figure 9–6 Glyoxylic bypass.

Role of the TCA Cycle in Intermediary Metabolism. The TCA cycle serves as a hub where many metabolic pathways converge. Some intermediates of the TCA cycle may also serve as intermediates in either the catabolism or anabolism of compounds other than those in the TCA cycle. Acetyl-CoA, for example, is a degradation product of a fatty acid and may also serve in the synthesis of new fatty acids. α-Ketoglutaric acid has a similar role in amino acid metabolism. (Metabolism of amino acids and fatty acids is considered in greater detail on pages 323 and 328). The existence of intermediates which are common to catabolic pathways and to the TCA cycle allow the degradation products of a number of compounds to enter the TCA cycle, where they are further dissimilated. An extremely important role of the TCA cycle is to make intermediates available that serve as precursors for biosynthesis of organic compounds.

Intermediates are drained from the TCA cycle for synthesis of other compounds, yet the cyclical nature of the TCA cycle continues. Some fungi are able to fix atmospheric carbon dioxide and use this as a source of carbon. This may be confirmed experimentally by providing radioactive carbon dioxide to the fungus in the air and later demonstrating that the radioactive carbon has been incorporated into organic compounds within the fungus. Fixation of carbon dioxide makes it possible for the TCA cycle to continue to operate because the necessary concentration of key intermediates is maintained. Although different mechanisms may operate, probably the most important in fungi is that in which pyruvic acid functions as an acceptor molecule that is carboxylated in the presence of pyruvate carboxylase:

$$\text{pyruvic acid} + \text{ATP} + \text{HCO}_3^- \rightleftharpoons \text{oxalacetic acid} + \text{ADP} + \text{P}_i$$

The enzyme pyruvate carboxylase contains biotin, and a monovalent ion, usually K^+, and a divalent ion such as Mn^{2+} or Mg^{2+} are required for the above reaction. The mechanism for carbon dioxide fixation in fungi is generally similar to that which occurs in mammalian tissues (Casselton, 1976; Utter et al., 1975).

Role of the TCA Cycle in Energy Release. Aerobic respiration is completed when the pairs of hydrogen and their electrons removed during the oxidation of glucose are passed along the *respiratory chain*, an electron transport system, finally reducing oxygen to water. Two respiratory chains occur in fungi that accomplish this task.

1. One respiratory chain contains *cytochromes*, enzymes with a heme prosthetic group, and also generates ATP as electrons are transported along it. This chain can be inhibited by cyanide. This chain, having cytochromes, is similar to that occurring widely in other organisms and is the principal chain involved in the terminal steps of respiration in fungi generally.
2. The second respiratory chain lacks cytochromes. This chain does not generate ATP as a result of its reactions, and is not inhibited by cyanide. It occurs in a number of fungi, particularly those having respiratory deficiencies. The role of this second chain is not known.

In further consideration of the respiratory chain, we consider the first type, the one with cytochromes. This is the principal respiratory chain in the fungi, as well as in other organisms, such as mammals. The order of components within it is

$$\text{flavin} \longrightarrow \text{ubiquinone} \longrightarrow \text{cytochrome } b \longrightarrow$$
$$\text{cytochromes } c_1, c \longrightarrow \text{cytochromes } a, a_3 \longrightarrow \text{oxygen}$$

NADH from glycolysis and the TCA cycle transfer their electrons and hydrogens to the flavoprotein, the yellow enzyme that is the first component of the respiratory chain. First the flavoprotein is reduced as it accepts hydrogen ions and electrons from the NADH, and then it is oxidized as the hydrogen ions and electrons are transferred to ubiquinone, a coenzyme (also called coenzyme Q). FADH from the TCA cycle enters at the ubiquinone step, also bringing about its reduction. The next constituents in the chain are groups of cytochromes. These cannot transfer hydrogen ions, which no longer participate directly in the redox reactions but instead are transferred across the membrane. The electrons are accepted sequentially by the cytochromes, which are reduced as they accept the electrons but oxidized as they are transferred to the next cytochrome. Finally, the electrons participate in the reduction of oxygen to water.

The precise composition of the cytochrome chain in the fungi is highly variable. The flavoprotein may be either FAD or FMN, and ubiquinone also varies from one fungus to another. Each cytochrome unit can be a complex of cytochromes; for example, cytochrome a may be subdivided into cytochromes a and a_3. The cytochromes from different fungi often have different physical properties, and coenzyme Q differs in composition among the fungi. The cytochromes can vary among genetic strains or if a single species is grown under different environmental conditions—a cytoplasmic mutant (*"petite colonie"*) of yeast and anaerobically grown normal yeast strains have the cytochromes a_1 and b_1, which are not normally present, but lack the normal cytochromes a, b, and c (Lindenmayer, 1965).

The net gain in energy as a result of respiration is of great importance, and is harvested as ATP. Each mole of ATP contains about 7.3 kilocalories of energy. A small amount of the ATP derived from respiration is formed directly by the phosphorylation of ADP during glycolysis, but a much larger amount of ATP is formed as a result of transport of hydrogen ions and their electrons down the cytochrome respiratory chain. The energy that is released as a result of this process is used to generate a proton gradient that drives the phosphorylation of ADP to ATP. For each pair of hydrogens from NADH entering this chain, three ATP molecules are generated. For each molecule of glucose entering glycolysis, there is the yield of two NADH molecules, and as indicated previously, these may be utilized in fermentation, leaving a net gain in energy from glycolysis of only two ATP molecules. Alternatively, if aerobic respiration occurs, these NADH molecules can enter the respiratory chain and generate an additional six molecules of ATP for each molecule of glucose that is converted to pyruvic acid. Additional pairs of hydrogen ions and their electrons are derived from the conversion of pyruvic acid to acetyl-CoA, and from the stepwise oxidation of citric acid. These are similarly transported through the respiratory chain and generate ATP. Altogether, the net yield of the complete oxidation of one molecule of glucose through aerobic respiration is 36 molecules of ATP, significantly greater than the two molecules that can be harvested as a result of fermentation.

Factors Influencing Respiration in Fungi. A number of environmental and intrinsic factors affect the course and rate of respiration. External factors influencing respiration are pH, temperature, oxygen pressure, carbon dioxide, and nutrients. The pH and temperature affect respiration in a manner similar to their effect on growth, discussed in Chapter 8.

Oxygen is a requirement for aerobic respiration to take place. The oxygen used in respiration is dissolved in liquid, and the amount of this dissolved oxygen is proportional to the amount of oxygen present in the gas phase. At very low concentrations of oxygen, the rate of respiration is directly correlated with the amount of oxygen available but, as the amount of oxygen increases, the rate of respiration becomes independent of the oxygen concentration. The point at which the respiration rate becomes independent of oxygen concentration is 0.022 millimoles per liter (about 16 millimeters Hg) for *Penicillium chrysogenum* and about 0.0046 millimoles per liter (about 4.0 millimeters Hg) for yeast cells (Finn, 1954).

Another extrinsic factor that may affect respiration is the sugar concentration. A high sugar concentration in the presence of adequate aeration may result in utilization of fermentative pathways in yeasts, as a glucose concentration over 5% totally inhibits synthesis of respiratory enzymes. As yeasts exhaust the sugar from their culture medium, there may be an increase in the activities of the enzymes of the TCA cycle and also of the glyoxylic acid cycle (Suomalainen and Oura, 1971).

Intrinsic factors include the enzyme complement of the cell, cell permeability, phase of development, and age. The most important of these is the enzyme complement of a species. This will ultimately dictate which substrates can be oxidized and the biochemical pathways to be followed. Failure of a fungus to oxidize a substrate may be due to inability of the substrate to penetrate the cell membrane, but a cell-free preparation may oxidize the substrate readily, indicating that the cell has the necessary enzymes to accomplish the oxidation. Those substrates most often excluded from the cell are ionizable compounds such as organic acids. A fungus passes through many developmenal stages (such as spore dormancy, spore germination, and mycelium development). Rates of respiration or the respiratory pathways may alter with the different stages or age. Respiration is greatest during periods of active growth and decreases with aging of the fungus and accompanying decrease in growth.

FERMENTATION, RESPIRATION, AND DEVELOPMENT

Fungal development involves a number of complex biochemical changes, and usually includes changes in respiration as well. These changes in respiration are not necessarily the determining factors in the developmental processes, but demonstrate the complexity of shifts in metabolism. Two examples are cited below.

1. A change from fermentative metabolism to oxidative metabolism occurs during perithecial development of *Neurospora tetrasperma* (Viswanath-Reddy and Turian, 1975). The respiratory rate is high in the early phase of vegetative growth but decreases as ascogonial differentiation begins. Ascogonial coils form under semianaerobic conditions (0.5% oxygen) or in the absence of carbon dioxide but fail to develop further under these conditions. Perithecial maturation requires an oxygen concentration of 5% and a carbon dioxide concentration of 1%. Inhibitors of glycolysis prevent formation of the ascogonial coils, while inhibitors of the TCA cycle prevent the subsequent development into protoperithecia. These data indicate that a shift toward oxidative metabolism occurs as differentiation proceeds.

2. An alteration in the TCA cycle occurs in *Blastocladiella emersonii* and exerts a profound effect on the course of differentiation. This Chytridiomycete forms colorless zoosporangia and pigmented resting spores. Numerous biochemical differences exist between the two morphological forms, and most of them appear to be triggered by alterations in respiration. As the resting spores differentiate, there is a marked decline in respiration and its enzymes disappear. Accompanying this decline of the TCA cycle is an increase of about 6500-fold in the activity of isocitric dehydrogenase, which in turn causes a reversal of the conversion of isocitric acid to α-ketoglutaric acid, an important step in the TCA cycle. The TCA is blocked and large quantities of isocitric acid are produced. The isocitric acid does not accumulate but is shunted into other metabolic pathways (Cantino and Turian, 1959).

Carbon Compounds Produced by Fungi

After carbon compounds are assimilated, a certain amount of the carbon is used by fungi in the biosynthesis of cellular materials. These include wall constituents, storage carbohydrates, and a wide variety of metabolites that are of no apparent value to the fungi.

POLYSACCHARIDES

A large amount of the carbon assimilated by fungi is used in the biosynthesis of wall material. As indicated in Chapter 1, the fungal walls consist of complex mixtures of carbohydrates, proteins, and lipids, and the principal carbohydrate is usually chitin. Biosynthesis of the wall polysaccharides and glycogen basically involves a repetition of the following basic reaction (Burnett, 1976):

$$\text{nucleoside diphosphate sugar} + \text{primer} \longrightarrow$$
$$\text{(primer-sugar)} + \text{nucleoside diphosphate}$$

Chitin biosynthesis, which follows the scheme shown above, is generally similar in the fungi, crustaceans, and insects (McMurrough et al., 1971). Chitin biosynthesis begins with the conversion of glucose-1-phosphate into N-acetylglucosamine-1-phosphate, which requires a series of reactions (Fig. 9–7). The N-acetylglucosamine-1-phosphate then reacts with UTP (the nucleotide uridine triphosphate) to form UDP-N-acetylglucosamine, which is the nucleoside diphosphate sugar. Finally the UDP-N-acetylglucosamine transfers the N-acetyl-

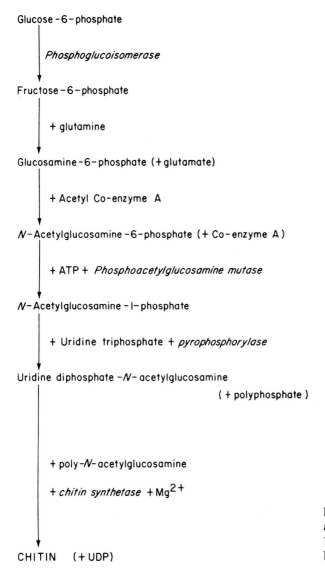

Figure 9–7 Synthesis of chitin as it probably occurs in fungi. [From J. H. Burnett, 1976, Fundamentals of mycology, 2nd ed., Edward Arnold, Ltd., London, 673 pp.]

glucosamine moiety to the growing chitin chain, functioning as a primer, and becomes one of its subunits. The enzyme chitin synthetase and Mg^{2+} ions are required for this polymerization step. The completed chitin molecule is a long chain of sugar subunits which are joined by β-1,4 links.

Fungi form polysaccharides that serve as an energy reserve. The sugar alcohol mannitol is formed and utilized as a food reserve and is especially common among the ascomycetes and basidiomycetes. Other sugar alcohols formed by fungi (and presumably serving as storage products) are D-sorbitol, D-volemitol, D-arabitol, and L-erythritol. The most common of the storage products is glycogen (Fig. 9–8), which usually comprises about 5% of the dry weight of the mycelium and spores (Cochrane, 1958). Like the glycogen formed by animals, glycogen is a branched glucose polymer. In the main portion of the chain, the glucose units are joined by α-1,4 linkages. In glycogen synthesis, UTP reacts with glucose-1-phosphate as follows (Burnett, 1976):

$$UTP + glucose\text{-}1\text{-}phosphate \rightleftharpoons UDP\text{-}glucose + pyrophosphate$$

The UDP-glucose (the nucleoside diphosphate sugar) donates D-glucose in the presence of the enzyme glycogen synthetase, and the D-glucose is added to the growing glycogen molecule, which functions as an acceptor. These last steps may be summarized as follows:

$$_x UDP\text{-}glucose + [\alpha\text{-}1, 4\text{-}glucose]_n \xrightarrow[\text{synthetase}]{\text{glycogen}} [\alpha\text{-}1, 4\text{-}glucose]_{n+x} + {}_x UDP$$

MISCELLANEOUS HYDROCARBON METABOLITES

Fungi produce a large variety of organic compounds. Many of these compounds are of no apparent value to the fungus as they are not directly involved in supplying energy to the cell nor are they involved in structural components of the cell. Instead, the compounds often arise either directly or indirectly from a surplus of some intermediate of normal metabolism and represent "overflow" products. Alternatively, they may arise from an elaborate biosynthetic pathway, which in reality is an error in metabolism since it does not provide the fungus with energy or a needed synthetic product but instead requires energy and intermediates to take place. Some of these compounds are of commercial interest (see Chapter 16).

ALIPHATIC METABOLITES

A number of saturated aliphatic compounds are formed by fungi. These include alcohols, acids, and esters. Some fairly common examples include acids and various unsaturated aliphatics.

Acids from respiratory pathways may accumulate in such large quantities that they may not be further metabolized. These include gluconic, citric, oxa-

Figure 9–8 Portion of a glycogen molecule.

loacetic, malic, fumaric, succinic, lactic, and acetic acids. Some acids may be derived from respiratory intermediates. These include oxalic acid, a common metabolic product of fungi generally and especially of the Agaricales, derived from oxaloacetic acid; itaconic acid, formed by strains of *Aspergillus terreus,* derived from *cis*-aconitic acid; and acetonedicarboxylic acids, produced by *Aspergillus niger* from citric acid metabolism. Acids which may be formed directly from sugars are mannonic acid (from mannose), galactonic acid (from galactose), and glucuronic acid (from glucose).

Unsaturated aliphatic compounds are also common, as an unsaturated double bond (–CH=CH–) and an unsaturated triple bond (–C≡C–) occur in a large number of fungal metabolites. The simplest compound of this type is ethylene (CH_2=CH_2), produced by some fungi that induce wilts of plants (Chapter 13). The unsaturated double bond may also occur in chains, either with saturated single bonds or with other double bonds. Many of these compounds are colored, a common example being β-carotene, a yellow pigment (Fig. 9–9). The unsaturated triple bond is especially common in metabolites from the higher fungi and almost always occurs in chains with other triple bonds or with double bonds.

Figure 9–9   β-Carotene.

AROMATIC METABOLITES

The aromatic nucleus is characterized by a ring of six carbon atoms and three unsaturated double bonds. There are two pathways by which the aromatic nucleus may be derived. These are the *shikimic acid* pathway and the *acetate* pathway. It is not known, however, whether all or even most aromatic compounds are derived from these pathways.

In the shikimic acid pathway, erythrose-4-phosphate (derived from the HMP pathway) condenses with phosphoenol pyruvic acid. The intermediate formed from this reaction undergoes successive rearrrangements (including formation of a ring and a double bond) to form shikimic acid. Shikimic acid condenses with a three-carbon unit and is converted to prephrenic acid, which has two double bonds. With the formation of a third double bond, prephrenic acid is converted to aromatic compounds (Fig. 9–10).

The two-carbon and one-oxygen skeleton of acetic acid (acetate) has long been recognized as a potential building block for fungal metabolites. Various combinations of acetate units or modified acetate units could account for the structure of many metabolites. A pathway involving condensation of acetate molecules is that in which acetyl-CoA combines with carbon dioxide to form malonyl-CoA, a three-carbon intermediate. Three malonyl-CoA molecules undergo condensation to form a six-carbon aromatic nucleus, meanwhile releasing three carbon dioxide molecules (Bu'Lock and Smalley, 1961). The intermediate reactions in this process are poorly understood.

There are large numbers of aromatic compounds in fungi, and they may occur in either monocyclic or polycyclic forms (Fig. 9–11). Most monocyclic aromatic metabolites are either aromatic alcohols or derivatives of these in which the alcoholic hydroxyl group is replaced by a methyl (–CH_3), methoxyl (–OCH_3),

Figure 9–10 Shikimic acid pathway for formation of aromatic compounds. Some of the steps shown are multiple.

or carboxyl (–CO₂H) group or by an organic radical. The aromatic alcohols are phenol, pyrocatechol, resorcinol, quinol, hydroxyquinol, and pyrogallol; these differ in the number and position of their hydroxyl groups. Derivatives of phenol, resorcinol, and quinol are especially common.

Examples of phenol derivatives are *p*-hydroxybenzoic acid and anisaldehyde which are derived via the shikimic acid pathway. Another member of this group, 6-methylsalicyclic acid, is an intermediate in the acetate pathway. The numerous resorcinol derivatives are presumably derived via the acetate pathway (Birkinshaw, 1965).

PHENOL DERIVATIVES

Figure 9–11 Some aromatic metabolites.

A quinol derivative may be oxidized by the fungus to form a *quinone*. There are several types of quinones, determined by the number of rings present and the manner in which they are joined. Of special interest are the *anthraquinones*, pigments that cause the mycelium to be brightly colored. The anthraquinones may account for as much as 30% of the dry weight of the mycelium. These compounds consist of three aromatic rings and may have from one to five hydroxyl groups, one of which may be methylated (Birkinshaw, 1965). Coenzyme Q in the cytochrome system is a quinone which is functional in the cell.

NITROGEN METABOLISM

Amino Acid Metabolism

DEAMINATION AND TRANSAMINATION REACTIONS

As noted in Chapter 8, proteins and peptides free amino acids upon hydrolysis, and these amino acids are a source of usable nitrogen. Before we can discuss utilization of amino acids by fungi, we must discuss two important types of reactions involving amino acids which are of general occurrence in organisms and are responsible for many of the transformations of amino acids. These are the oxidative deamination and transamination reactions.

A deamination reaction is one in which the amino group is removed from the amino acid, leaving a keto acid:

$$NH_2 - \underset{\underset{H}{|}}{\overset{\overset{R}{|}}{C}} - COOH + \tfrac{1}{2}O_2 \longrightarrow O = \underset{}{\overset{\overset{R}{|}}{C}} - COOH + NH_3$$

amino acid   keto acid

The enzymes involved in the deamination reaction are D- or L-amino acid oxidases, which deaminate D- or L-amino acids, respectively. The amino acid oxidases of fungi are generally similar to those of animal systems, but they may differ in the rates with which they affect the deamination reaction and also in their degree of substrate specificity.

Transamination involves transfer of the amino group from one compound to another. It takes place between an amino acid and a keto acid, converting the keto acid into the corresponding amino acid (see below). Transamination allows for the synthesis of an amino acid from any appropriate carbon skeleton, as long as a donor amino acid is available.

$$NH_2 - \underset{\underset{H}{|}}{\overset{\overset{R}{|}}{C}} - COOH + O = \underset{}{\overset{\overset{R'}{|}}{C}} - COOH \xrightarrow{\text{transaminase}}$$

amino acid A   keto acid B

$$O = \underset{}{\overset{\overset{R}{|}}{C}} - COOH + NH_2 - \underset{\underset{H}{|}}{\overset{\overset{R'}{|}}{C}} - COOH$$

keto acid A   amino acid B

AMINO ACID CATABOLISM

Most of the amino acids assimilated by the fungus are used directly without being dissimilated. Transamination may convert amino acids into other required amino acids, which are then used directly in protein synthesis or in other syntheses. *Neurospora crassa* mutants, deficient for glutamic dehydrogenase and therefore unable to directly synthesize amino acids via α-glutaric acid (see the next section), are able to utilize a wide variety of amino acids, indicating that an extensive transaminase system exists in that organism (Nicholas, 1965).

The principal route of amino acid catabolism in fungi is oxidative deamination, but a number of other mechanisms exist. For example, asparagine may be hydrolyzed to aspartic acid and ammonia in *Aspergillus niger* and other fungi; threonine is cleaved to glycine and acetaldehyde by *Neurospora crassa*; and arginine is degraded via the ornithine cycle (discussed in the next section) (Cochrane, 1958).

THE GLUTAMIC ACID CROSSROADS

Both the ammonia freed by the deamination reaction and inorganic nitrogen (nitrates, nitrites, and ammonia) which are directly taken up by the fungi enter into synthetic reactions. It is probable that all inorganic forms of nitrogen are reduced to ammonia and that ammonia enters amino acid biosynthetic pathways by combining with α-ketoglutaric acid, an intermediate in the TCA cycle:

$$HOOC-(CH_2)_2-CO-COOH + NADH + H^+ + NH_3 \rightleftharpoons$$
α-ketoglutaric acid

$$HOOC-(CH_2)_2-CH(NH_2)-COOH + NAD$$
L-glutamic acid

The enzyme that mediates this reaction is glutamic dehydrogenase. The amino acid formed in this reaction, L-glutamic acid, is actively involved in transamination reactions and in specific synthetic pathways that give rise to other amino acids. The formation of glutamic acid from ammonia and its keto acid and the subsequent shunt of the amino group to other amino acids provides an important mechanism for introducing inorganic nitrogen into amino acid biosyntheses (Fig. 9–12).

Just as glutamic acid provides a bridge leading to amino acid biosyntheses, it may function in reverse. The reaction above is reversible, and through glutamic acid the amino group may be lost as ammonia and the carbon skeleton of the amino acid introduced into the TCA cycle as α-ketoglutaric acid, which is then oxidized. A similar bridge involving aspartic acid and fumaric acid occurs in the bacteria, but it is apparently of minor importance in the fungi (Cochrane, 1958).

AMINO ACID BIOSYNTHESIS

Utilization of Nutrient-Deficient Mutants. The use of genetic mutants has contributed a great deal to our understanding of amino acid biosynthetic pathways and, in many cases, has provided the bulk of information about a pathway which is then supplemented by chemical and isotopic analyses. *Neurospora crassa* has been widely used in these studies.

For these studies, mutants are induced, usually by ultraviolet irradiation of the spores. The spores are inoculated onto a minimal medium (containing only the minimum amount of inorganic nutrients and vitamins that support growth of a normal strain). Any spore that fails to develop a germ tube is transferred to a medium which contains supplementary nutrients. If an arginine-

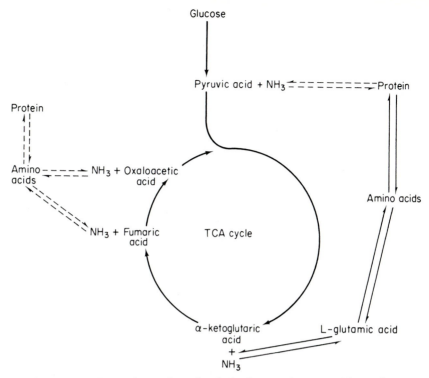

Figure 9–12 Interrelationship of carbohydrate and amino acid metabolism. Those conversions which are important in fungi are indicated by solid lines, while those in other organisms are indicated by a dashed line.

deficient mutant is being sought, for example, the supplemented medium would contain arginine. If the spore then germinates and a mycelium develops on the supplemented medium, it indicates that the strain is deficient for that nutritional factor provided in the supplemented medium.

Not all mutants deficient for a single nutrient are genetically identical and may show quite different responses to chemically related nutrients that may or may not support growth. Each mutant has a genetic deficiency which does not allow it to convert one precursor to another compound in a biosynthetic sequence and this leads to the formation of the deficient nutrient. Any compound normally coming after the genetic block will support growth if it is provided in the medium, since the necessary end product can then be synthesized. Also, if a genetic block occurs, that precursor occurring immediately before the block will not be utilized and will accumulate. The nutritional differences among the mutants are presumably due to genetic blocks at different points along the biosynthetic route.

We illustrate the elucidation of a biochemical pathway with methionine-requiring mutants (Horowitz, 1947). Both cysteine and homocysteine were recognized as probable precursors of methionine, and analysis of an accumulated substance in a group of mutants indicated that cystathionine might also be a precursor. The methionine-deficient mutants analyzed comprised four nutritional groups, each controlled by a single gene. These groups were

1. Strain me-1, which can only use methionine
2. Strain me-2, which can use homocysteine and methionine
3. Strain me-3, which can use cystathionine, homocysteine, and methionine

4. Strain me-4, which can use cysteine, cystathionine, homocysteine, and methionine

It is assumed that those mutants in group 4 have a genetically controlled block that does not permit the synthesis of the first precursor in this sequence. Cysteine does not support growth of group 3, although all other compounds will, and this indicates that the genetic block in group 3 interrupts the conversion of cysteine to the next precursor, which places cystine as the first precursor in this sequence. Similar reasoning shows us that cystathionine is the next precursor since group 2 cannot utilize either cysteine or cystathionine, both of which occur before its point of blockage. Since mutants exist which can use only methionine, it is apparently the last compound in the sequence. Homocysteine is then placed in the penultimate position. The sequence can then be constructed:

gene me-4 ──┼─→ cysteine gene me-3 ──┼─→ cystathionine gene me-2 ──┼─→ homocysteine gene me-1 ──┼─→ methionine

Biosynthesis of Other Amino Acids. The ornithine cycle was first established in mammalian liver, but it was later found to occur in *Neurospora crassa* through the use of single-gene nutritional mutants (Fincham and Boylen, 1957; Srb and Horowitz, 1944). In this cycle, glutamic acid is converted to ornithine by reduction and transamination (Fig. 9–13). Ornithine is converted to citrulline by the addition of a $O=C-NH_2$ group and citrulline is converted to arginine (via arginosuccinic acid) by the replacement of this keto oxygen with $=NH$. Arginine may be degraded by the enzyme arginase in animals to give urea, an end product of metabolism, and ornithine, which can initiate the cycle again. Like animals, *Neurospora* also has arginase but the urea formed is broken down by an additional enzyme, urease, to form carbon dioxide and ammonia. Apparently the role of the ornithine cycle in *Neurospora* and other fungi is to supply arginine for protein synthesis.

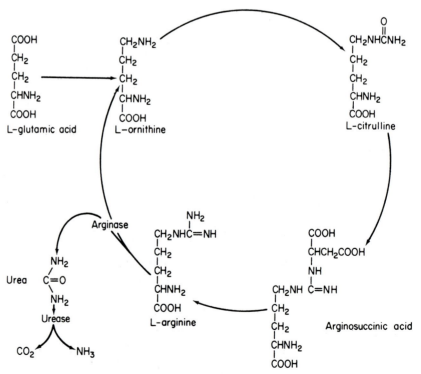

Figure 9–13 Ornithine cycle in *Neurospora*.

We know from page 321 that the shikimic acid pathway gives rise to aromatic intermediates that may be parent compounds for aromatic metabolites. The shikimic acid pathway also gives rise to the amino acids tryptophan, tyrosine, and phenylalanine, in addition to the vitamin p-aminobenzoic acid. All of these have an aromatic nucleus. A *Neurospora crassa* strain deficient for p-aminobenzoic acid, tryptophan, tyrosine, and phenylalanine will grow if shikimic acid is available (Tatum et al., 1954).

Nitrogen Compounds of Fungi

FUNCTIONAL NITROGEN COMPOUNDS

As in other organisms, nitrogen is an important component of the protoplasm of fungi. In a survey of cultivated fungi, the percentage of the dry weight attributable to nitrogen was found to vary between 2.27% in *Coprinus radicans* to about 5.13% in *Trichoderma lignorum* (Heck, 1929). The amount of nitrogen incorporated into the protoplasm varies with environmental conditions and with age of the mycelium. The nitrogen content is greatest when nitrogen is readily available in the medium, especially as ammonium salts. A young mycelium characteristically has a higher content than an older mycelium.

About 60 to 70% of the total nitrogen of the fungal cell is protein. The nitrogen which does not exist as protein occurs as nucleic acids, chitin, and nonessential metabolites.

MISCELLANEOUS NITROGEN METABOLITES

Fungi produce a large number of nitrogen metabolites that are of no apparent value to the fungus. We discuss some specific examples of these, which are pertinent to discussions elsewhere in this book and are indicative of the wide variety of these compounds which are formed by fungi.

Ammonia and its derivatives in which one or more of the hydrogen atoms are replaced by an alkyl group (the amines) are of frequent occurrence in the fungi. Many of the amines are apparently derived by decarboxylation of an amino acid. The simplest of these is methylamine (CH_3NH_2), which was found in 22 fungi in a survey of 105 species (as many as eight amines were found in a single species) (Birkinshaw, 1965). A hydroxylated derivative of an amine, ethanolamine ($H_2NCH_2–CH_2OH$), may be converted into an ionic compound by substitution with a fourth group. This ionic compound is choline:

$$(CH_3)_3N^+CH_2CH_2OH(Cl^-)$$

Choline is the parent compound for muscarine and muscaridine, poisons from *Amanita muscaria* (Chapter 16).

Other acyclic nitrogen metabolites are complex molecules consisting of many units. These include short peptide chains made up entirely of amino acids and molecules with mixtures of amino acids and nitrogen-free organic acids. Lycomarasmin, which induces wilting in plant diseases caused by *Fusarium*, is a molecule of this latter type (Chapter 13).

A number of nitrogen metabolites are heterocyclic compounds. These include five- and six-membered rings. Those most important to us are the metabolites which have the indole nucleus. Metabolites with the indole nucleus include the active components of hallucinogenic fungi, psilocin and psilocybin; the ergot alkaloids; and the poisons phalloidin and phalloin produced by *Amanita phalloides* (Chapter 16). Many of the antibiotics are heterocyclic nitrogen compounds which include a sulfur or chlorine atom. Among these are penicillin and griseofulvin (see Chapters 12, 13, and 16).

Lipids include a diverse variety of compounds that are soluble in nonpolar organic solvents. Many of the lipids contain glycerol, which may be esterified with one, two, or three fatty acid molecules to form mono-, di-, or tri-glycerides, respectively (these are called the neutral fats). The phospholipids also contain glycerol, as these compounds are derivatives of glycerol phosphate. Other lipids, including the sterols, may lack glycerol. Lipids may also be combined in nature with other compounds such as proteins, amino acids, or polysaccharides.

Lipid Catabolism

The lipids sometimes serve as a source of energy and carbon for fungi. Fungi do not usually require an exogenous source of lipids, as they are capable of synthesizing their own, and lipids are not ordinarily added to culture medium. Addition of lipids or fatty acids to the culture medium may enhance the growth of some fungi; for example, some boletes that are ordinarily difficult to grow in culture may grow quite well if vegetable oil is added (Schisler and Volkoff, 1977). The yeast *Pityrosporum ovale* actually requires an exogenous source of lipid (Hunter and Rose, 1971). Fungi produce triglycerides, which serve as an energy source when required (for example, upon spore germination). Triglycerides are also especially abundant in terms of quantity in natural substrate encountered by fungi.

Triglycerides are hydrolyzed by lipases that release glycerol and the three fatty acids. Glycerol may enter glycolysis after its phosphorylation and oxidation to 3-phosphoglyceraldehyde and fatty acids. The fatty acids must undergo further degradation before they can be utilized.

The naturally occurring fatty acids have a straight chain with an even number of carbon atoms. That carbon atom next to the carboxyl group is the α-carbon, while the second one is the β-carbon. Fatty acid catabolism is primarily a sequence of oxidations of the β-carbon atoms (Fig. 9–14).

In animals, fatty acid catabolism begins with the activation of the fatty acid molecule by a linkage with a molecule of coenzyme A in the presence of ATP. This activated molecule is then dehydrogenated (with FAD as the hydrogen carrier) with one hydrogen atom being removed from both the α- and β-carbons, creating an unsaturated double bond (–C=CH–). The next step is a hydration, producing an alcoholic group on the β-carbon, which is then oxidized by a β-keto fatty acid (β-oxidation). The hydrogen acceptor in this last reaction is NAD. The β-keto fatty acid then reacts with a second molecule of coenzyme A and yields an acetyl-CoA and an activated fatty acid that has two fewer carbons than the original fatty acid. Successive β oxidations occur until only a two-carbon fragment remains. The acetyl-CoA molecules which are formed enter the TCA cycle.

In fungi, the catabolism of fatty acid often proceeds in a similar fashion to the point where a β-keto fatty acid is formed, but the β-keto fatty acid may be decarboxylated to form a methyl ketone having one less carbon atom than the original fatty acid, that is,

$$\longrightarrow \; R\!-\!CO\!-\!CH_2\!-\!COOH \; \longrightarrow \; R\!-\!CO\!-\!CH_3 + CO_2$$

ß-keto fatty acid   methyl ketone

These methyl ketones are formed in cheeses processed with fungi and in edible oils attacked by fungi, perhaps by a detoxification mechanism. If complete oxidation to acetyl fragments occurs, the fragments may be utilized in the glyoxylic acid cycle.

Figure 9–14 β-Oxidation of a fatty acid as it occurs in animals. In fungi, the pathway occurs up to and including β-keto fatty acid. See the text.

Enzymes for the β-oxidation of fatty acids, as well as those for the glyoxylic acid cycle, are contained in specialized microbodies, the glyoxysomes. This close association is functionally useful because the acetyl units derived from the oxidation to fatty acids are further metabolized via the glyoxylic acid cycle.

Not unexpectedly, there is sometimes a close association between particular phases of the life cycle in which a fungus is preferentially utilizing the glyoxylic acid cycle and in which it is metabolizing its lipid reserves. This has been observed in the motile zoospores of *Phytophthora erythroseptica* and upon spore germination (Brennan and Lösel, 1978).

Lipids Produced by Fungi

Like other organisms, fungi can synthesize a wide variety of lipids. These include the neutral fats, phospholipids, and sterols, as well as others. The lipid content of fungi fluctuates greatly in response to age, stage of development, and cultural conditions. The lipid content of yeasts usually ranges from approximately 7 to 15% of the dry weight, while that of filamentous fungi ranges from 6 to 9% when grown under favorable conditions (Brennan and Lösel, 1978). However, in fungi a lipid content of 20% of the dry weight is not uncommon, and the lipid content may be as great as 50% of the dry weight (Cochrane, 1958). Some yeasts that accumulate unusually high concentrations of lipid (up to 60%

of their dry weight) may be commercially useful for the production of these fats (Hunter and Rose, 1971). Generally, the triglycerides and their contained fatty acids comprise a rather large fraction of the lipids produced by fungi. Fatty acids may comprise as much as 88% of the lipid fraction, and among these, the 16- and 18-carbon acids are most abundant, the unsaturated fatty acids (especially oleic and linoleic acids) predominating. Sterols are probably produced by all fungi, but the amount of sterol is usually about 1% or less of the dry weight. The sterol usually present in the greatest quantity is ergosterol, first detected in the sclerotia of *Claviceps purpurea* (the ergots) (Weete, 1973):

Lipids serve diverse functions within the fungi. Lipids may serve as an energy reserve, and lipid droplets are often especially prominent in rapidly growing cells that are provided with abundant sugars and in various types of reproductive structures and their spores. The major component of these reserve lipids probably consists of triglycerides. Lipids are also an important component of the cellular membranes, which include the plasmalemma, endoplasmic reticulum, and organelle membranes. Lipids may comprise as much as 30 to 50% of these membranes (Weete, 1974). Most of the lipid within the membranes is phospholipid, but there are also sterols, sterol esters, and neutral lipids. Phospholipids are major constituents of cellular membranes, function in the transport of ions across membranes, and are involved in the functional organization of a number of complex systems within the membranes (e.g., the electron transport system within the mitochondrial membranes) (Brennan and Lösel, 1978). In experiments with various species of yeasts subjected to ultraviolet radiation, it was demonstrated that long-chain fatty acids or sterols would induce the recovery of some, but not all, species. This led to the suggestion that sterols that naturally occur in the membranes may induce recovery from ultraviolet radiation (Sarachek and Higgins, 1972). Lipids also occur in fungal cell walls and form a coating on spores; they may have a protective, waterproofing function. As outlined in the next chapter, some fungi produce sexual hormones. Three of the best known hormones (sirenin, antheridiol, and trisporic acid) are lipids. As the name implies, antheridiol is a sterol (Brennan and Lösel, 1978).

A number of lipid metabolites with no apparent function are produced by fungi. These include fatty acids that are not utilized in fat synthesis, the well-known phosphatides cephalin and lecithin (lipids containing nitrogen and phosphorus), cerebrins (the nitrogenous lipids), and some of the sterols. The cerebrins may constitute 0.2 to 0.3% of the dry weight and are end products of the autodigestion of some complex membranous lipids. The yellow pigments, carotenes, are often abundant in fungi.

REFERENCES CITED

BIRKINSHAW, J. H. 1965. Chemical constituents of the fungal cell. 2. Special chemical products. *In:* Ainsworth, G. C., and A. S. Sussman, Eds. The fungi: an advanced treatise. Academic Press, New York, Vol. 1, pp. 179–228.

BLUMENTHAL, H. J. 1965. Carbohydrate metabolism. 1. Glycolysis. *In:* Ainsworth, G. C., and A. S. Sussman, Eds. The fungi: an advanced treatise. Academic Press, New York, Vol. 1, pp. 229–268.

BRENNAN, P. J., and D. M. LÖSEL. 1978. Physiology of fungal lipids: selected topics. Advan. Microb. Physiol. 17:47–179.

BU'LOCK, J. D., and H. M. SMALLEY. 1961. Biosynthesis of aromatic substances from acetyl- and malonyl-coenzyme A. Proc. Chem. Soc. 1961:209–211

BURNETT, J. H. 1976. Fundamentals of mycology, 2nd ed. Crane, Russak, New York, 673 pp.

CANTINO, C., and G. TURIAN. 1959. Physiology and development of lower fungi (Phycomycetes). Ann. Rev. Microbiol. 13:97–124.

CASSELTON, P. J. 1976. Anaplerotic pathways. In: Smith, J. E., and D. R. Berry, Eds. The filamentous fungi: biosynthesis and metabolism. Edward Arnold, London, Vol. 2, pp. 121–136.

COCHRANE, V. W. 1958. Physiology of fungi. Wiley, New York, 524 pp.

——. 1976. Glycolysis. In: Smith, J. E., and D. R. Berry, Eds. The filamentous fungi: biosynthesis and metabolism. Edward Arnold, London, Vol. 2, pp. 65–91.

COLLINS, J. F., and H. L. KORNBERG. 1960. The metabolism of C_2 compounds in microorganisms. 4. Synthesis of cell materials from acetate by Aspergillus niger. Biochem. J. 77:430–438.

FINCHAM, J. R. S., and J. B. BOYLEN. 1957. Neurospora crassa mutants lacking arginosuccinase. J. Gen. Microbiol. 16:438–448.

FINN, R. K. 1954. Agitation-aeration in the laboratory and in industry. Bacteriol. Rev. 18:254–274.

HECK, F. I. 1929. A study of the nitrogenous compounds in fungous tissue and their decomposition in the soil. Soil Sci. 27:1–45.

HOROWITZ, N. H. 1947. Methionine synthesis in Neurospora. The isolation of cystathionine. J. Biol. Chem. 171:255–264.

HUNTER, K., and A. H. ROSE. 1971. Yeast lipids and membranes. In: Rose, A. H., and J. S. Harrison, Eds. The yeasts. Academic Press, New York, Vol. 2, pp. 211–270.

KUBICEK, C. P. 1988. Regulatory aspects of the tricarboxylic acid cycle in filamentous fungi—a review. Trans. Brit. Mycol. Soc. 90:339–349.

LINDENMAYER, A. 1965. Carbohydrate metabolism. 3. Terminal oxidation and electron transport. In: Ainsworth, G. C., and A. S. Sussman, Eds. The fungi: an advanced treatise. Academic Press, New York, Vol. 1, pp. 301–348.

McMURROUGH, I., A. FLORES-CARREON, and S. BARTNICKI-GARCIA. 1971. Pathway of chitin synthesis and cellular localization of chitin synthetase in Mucor rouxii. J. Biol. Chem. 246:3099–4007.

NICHOLAS, D. J. D. 1965. Utilization of inorganic nitrogen compounds and amino acids by fungi. In: Ainsworth, G. C., and A. S. Sussman, Eds. The fungi: an advanced treatise. Academic Press, New York, Vol. 1, pp. 349–376.

NIEDERPRUEM, D. J. 1965. Carbohydrate metabolism. 2. Tricarboxylic acid cycle. In: Ainsworth, G. C., and A. S. Sussman, Eds. The fungi: an advanced treatise. Academic Press, New York, Vol. 1, pp. 269–300.

SARACHEK, A., and N. P. HIGGINS. 1972. Effects of ergosterol, palmitic acid and related simple lipids on the recovery of Candida albicans from ultraviolet irradiation. Arch. Mikrobiol. 82:38–54.

SCHISLER, L. C., and O. VOLKOFF. 1977. The effect of safflower oil on mycelial growth of Boletaceae in submerged liquid cultures. Mycologia 69:118–125.

SRB, A. M., and N. H. HOROWITZ. 1944. The ornithine cycle in Neurospora and its genetic control. J. Biol. Chem. 154:129–139.

SUOMALAINEN, H., and E. OURA. 1971. Yeast nutrition and solute uptake. In: Rose, A. H., and J. S. Harrison, Eds. The yeasts. Academic Press, New York, Vol. 2, pp. 3–74.

TATUM, E. L., S. R. GROSS, G. EHRENSVAARD, and L. GARNJORBST. 1954. Synthesis of aromatic compounds by Neurospora. Proc. Natl. Acad. Sci. USA 40:271–276.

UTTER, M. E., R. E. BARDEN, and B. L. TAYLOR. 1975. Pyruvate carboxylase: an evaluation of the relationships between structure and mechanism and between structure and catalytic activity. Advan. Enzymol. 42:1–72.

VISWANATH-REDDY, M., and G. TURIAN. 1975. Physiological changes during protoperithecial differentiation in *Neurospora tetrasperma*. Physiol. Plantarum 35:166–174.

WEETE, J. D. 1973. Sterols of the fungi: distribution and biosynthesis. Phytochemistry 12:1843–1846.

———1974. Fungal lipid biochemistry: distribution and biosynthesis. Plenum Press, New York, 393 pp.

Additional Recommended References

BLUMENTHAL, H. J., and S. ROSEMAN. 1957. Quantitative estimation of chitin in fungi. J. Bacteriol. 74:222–224.

ELLIOTT, C. G. 1977. Sterols in fungi: their functions in growth and reproduction. Advan. Microb. Physiol. 15:121–173.

GARRAWAY, M. O., and R. C. EVANS. 1984. Fungal nutrition and physiology. Wiley, New York, 401 pp.

RATTRAY, J. B. M., A. SCHIBECI, and D. K. KIDBY. 1975. Lipids of yeasts. Bacteriol. Rev. 39:197–231.

SMITH, J. E., and D. R. BERRY, Eds. 1976. The filamentous fungi: biosynthesis and metabolism. Edward Arnold, London, Vol. 2, 520 pp.

SOLS, A., C. GANCEDO, and G. DELA FUENTE. 1971. Energy-yielding metabolism in yeasts. *In:* Rose, A. H., and J. S. Harrison, Eds. The yeasts. Academic Press, New York, Vol. 2, pp. 272–307.

STRATHERN, J. N., E. W. JONES, and J. R. BROACH, Eds. 1982. The molecular biology of the yeast *Saccharomyces:* metabolism and gene expression. Cold Spring Harbor Laboratory, Cold Spring Harbor, N.Y. 680 pp.

10 *Sporulation*

Sporulation involves the development of sexual or nonsexual spores and their associated structures. As we have seen in earlier chapters, the fungi form a wide variety of reproductive structures associated with spore production. Nonsexual spores include zoospores produced in zoosporangia, nonmotile sporangiospores produced in sporangia, and condia on condiophores. Sexual reproduction may involve the differentiation of gametes and gametangia and their mating before spore production occurs, as is the case with zygospore production. In the Ascomycotina and Basidiomycotina, sexual spores are borne by complex multicellular sporocarps. The formation of these structures and their behavior are controlled by genetic factors, discussed in Chapter 8, as well as by hormonal, nutritional, and environmental factors, which are considered in this chapter.

HORMONAL CONTROL

The regulatory systems of higher plants and animals involve hormonal control. Hormones are produced in minute quantities by a specific animal organ or region of the plant, and are transported to another part of the organism, where they exert their effects. Hormones have been detected in some fungi. Like the hormones of animals and plants, these hormones are produced in minute quantities, diffuse from their site of production, and finally, exert a highly specific effect on a particular organ of the fungus. This description of hormones excludes substances such as vitamins, purines, and other growth factors which usually exert a more generalized effect on reproduction and growth. Fungal hormones can regulate differentiation of gametangia, serve as attractants in the mating process, and have a role during sporocarp development. We do not know how widely hormones are distributed among the fungi, and comparatively few examples of hormonal action are well documented. In part, our comparative state of ignorance is due to the experimental difficulties encountered in actually demonstrating that a hormone exists.

Hormones are known to serve as attractants for gametes, to initiate development of gametangia and also serve as attractants during mating, and to have a role in ascocarp (and possibly basidiocarp) development.

Role of Hormones in Gamete Attraction

One of the most important stages in many fungal life cycles is the coming together of gametes. In aquatic organisms, the gametes may be distant from each other and an attractant may be required to bring the gametes together.

Male gametes of *Allomyces*, a Chytridiomycete, are attracted to female gametangia before and during the release of female gametes. The female gametes release the hormone *sirenin*, which attracts the male gametes, but the release of this hormone ceases after the male gametes reach the female. Sirenin can attract male gametes in concentrations as low as $10^{-10} M$, but fails to attract them at higher concentrations ($10^{-4} M$). Female gametes of *Allomyces* are not attracted to sirenin (Carlile and Machlis, 1965-a, b). Sirenin is a sesquiterpene (Fig. 10–1) (Nutting et al., 1968). The particular chemical structure is critical in eliciting a response from the male gametes, as they fail to respond to closely related chemical analogues of sirenin (Machlis, 1973). Sirenin is of historical interest, as it was the first fungal hormone to be chemically isolated and characterized.

Sirenin

Trisporic acid C

Antheridiol

I, R = (CH₃)₂CHC=O

Oogoniol–I

Figure 10–1 Fungal hormones.

Hormones are also involved in mating of gametes in yeasts, such as *Saccharomyces cerevisiae*, which have two-allele heterothallism involving mating between cells having the mating factors *A* and *a*. Mating depends on chance collisions in a medium and continued close cell-to-cell contact. Cells of each mating type produce a diffusible proteinaceous hormone which causes changes in the cells of the opposite mating type. The hormone increases the selectivity of the cells for the opposite mating type as well as arresting cellular growth at the G_1 stage (the transient phase between nuclear division and DNA synthesis), coordinating the cell cycle. As a result of the hormonal activity, the cells agglutinate and localized autolysis of the cell wall and plasmalemma occurs, allowing the protoplasts to fuse. Once the gametes join to form a zygote, the new diploid cell is incapable of responding to the hormones produced by other haploid cells (Thorner, 1981).

In many fungi, mating may occur between two gametangia or a gametangium and a unicellular spermatium. Hormonal control of gametangial development and attraction is known to occur in *Achlya,* the Mucorales, and in some members of the Ascomyctotina.

ACHLYA

Some species of the aquatic fungus, *Achlya,* bear the male organs (antheridia) and female organs (oogonia) on separate thalli. Raper (1951) ingeniously demonstrated hormonal control of sexual differentiation in *Achlya.* Potentially male and female thalli were alternated in sequential microaquaria. Water was then allowed to pass through the aquaria in one direction only. If growth factors or hormones were released into the medium, then theoretically those thalli occupying a stage in the sequence after hormonal release would respond. It was discovered through this arrangement that a series of highly specific hormones are formed that control the sexual process. The female vegetative plant produces hormones A and A^2, which induce the formation of the antheridial hyphae on the male plant. The male plant then produces hormone A^1, which passes back to the female plant, augmenting the continued production of hormones A and A^2. Eventually, the male plant produces hormone A^3, which depresses the action of A and A^2; then the antheridial hyphae produce hormone B, which passes to the female plant, where it induces the formation of the oogonial initials. The oogonial initials then produce hormone C, which (1) induces the antheridial hyphae to grow along a concentration gradient toward the oogonium until physical contact is made between the organs, and (2) induces the formation and delimitation of the antheridium in the male plant. The antheridia may then produce hormone D, which induces (1) the delimitation of the oogonium, and (2) the differentiation of the oospheres. Therefore, as interpreted by Raper, sexual differentiation in *Achlya* is under the control of at least seven distinct hormones, four secreted by the male thallus and three secreted by the female thallus. The sequence is clarified further in Fig. 10-2. The hormones are effective in extremely minute quantities, and hormone A induced the formation of antheridial hyphae when the culture medium was tested in a dilution as low as 10^{-13} M.

Both hormones A and B have been isolated and were found to be steroids (Fig. 10-1). Hormone A is now called *antheridiol* and the three steroids that give the effects of hormone B are called *oogoniol* (McMorris et al., 1975). Because the hormones are effective in extremely minute dosages, large quantities of culture medium must be processed to obtain the hormones in pure form. Barksdale (1967) reported that 6000 liters of the culture medium are required to yield only 20 milligrams of crystalline antheridiol. Action of the hormones is highly specific, and even steroids that are closely related in the structure typically fail to elicit a morphogenetic response when tested on cultures of *Achlya* (Barksdale et al., 1974). The mode of action of antheridiol has been investigated. When antheridiol is added to the culture medium in which a vegetative thallus is growing, the antheridiol is removed from the culture medium by the thallus and metabolized (Fig. 10-3). This uptake of antheridiol requires about 60 minutes and initiates an increase in the level of both messenger and ribosomal RNA as well as of the protein within the thallus. These metabolic changes may be detected in advance of the visible morphogenetic changes (initiation of the antheridial branches), which begin about 60 to 90 minutes after administration of the antheridiol but may require about 300 minutes for completion. Both the RNA and protein are required to support the growth of the antheridial branches.

Figure 10–2 Interrelationships between sequential hormonal secretions and morphology in heterothallic species of *Achlya*. [From J. R. Raper, 1955 *in:* Butler, E. G., Ed., Biological specificity and growth, Princeton University Press, Princeton, N.J.]

After about 180 minutes, no additional RNA synthesis is necessary, but protein synthesis must continue. There is also an increase in the enzyme cellulase, required to soften the parent walls to allow branching to take place (Timberlake, 1976), and an increase of catalase in microbodies (Choinski and Mullins, 1977).

Nutritional conditions can influence the activity of antheridiol, as an adequate concentration of a nitrogen and carbon source must be present for the initiation of antheridial branches and also influences their number (Barksdale, 1970).

MUCORALES

Several members of the Mucorales are heterothallic, and as discussed in Chapter 3, gametangia are produced by hyphae of different mating types. Hyphae of each mating type produce a volatile compound that passes through the air and induces a response in hyphae of the opposite mating type. The active volatiles produced by the plus strain are 4-hydroxymethyltrisporates, while those produced by the minus strains are trisporins (Nieuwenhuis and van den Ende, 1975). The production of trisporic acids acts as a positive feedback mechanism and continues to stimulate production of the volatile precursors (Werkman

Figure 10–3 Summary of events associated with antheridiol-induced differentiation in *Achlya*. The brackets indicate the time periods during development in which the process occurs. The dashed lines indicate relative morphological differentiation. [From W. E. Timberlake, 1976, Develop. Biol. 51:202–214.]

and van den Ende, 1973). These active volatiles induce progametangia formation and synthesis of both carotene and trisporic acid. Mated cultures produce about 15 to 20 times as much carotene as separate cultures of each mating type and also produce trisporic acid, which can be chemically derived from β-carotene. Trisporic acid (Fig. 10–2) is formed by both the plus and minus strains, and if either partner is removed, production will cease. Purified trisporic acid in small concentrations ($10^{-8}M$) will induce formation of progametangia in unmated plus or minus cultures (Gooday, 1973). Addition of trisporic acid to minus cultures of *Blakeslea trispora* will cause a massive (about 73-fold) increase in production of carotenoid pigments (Thomas and Goodwin, 1967). Although trisporic acids greatly enhance sexual reproduction, they apparently are not essential. Mutants of *Mucor mucedo* unable to respond to the addition of trisporic acids to the culture have been found capable of mating and producing progametangia (Wurtz and Jockusch, 1975).

ASCOMYCOTINA

Hormonal control in the Ascomycotina was first demonstrated in *Ascobolus stercorarius*, a discomycete (Bistis, 1956, 1957). Plasmogamy occurs by contact of a spermatium and ascogonium of different mating types. If a spermatium is placed on the agar near (but not in physical contact with) the hyphae of the opposite mating type, the ascogonium-producing hyphae will differentiate. Evidently, an ascogonium-inducing hormone is being produced by the spermatium, as this effect occurs only within a radius of approximately 500 micrometers of that cell. There is no apparent control by the spermatium during the initial stages of ascogonium development in which it is coiling. The trichogyne eventually appears and exhibits a strongly positive chemotropic response to the sper-

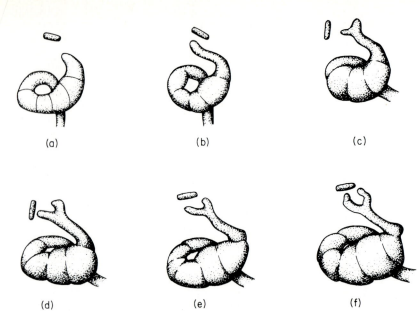

(a)   (b)   (c)

(d)   (e)   (f)

Figure 10–4 Hormonal attraction of the spermatium for the trichogyne of *Ascobolus stercorarius:* (a) at 6:00 P.M., the spermatium was placed in the vicinity of the ascogonium; (b) at 6:12 P.M., the ascogonial tip now shows a directional response to the spermatium; (c) 6:25 P.M., there is now the development of a lateral, subterminal projection as a response to the relocation of the spermatium, which was made at 6:16 P.M.; (d) 6:37 P.M., growth of the projection directly to the spermatium and a change in direction of growth of the trichogyne tip; (e) 6:41 P.M., the spermatium has been moved to a new position; (f) 6:48 P.M., both apices have now responded to the new location of the spermatium. [Adapted from G. Bistis, 1957, Am. J. Botany 44:436–443.]

matium. If the spermatium is moved about on the agar surface, the trichogyne will change its direction of coiling and will always point toward the spermatium (Fig. 10–4). If the spermatium is removed at this point, growth of the trichogyne will cease, and in some instances the trichogyne will revert to the vegetative condition. Plasmogamy occurs, and development of the ascogonium is completed. The sheathing hyphae develop below the ascogonium, surround it, and proliferate to form the sterile tissue of the apothecium. These ensheathing hyphae exhibit a positive chemotropic response, which appears to be due to hormone(s) produced by the ascogonium.

Hormonal control in the Ascomycotina is also known to occur in *Nectria haematococca* var. *curcubitae* (Bistis and Georgopoulos, 1979) and *Neurospora crassa* (Bistis, 1981, 1983). These two fungi also have hormones mediating the attraction between the trichogynes and spermatia.

A Hormonal Role in Basidiocarp Development?

In some agarics stipe elongation is regulated by a growth factor, perhaps a hormone, which is produced in the gills and diffuses downward to the stipe. If the gills are removed from the pileus of *Agaricus brunnescens,* the stipe will not elongate normally. If a portion of the pileus is unilaterally removed, curvature of the stipe will occur in the direction opposite from that of the remaining pileus. Curvature of the stipe can also be caused by placing agar blocks containing diffusate from the gills on one side only of the decapitated stipe (Gruen, 1963). Although the gills are required for the elongation of the stipe of *Flammulina velutipes,* the agar block test does not produce a curvature (Gruen, 1969). These results suggest that some diffusible material effective in small co█ passing from the pileus to the stipe. However, no specific materia█ isolated.

ENVIRONMENTAL FACTORS AFFECTING SPORULATION

Survival of a fungus species may depend on its ability to mate with a compatible partner, but it is equally dependent on the ability of the fungus to produce large quantities of viable spores under the prevailing environmental conditions. Fungi require a period of vegetative growth before sporulation will occur (Fig. 10–5). As the mycelium enters the reproductive stage, materials produced by the mycelium are transported to the developing reproductive structures. Compounds in the mycelium that can serve as carbon and energy sources for sporulation include proteins and various carbohydrates, such as trehalose (a disaccharide), glycogen (a reserve polysaccharide that accumulates in the cytoplasm), and glucan (a polysaccharide occurring in the cell wall). As much as 40% of the dry weight of the mycelium consists of a glucan, which is hydrolyzed and the glucose monomers utilized in cleistothecial production in *Aspergillus nidulans*. If physiological conditions prevent glucan digestion, the proteins in the mycelium of *A. nidulans* will be used instead (Zonneveld, 1972, 1980).

Reproduction is associated with the decline or cessation of vegetative growth, and a number of factors interact to induce the shift from vegetative

Figure 10–5 Basidiocarps of *Flammulina velutipes* arising from vegetative mycelium in culture. [Courtesy Plant Pathology Department, Cornell University.]

owth to sporulation. Even the presence of a simple barrier can induce sporula-
on in some fungi. Frequently cultures of fungi in petri dishes will sporulate
around the periphery of the dish, and experiments with one such fungus, *Sordaria brevicollis,* confirm that it is the actual physical contact with the glass that
is the stimulus for perithecial production (MacDonald and Bond, 1976). Similar
observations can be made in nature. The hypogeous fungus *Hymenogaster luteus*
tends to form its basidiocarps in contact with hard surfaces such as a layer of
clay, stones, or tree roots (Hawker, 1954). Sporulation is particularly affected by
other environmental factors, including nutrition, light, and temperature.

The physical and nutritional requirements for sporulation are usually more
precise, or more restricted, than those permitting vegetative growth. Vegetative
growth will occur over a much wider range of conditions than will reproduction.
Simple spore-bearing structures such as sporangiophores and conidiophores,
in turn, form under a wider range of conditions than do those permitting formation of the more complex sporocarps.

There is no single set of optimum conditions for reproduction, as virtually
each species responds uniquely to environmental conditions. In addition, optimal conditions for nonsexual reproduction may differ from those for sexual reproduction in the same species. Sometimes the effect of a particular physical or
nutritional factor can be modified by other cultural conditions—light requirements, for example, can be modified by temperature, and vitamin requirements
by the available carbohydrate.

The majority of nutritional and physical effects on reproduction are quantitative, either decreasing or increasing the number of spores and other reproductive structures. Relatively few external effects on reproduction are of a qualitative all-or-none nature. Cultural conditions sometimes affect morphology by
causing a change in size, shape, septation, or pigmentation of reproductive
structures. Even the dynamic aspects of reproduction, such as heterokaryon formation and plasmogamy, may be affected.

The environmental factors exert their effects by altering biochemical and
developmental pathways. The precise manner in which this occurs is, for the
large part, unknown. On pages 350–352, a few examples of altered metabolism
associated with reproduction are discussed.

Nutrition

MEDIUM CONCENTRATION

In classic experiments, Klebs (1899) cultured *Saprolegnia mixta* in the vegetative state for $2\frac{1}{2}$ years by replenishing the enriched culture medium. When the
fungus was transferred to water, zoospore production began in a few days. Similar results were obtained in experiments with other fungi, and Klebs concluded
that a fungus is most likely to sporulate after the vegetative mycelium exhausts
the nutrient supply in the medium. Mycologists have repeated Klebs' observations with a wide variety of fungi—reproduction generally occurs when vigorous
mycelial growth depletes the nutrients or when the mycelium is transferred to a
more dilute medium (Fig. 10–6). Culture medium favoring reproduction is often
formulated to provide a weak nutrient concentration that will support scant mycelial growth, particularly for fungi such as basidiomycetes that produce large
sporocarps and are favored by relatively large quantities of dilute medium. In
some cases, heavy growth of a fungus in culture will induce earlier sporulation
because the nutrients are used more rapidly.

The reasons for inhibition of reproduction by high nutrient levels are not
understood. It has often been suggested that cessation of vegetative growth may
be the actual stimulus initiating reproduction and that the high nutrient levels
simply postpone starvation. Some insight into a physiological mechanism may

Figure 10–6 Relationships between growth, sporulation and glucose content of the medium of cultures of *Rhizopus arrhizus*. Squares indicate glucose concentration, circles the mycelial dry weight, and triangles the number of spores per culture. [From G. C. Lawler and D. J. Weber, 1980, J. Gen. Microbiol. 117:465–474.]

be obtained by examining the yeast *Saccharomyces cerevisiae*, in which meiosis and sporulation may be induced by transfer from a nutrient-rich medium to a nutrient-poor medium (Croes, 1967-a, b). During vegetative growth, *S. cervisiae* utilizes the glucose in its medium in fermentation pathways, which results in a decrease of glucose but an increase of ethanol in the medium (Fig. 10–7). Mitotic

Figure 10–7 Comparison of the pathways involving carbohydrate and protein metabolism during active growth (left), the late phase of growth (center), and early sporogenesis (right) in *Saccharomyces cerevisiae*. Lines of different thickness mark the relative reaction rates of the processes. See the text for additional details. [From A. F. Croes, 1967, Planta 76:227–237.]

divisions occur in the presence of glucose. Near the end of the logarithmic phase of growth, a shift in the respiratory pathway occurs so that acetate and ethanol may be metabolized in an aerobic pathway (glyoxylic acid bypass of the TCA cycle). The utilization of acetate or ethanol is essential to sporogenesis and appears to induce meiosis, as the addition of a small amount of glucose at this point will cause a reversion both to fermentation and to vegetative growth. Immediately after meiosis is induced, there is an initial sharp rise in aerobic respiration followed by its steady decline, reflecting the initial period of high metabolic activity followed by low metabolic activity. Metabolism of acetate makes the accumulation of glycogen and fats possible, which increases the dry weight. There is a steady decline in the RNA levels, which reflects the lowering of metabolic activity in later stages of sporogenesis. The level of protein is not altered greatly during sporulation, which indicates that the fungus is utilizing reserves that had been accumulated during growth. As we can see in the account above, the metabolic changes involved in reproduction in *S. cerevisiae* are directly related to glucose starvation.

Reproduction at high nutrient concentrations can sometimes be inhibited by metabolite excretion by the mycelium that causes staling of the medium or alters the pH. Asthana and Hawker (1936) and Barksdale (1962) demonstrated that staling of the medium can inhibit reproduction by adding fresh medium to cultures in which sporulation had declined. The staling products were diluted without an accompanying reduction in nutrients, and the formation of reproductive structures resumed.

CARBON NUTRITION

Fungi differ in their ability to use different carbon sources for reproduction. Mycelial growth is often favored by a monosaccharide (particularly by glucose or fructose), and nonsexual reproduction (such as production of conidia) is often similarly favored by a monosaccharide. However, sexual reproduction in the same fungus is usually favored by a disaccharide or polysaccharide. *Cordyceps militaris* provides such an example—perithecial stromata are produced earlier and more abundantly on medium containing sucrose or starch than on glucose (Basith and Madelin, 1968). Cellulose, which has limited availability for most fungi and is not generally favorable for vegetative growth, may produce abundant sporulation in some fungi. Glucose is often an unsatisfactory carbon source for sexual reproduction even though other hexoses, pentoses, disaccharides, or polysaccharides may be satisfactory. Sporulation may even be repressed by glucose. Concentrations of glucose as low as 0.2% decrease the number of perithecia formed by *Magnaporthe salvinii* and prevent ascus formation (Tsuda et al., 1982) as well as completely suppressing apothecial production by *Pyronema domesticum* (Moore-Landecker, 1975).

Not only is the source of carbon very important, but its concentration in the medium can be critical in determining whether or not sporulation will occur, as outlined previously. Typically sexual reproduction is favored by lower concentrations than those supporting maximum vegetative growth and is completely inhibited by high concentrations. Fertile perithecia of *Ophiobolus graminis* require a minimum carbohydrate concentration of 0.75% glucose, becoming more numerous as the glucose concentration is increased to 1.5%, the optimum concentration. No perithecia are produced on media with 10% glucose, although maximum vegetative growth and pigmentation occur at this concentration. In contrast, numerous sterile perithecia form if the medium contains sucrose rather than glucose (Weste, 1970). More perithecial stromata are produced by *Cordyceps militaris* on medium containing 5% rather than 10% sucrose but form perithecia only on medium containing no more than 1% glucose (Basith and Madelin, 1968). The primary effect of polysaccharides is that of concentration. The fungus utilizes the complex molecules at a slower rate than the simple hexoses; therefore, the amount of available carbohydrate is less but becomes available as hydrolysis of the molecule occurs over time. At any one time, this approximates a dilute medium with low hexose concentrations.

Both the source and concentration of the carbon source can affect the behavior or morphology of the reproductive structures. Zoosporangia of the chytrid *Chytriomyces hyalinus* are larger when grown on cellobiose than on fructose (starch supports the best vegetative growth) (Hasija and Miller, 1971). Hyphal fusions leading to dikaryon formation in *Schizophyllum commune* occur with a much greater frequency in the absence of a carbohydrate than in the presence of even small concentrations (0.5% or higher) (Ahmad and Miles, 1970). An increase in carbohydrate causes a decrease in the pycnidial size in some deuteromycetes and may also increase the length and curvature of the spore (Nitimargi, 1937).

NITROGEN NUTRITION

Fungi differ in their ability to utilize different nitrogen sources for reproduction, and a nitrogen source that gives good vegetative growth may not favor reproduction. For example, asparagine and ammonium compounds are usually good nitrogen sources for vegetative growth but generally suppress reproduction. The inhibitory effect of these compounds probably is due to the accumulation of ammonia during the vegetative growth phase, which makes the medium alkaline and inhibits reproduction, or alternatively, the ammonia that is present in the culture as a gas may be toxic. *Venturia inaequalis* fails to form perithecia on medium containing ammonium salts but will do so if calcium carbonate is added to control the pH (Ross and Bremner, 1971). In their study of species of *Phytophthora*, Leal et al. (1970) determined that the accumulation of ammonia as a gas is toxic and inhibits oogonium formation but that the increased pH and accumulation of ammonium ions in the medium apparently do not affect reproduction.

Nitrates, urea, and particularly the amino acids are likely to be favorable nitrogen sources for reproduction. Fungi often show a great deal of specificity and respond uniquely to a particular nitrogen source, and may be able to utilize all of these nitrogen sources or only some of them. They are often best able to utilize a particular amino acid or only a few amino acids. The particular amino acids that are favorable vary from one fungus to another, and a mixture of amino acids is usually better than a single amino acid. Examples of specificity for a particular nitrogen source follow. Some species of *Sporomiella* are able to utilize nitrates and certain amino acids but not urea for perithecial maturation (Asina et al., 1977). *Cordyceps militaris*, an insect parasite, is unusual because perithecial stromata are favored by complex nitrogen compounds (hemoglobin, casein, or peptone), while amino acids are not suitable nitrogen sources. With some amino acids, stromatal initials form but then abort (Basith and Madelin, 1968).

Nitrogen source may also affect specific aspects of reproduction or morphology of reproductive structures. Diploidization, required for the parasexual cycle, is greatly favored by nitrogen supplied as a nitrate but not by asparagine or ammonium in *Verticillium albo-atrum* (Ingle and Hastie, 1974). Pycnidia of *Phyllosticta antirrhini* are almost twice as large when grown on a medium containing an ammonium salt than when grown on a medium containing alanine (Maiello and Cappellini, 1976).

The minimum amount of nitrogen permitting sporulation is generally above that at which sparse vegetative growth can take place, but at high nitrogen concentrations growth becomes vigorous and inhibits sporulation. Casein hydrolysate, which provides a mixture of amino acids, favors apothecial maturation by *Pyronema domesticum* in concentrations up to 0.1%, but higher concentrations depress maturation of the ascogonia and ascogenous hyphae. At a concentration of 0.35% numerous apothecia with a well-developed sterile system form, but they lack asci (Moore-Landecker, 1975). The optimum nitrogen concentration can also vary with the nitrogen source, as demonstrated with *Venturia inaequalis* (Ross and Hamlin, 1965). Nitrates and urea are most effective for reproduction of *V. inaequalis* at a concentration of about 30 parts per million (ppm) nitrogen, while the most effective concentration of amino acids varies. Perithecial production is abundant at 100 ppm nitrogen if glutamic acid and valine are used as nitrogen sources, while this concentration is too great to permit normal perithecial maturation if aspartic acid is provided.

MINERALS, VITAMINS, AND GROWTH FACTORS

Essential mineral elements and vitamins were discussed in Chapter 8. Formation of reproductive structures typically requires a higher concentration of minerals or vitamins than concentrations that permit vegetative growth.

Sporulation may not occur when the fungus is grown on medium containing a mineral concentration sufficient to support scant or even optimal vegetative growth (Fig.10–8). A particular mineral may be required for reproduction but not for growth. This is the case for the bird's nest fungus *Cyathus stercoreus*, which can grow vegetatively on a medium lacking calcium ions but requires their addition for basidiocarp formation (Lu, 1973).

Similarly, a fungus may be partially deficient for a vitamin, and this partial deficiency may be expressed as an inability to sporulate although vegetative growth is good. For example, *Sordaria fimicola* will grow vegetatively on a biotin-free medium but will not form perithecia unless biotin is added (Barnett and Lilly, 1947-a). Generally, vitamins that favor mycelial growth also favor reproduction. The optimum level of the deficient vitamin may be determined by the concentration of the other nutrients available or by the type of carbohydrate. The amount of thiamine required for perithecium development by *Ceratostomella fimbriata* increases as the available amount of carbohydrate increases (Barnett and Lilly, 1947-b). In a similar manner, the amount of available thiamine raises the optimum carbohydrate level for sporulation in other fungi.

Reproduction is sometimes enhanced by adding growth factors to the medium. Species of *Pythium* and *Phytophthora* and some other fungi require sterols for both nonsexual and sexual reproduction but not for vegetative growth (Hendrix, 1970). Fatty acids promote perithecial formation by species of *Ceratocystis* (Dalpe and Neumann, 1976; Marshall et al., 1982). Reproduction by many plant

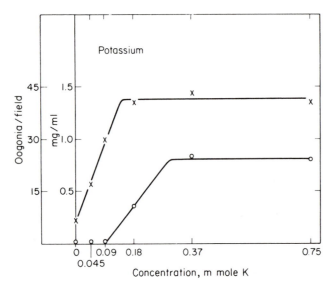

Figure 10–8 Effect of the potassium concentration on vegetative growth (*x*) and oogonium formation (*o*) in *Achlya ambisexualis*. Note that the optimum concentration of potassium for vegetative growth is barely sufficient for reproduction. [From A. W. Barksdale, 1962, Am. J. Botany 49:633–638.]

parasitic fungi either requires or is greatly enhanced by the addition of plant materials (containing vitamins and unidentified growth factors) to the medium.

Physical Factors

Physical factors affecting growth include temperature, acidity, light, moisture, carbon dioxide, and oxygen (Chapter 8). The mycelium can tolerate and grow over a relatively wide range of conditions, but reproduction requires a more precise set of conditions. This general rule can be illustrated by examining temperature. There is a fairly broad range of temperatures permitting growth, but this range usually includes a low temperature range that allows vegetative growth to occur but is too low for reproduction, and a range too high to allow reproduction but still permitting vegetative growth (Fig. 10–9). The optima for vegetative growth and reproduction may or may not coincide. If they do not coincide, the optimum for reproduction may be either higher or lower than for vegetative growth.

Reproduction may require some physical factor that is not required for vegetative growth. Light affects reproduction of many fungi but rarely affects vegetative growth (Chapter 8). Basidiocarps of *Gloeophyllum saepiarium* will form in culture only if the vegetative mycelium is exposed to periods of drying (States, 1975).

TEMPERATURE

Temperature is one of the most important physical factors affecting reproduction, as it may influence a number of metabolic and physiological processes, any one of which may regulate reproduction. An interesting example may be found in *Schizophyllum commune,* in which the rate of nuclear migration in the dikaryon and recombination within the *A* mating factor can be increased by raising the temperature to about 32°C (Raper, 1966). Culturing *Neurospora tetrasperma* at a high temperature (37°C) prevents perithecial formation but also results in about six times the normal levels of amino acids, indicating that a disturbance in nitrogen metabolism occurs at high temperatures (Viswanath-

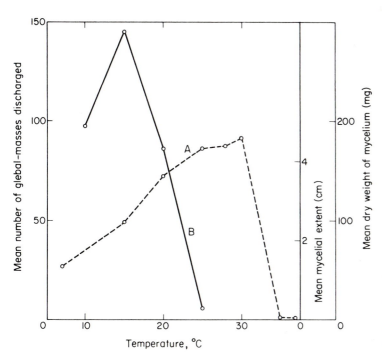

Figure 10–9 Comparison of vegetative growth and basidiocarp production of an isolate of *Sphaerobolus* as affected by temperature. A, vegetative growth; B, glebal masses discharged after 38 days. [Adapted from S. O. Alasoadura, 1963, Ann. Botany (London) II 27:123–145.]

Environmental Factors Affecting Sporulation

Reddy and Turian, 1975). The effect of temperature on reproduction reflects the summation of its effects on many temperature-dependent processes such as these.

Different fungal species as well as different strains of the same species often have unique temperature requirements. Even the nonsexual spores and sexual spores produced by the same fungus can respond differently to temperature. The temperature response for production of nonsexual spores often does not differ greatly from that of the vegetative mycelium, while sexual reproduction in the same species often has very precise temperature requirements.

The vast majority of fungi are mesophilic in their temperature requirements, thriving at temperatures somewhere between 10 and 40°C. Although there is considerable difference among the fungi and exceptions can be found, the optimum temperature for sexual reproduction of mesophilic fungi frequently lies somewhere between 20 and 25°C and drops off sharply with either lower or higher temperatures, having a minimum and maximum temperature for reproduction within about 5 to 10°C on either side of the optimum range (Fig. 10-9). Optimum temperatures as low as 10 to 15°C or as high as 30 to 35°C occur but are unusual. Some discomycetes in the Helotiales have an unusually low temperature optimum—apothecial development by *Monilinia* is favored by cool temperatures (10 to 15°C) (Willetts and Harada, 1984). Species of *Thelebolus* are unusually tolerant to cold. Fertile apothecia can be produced at 0°C, although the optimum range is 15 to 20°C (Wicklow and Malloch, 1971).

Some Pyrenomycetes require a long period of cold to induce perithecial formation. An example is *Pleospora herbarum*, a plant parasite that forms mature perithecia in the spring (Leach, 1971). Actively growing mycelium must be exposed to uninterrupted temperatures of 5 to 10°C for several days to stimulate maximum production of protoperithecia. Although the actual differentiation of protoperithecia is favored by relatively high temperatures (24 to 27°C), perithecial maturation is favored by low temperatures (10 to 15°C).

Tyrosinase activity apparently regulates the induction of perithecia in fungi requiring cold. Tyrosinase oxidizes tyrosine to form melanin, a dark pigment occurring in perithecial walls. In *Hypomyces solani* var. *curcurbitae*, an increase in tyrosinase activity accompanies formation of the perithecium primordium. Exposure to a cold temperature (4°C) increases both primordium development and tryosinase activity. The addition of exogenous tyrosinase also stimulates the abundant formation of primordia (Wilson, 1968). Using different genetic strains, Wilson and Baker (1969) determined that the response to cold is controlled by a gene having two alleles. Strains having one allele required exposure to the cold or the addition of tyrosinase to induce perithecial primordia, while strains having the second allele produced perithecial primordia without exposure to cold or the addition of tyrosinase. However, the latter strains had extracellular tyrosinase activity even without cold treatment.

LIGHT

Light affects sporulation by numerous fungi of every major class, and may be either stimulatory (inductive) or inhibitory to the formation of reproductive structures and spores. The fungi may be divided into five groups on the basis of their response to light: (1) those that are apparently indifferent to light; (2) those in which sporulation is decreased or prevented on exposure to light; (3) those that require alternating light and darkness to sporulate; (4) those that are able to produce viable spores in complete darkness but sporulate more abundantly with exposure to light; and (5) those that require light in order to produce reproductive structures and spores.

Effects of light on reproduction in fungi are extremely complex. Closely related species or different isolates of the same species may differ in their response to light. Different developmental stages or spore forms of the same isolate may have varying light requirements. The intensity, duration, and quality of light all play a role in the overall effect of light on reproduction.

Effective Wavelengths. The wavelengths affecting reproduction are most often in the near ultraviolet (NUV), violet, or blue regions of the spectrum (about 320 to 490 nanometers) in length (Fig. 10–10). Occasionally, longer wavelengths in the yellow, red, or far-red ranges may affect reproduction. Moreover, a fungus may have more than one photoresponse that may or may not respond to other wavelengths. For example, NUV promotes pycnidial formation in *Mycosphaerella ligulicola*, while far red promotes perithecial formation (McCoy et al., 1975). Some cases are known where a photoresponse to one wavelength may be reversed by subsequent exposure to another wavelength. Conidial formation by *Botrytis cinerea* is inhibited by blue light, but this inhibition can be reversed by exposure to far-red light (Tan, 1975).

The Photoreceptor and the Initial Responses. Although mycelial growth is usually not affected by light, it is often the growing mycelium that receives the light that will later affect reproduction. The light must be absorbed by a special pigment, the *photoreceptor,* which occurs within the cells. The absorption spectrum of the pigment and the action spectrum for the photoresponse should correlate closely with each other. Probably more than one photoreceptor exists in the fungi, because it is unlikely that any one pigment could absorb all wavelengths known to exert biological effects. Because the majority of photoresponses are controlled by NUV and blue light, scientists have been particularly interested in identifying the photoreceptor for these responses. Current evidence supports the view that flavins are photoreceptors for NUV/blue photoresponses. The flavins are yellow pigments that strongly absorb both NUV and blue light. The flavin photoreceptor apparently exerts its primary effect by participating in a redox reaction. Supporting evidence includes (1) a requirement for molecular oxygen, and (2) inhibition of the photoresponse in the light by flavin inhibitors. Such redox reactions have been demonstrated in an albino mu-

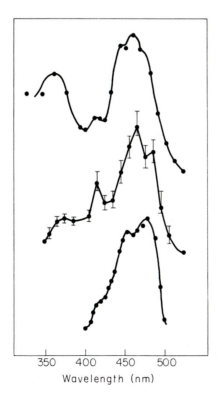

Figure 10–10 Action spectra in *Neurospora* for photoreduction of cytochrome *b* (top curve), photoinhibition of the circadian rhythm of conidiation (middle curve), and photoactivation of carotene synthesis (bottom curve). [From V. Mūnoz and W. L. Butler, 1975, Plant Physiol. 55:421–426.]

Wavelength (nm)

tant of *Neurospora crassa* which has a circadian rhythm (page 353). Light can cause a phase shift in the rhythm and also promotes conidial formation under certain conditions. If flavins are added to a culture, the reduction of cytochrome *b* can be detected when the mycelium is exposed to light. Further *in vitro* studies confirmed that the photoresponses are proportional to the amount of exogenous flavin added, suggesting that the free flavins are the primary photoreceptors that donate electrons to cytochrome *b* (Klemm and Ninnemann, 1978, 1979; Fritz and Ninnemann, 1985).

Dosage. The fungus must receive the effective wavelengths as well as a sufficient dose of light. The exposure time required for the photoresponse in some fungi will decrease as the light intensity is increased, indicating that the required light dosage at a given wavelength remains constant. *Neurospora crassa*, for example, requires 12 seconds of exposure to 1.05 W$\cdot$m^{-2} (watts per square meter) of energy from blue light to produce perithecia. However, if the light intensity is decreased to 5.25×10^{-2} W$\cdot$m^{-2}, the length of the exposure must be increased 20 times for the same effect (Innocenti et al., 1983). *N. crassa* is like a number of other fungi that require exposure to light for only a few seconds or minutes. Other fungi may require several hours or even days of exposure. Perithecia of *Pyricularia* are induced only by exposure to light for at least 6 hours daily for 5 to 10 days (Yaegashi and Hebert, 1976).

Developmental Phase. Although successful reproduction may require light, only a certain developmental phase may be light dependent. The discomycete *Sclerotinia trifoliorum* can produce elongated immature apothecia in the dark. In the presence of red or green light, further expansion of the apothecium can occur but asci are absent. Exposure to NUV is required for the initiation of basidiocarps of *Corpinus lagopus* and for gill expansion, but stipe elongation can occur in the dark (Lu, 1974). Light has differing effects on the various apothecial components of *Pyronema domesticum* (Moore-Landecker, 1979-a, b). The gametangia require about 12 hours of exposure to light for their full development. A few ascogenous hyphae and asci can subsequently develop in the dark, but light is required for optimum development of the ascogenous hyphae, asci, and ascospores. The ascospores mature best if the light is alternating at the time of their development, and exposure to constant light inhibits ascospore maturation. Development of the sterile tissues and paraphyses, which occurs concurrently, is not light dependent.

Photoinduction in Combination with Photoinhibition. Light can sometimes have both inhibitory and stimulatory effects on a fungus. This phenomenon is particularly notable in *diurnal sporulators*, fungi that require the normal alternation of day and night for sporulation and whose development is synchronized with light and dark cycles in nature. The agaric *Coprinus* responds in this manner. Young primordia in the "button" stage begin to expand about 4:30 P.M. and reach full maturity early the next morning (Johnson and Jones, 1941). In laboratory studies of *Corprinus congregatus*, Manachère (1980) determined that this agaric requires a period of light that initiates morphogenesis. A dark period of at least 4 hours but up to 12 hours is required for stipe elongation and basidium development, which are inhibited by light. Following the dark period, a second light period is required for normal maturation of basidiospores and opening of the pileus, required for basidiospore discharge. The deuteromycete, *Stemphylium botryosum*, also is a diurnal sporulator and produces abundant spores only in alternating light and dark. It does not sporulate at all or forms only a few spores in constant darkness, and in constant light it forms sterile conidiophores. This fungus requires ultraviolet light for the production of conidiophores, while the second developmental phase, during which the conidia are formed, requires darkness. The particular length of the dark period required is temperature dependent (Leach, 1968). Laboratory cultures of diurnal sporulators often develop concentric rings or zones of reproductive structures that alternate with sterile mycelium (Fig. 10–11) (see also page 352).

Light can either inhibit or promote perithecial production by *Gelasinospora reticulospora*. The effects of light or darkness depend on the developmental phase of the fungus (Inoue and Furuya, 1974-a, b, 1975; Inoue and Watanabe, 1984). The growing hyphae require at least 30 hours of uninterrupted dark, and brief exposure to NUV or blue light during this dark phase will prevent subsequent perithecial formation. As the dark period increases in length, the most effective wavelengths shift from NUV to blue and the sensitivity to light increases. After the required dark inductive period is completed, blue light (350 to 520 nanometers) is required to induce perithecial formation. However, if the mycelium is exposed to the blue light ordinarily required for perithecial formation and then to NUV light, perithecial formation does not occur.

ATMOSPHERIC CONDITIONS

Reproduction may be affected by the oxygen, carbon dioxide, or volatiles that occur in air (see discussion in Chapter 8). The importance of oxygen is not unexpected because most fungi are aerobes, requiring oxygen for respiration as a terminal electron acceptor. Reproduction of some species is known to accom-

Figure 10–11 Effect of light on sporulation of *Trichoderma lignorum*. The culture at the top left was grown in constant light, and is dark owing to the uniform distribution of spores. The culture at the top right was grown in constant darkness and has no spores. The bottom culture was grown in alternating light and darkness, resulting in zonation. [Courtesy Plant Pathology Department, Cornell University.]

pany a shift from anaerobic to aerobic pathways (e.g., *Neurospora tetrasperma*, page 318, and *Saccharomyces cerevisiae*, page 341). In addition, aerobic conditions are required for the initial redox reactions occurring in the photoresponses, indicating one of the significant roles of oxygen for sporocarp development. Aeration can substitute for the light requirement in *Pyronema domesticum*, which readily forms apothecia in darkness if cultures are placed in a relatively strong air current (Moore-Landecker and Shropshire, 1982). *Leptosphaeria typhae* produces numerous perithecia only with photoperiods that strongly promote oxidative metabolism, also indicating the importance of oxygen in the photoresponses (Vidal and Viala, 1979). Light also causes an increase in the concentration of ATP in the mycelium of *Trichoderma viride* (ATP is generated as a result of aerobic respiration) (Farkaš et al., 1985).

Although air normally has a carbon dioxide concentration of about 0.3%, the mycelium of many species can tolerate much higher levels and may even be stimulated by the excess carbon dioxide. In contrast, sporulation is very sensitive to carbon dioxide and is generally suppressed by high levels. Sometimes inhibitory levels of carbon dioxide are only slightly greater than those found normally in the air. Species of *Pleurotus*, an agaric, serve as such an example. A carbon dioxide concentration of 16 to 22% is optimum for mycelial growth, but only 0.4% causes abnormal development of the basidiocarp (Manachère, 1980).

Volatiles occurring in the habitat undoubtedly affect reproduction by fungi, but there is little direct evidence that this occurs. Apothecial development by *Pyronema domesticum* is very sensitive to the volatiles to which it is exposed. Volatiles emitted by some bacteria and fungi can either prevent apothecial development or impair it in laboratory cultures where the two organisms share a common air space. In contrast, the volatiles emitted by some other microorganisms have no apparent effect. Even volatiles emanating from sterilized soil and wood have some deleterious effects on *P. domesticum* (Moore-Landecker, 1988).

MOLECULAR AND BIOCHEMICAL CHANGES ACCOMPANYING REPRODUCTION

The nutritional and environmental factors are all important in reproduction, and no single environmental factor itself can be said to be *responsible* for reproduction. All factors are involved, and a deficiency of any one is sufficient to maintain the fungus in a vegetative state.

The shift from vegetative growth to reproduction itself is complex, is affected by all cultural conditions, and is subject to genetic regulation. Numerous molecular and biochemical changes occur as the fungus produces sporocarps and spores. These often involve the respiratory pathway and rate; rates of DNA, RNA, and protein synthesis; and changes in carbohydrate metabolism. Many of these molecular and biochemical changes normally accompany growth and development—increased DNA synthesis, for example, is to be expected when nuclear division occurs.

There are some metabolic changes that are not of a general nature and are associated specifically with reproduction. Particular compounds increase significantly only in the reproductive structures themselves, or when a mycelium is actually producing reproductive structures. It is often difficult to determine whether such changes are secondary to the reproductive process, induced by the same nutritional and environmental conditions and merely accompanying it, or are actually required for reproduction to occur. An example of such a specific change was given on page 346 in connection with the role of tyrosinase in cold induction. In addition to the positive correlation often noted between tyrosinase production and that of related enzymes (the laccases) involved in melanin production, we examine three other examples in which there appears to be more than a coincidental relationship between cellular events and reproduction.

Transcription and Translation Associated with Reproduction

Satisfactory development of reproductive structures depends on the expression of some genes, which is reflected in both the transcription of messenger RNA and the translation into polypeptides specifically required for reproduction. Comparison of the mRNA content of vegetative mycelium, of mycelium producing reproductive structures, and of the reproductive structures themselves can provide valuable clues about the genetic control of reproduction. Studies with a variety of fungi have demonstrated that changes in transcription and translation accompany reproduction. Sexual reproduction by *Schizophyllum commune* has been analyzed in this manner (Hoge et al., 1982; Mulder and Wessels, 1986; Raper and Timberlake, 1985). The total RNA was compared in monokaryotic hyphae, dikaryotic hyphae prior to initiation and during production of basidiocarps, the young primordia, and expanded basidiocarps. The transition from the monokaryotic to the dikaryotic phase is marked by the production of a class of mRNAs associated with reproduction only. Nine of these specific RNAs have been identified and their production appears to be under the control of the incompatibility factors. Formation of basidiocarp initials is accompanied by a marked increase in the transcription of these specific RNAs in the dikaryon, although they total only about 5% of the total RNA. These reproduction-specific RNAs increase in concentration in the developing primordium and they also occur in the mature basidiocarp. The translation of many polypeptides occurring during morphogenesis is controlled exclusively by these specific mRNAs (Mulder and Wessels, 1986).

Cyclic AMP

Levels of the nucleotide cyclic AMP are known to play a regulatory role in the reproduction of some fungi. These include the morphogenesis of cleistothecia of *Aspergillus nidulans* (Zonneveld, 1980) as well as oospore and zoospore formation in *Lagenidium giganteum* and some other Oomycetes (Kerwin and Washino, 1984). Cyclic AMP influences the development of basidiocarps of some fungi in which endogenous levels of this nucleotide rise as reproduction is initiated. Evidence correlating the synthesis of cyclic AMP with basidiocarp morphogenesis was first found in studies of *Coprinus cinereus* (= *C. macrorhizus*). An increase in adenyl cyclase (responsible for synthesis of cyclic AMP) occurs in dikaryotic mycelia preceding primordium formation, but then decreases as the basidiocarps develop. A similar increase does not occur in strains lacking the ability to form basidiocarps. A monokaryotic mycelium can be induced to form basidiocarps if exogenous cyclic AMP or inhibitors of phosphodiesterase activity (responsible for breaking down endogenous cyclic AMP) are added to the culture (Uno and Ishikawa, 1973). *C. cinereus* normally requires light for basidiocarp formation, and exposure to inductive light also results in an increase in cyclic AMP (Uno et al., 1974). In *C. cinereus*, the cyclic AMP apparently regulates glycogen metabolism and is involved in the breakdown of reserve glycogen in the mycelium, which is transported to the developing pileus, where it provides energy (Kuhad et al., 1987).

Cyclic AMP may also be involved in photoinduction. Light is required for induction of apothecia by *Saccobolus platensis*, but apothecia are produced in the dark if cyclic AMP is added to the culture. The activity of enzymes associated with cyclic AMP also show a response to light exposure. Adenylate cyclase activity is twice as high in extracts prepared from irradiated mycelia than in dark-grown mycelia. The reverse is true for cyclic AMP phosphodiesterase activity, which is about half the level in the light-grown mycelium than in the dark-grown mycelium (Galvagno et al., 1984).

Nitrogen Metabolism in Coprinus cinereus

Metabolic control of morphogenesis in *Coprinus cinereus* involves nitrogen metabolism (Moore, 1984; Rosin and Moore, 1985). Four enzymes involved in nitrogen metabolism increase greatly in activity in the expanding pileus while remaining at low levels or even declining in the stipe. One of these enzymes is NADP-linked glutamate dehydrogenase. Acetyl-CoA accumulates in tissues where ammonia is limiting, and in so doing induces NADP-glutamate dehydrogenase and results in a shift in metabolism. The urea cycle increases in activity, resulting in the accumulation of urea in the pileus. The urea increases the osmotic potential of the cells and serves to draw water into the cells by osmosis, resulting in the expansion of cells in the hymenium and in gill enlargement. These changes are accompanied by increased activity of the tricarboxylic acid cycle in the pileus, utilizing stored glycogen as a substrate. In contrast, urease is absent in the pileus but present in the stipe; consequently, urea does not accumulate in the stipe.

RHYTHMS

Many events occurring in nature take place in regular cycles. These may be daily, perhaps synchronized with the alternation of day and night or with lunar or tidal cycles. Timing mechanisms are often called biological clocks, and serve to coordinate the activities of an organism with its environment.

Both sporulation and spore discharge by fungi can occur with a predictable periodicity, perhaps forming spores (Fig. 10–12) or discharging them only at certain times of the day (Chapter 11). The effects of these *rhythms* may be most readily observed in some cultures in which spores are formed in concentric rings or bands separated by a vegetative mycelium. Similar concentric rings formed

Figure 10–12 *Pilobolus* requires alternating light and dark for the production of sporangiophores and sporangia, which mature in the morning. [Courtesy Carolina Biological Supply.]

Figure 10–13 Formation of basidiocarps in nature. These mushrooms are forming at the edge of an underground mycelium, and are maturing simultaneously because all environmental conditions are suitable for reproduction. Such a ring of mushrooms is popularly called a "fairy ring." [From E. J. Moore and V. N. Rockcastle, 1963, Fungi, Cornell Science Leaflet.]

in nature include rings of perithecia formed by *Nectria* on wood, sporodochia of *Sclerotinia* on a decaying peach, or "fairy rings" of mushrooms (Fig. 10–13).

The timing of sporulation and spore discharge are produced as a result of physiological and developmental events that occur in a rhythmic manner. Three principal types of rhythms may be distinguished (Lysek, 1984):

1. *Diurnal rhythms.* These are rhythms created by the alternation of light, temperature, or humidity and have a period of about 24 hours. These rhythms are actually caused exogenously by the environmental variables that affect the developing fungus. The rhythmic response does not occur if the environmental condition is held constant (Fig. 10–11).

2. *Circadian rhythms.* Circadian rhythms occur widely among plants and animals and occur with a periodicity corresponding with day–night cycles in nature. A number of specific factors characterize circadian rhythms. They have a periodicity of 24 hours, which is synchronized with light–dark cycles. The phase can be shifted with single disturbances involving light, temperature, or chemicals. Once the rhythm is established, it will continue to occur in continuous dark but often disappears in continuous light.

3. *Endogenous rhythms.* A particular event occurs with great regularity, perhaps every 24 hours, but the rhythms are neither triggered by nor synchronized with alternating environmental factors. Unlike circadian rhythms, they can occur even in darkness or at constant temperatures, and their phase cannot be shifted.

The vast majority of the rhythms observed in the fungi are diurnal rhythms, imposed by fluctuation of environmental factors such as temperature and light. On page 348, the effects of light in controlling diurnal sporulation were discussed. Growth rhythms induced by light or temperature are commonly observed in laboratory cultures and probably account for most instances where concentric rings of reproductive structures are observed in nature. Diurnal rhythms occur in spore discharge as well (Chapter 11).

True circadian rhythms are known in only a few fungi. As indicated above, rigorous testing is required to determine whether or not an observed rhythm is actually circadian, and additional examples of circadian rhythms may be discov-

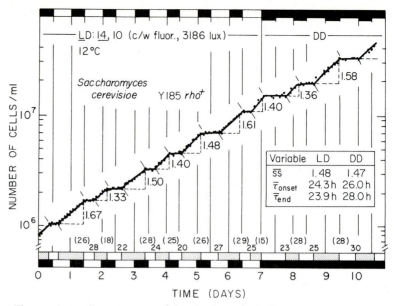

Figure 10–14 Entrainment of the rhythm of cell division of *Saccharomyces cerevisiae* by light and dark cycles of 14 hours and 10 hours, respectively. Once the rhythm is entrained, it endures in constant darkness, a characteristic of circadian rhythms. [From L. N. Edmunds, Jr., R. I. Apter, P. J. Rosenthal, W.-K. Shen, and J. R. Woodward, 1979, Photochem. Photobiol. 30:595–601.]

ered. Circadian rhythms time cellular division in *Saccharomyces cerevisiae* (Fig. 10–14) and production of conidia in bands in some isolates of *Sclerotinia fruticola* and in *Neurospora crassa* (Fig. 10–10). Biochemical properties that are possibly correlated with condia development in *N. crassa* may also be rhythmic. These include a number of enzymes—in particular, the enzymes associated with the tricarboxylic acid cycle and the glyoxylate cycle are more active while the condidia are forming, while those associated with glycolysis and the HMP pathway are simultaneously reduced (Hochberg and Sargent, 1974). Additionally, circadian rhythms of carbon dioxide production, DNA and RNA synthesis, and levels of cofactors occur (Feldman and Dunlap, 1983).

Endogenous rhythms are principally known in "clock" mutants, which differ from the wild type by exhibiting rhythmic behavior. These mutants are known only in the fungi, and include *Podospora anserina*, which has rhythmic growth of hyphae, in turn producing concentric rings of perithecia. In the clock mutant of *P. anserina*, the hyphae on the surface of the agar begin to develop more branches and to grow at a slower rate, responding to staling conditions. This retards the advance of the superficial part of the mycelium while not affecting the growth rate of hyphae submerged within the agar. These submerged hyphae extend beyond the previous retarded front, penetrate the agar and become the new advancing front. Repeated cycles of this differential growth form the bands. Gradients in oxygen availability seem to be involved in the *P. anserina*. The surface hyphae experience a slowdown in their growth due to enhanced oxygen consumption, while the submerged hyphae are not affected in this manner. When the submerged hyphae penetrate the agar and become superficial, they, too, have enhanced oxygen consumption (Lysek, 1984). Once hyphal growth has been retarded, initiation of perithecial development occurs.

The underlying mechanism actually regulating the endogenous and circadian rhythms is unknown, but a number of hypotheses have been put forth. These models require that two alternative physiological states exist, and that the oscillation from one state to the other drives additional physiological and biochemical factors that lead to the expression of the rhythm (Edmunds, 1984).

REPRODUCTION IN THE NATURAL HABITAT

Even a casual observer of fungi in their natural habitats will note that fungi are often very specific in their occurrence on particular types of substrata; often they occur only on plants or plant debris of certain species and sometimes only on a specific part of that plant, on insects and also sometimes only on specific parts of certain insect species, on dung of certain species of animals, or on freshly burned material. Also fungi may occur only in a specific type of ecological niche, such as a bog, woodland, or open field. In addition, fungi fruit during limited parts of the growing season and may occur only sporadically. It is sometimes impossible to collect a given fungus species for years at a time because environmental conditions do not allow reproduction, but during a satisfactory season the fungus will suddenly reappear (Fig. 10–13).

Specificity of fungi for particular ecological niches or seasonal distribution is determined by any specificity for hosts that have a narrow ecological or seasonal distribution and by an ability to grow vegetatively or reproduce under the normally prevailing environmental conditions. A fungus in either nature or the laboratory will form sporocarps only if nutritional requirements and all environmental conditions (temperature, light, pH, moisture, and proper aeration of the substratum) are within the narrow tolerance ranges simultaneously. Therefore, it is to be expected that a fungus will be somewhat restricted in its reproductive season and that its rather physiologically flexible mycelium will remain vegetative until the prevailing climate and conditions become satisfactory for reproduction.

Those environmental factors most subject to daily and seasonal fluctuation are light, temperature, and moisture. Light-dependent fungi usually require only relatively small amounts of light, and it is doubtful that light intensity becomes too low to permit reproduction in nature. It has been determined (Wilkins and Patrick, 1940; Wilkins and Harris, 1946) that the seasonal and ecological distribution of fungi is determined by temperature and moisture content of the substratum through the season and that a certain period of suitable environmental conditions must prevail before a sporocarp is produced. This period of time, the lag phase, varies with the fungus species but is apparently proportional to the size of the fungus. Therefore a large sporocarp requires a longer period of suitable conditions than a very small one. Accordingly, the rapid rise of basidiocarps in the autumn may be partially a response to high maximum temperatures during the summer. Moisture is perhaps the most important single factor controlling reproduction, as all fungi require a certain amount of moisture but many fungi are apparently indifferent to temperature (Wilkins and Harris, 1946).

REFERENCES CITED

AHMAD, S. S., and P. G. MILES. 1970. Hyphal fusions in *Schizophyllum commune*. 2. Effects of environmental and chemical factors. Mycologia 62:1008–1017.

ASINA, S., JAIN, K., and R. F. CAIN. 1977. Factors influencing growth and ascocarp production in three species of *Sporomiella*. Can. J. Botany 55:1915–1925.

ASTHANA, R. P., and L. E. HAWKER. 1936. The influence of certain fungi on the sporulation of *Melanospora destruens* Shear, and some of other ascomycetes. Ann. Botany (London) 50:325–344.

BARKSDALE, A. W. 1962. Effect of nutritional deficiency on growth and sexual reproduction of *Achlya ambisexualis*. Am. J. Botany 49:633–638.

BARKSDALE, A. 1967. The sexual hormones of the fungus *Achlya*. Ann. N.Y. Acad. Sci. 144:313–319.

——. 1970. Nutrition and antheridiol-induced branching in *Achlya ambisexualis*. Mycologia 62:411–420.

——, T. C. McMORRIS, R. SESHADRI, T. ARUNACHALAM, J. A. EDWARDS, J. SUNDEEN, and D. M. GREEN. 1974. Response of *Achyla ambisexualis* E 87 to the hormones antheridiol and certain other steroids. J. Gen. Microbiol. 82:295–299.

BARNETT, H. L., and V. G. LILLY. 1947-a. The effects of biotin upon the formation and development of perithecia, asci, and ascospores by *Sordaria fimicola* Ces. and DeNot. Am. J. Botany 34:196–204.

——, and ——. 1947-b. The relationship of thiamin to the production of perithecia by *Ceratostomella fimbriata*. Mycologia 39:699–708.

BASITH, M., and M. F. MADELIN. 1968. Studies on the production of perithecial stromata by *Cordyceps militaris* in artificial culture. Can. J. Botany 46:473–480.

BISTIS, G. N. 1956. Sexuality in *Ascobolus stercorius*. I. Morphology of the ascogonium; plasmogamy; evidence for a sexual hormonal mechanism. Am. J. Botany 44:436–443.

——. 1957. Sexuality in *Ascobolus sterocrarius*. II. Preliminary experiments on various aspects of the sexual process. Am. J. Botany 44:436–443.

——. 1981. Chemotropic interactions between trichogynes and conidia of opposite mating-type in *Neurospora crassa*. Mycologia 73:959–975.

——. 1983. Evidence for diffusible, mating-type-specific trichogyne attractants in *Neurospora crassa*. Exptl. Mycol. 7:292–295.

——, and S. G. GEORGOPOULOS. 1979. Some aspects of sexual reproduction in *Nectria haematococca* var. *curcurbitae*. Mycologia 71:127–143.

CARLILE, M. J., and J. MACHLIS. 1965-a. The response of male gametes of *Allomyces* to the sexual hormone sirenin. Am. J. Botany 52:478–483.

——, and ——. 1965-b. A comparative study of the chemotaxis of the motile phases of *Allomyces*. Am. J. Botany 52:484–486.

CHOINSKI, J. S., and J. T. MULLINS. 1977. Ultrastructural and enzymatic evidence for the presence of microbodies in the fungus *Achlya*. Am. J. Botany 64:593–599.

CROES, A. F. 1967-a. Induction of meiosis in yeast. I. Timing of cytological and biochemical events. Planta 76:209–226.

——. 1967-b. Induction of meiosis in yeast. II. Metabolic factors leading to meiosis. Planta 76:227–237.

DALPE, Y., and P. NEUMANN. 1976. L'effet d'acides gras sur la stimulation des peritheces de *Ceratocystis ips*, *C. minor* and *C. capillifera*. European J. Forest Pathol. 6:335–342.

EDMUNDS, L. N., Jr. 1984. Physiology of circadian rhythms in micro-organisms. Advan. Microb. Physiol. 25:61–148.

FARKAŠ, V., Z. SULOVÁ, and J. LEHOTSKÝ. 1985. Effect of light on concentration of adenine nucleotides in *Trichoderma viride*. J. Gen. Microbiol. 131:317–320.

FELDMAN, J. F., and J. C. DUNLAP. 1983. *Neurospora crassa*: a unique system for studying circadian rhythms. Photochem. Photobiol. Rev. 7:319–368.

FRITZ, B. J., and H. NINNEMANN. 1985. Photoreactivation by triplet flavin and photoinactivation by singlet oxygen of *Neurospora crassa* nitrate reductase. Photochem. Photobiol. 41:39–45.

GALVAGNO, M. A., F. FORCHIASSIN, M. L. CANTORE, and S. PASSERON. 1984. The effect of light and cyclic AMP metabolism on fruiting body formation in *Saccobolus platensis*. Exptl. Mycol. 8:334–341.

GOODAY, G. W. 1973. Differentiation in the Mucorales. Symp. Soc. Gen. Microbiol. 23:269–294.

GRUEN, H. E. 1963. Endogenous growth regulation in carpophores of *Agaricus bisporus*. Plant Physiol. 38:652–666.

——. 1969. Growth and rotation of *Flammulina velutipes* fruit bodies and the dependence of stipe elongation on the cap. Mycologia 61:149–166.

HASIJA, S. K., and C. E. MILLER. 1971. Nutrition of *Chytriomyces* and its influence on morphology. Am. J. Botany 58:939–944.

HAWKER, L. E. 1954. British hypogeous fungi. Phil. Trans. Roy. Soc. London Ser. B 237:429–546.

HENDRIX, J. W. 1970. Sterols in growth and reproduction of fungi. Ann. Rev. Phytopathol. 8:111–130.

HOCHBERG, M. L., and M. L. SARGENT. 1974. Rhythms of enzyme activity associated with circadian conidiation in *Neurospora crassa*. J. Bacteriol. 120:1164–1175.

HOGE, J. H. C., J. SPRINGER, and J. G. H. WESSELS. 1982. Changes in complex RNA during fruit-body initiation in the fungus *Schizophyllum commune*. Exptl. Mycol. 6:233–243.

INGLE, M. R., and A. C. HASTIE. 1974. Environmental factors affecting the formation of diploids in *Verticillium albo-atrum*. Trans. Brit. Mycol. Soc. 62:313–321.

INNOCENTI, F. G., U. POHL, and V. E. A. RUSSO. 1983. Photoinduction of protoperithecia in *Neurospora crassa* by blue light. Photochem. Photobiol. 37:49–51.

INOUE, Y., and M. FURUYA. 1974-a. Perithecial formation in *Gelasinospora reticulispora*. II. Promotive effects of near-ultraviolet and blue light after dark incubation. Plant and Cell Physiol. (Tokyo) 15:195–204.

———, and ———. 1974-b. Perithecial formation in *Gelasinospora reticulispora*. III. Inhibitory effects of near-UV and blue light during the inductive dark period. Plant and Cell Physiol. (Tokyo) 15:469–475.

———, and ———. 1975. Perithecial formation in *Gelasinospora reticulispora*. IV. Action spectra for the photoinduction. Plant Physiol. 55:1098–1101.

INOUE, Y., and M. WATANABE. 1984. Perithecial formation in *Gelasinospora reticulispora*. VII. Action spectra in UV region for the photoinduction and the photoinhibition of photoinductive effect brought by blue light. Plant and Cell Physiol. (Tokyo) 25:107–113.

JOHNSON, G. T., and A. C. JONES. 1941. Data on the cultural characteristics of a species of *Corprinus*. Mycologia 33:424–433.

KERWIN, J. L., and R. K. WASHINO. 1984. Cyclic nucleotide regulation of oosporogenesis by *Lagenidium giganteum* and related fungi. Exptl. Mycol. 8:215–224.

KLEBS, G. 1899. Zur Physiologie der Fortpflanzung einiger Pilze. II. Jahrb. Eiss. Botan. 33:513–593.

KLEMM, E., and H. NINNEMANN. 1978. Correlation between absorbance changes and a physiological response induced by blue light in *Neurospora*. Photochem. Photobiol. 28:227–230.

———, and ———. 1979. Nitrate reductase: a key enzyme in blue light-promoted conidiation and absorbance change of *Neurospora*. Photochem. Photobiol. 29:629–632.

KUHAD, R. C., I. V. ROSIN, and D. MOORE. 1987. A possible relation between cyclic-AMP levels and glycogen mobilization in *Coprinus cinereus*. Trans. Brit. Mycol. Soc. 88:229–236.

LEACH, C. M. 1968. An action spectrum for light inhibition of the "terminal phase" of photosporogenesis in the fungus *Stemphylium botryosum*. Mycologia 60:532–546.

———. 1971. Regulation of perithecium development and maturation in *Pleospora herbarum* by light and temperature. Trans. Brit. Mycol. Soc. 57:295–315.

LEAL, J. A., V. G. LILLY, and M. E. GALLEGLY. 1970. Some effects of ammonia on species of *Phytophthora*. Mycologia 62:1041–1056.

LU, B. C. 1973. Effect of calcium on fruiting of *Cyathus stercoreus*. Mycologia 65:329–334.

———. 1974. Meiosis in Coprinus. V. The role of light on basidiocarp initiation, mitosis, and hymenium differentiation in *Coprinus lagopus*. Can. J. Botany 52:299–305.

LYSEK, G. 1984. Physiology and ecology of rhythmic growth and sporulation in fungi: *In:* Jennings, D. H., and A. D. M. Rayner, Eds. The ecology and physiology of the fungal mycelium. Cambridge University Press, Cambridge, pp. 323–342.

MacDONALD, D. J., and D. J. BOND. 1976. Genetic and environmental factors influencing the production and distribution of protoperithecia in *Sordaria brevicollis*. J. Gen. Microbiol. 95:375–380.

MACHLIS, L. 1973. The chemotactic activity of various sirenins and analogues and the uptake of sirenin by the sperm of *Allomyces*. Plant Physiol. 52:527–530.

MAIELLO, J. M., and R. A. CAPPELLINI. 1976. The influence of carbon and nitrogen nutrition on pycnidium development in *Phyllosticta antirrhini*. Mycologia 68:1174–1180.

MANACHÈRE, G. 1980. Conditions essential for controlled fruiting of macromycetes—a review. Trans. Brit. Mycol. Soc. 75:255–270.

MARSHALL, M. R., D. F. HINDAL, and W. L. McDONALD. 1982. Production of perithecia in culture by *Ceratocystis ulmi*. Mycologia 74:376–381.

McCOY, R. E., R. K. HORST, and A. W. DIMOCK. 1975. Environmental factors regulating sexual and asexual reproduction by *Mycosphaerella ligulicola*. Phytopathology 62:1188–1195.

McMORRIS, T. C., R. SESHADRI, G. R. WEIHE, G. P. ARSENAULT, and A. W. BARKSDALE. 1975. Structures of oogoniol-1, -2, and -3, steroidal sex hormones of the water mold, *Achlya*. J. Am. Chem. Soc. 97:2544–2545.

MOORE, D. 1984. Developmental biology of the *Coprinus cinereus* carpophore: metabolic regulation in relation to cap morphogenesis. Exptl. Mycol. 8:283–297.

MOORE-LANDECKER, E. 1975. Effect of cultural conditions on apothecial morphogenesis in *Pyronema domesticum*. Can. J. Botany 53:2759–2769.

——. 1979-a. Effect of light regimens and intensities on morphogenesis of the discomycete *Pyronema domesticum*. Mycologia 71:699–712.

——. 1979-b. Effect of cultural age and a single photoperiod on morphogenesis of the discomycete *Pyronema domesticum*. Can. J. Botany 57:1541–1549.

——. 1988. Response of *Pyronema domesticum* to volatiles from microbes, seeds, and natural substrata. Can. J. Botany 66:194–198.

——, and W. SHROPSHIRE, JR. 1982. Effects of aeration and light on apothecia, sclerotia, and mycelial growth in the discomycete *Pyronema domesticum*. Mycologia 74:1000–1013.

MULDER, G. H., and J. G. H. WESSELS. 1986. Molecular cloning of RNAs differentially expressed in monokaryons and dikaryons of *Schizophyllum commune* in relation to fruiting. Exptl. Mycol. 10:214–277.

NIEUWENHUIS, M., and H. VAN DEN ENDE. 1975. Sex specificity of hormone synthesis in *Mucor mucedo*. Arch. Mikrobiol. 102:167–169.

NITIMARGI, N. M. 1937. Studies in the genera *Cytosporina, Phomopsis,* and *Diaporthe*. VII. Chemical factors influencing sporing characteristics. Ann. Botany (London) 49:19–40.

NUTTING, W. H., H. RAPOPORT, and L. MACHLIS. 1968. The structure of sirenin. J. Am. Chem. Soc. 90:6434–6438.

RAPER, C., and W. E. TIMBERLAKE. 1985. Identification of a polyadenylated transcript correlated with development in the sexual cycle of *Schizophyllum*. Exptl. Mycol. 9:269–274.

RAPER, J. R. 1951. Sexual hormones in *Achlya*. Am. Scientist 39:110–120, 130.

——. 1966. Genetics of sexuality in higher fungi. Ronald Press, New York, 283 pp.

ROSIN, I. V., and D. MOORE. 1985. Differentiation of the hymenium in *Coprinus cinereus*. Trans. Brit. Mycol. Soc. 84:621–628.

ROSS, R. G., and D. J. BREMNER. 1971. Effect of ammonium nitrogen and amino acids on perithecial formation of *Venturia inaequalis*. Can. J. Plant Sci. 51:29–33.

——, and S. A. HAMLIN. 1965. Influence of nutrients on perithecial production of *Venturia inaequalis* (Cke.) Wint. Can. J. Botany 43:959–965.

STATES, J. S. 1975. Normal basidiocarp development of *Gloeophyllum (Lenzites) saepiarium* in culture. Mycologia 67:1166–1175.

TAN, K. K. 1975. Interaction of near-ultraviolet, blue, red and far red light in sporulation of *Botrytis cinerea*. Trans. Brit. Mycol. Soc. 64:215–222.

THOMAS, D. M., and T. W. GOODWIN. 1967. Studies on carotenogenesis in *Blakeslea trispora*. I. General observations on synthesis in mated and unmated strains. Phytochemistry 6:355–360.

THORNER, J. 1981. Pheromonal regulation of development in *Saccharomyces cerevisiae*. *In:* Strathern, J. N., E. W. Jones, and J. R. Broach, Eds. The molecular biology of the yeast *Saccharomyces:* life cycle and inheritance. Cold Spring Harbor Laboratory, Cold Spring Harbor, N.Y., pp. 143–180.

TIMBERLAKE, W. E. 1976. Alterations in RNA and protein synthesis associated with steroid hormone-induced sexual morphogenesis in the water mold *Achlya*. Develop. Biol. 51:202–214.

TSUDA, M., T. WAKI, M. TAGA, and A. UEYAMA. 1982. Ascocarp production of *Magnaporthe salvinii* in culture. Trans. Brit. Mycol. Soc. 78:515–519.

UNO, I., and T. ISHIKAWA. 1973. Metabolism of adenosine 3′, 5′-cyclic monophosphate and induction of fruiting bodies in *Coprinus macrorhizus*. J. Bacteriol. 113:1249–1255.

——, M. YAMAGUCHI, and T. ISHIKAWA. 1974. The effect of light on fruiting body formation and adenosine 3′:5′-cyclic monophosphate metabolism in *Coprinus macrorhizus*. Proc. Natl. Acad. Sci. USA 71:479–483.

VIDAL, G., and G. VIALA. 1979. Influence de l'éclairement continu sur la croissance, la reproduction sexuée et le métabolisme du *Leptosphaeria typhae*. Ann. Sci. Nat. Botan. Biol. Vegetale 13 Ser. 1:283–288.

VISWANATH-REDDY, M., and G. TURIAN. 1975. Physiological changes during protoperithecial differentiation in *Neurospora tetrasperma*. Physiol. Plantarum 35:166–174.

WERKMAN, T. A., and H. VAN DEN ENDE. 1973. Trisporic acid synthesis in *Blakeslea trispora*—interaction between plus and minus mating types. Arch. Mikrobiol. 90:365–374.

WESTE, G. 1970. Factors affecting vegetative growth and the production of perithecia in culture by *Ophiobolus graminis*. I. Variations in media and age of mycelium. Australian J. Botany 18:1–10.

WICKLOW, D., and D. MALLOCH. 1971. Studies in the genus *Theobolus*: temperature optima for growth and ascocarp development. Mycologia 63:118–131.

WILKINS, W. H., and G. C. M. HARRIS. 1946. The ecology of the larger fungi. V. An investigation into the influence of rainfall and temperature on the seasonal production of fungi in a beechwood and a pinewood. Ann. Appl. Biol. 33:179–188.

——, and S. H. M. PATRICK. 1940. The ecology of the larger fungi. IV. The seasonal frequency of grassland fungi with special reference to the influence of environmental factors. Ann. Appl. Biol. 27:17–34.

WILLETTS, H. J., and Y. HARADA. 1984. A review of apothecial production by *Monilinia* fungi in Japan. Mycologia 76:314–325.

WILSON, D. M. 1968. Physiology of sexual reproduction in *Hypomyces solani* f. sp. *cucurbitae*. V. Influence of tyrosinase on perithecial primordium formation. Phytopathology 58:1697–1699.

——, and R. BAKER. 1969. Physiology of production of perithecia and microconidia in *Hypomyces solani* f. sp. *cucurbitae*. Trans. Brit. Mycol. Soc. 53:229–236.

WURTZ, T., and J. JOCKUSCH. 1975. Sexual differentiation in *Mucor*: trisporic acid response mutants and mutants blocked in zygospore development. Develop. Biol. 43:213–220.

YAEGASHI, H., and T. T. HEBERT. 1976. Effect of temperature, light, and nutrients on production of perithecia by *Pyricularia*. Ann. Phytopathol. Soc. Japan 42:556–562.

ZONNEVELD, B. J. M. 1972. Morphogenesis in *Aspergillus nidulans*. The significance of α-1,3-glucan of the cell wall and α-1,3-glucanase for cleistothecium development. Biochim. Biophys. Acta 273:174–187.

——. 1980. Protease levels in relation to cAMP and a reserve polymer during growth and sexual differentiation of *Aspergillus nidulans*. Exptl. Mycol. 4:140–146.

Additional Recommended References

CANTINO, E. C. 1966. Morphogenesis in aquatic fungi. *In*: Ainsworth, G. C., and A. S. Sussman, Eds. The fungi: an advanced treatise, Vol. 2, pp. 283–337.

CARLILE, M. J. 1965. The photobiology of fungi. Ann. Rev. Plant Physiol. 16:175–202.

ESSER, K. 1968. Phenol oxidases and morphogenesis in *Podospora anserina*. Genetics 60:281–288.

FELDMAN, J. F. 1982. Genetic approaches to circadian clocks. Ann. Rev. Plant Physiol. 33:583–608.

FELDMAN, J. F., and J. P. THAYER. 1974. Cyclic AMP-induced tyrosinase synthesis in *Neurospora crassa*. Biochem. Biophys. Res. Commun. 61:977–981.

GOODAY, G. W. 1983. Hormones and sexuality in fungi. *In*: Bennett, J. W., and A. Ciegler, Eds. Secondary metabolism and differentiation in fungi. Marcel Dekker, New York, Vol. 5, pp. 239–266.

——. 1986. Elongation of the stipe of *Coprinus cinereus*. *In:* Moore, D., L. A. Casselton, D. A. Wood, and J. C. Frankland, Eds. Developmental biology of higher fungi. Cambridge University Press, Cambridge, pp. 311–331.

GRUEN, H. E. 1982. Control of stipe elongation by the pileus and mycelium in fruitbodies of *Flammulina velutipes* and other Agaricales. *In:* Wells, K., and E. K. Wells, Eds. Basidium and basidiocarp: evolution, cytology, function, and development. Springer-Verlag, New York, pp. 125–155.

HAMMOND, J. B. W. 1984. The biochemistry of *Agaricus* fructification. *In:* Moore, D., L. A. Casselton, D. A. Wood, and J. C. Frankland, Eds. Developmental biology of higher fungi. Cambridge University Press, Cambridge, pp. 389–401.

HERMANN, T. E., M. B. KURTZ, and S. P. CHAMPE. 1983. Laccase localized in hülle cells and cleistothecial primordia of *Aspergillus nidulans*. J. Bacteriol. 154:955–964.

HAWKER, L. E. 1966. Environmental influences on reproduction. *In:* Ainsworth, G. C., and A. S. Sussman, Eds. The fungi: an advanced treatise. Academic Press, New York, Vol. 2, pp. 435–469.

JEREBZOFF, S. 1965. Growth rhythms. *In:* Ainsworth, G. C., and A. S. Sussman, Eds. The fungi: an advanced treatise. Academic Press, New York, Vol. 2, pp. 625–645.

KOCHERT, G. 1978. Sexual pheromones in algae and fungi. Ann. Rev. Plant Physiol. 29:461–486.

LEATHAM, G. F., and M. A. STAHMANN. 1981. Studies on the laccase of *Lentinus edodes*: specificity, localization and association with the development of fruiting bodies. J. Gen. Microbiol. 125:147–157.

MADELIN, M. F. 1956-a. Studies on the nutrition of *Coprinus lagopus* Fr., especially as affecting fruiting. Ann. Botany (London) II 20:307–330.

——. 1956-b. The influence of light and temperature on fruiting of *Coprinus lagopus* Fr. in pure culture. Ann. Botany (London) II 20:467–480.

PALL, M. L. 1981. Adenosine 3′,5′-phosphate in fungi. Microbiol. Rev. 45:462–480.

SCHWALB, M. N. 1978. Regulation of fruiting. *In:* Schwalb, M. N., and P. G. Miles, Eds. Genetics and morphogenesis in the basidiomycetes. Academic Press, New York, pp. 135–165.

SMITH, J. E., and J. C. GALBRAITH. 1971. Biochemical and physiological aspects of differentiation in the fungi. Advan. Microb. Physiol. 5:45–134.

TABER, W. A. 1966. Morphogenesis in basidiomycetes. *In:* Ainsworth, G. C., and A. S. Sussman, Eds. The fungi: an advanced treatise. Academic Press, New York, Vol. 2, pp. 387–412.

TAN, K. K. 1978. Light-induced fungal development. *In:* Smith, J. E., and D. R. Berry, Eds. The filamentous fungi. Edward Arnold, London, Vol. 3, pp. 334–357.

TURIAN, G. 1966. Morphogenesis in Ascomycetes. *In:* Ainsworth, G. C., and A. S. Sussman, Eds. The fungi: an advanced treatise. Academic Press, New York, Vol. 2, pp. 339–385.

11

Spores

Fungal spores occupy a unique position in the life cycle as they terminate both the reproductive and developmental cycles and have the inherent potentialities to develop into a new generation. The spore is much less susceptible to adverse environmental conditions than the vegetative thallus is, and it often enables the fungus to endure unfavorable conditions. More important, however, the fungus spore is a reproductive unit capable of being dispersed for great distances.

Events directly involving spores are their formation by the parent sporocarp or mycelium, discharge or release from the parent, dispersal to new locales, possible period of dormancy, and finally germination into a new thallus.

There are many hazards met by the spores along the way—discharge or dispersal mechanisms may not be operating effectively, a suitable substratum for germination and growth may not be encountered, and finally, environmental conditions may not be suitable for either germination or further development. The laws of probability work against the success of any single spore, and the majority of the spores are wasted. In order to realize this, we need only consider that the giant puffball, *Calvatia gigantea*, produces 7×10^{12} basidiospores in a basidiocarp of 232 grams; if each spore were to germinate and form an equally large basidiocarp, their combined mass would be 800 times the size of our planet (Buller, 1924).

SPORE RELEASE

After spores are produced by the parent, they must be released in order to reach a new locale. Spores may be freed through the intervention of physical aspects of the environment during which the fungus plays a passive role, or they may be forcibly discharged by the fungus.

Passive Mechanisms of Spore Release

A large number of fungi, especially conidial forms, produce dry spores that are freely exposed to the surrounding air. Spores of this type are especially adapted to removal from the parent by air currents which blow the spores off. Removal of spores by air currents is much more effective in a cupulate sporocarp than on a flat surface because an eddy system is produced within the cup which carries the spores upward where they can join a second eddy stream which carries the spores away from the cup (Brodie and Gregory, 1953). An equivalent type of eddy system is not produced over a flat surface.

Falling raindrops fragment into smaller drops upon impact and, if they land on a fungus, will dislodge the spores and carry them away in their splatter. This type of spore release may occur in dry-spored types in which wind dislodgement occurs, but it is especially significant for fungi which produce their spores in a slime, making wind dislodgement impossible.

Fungi bearing spores within a slime include conidial fungi and some ascomycetes that lack an explosive ascus. Ascomycetes of this nature include some bitunicate members in which ascospore liberation occurs through gelatinization of the ascocarp and in some members of the Ophiostomatales (such as *Chaetomium*) in which the asci deliquesce. In the latter type, the disintegrating asci form a slime which contains the freed ascospores. The slime oozes out of the perithecial neck, much as paste would ooze out of a squeezed tube. Liberation of the spores is accomplished by splashing rain. In some instances, the entire sporocarp may be dislodged by the falling raindrop and splashed away. Perithecia of *Chaetomium* were found to be dispersed to a distance of 33 centimeters in this manner (Dixon, 1961).

Raindrops may also dislodge spores through bombardment of the sporocarp. Raindrops landing on a Gasteromycete peridium may depress it like a bellows, forcing air and spores out through the ostiole in a puff that rises to a height of 2 to 3 centimeters (Gregory, 1949).

Structures especially adapted to utilize the force of landing raindrops are the "splash cups" found in the Nidulariales, the bird's nest fungi (see Chapter 5) (Brodie, 1956). We consider *Cyathus striatus* as an example of this mechanism. Large raindrops are about 4 millimeters in diameter, only slightly smaller than the 6- to 8-millimeter diameter of the mouth of the funnel-shaped basidiocarp of *C. striatus*. The large drops have a terminal velocity of 6 meters per second, and the displacement of water in the funnel creates a strong thrust along the sides, inclined 60 to 70° horizontally (through the use of models, this was demonstrated to be the most effective inclination). The strong upward thrust of the water dislodges the peridioles and ejects them to a distance of up to 1 meter. While in flight, the cablelike funiculus trails along behind the peridiole until its adhesive ending, the hapteron, attaches itself to an object. The peridiole is suddenly checked in its flight and is jerked backward, and the funiculus becomes wound around the retaining object (Fig. 11–1). This object is frequently some form of vegetation that may be eaten by an animal, enabling the fungus to complete its life cycle.

Active Mechanisms of Spore Release

All active mechanisms of spore release share the characteristic of flinging the spore into the air with considerable force. The distance to which the spore will travel is determined by its initial velocity, shape, size, and density. A simplified formula for the distance (d) that a spore will travel may be expressed as $d = Kr^2$, where K is a constant and r is the radius of the spore. As the radius of the spore increases, so does the distance to which it will travel as a result of the forceful discharge. Therefore we find that the smaller spores can be thrown only short

Figure 11–1 Splashing of a peridiole from a basidiocarp of *Cyathus striatus:* (a, b) raindrop lands in a cup (c) the peridiole is splashed out with the hapteron extended; (d) hapteron sticks to a plant as the peridiole is carried forward by its own momentum and the funiculus is extended by a pull; (e) peridiole is jerked backwards when the funiculus is extended to its full length; (f) jerk causes the peridiole to swing around its point of attachment and the cord is wrapped around the plant stem as another raindrop lands in the splash cup. [Reproduced by permission of the National Research Council of Canada from H. J. Brodie, Can. J. Botany 29:1951, 224–234.]

distances, while the larger ones can be thrown for comparatively large distances (Ingold, 1971).

EXPLOSIVE MECHANISMS

A common means of forcible spore discharge is that in which a structure containing a large vacuole becomes turgid through an increase in osmotic concentration and suddenly bursts, carrying the spores away in a jet of water. This mechanism occurs in *Pilobolus* (Mucorales), in *Basidiobolus ranarum* and *Entomophthora muscae* (Entomophthorales), and in the majority of the ascomycetes.

Pilobolus.   Species of *Pilobolus* forcibly eject an entire sporangium, thus earning their name which literally means "hat-thrower" (Greek). These fungi occur in abundance on horse dung. If fresh horse dung is placed in a moist chamber which is exposed to the light, numerous sporangia of *Pilobolus* will usually appear within a few days. Successive crops of sporangia appear each morning and complete their discharge by early afternoon. *Pilobolus* bears a single sporangium on each sporangiophore. The sporangium has a dense, black wall, is lenticular in shape, and contains from 15,000 to 30,000 spores. The sporangiophore is a single cell consisting of three parts: (1) a basal swelling that attaches it firmly to the substratum, (2) an elongate stipe, and (3) a subsporangial vesicle (Fig. 11–2). The subsporangial vesicle is pyriform and considerably larger than the sporangium, which it supports on a nipple-shaped columella. The subsporangial vesicle contains a single vacuole which completely fills it except for a thin layer of protoplasm adjacent to the wall. This protoplasm is specialized in two sites where it is thickened and reddish, owing to the presence of carotene. One such site occurs near the top of the subsporangial vesicle, and the second occurs

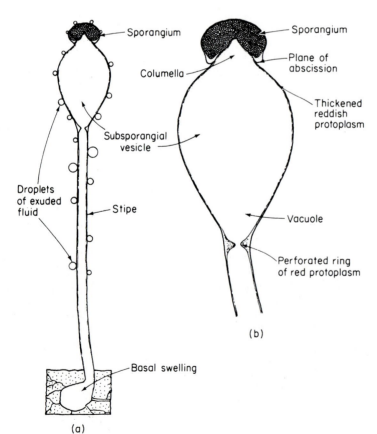

(b)

Figure 11–2   *Pilobolus kleinii:* (a) longitudinal section through the entire sporangiophore and sporangium × 3.3; (b) detail of the subsporangial vesicle and sporangium, approximately × 10. [Adapted from A. H. R. Buller, 1909, Researches on fungi, Vol. 1, Longmans, Green & Co. Ltd., London.]

at its base where the reddish protoplasm forms a perforated ring delimiting the vesicle from the stipe.

The sporangiophore is phototropic, and the stipe will bend toward a source of light until the sporangium is aimed directly at the light. This phototropic response is controlled by the subsporangial vesicle. The black sporangium blocks any light rays which strike it, but the protruding subsporangial vesicle is fully exposed to the light that strikes the thickened, reddish protoplasm near the top of the vesicle. This mass of protoplasm functions in the same manner as a biconcave lens does and refracts the light rays so that they converge in a spot on the opposite side of the vesicle. If the sporangium is already pointing directly toward the source of light (i.e., light rays are parallel to the long axis of the subsporangial swelling), the refracted light will strike the light-sensitive, reddish protoplasmic ring at the base of the subsporangial vesicle in a symmetrical manner. If the light strikes in a symmetrical fashion, no change will occur. If the sporangium is not pointing toward the source of light, however, the refracted light will strike the subsporangial wall (and not the ring) and will be asymmetric for the vesicle as a whole (Fig. 11–3). A stimulus is then exerted on the protoplasm which is conveyed to the motor region of the stipe (beneath the light-sensitive ring), causing the stipe wall on the side away from the light to

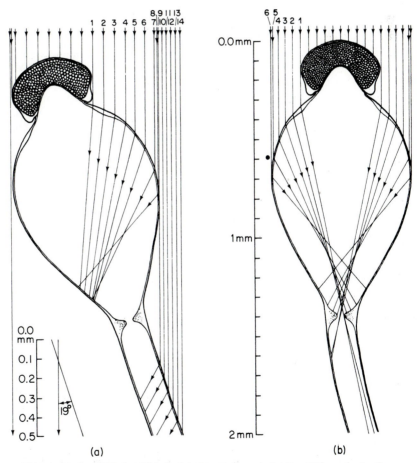

Figure 11–3 *Pilobolus kleinii:* (a) the main axis is at a 19° angle to the source of strongest illumination; this is a physiologically unstable position; (b) the main axis of the subsporangial vesicle and sporangium is now redirected toward the source of strongest illumination, a physiologically stable position. [From A. H. R. Buller, 1934, Researches on fungi, Vol. 6, Longmans, Green & Co. Ltd., London.]

grow faster than the wall toward the light. Growth of the wall continues until the sporangiophore is parallel to the light rays, which is registered when the refracted light strikes the light-sensitive ring.

Meanwhile, certain changes are occurring as a prelude to sporangium discharge. The lower wall of the sporangium is colorless and breaks away from the subsporangial vesicle. A ring of mucilage surrounding the base of the sporangium is thus exposed. The subsporangial vesicle is extremely turgid (about 5.5 atmospheres of pressure) and finally a weak abscission zone (surrounding the columella) ruptures. The pressure within the subsporangial vesicle is suddenly relieved, and the vacuolar fluid surges upward as a forceful squirt, carrying the sporangium with it. The speed of the discharge is about 14 meters per second, and the distance of the horizontal throw is as great as 2 meters (Buller, 1934). The ring of mucilage is wetted by the fluid and causes the sporangium to adhere to any object on which it lands (Fig. 11–4). This object may be a blade of grass that will be eaten by a horse.

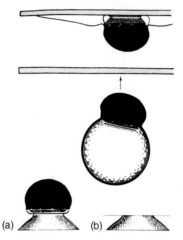

Figure 11–4 *Pilobolus kleinii:* (a) sporangium shortly after dehiscence of the lower sporangial wall; the base of the sporangium is surrounded by a gelatinous band through which the sporangiospores can be distinguished; (b) flight and landing of the sporangium after discharge. Part of the subsporangial wall remains attached to the sporangium, and it is accompanied by a large drop of fluid. This fluid enables the sporangium to adhere to a surface upon landing (top). [Adapted from A. H. R. Buller, 1909, Researches on fungi, Vol. 1, Longmans, Green & Co. Ltd., London.]

The phototropic response and forcible discharge of the sporangium are adaptations to the environment in which species of *Pilobolus* occur. They sporulate on horse dung, which is deposited in irregular and overlapping piles. It is necessary that the sporangia reach open spaces and blades of grass to again be consumed by a horse: the sporangia would be wasted if they were discharged onto overhanging dung. *Pilobolus* species are well adapted to these constraints because their sporangiophores are directed towards the open spaces (detected by the direction of the light), and their sporangia are forcibly ejected in that direction. The *Pilobolus* gun is deadly accurate, as can be determined by placing dung bearing sporangia within a dark box with a target of white paper with a small illuminated hole at the top of the box.

Basidiobolus ranarum and Entomophthora muscae. *Basidiobolus ranarum*, which occurs on frog excrement, and *Entomophthora muscae*, a parasite of flies, rely upon the sudden rupture of the sporangiophore to eject the single sporangium (Fig. 11–5).

Basidiobolus ranarum forms minute sporangiophores from either the mycelium or a discharged sporangium. The sporangiophore is enlarged at the apex into a subsporangial bulb that supports the single sporangium on a columella. There is a weak circumsessile zone near the base of the subsporangial bulb, which suddenly ruptures when the turgor increases. The wall of the subsporangial bulb above the line of dehiscence is elastic, while that below the line is rigid. When dehiscence occurs, the elastic upper portion contracts and squirts the vacuolar fluid backward. The upper portion of the subsporangial bulb is shot off

Chap. 11 / Spores

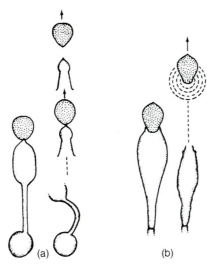

Figure 11–5 Sporangium discharge by *Basidiobolus ranarum* (a) and *Entomophothora muscae* (b). [From C. T. Ingold, 1934, New Phytologist 33:274–277.]

together with the sporangium, which is shot to a distance of 1 to 2 centimeters (Ingold, 1934).

The sporangium of *Entomophthora muscae* is also borne on an inflated sporangiophore, but the sporangiophore, unlike that of *Basidiobolus ranarum*, ruptures immediately beneath the sporangium. The resulting jet of sap then propels the sporangium to a distance of 1 to 1.5 centimeters.

Ascomycotina. An almost universal characteristic of the ascus is its explosive discharge of the ascospores. The ascus is lined with a thin layer of cytoplasm which surrounds a large vacuole. The ascospores are suspended in the vacuole, hanging together by their own adhesiveness and apically supported by a strand of mucilage. A great deal of pressure is then built up in the ascus. This buildup of pressure is usually attributed to a large quantity of glycogen (detectable by chestnut-brown staining in iodine) in the vacuole and its subsequent conversion to soluble sugars. The soluble sugars, in combination with any salts which may be present, increase the osmotic concentration of the ascus contents. It has been estimated that the concentration of the fluid within the ascus of *Sordaria fimicola* is about 10 to 30 atmospheres (Ingold, 1971). In some species, pressure in the asci may be generated primarily by the direct absorption of rain water (Zoberi, 1973), or by the swelling of a mucilage (Ingold, 1971). Regardless of the mechanism for pressure buildup, the resulting pressure causes the ascus tip to open at a predetermined pore or slit and forces the ascospores out violently. The asci within an ascocarp burst singly, in succession, or simultaneously. The distance to which the ascospores may be discharged may be less than a millimeter or up to 40 centimeters but is usually 0.5 to 2.0 centimeters (Ingold, 1971).

The morphology of the ascocarp bearing explosive asci may place special requirements on ascus behavior before ascospore discharge, and we may distinguish three principal types of situations: those encountered in a cleistothecium with a nondisintegrating wall, in an ostiolate ascocarp, and in an apothecium. These types are considered in greater detail.

Cleistothecia with Nondisintegrating Walls. Cleistothecia with nondisintegrating walls are found in the Erysiphales. In these fungi, the ascus enlarges and protrudes through the confining ascocarp wall. The wall which is forced open in this way is under some strain and suddenly snaps shut, ejecting the entire ascus to a distance of several centimeters. The ascus then later forcibly ejects its ascospores (Woodward, 1927).

Ostiolate Ascocarps. Enclosed ascocarps opening by only a small ostiole occur in the locular and perithecial fungi. They can release their ascospores only if the asci can protrude through the ostiole. Some species have a neck that is positively phototropic, bending towards the light (Ingold, 1971). The ostiole is so narrow that only one ascus can discharge its ascospores at a time.

In most ostiolate ascomycetes, the asci remain attached at their bases and reach the ostiole by elongation. If the ascus is unitunicate, the entire ascus elongates, whereas if the ascus is bitunicate, only the inner ascus wall elongates. The asci apparently elongate by their own growth and by pressure of the surrounding asci. Each ascus elongates, reaches the ostiole, and discharges its ascospores. The empty ascus shrinks back into the ascocarp and disintegrates (Ingold, 1971).

Alternatively, some asci become detached at their bases, float upward to the ostiole in a mucilage, individually lodge between the lips of the ostiole, and then forcibly discharge their ascospores. Those Pyrenomycetes having detached asci are found in some members of the Sphaeriales with very long perithecial necks. In these fungi, an elongating ascus could not reach the exterior (Ingold, 1933).

Physiological aspects of ascospore discharge in ostiolate ascocarps with elongating asci have been studied, and some interesting periodic responses to light have been demonstrated. Ascospore discharge by *Sordaria fimicola* occurs primarily during the light if the fungus is cultured in alternating cycles of light and dark (Ingold and Dring, 1957). The perithecial necks of *S. fimicola* are positively phototropic, and both the ascospore discharge and phototropism respond to wavelengths of light in the blue range (less than 520 nanometers). In contrast, ascospore discharge by *Hypoxylon fuscum* and *Daldinia concentrica* is inhibited by light and therefore is greater in darkness (Ingold and Marshall, 1963; Ingold, 1956, respectively). A pronounced endogenous rhythm exists in *Daldinia concentrica*, which will continue to release its ascospores during the night even if perithecia are maintained in total darkness (Ingold and Cox, 1955).

Returning to *Sordaria fimicola*, it has been demonstrated that light and temperature may interact in controlling the rate of ascospore discharge. In some experiments, cultures with perithecia were subjected to three alternating 24-hour cycles of low and high temperatures (8 and 20°C, 20 and 25°C, or 15 and 25°C) in the dark. In all three experiments, the high temperatures stimulated ascospore discharge during the initial 8 to 10 hours of that temperature period, but the rate of ascospore discharge subsequently declined, becoming equivalent, or nearly so, to that occurring at the lower temperature. When the temperature again rose, ascospore discharge again peaked (Fig. 11–6). This temperature-dependent cycle could be repeated several times in the dark. However, if alternating periods of light and dark were provided, the fungus would respond to the light cycles and not to the temperature in those cycles utilizing 15 and 25°C or 20 and 25°C. Light was incapable of stimulating discharge in cultures maintained at 8°C. Ingold (1971) suggests that light is the limiting factor at high temperatures, while temperature itself is limiting under cold conditions. In nature, both higher temperatures and maximum light occur during the day, presumably contributing to maximum ascospore discharge by *S. fimicola*.

Apothecia. Ascospore discharge from apothecia is more easily accomplished than from perithecia because all asci are exposed to the exterior. If the apothecium has a flat or convex hymenium, discharge of the ascospores is completely unhindered. There are problems encountered, however, in a deeply cupulate apothecium. In such an apothecium, the asci line the inside of the cup and asci on the sides would aim their ascospores toward the opposite hymenium if the line of discharge were at right angles to the hymenium, as we would normally expect. An adaptation avoiding this potential problem exists in the form of positive phototropism, bending of the individual asci toward the strongest source of light. Phototropic asci in a deeply cupulate apothecium curve toward the light, resulting in a different degree of curvature in the different

Figure 11–6 *Sordaria fimicola.* Upper curve: rate of spore discharge from culture grown in the dark at 20°C and then subjected to alternate days at 25 and 20°C in dark; lower two curves: two parallel experiments on cultures reared in light at 20°C and then treated as shown with alternating days at 20 and 25°C and alternating days of darkness and light (1000 lux, "daylight" fluorescent): in one relatively high temperature and darkness coincide, in the other relatively high temperature and light. In the lower two curves note change of scale above 1000 spores per hour. [Figure and legend from C. T. Ingold, 1971, *Fungal spores: their liberation and dispersal,* Clarendon Press, Oxford, 302 pp.]

parts of the apothecium (Fig. 11–7). This orients the asci in such a way that each is pointing toward open space, with the result that the ascospores will be discharged into open air.

As the entire hymenium is exposed in the Discomycetes, there is no restriction on the number of asci that can simultaneously discharge their ascospores (unlike the former types). Simultaneous discharge of ascospores is a common occurrence in the Discomycetes and is especially likely to occur when there is a change in humidity or temperature or if the apothecium is touched. In *Ascobolus viridulus,* this discharge may be induced by transferring the apothecium from darkness to high-intensity light in the blue range (400 to 460 nanometers); however, if red light (greater than 600 nanometers) is provided along with the blue light, no discharge occurs (Ingold and Oso, 1968). The simultaneous discharge may be detected as a sudden puffing forth of a cloud of spores (Fig. 11–8) and produces an audible hiss if the puffing apothecium is placed closed to one's ear. The sudden puffing of the spores creates a minute blast of air above the apothecium that increases the normal turbulence and aids in the dispersal of the ascospores. In *Ascobolus viridulus,* the increased distance to which spores are discharged during puffing is five times that achieved without puffing.

Ascospore discharge in the Discomycete *Cookeina sulcipes* is favored by rain.

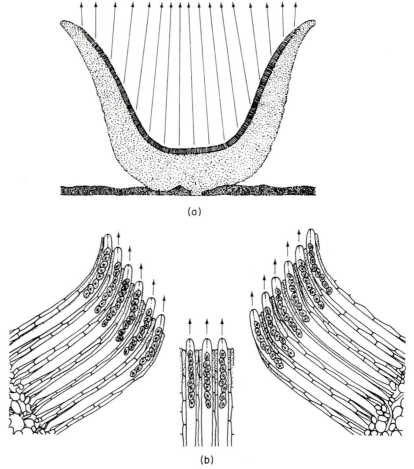

(a)

(b)

Figure 11–7 Phototropism of asci in the Discomycete *Peziza vesiculosa:* (a) direction of spore discharge from the apothecium, × 1.7; (b) phototropic curvature of asci from three sections of the apothecium, × 71. [From A. H. R. Buller, 1934, Researches on fungi, Vol. 6, Longmans, Green & Co. Ltd., London.]

Figure 11–8 Ascospore discharge by apothecia of *Ciboria aestivalis.* Ejection of the ascospore cloud was stimulated by a sudden change in atmospheric conditions. [Courtesy Plant Pathology Department, Cornell University.]

The apothecia are oriented toward the light, owing to positive phototropism of their asci. The cuplike apothecium catches the falling raindrops and retains the water until the cup is full. The asci absorb the water, becoming turgid and developing hydrostatic pressure. As the water in the cup evaporates, the ascus tips above the water's surface become desiccated, producing a negative vapor pressure which causes the ascus tips to be ejected. The ascus tips dry in succession as the water level goes down, and finally there is a big puff when all the water has evaporated (Zoberi, 1973).

ROUNDING-OFF MECHANISM

The normal shape of a turgid cell would be spherical if it were completely unconfined. An analogy may be found in soap bubbles, which are spherical when independently floating through air but assume polyhedral shapes when compressed by neighboring bubbles. Sudden rounding out of a previously compressed cell to discharge spores occurs in some fungi.

A fungus of this type is *Entomophthora coronata*, a member of the Entomophthorales and a parasite of aphids and termites. A single sporangium is borne on each sporangiophore. Both have a double wall, with the outer wall of the sporangium continuous with that of the sporangiophore. In addition, the sporangium is separated from the sporangiophore by a double wall: one layer belongs to the sporangium and the second layer belongs to the sporangiophore. This cross wall forms a columella and bulges into the sporangium. Prior to spore discharge, the outer wall joining the sporangiophore with the sporangium ruptures, and there is a sudden eversion or rounding out of the depressed portion of the sporangium (Fig. 11–9). The sporangium is suddenly shot away when the tension on it is released and travels a distance of 0.5 to 4.0 centimeters. Spore discharge is favored by high relative humidity, and transfer of the fungus from high to low relative humidity will greatly reduce spore discharge (Ingold, 1971).

A similar type of discharge mechanism occurs in the downy mildews when both the sporangiophore tips and the sporangium suddenly round out and is also observed in aeciospore release in rusts. Rust aeciospores are tightly packed in the aecium and tend to assume a polyhedral form from the pressure of adjacent spores. When damp conditions prevail, the cells become more turgid and round out, and the terminal cells are shot off.

FORCIBLE SPORE DISCHARGE IN THE BASIDIOMYCOTINA

Violent discharge of basidiospores occurs in the majority of the basidiomycetes, probably occurring in all the Hymenomycetes, but being absent in the

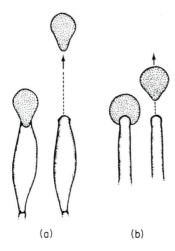

(a)   (b)

Figure 11–9 Sporangium discharge by the rounding out mechanism in *Entomophthora grylii* (a) and *Entomophthora coronata* (b). [From C. T. Ingold, 1934, New Phytologist 33:274–277.]

Gasteromycetes. A marked periodicity in spore discharge in response to day and night cycles has been noted under natural conditions in many Hymenomycetes (Haard and Kramer, 1970).

Basidia that forcibly eject their spores have well-differentiated sterigmata that bear the basidiospores on a 45° angle at the hilar appendix, a kneelike projection (Fig. 11–10). A droplet of fluid begins to accumulate before spore discharge. At first, the small amount of fluid can be discerned only in electron micrographs of *Boletus rubinellus* as granular material enclosed by a pellicle, which also covers the basidiospore. Appearance of the drop occurs concurrently with the formation of separate plugs that seal both the basidiospore and sterigma. As the droplet expands, the pellicle disappears (Yoon and McLaughlin, 1986). When observed with the light microscope, the droplet appears to enlarge rapidly and may become 10 micrometers or more in size within about 40 seconds after becoming visible in *Intersonilia perplexans*. The spores do not decrease in size as the droplet appears, and are capable of developing droplets even after being detached from the sterigmata by a micromanipulator. These observations indicate that the water is not extruded by either the spores or sterigmata. Webster et al. (1984-b) suggest that the water accumulates because the spore surface is hygroscopic and absorbs liquid from the atmosphere. After reaching its full size, the spore is suddenly shot away violently, carrying the droplet with it. The droplet quickly disappears as the liquid is distributed over the surface of the spore.

(a)   (b)

Figure 11–10 Diagram of a longitudinal section through the basal region of a basidiospore, showing two stages in droplet expansion: (a) early and late (b) in basidiospore maturation. (Ab, abscission layer; D, droplet; HP, hilar plug; Sp, sterigmal plug; PM, plamalemma; wall layers are numbered. Note that the hilum and sterigma become plugged as the droplet enlarges, and that the membrane that originally surrounded the droplet disappears. [From K. S. Yoon and D. J. McLaughlin, 1986, Mycologia 78:185–197, The New York Botanical Garden.]

The spore is propelled to a distance of about 1 to 1.5 millimeters (Webster et al., 1984-a), a very short distance compared with the distance to which an ascus can eject its ascospores. This distance is sufficient to eject the spore to approximately the midway point of the free passage between the hymenia lining either the teeth, lamellae, or tubes; then the spore changes its course and begins to fall in response to gravity (Fig. 11–11).

The actual mechanism responsible for spore discharge is unknown, but has been the subject of a great deal of discussion and debate ever since it was originally described by Buller in 1909. Recently Webster et al. (1984-a) suggested that one of the following two mechanisms may be involved, either alone or in

Figure 11–11 Basidiospore discharge; (a) successive stages in basidi-ospore discharge; a basidiospore is on a sterigma (left), and a bubble then appears (center); both the basidiospore and bubble are lost, leaving the basidium behind (right); (b) section of a hymenium showing the relative distance of the horizontal distance and path (sporobola) followed by a discharged basidiospore; (c) gills in their normal, vertical position which allows basidiospores to escape (left), and gills tilted in a position which would cause some of the basidiospores to be trapped. [Adapted from A. H. R. Buller, (a, b) 1924, (c) 1909, Researches on fungi, Longmans, Green & Co. Ltd., London.]

combination. (1) The droplet spreads quickly over the surface of the spore, wetting it and providing the kinetic energy required to eject the spore from the sterigma. This proposal is based on mathematical models but direct experimental evidence is lacking. (2) The sudden wetting of the spore by the spreading of liquid over its surface might change the electrical charge on the spore's surface, resulting in an electrostatic repulsion from the sterigma. There is some experimental support for this concept, as it can be demonstrated that discharged basidiospores bear an electrical charge. Webster and his coworkers proposed a model in which both mechanisms interact (Fig. 11–12).

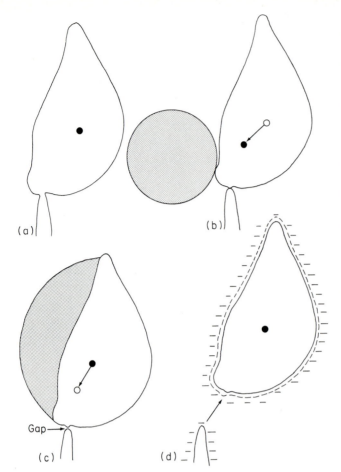

(a)

(b)

(c)

Gap →

(d)

Figure 11–12 Diagrammatic representation of the interpretation of the events associated with basidiospore discharge. (A) Basidiospore attached to its sterigma before drop formation. The closed circle inside the spore indicates the position of the center of mass of the spore. There is still firm attachment between the spore and its sterigma. (B) Basidiospore with a drop of liquid at the hilar appendix. The drop has attained maximal size but has not yet made contact with the adjacent wall of the spore. The extrusion of the drop has caused a shift in the center of mass to a new position indicated by the arrow. The spore is now shown perched on the sterigma. (C) Contact between the drop and the spore wall has resulted in sudden wetting of the spore wall and a rapid redistribution of the center of mass of the spore. The rapid redistribution of mass has created a linear momentum which has caused the separation of the spore from its sterigma. (D) Liquid from the drop has spread over the whole of the spore wall and this has caused the spore wall to become negatively charged. Electrical repulsion between the spore and its sterigma, which has a narrow radius of curvature, causes the spore to be propelled. [From J. Webster, R. A. Davey, G. A. Duller, and C. T. Ingold, 1984, Trans. Brit. Mycol. Soc. 82:13–29.]

For the mechanism to be effective in dispersal of spores from a pileus, the hymenophores must be vertically oriented. Then the basidiospore, once shot horizontally clear of the hymenium, is free to fall through the narrow free space between the hymenia (Fig. 11–11) until it reaches open air, where it will be caught up in the turbulent air currents and disseminated. If the hymenophore were tilted, the falling spore would be captured on a neighboring hymenium that obstructs its path. The vertical orientation of the hymenium is maintained by positive *geotropism* of the hymenophore and negative geotropism of the stipe. The stipe may grow in various directions, although the lamella, tubes, or teeth will remain on the required vertical axis. For example, an agaric growing on an inclined surface produces a curved stipe that holds the gills vertically (Fig. 11–13), and a bracket fungus that is growing on a log that is accidentally overturned reorients its growth (no matter how grotesque the final result) (Fig. 11–14) and forms its tubes vertically. In addition to exhibiting positive geotropism, the hymenia must be separated from the ground by open air into which the spores can fall. This is accomplished either by forming a stipe which supports the pileus above the ground or by growing on a tree or similar object which serves the same function. Further, the pileus protects the hymenium from wetting, which would make it impossible for the spores to be discharged (Ingold, 1971).

EVERSION MECHANISM

Sphaerobolus is a Gasteromycete that may occur in large groups on rotting wood or on dung of herbivorous animals: it is usually 1 to 2 millimeters in diameter. During the first half of its development, exposure to blue light is required, while during maturation exposure to red light is required. The fungus is not

Figure 11–13 Geotropic swinging and adjustment of the pileus in space. A basidiocarp was placed in the position shown in (A), and after 2½ hours it assumed the position shown in (B). The remaining drawings were made at intervals of 1 hour, until finally the basidiocarp assumed the final position in (H) in which the gills are perpendicular to the ground. [Adapted from A. H. R. Buller, 1909, Researches on fungi, vol. 1, Longmans, Green & Co. Ltd., London.]

responsive to light during its final day of development. Meanwhile the basidiocarp is becoming oriented towards the light; this positive phototropism is controlled by blue light (Alasodura, 1963; Ingold and Nawaz, 1967; Nawaz, 1967).

At maturity, the gleba is exposed when the apex of the peridium splits in

Figure 11–14 Distorted basidiocarp of the bracket fungus *Ganoderma applanatum*. The log on which this fungus was growing was moved several times, and the newly forming tissue always reoriented itself in such a manner that the tubes would be perpendicular to the ground. About seven-eighteenths natural size. [Courtesy Plant Pathology Department, Cornell University.]

a stellate fashion and folds backward. The glebal mass is suddenly ejected with a catapultlike mechanism (Fig. 11–15) and is the largest fungal projectile known. Opening of the peridium and ejection of the gleba occur in the early daylight hours.

To more fully explore this mechanism, we discuss *Sphaerobolus stellatus*. The gleba of *S. stellatus* is surrounded by six distinct peridial layers, the inner three of which are of particular interest in this discussion (Fig. 11–15). The innermost peridial layer (6) which surrounds the gleba is a thin layer of pseudoparen-

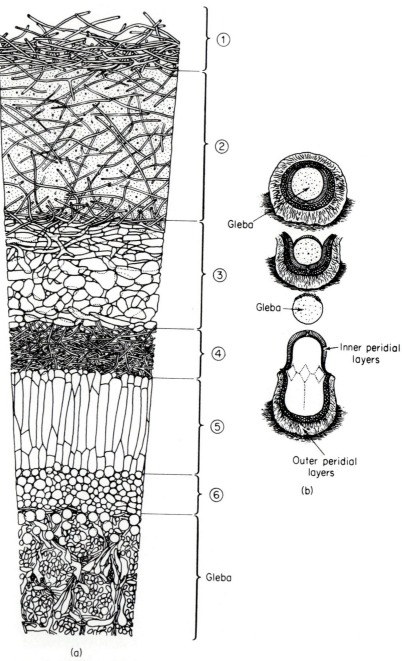

Figure 11–15 *Sphaerobolus stellatus*: (a) detail of the peridial layers, × 63; (b) successive stages in discharge of the gleba top to bottom. See the text for additional explanation. [(a) From A. H. R. Buller, 1933, Researches on fungi, Vol. 1, Longmans, Green & Co., Ltd., London; (b) from A. Engler and K. Prantl, 1900, Die natürlichen Pflanzenfamilien, Engelmann, Leipzig.]

chyma; the adjacent layer (5) is a palisade layer made up of prismatic cells; number 5 is surrounded by a layer (4) of fibrous cells. Just before the peridium opens, the thin pseudoparenchymatous layer (6) adjacent to the gleba deliquesces, separating the gleba from the outer peridial layers and forming a watery pool which serves as a lubricant. The outermost portion of the fibrous layer (4) also deliquesces, producing a space between the remainder of layer 4 and the outer layers of the peridium (3, 2, and 1) but remaining solidly affixed to these outer peridial layers at the teeth of the stellate fissures. The prismatic cells of the fifth layer then absorb water made available from the deliquescence that has occurred and become increasingly turgid. At this time, there is a decrease in glycogen content of the prismatic cells, and an increase in soluble sugars, greatly increasing the osmotic concentration of these cells. As the cells of layer 5 increase in turgidity, the rim of the opened peridium bends outward as a result of the continual expansion of this turgid layer. Only the upper portion of the layer 5 is free to expand, as the fibrous layer (4) is attached to the bottom and is comparatively inelastic. A tension is exerted on these layers, and a point is reached when the layers suddenly yield to the strain and turn inside out, remaining attached at the teeth to the rigid outer three peridial layers, and sling away the gleba with great force. The actual discharge of the gleba is too rapid to perceive, but the initial velocity may be calculated to be as great as 9.3 meters per second. The horizontal distance to which a gleba may be ejected has been observed to be as great as 6.1 meters, the greatest distance known for any fungus projectile (Buller, 1933).

TENSILE WATER MECHANISM

A mechanism of apparently common occurrence in the Hyphomycetes operates when the atmospheric humidity is decreasing. Evaporation of cellular moisture creates a negative pressure within the cell (conidium, conidiophore, or both). The cohesion of water molecules to each other and the adhesion of the water molecules to the cell wall pull the cell walls inward, placing them under considerable tension. Suddenly a rupture of the tensile water occurs, possibly because either the adhesion or cohesion can no longer be maintained. This rupture is marked by the sudden appearance of a gas bubble and by the sudden relief of the tension, allowing the walls to spring back to their original position. The conidium is catapulted away (Fig.11–16).

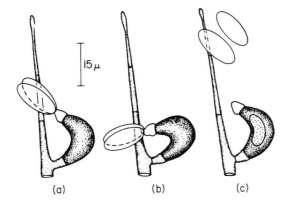

(a)   (b)   (c)

Figure 11–16 Conidium discharge by *Zygosporium oscheoides* by rupture of tensile water: (a) conidia borne on swollen, curved vesicle, and in its normal form before its first movement; (b) curvature of the vesicle in response to internal negative pressure; (c) the sudden return of vesicle to its normal position after catapulting the conidia. A gas bubble has appeared. [From D. S. Meredith, 1962, Ann. Botany (London) II 26:233–241.]

SPORE DISPERSAL

Spores are passively disseminated by air, water, and animals or by transport of their plant hosts. The fungi depend on these mechanisms for the eventual deposit of their spores in a satisfactory environment and for their continued existence. Dispersal is important because it enables a fungus species to occupy diverse habitats.

Dispersal by Air

The majority of fungus spores are released (or simply dislodged by wind) into and disseminated by the air. A sample of air may contain as many as 200,000 fungus spores per cubic meter, although 10,000 spores per cubic meter would be a more usual figure (Gregory, 1952). Hyphal fragments are also found in the air. The number of spores in the air fluctuates with prevailing conditions and is particularly great after a rainfall.

In some species the spores that may be collected from the air may differ markedly at various times of the day (e.g., morning as opposed to the afternoon). This periodicity may be in response to rhythms in the environment: for example, winds are often stronger or gustier near noon than at other times of the day, which would in turn dislodge more mature spores. Some fungi respond to light and dark cycles, and their spores may follow a periodicity in ripening. The urediniospores of many rusts are diurnal, maturing in the day. Therefore at midday when the winds are the highest, there will be an exceptionally large number of urediniospores in the air. In contrast, discharge of aeciospores tends to be nocturnal because the high relative humidity favors the buildup of turgor required for their discharge. Numbers of spores in the air at a particular time may be affected by the rhythmic response of a particular fungus to light and dark cycles (Ingold, 1971) or to relative humidity. As an example of the latter, *Drechslera turcica* discharges its spores during the morning when the high relative humidity generated by dew deposition is decreasing. Leach (1976) measured large voltage changes in the diseased portion of the maize leaf that were a result of changes in relative humidity and were correlated with spore discharge. He suggests that the conidiophores and conidia become charged with similar electrical charges during periods of high relative humidity, thereby creating an electrostatic repulsion between the conidia and conidiophores and resulting in the discharge of the spore as the relative humidity decreases.

The number and kind of spores in the air may be determined by a device such as the Hirst automatic volumetric suction trap, through which a known volume of air is drawn and directed onto a slide made sticky with glycerol or petroleum jelly. The spores then become impacted on the slide and can be counted microscopically. Simpler methods (1) exposure of a similar sticky slide to the air, (perhaps for 24 hours), after which the slide can be examined for spores, and (2) exposure of a petri dish with nutrient medium to the air, which allows some of the spores to germinate on the medium. Disadvantages of these methods are that they measure spore deposition and do not accurately account for the total spore flora of the air. A disadvantage of the second method using nutrient medium is that not all spores will germinate on any single medium, and only those spores actually germinating are accounted for.

In perfectly still air, a spore would fall in response to gravity. The rate of sedimentation of the spore fall may be calculated using Stokes' law, which describes the rate of fall of a small sphere in a viscous fluid.* (For the purposes of

---

* $V + \frac{2}{9}\left(\frac{\sigma - P}{\mu}\right)gr^2$

where V = terminal velocity (velocity of sedimentation), centimeters/second
      σ = density of sphere, grams/cubic centimeter (water = 1.00) (approximately equal to spore)
      P = density of medium (air = 1.27×10^{-3} gram/cubic centimeter)
      g = acceleration of gravity (981 centimeters/square second)
      μ = viscosity of medium (air at 18°C = 1.8×10^{-4} gram/square centimeter)
      r = radius of sphere, centimeters

this calculation, the density of spores is assumed to be approximately equal to that of water.) The rate of sedimentation of fungus spores ranges from approximately 0.003 to 2.78 centimeters per second with the usual order of magnitude about 1.0 centimeter per second. Asymmetry and roughness of the spore would presumably affect the rate of fall (Gregory, 1973).

Although the rate of fall of a spore in still air may be determined, still air is virtually never met in nature. A microscopically thin layer of perfectly still air does exist at the surface of the ground, but immediately above this still layer is the *lamellar layer* of the air. The lamellar layer is the lowest layer of moving air, and the air moves parallel to the nearest surface. The thickness of the lamellar layer is usually about 1 millimeter but varies with the wind speed and with the roughness of the adjacent surface. It is thin in high wind speeds and much thicker under calm conditions. Above the lamellar layer is the *turbulent layer*, where airflow is complex, the result both of wind and local eddies. One can detect turbulence by watching smoke from a bonfire or chimney. The turbulent layer increases in thickness with increasing wind.

The actual rate of sedimentation of a spore, and the path it will travel, is determined by its velocity in response to gravitation (Stokes' law) and the direction and rate of airflow. A spore in the lamellar layer falls in response to gravity, but its path is determined by both its velocity and the rate of lamellar air movement. A spore in the turbulent layer may be carried both upward and laterally by the air currents and may remain in the turbulent layer indefinitely. The flight of a spore in the turbulent layer may last hours, days, or even years, and the distance traveled may be tens or hundreds of miles.

It is evident that if a spore is to be disseminated for great distances, it must be discharged directly into the turbulent layer and not into the lamellar layer. Those fungi which have become adapted to air transport of their spores have mechanisms which release their spores directly into the turbulent layer. This may be seen in Discomycetes which expel their spores with such force that they are shot through the lamellar layer and into the turbulent layer and in basidiomycetes which form the hymenium well above the ground so that spores drop directly into the turbulent layer (Fig. 11–17).

Air dissemination of the spore ends with deposition, or the landing of the spore, on a substratum that may or may not be favorable for its germination. Spore deposition may occur (1) by the impaction of a spore on a protruding surface (such as a blade of grass) which interrupts the flow of air, (2) by sedimentation from the air in response to gravity, (3) by washing out of the air by rain or snow, (4) by electrostatic deposition, which may occur if a spore bearing a minute charge (either positive or negative) encounters a substratum bearing an

Turbulent layer

Lamellar layer

Ground

Figure 11–17 Discharge of spores of two Hymenomycetes and a Discomycete directly into the turbulent air layer.

Spore Dispersal

opposite charge, or (5) by boundary-layer exchange in which spores pass from the turbulent layer into the lamellar layer, whereupon they sediment out. Spore deposition is a comparatively inefficient process as it is random. Many spores must be produced, but only a few accidentally land on a suitable substratum. There is a great wastage of spores.

Dispersal by Water

Aquatic fungi include those which form zoospores, such as chytrids, their allies, and the Oomycetes. Fewer than 2% of the fungi are known to be aquatic (Ingold, 1975). The zoospores of these fungi swim feebly and randomly about and are also carried by strong water currents or by any motion of the water. Passive transport by water currents is undoubtedly of greater significance in the dispersal of these fungi than their own rather limited locomotion.

A second type of aquatic fungi are those which do not produce a motile zoospore stage. Abundantly occurring fungi of this latter type include numerous species of hyphomycetous deuteromycetes growing on leaves and debris at the bottom of a swiftly moving, well-aerated stream. The majority of these fungi produce four-armed (tetraradiate) spores (Fig. 11–18), which may be collected in

Figure 11–18 Spores collected in a single foam sample below the falls in a river. Spores *B, H, K, M, N, O, P,* and *Q* are tetradiate. [From C. T. Ingold, 1956, Mycologia 58:43–56.]

abundance in the foam and scum that collects behind barriers in the stream, especially in the late autumn or winter. The spores are transported in swiftly moving currents. For spores in swiftly moving water, coming to rest on a substratum can pose a problem, and it is probable that the irregular form of the tetraradiate spore enables it to anchor or entangle itself on the substratum more easily than a rounded spore. It has been demonstrated experimentally that a tetraradiate spore is caught more efficiently on a flat surface than a rounded spore is, thus tending to confirm the hypothesis. Such a spore makes contact at three points, like a tripod, which gives a stable form of attachment (much more so than one point). Further, a branched spore can be trapped in air bubbles better than nonbranched spores, thus favoring the capture of the branched spores in the foam which removes them from the swiftly moving current (Iqbal and Webster, 1973). Although most of the tetraradiate spores are produced by Hyphomycetes, a few are produced by basidiomycetes. Ingold (1975) suggests that this is an adaptive morphological form which was achieved by convergent evolution.

Spores of many terrestrial fungus species are distributed by water, even though the fungi may not be primarily adapted for water dispersal of their spores. Such a situation may be found when raindrops strike a spore-bearing structure and splatter, carrying the spores in the splash droplets (as we have seen in the slime-spored fungi). This provides an effective means of dispersal as a water drop of 5 millimeters (the size of a large raindrop) falling from a height of 7.4 centimeters may produce over 5000 splash droplets which travel an average distance of 20 centimeters (Gregory et al., 1959). Running water that passes from one locale to another may carry spores with it. This may occur in the trickling of rain down an object bearing fungi, such as a tree; the seepage of water through soil; and the overflowing of creeks or rivers over their normally dry banks.

Dispersal by Animals

Animals of all types, including humans, are often important agents of spore dispersal. The animal may accidentally come into contact with fungus spores by brushing against the plants or other substrata bearing fungi and thereby pick up spores on the integument, hair, claws, or clothing. Dispersal is completed when the spores are again deposited. A survey of 36 birds shot while resting in chestnut trees revealed that 19 of them were carrying spores of the chestnut blight fungus, *Cryphonectria parasitica* (Heald and Studhalter, 1914). Also the animal may consume vegetation bearing spores of the fungus or consume a fleshy sporocarp, and the spores then pass through the alimentary canal of the animal, sometimes unharmed and capable of germination when defecated. Talbot (1952) made a survey of 72 small animals (such as wood lice, mites, grubs, centipedes, slugs, springtails, and worms) and found that a large portion of the contents of the gut and feces were of fungal origin and that only one animal did not contain fungus spores. Many of the spores were viable. In a study in which the shoes of airplane passengers were swabbed and nutrient agar streaked, at least 65 species of fungi were recovered (Baker, 1966).

Virtually any fungus spore may be transported by an animal if the animal picks up the spore on its body or ingests it. The relative importance of this type of dispersal depends on the distance traveled by the animal, the viability of the spores under the conditions imposed, and the extent to which the fungus is dependent on animal dispersal in comparison with wind and water dispersal.

Numerous fungi depend almost entirely on dispersal by animals and utilize a number of devices to ensure transport by animals. This increases efficiency of dispersal and decreases spore wastage. Examples of these include the following:

1. *Occupation of a mutual habitat.* Phloem-feeding bark beetles of the Scolytidae bore out brood galleries under the bark, where eggs are laid and the larvae develop (Fig. 11–19). Phloem-feeding fungi such as species of *Ceratocystis* (which cause a blue staining of the wood) and yeasts occupy these brood galleries. The bark beetles feed on the same plant material on which the fungi feed and also feed on the fungi; so it is inevitable that the slimy spores of the fungi adhere to the surface of the beetle. The beetle migrates to new substrata, carrying the spores with it, and prepares a new tunnel in which the spores are rubbed off or deposited in the feces where they remain viable. This process not only provides for the dispersal of the fungus but also causes it to be inoculated directly into a suitable substratum. This greatly eliminates the hazards of a random spore flight. Many similar situations exist whereby a plant pathogenic fungus is regularly transported by insects.

Figure 11–19 Brood chambers beneath the bark of an elm tree formed by the beetle *Scolytus multistriatus* (top). [Courtesy Plant Pathology Department, Cornell University.]

An example is the transportation of smut spores by insects which are involved in pollination (Chapter 5).

2. *Odor attracting animals.* A rather conspicuous example is found in the stinkhorns (Phallales) which produce a foul-smelling, slimy gleba containing the basidiospores. Flies (especially blowflies and green bottle flies) and slugs visit the basidiocarp, feed on the foul-smelling liquid, and leave, carrying basidiospores on the outside of their bodies.

A similar phenomenon may be found in the rust fungi which produce their spermatia, pycniospores, in pycnia. The pycnia exude droplets of a sweet fluid, which is attractive to flies. Flies feed on the fluid, pick up pycniospores, and, in subsequent visits to other pycnia, leave behind pycniospores which they have accumulated. This is the means by which cross-fertilization is accomplished in rusts, and the sexual nature of pycnia was discovered only after Craigie (1927) observed the regular visitation by flies and was prompted to carry out cross-inoculation studies.

Hypogeous fungi such as *Endogone* and members of the Tuberales and some Gasteromycetes produce an odor which attracts animals (especially rodents) that dig up the sporocarps, consume them, and later deposit dung containing spores. Spores of *Endogone* can be recovered regularly from the alimentary canals of rodents that have fed on the sporocarps (Dowding, 1955).

3. *Ejection of spores onto edible vegetation.* Coprophilous fungi must have dung as a substratum on which to sporulate and, therefore, both their dispersal and further livelihood depend on the deposition of their spores onto vegetation which will be tempting food for an herbivorous animal. Assuming that the spore is consumed along with the supporting vegetation, it will pass through the alimentary canal where it is activated and then be deposited in dung where it can again sporulate.

Adaptations which help ensure the ejection of spores onto vegetation include phototropism (discussed in connection with *Pilobolus* but also occurring in the ascomycetes *Dasyobolus immersus*, *Saccobolus* spp., and *Sordaria* spp.) and forceful discharge mechanisms that fling the projectile to great distances (as we have seen in the coprophilous fungi *Basidiobolus ranarum*, *Sphaerobolus* spp., bird's nest fungi, and *Pilobolus* spp.). In addition, coprophilous fungi often have a very large projectile (a large sporangium, gleba, peridiole, or all eight ascospores stuck together with mucilage) which makes the distance traveled upon ejection rather large (recall that $d = Kr^2$). Those ascomycetes which fling their spores the greatest distance are coprophilous species in which the ascospores cling together (25 centimeters in *Sordaria curvula* and 60 centimeters in *Dasyobolus immersus*).

SPORE DORMANCY

Some spores may be mature and able to germinate immediately upon their introduction into a suitable environment. Many spores, however, undergo a *dormancy* of varying lengths during which they remain inactive, and germination occurs only when dormancy is broken. Dormancy is of varying significance to different fungi and is controlled by different factors.

Dormancy may be *constitutive* and under the innate control of the spore. This dormancy may be an enforced rest period during which the spore completes its maturation, or it may occur in spores which are already mature.

If spores are deposited in an environment unsuitable for germination and growth of the thallus, some spores may die while others are forced into dormancy that may tide the spore over until the environmental conditions improve to the extent that the vegetative stage could survive. This type of dormancy is of an *exogenous* nature and is under control of external factors (Sussman, 1966; Sussman and Halvorson, 1966).

Constitutive Dormancy

Constitutive dormancy may be under the control of self-inhibitors produced by the spores or even by the metabolism of the spore itself.

SELF-INHIBITORS

The role of self-inhibitors in imposing dormancy is frequently apparent when high concentrations of spores fail to germinate, while the frequency of germination increases if the spore concentration is decreased. One of the earliest observations of this phenomenon was that of Edgerton (1910) who noted that when more than 12 to 15 conidia of *Colletotrichum lindemuthianum* were included in 1 cubic millimeter of medium, the percentage of germination was reduced. In some cases, this inhibition is due to the effects of specific chemical inhibitors, but in other cases, this inhibition may simply be due to an inadequate supply of oxygen or nutrients.

Although the chemical nature of the inhibitor(s) is not known for many fungal species, we are learning that there is a great deal of variability among the different inhibitors in terms of their chemical nature and mode of action. Some of the inhibitors are volatile compounds that accumulate in high concentrations as a result of the crowded conditions. Volatile inhibitors include ethanol and acetaldehyde, which may be the major inhibitory compounds (Robinson et al., 1968). Other inhibitors may be removed from the spores simply by washing them in water, which suggests that these particular inhibitors are small molecules occurring on or near the surface of the spores. Inhibitors of urediniospore germination in some rust fungi have been extensively studied (Macko et al., 1976). In the rust fungi, the inhibitor present in many species is methyl *cis*-3,4-dimethoxycinnamate, while that formed by *Puccinia graminis* is methyl *cis*-ferulate (Fig. 11–20). These inhibitors are probably located either on or next to the surface of the urediniospores and can be extracted by floating the urediniospores in water. The inhibitory action is exerted against the initiation and growth of the germ tube prior to its emergence from the spore, but no inhibition is exerted after the emergence from the spore. Methyl *cis*-ferulate prevents digestion of the wall material in the pore area where the germ tube normally emerges upon germination. The effects of this inhibitor are reversible, and upon its removal, digestion of the wall to form the pore will occur. Self-inhibitors of a different chemical nature occur in the aeciospores of the rust *Cronartium fusiforme*. These inhibitors were effective against the aeciospores of other rust species but not of other spores produced by the rusts (Foudin et al., 1977).

(a)

CH_3O — — $\overset{\overset{H}{|}}{C} = \overset{\overset{H}{|}}{C} - COOCH_3$

CH_3O —

(b)

CH_3O — — $\overset{\overset{H}{|}}{C} = \overset{\overset{H}{|}}{C} - COOCH_3$

HO —

Figure 11–20 (a) Methyl *cis*-3,4-dimethoxycinnamate; (b) methyl *cis*-ferulate.

Dormancy of some spores may be terminated and germination initiated *(activated)* by certain treatments or the addition of nutrients which are *not* required to support vegetative growth once initiated. These activators serve as a "trigger," which modifies the spore metabolism in such a way as to cause germination to take place. Activators may include any aspect of either the physical or chemical environment, such as temperature, light, moisture, composition of the substratum, or any combination of these. Examples of activating treatments include those discussed below.

1. *Exposure of spores to high heats.* A requirement for exposure of spores to high heats is especially common among saprophytic ascomycetes. For example, ascospores of species of *Neurospora* require exposure to temperatures in the range of 50 to 60°C for about 20 minutes to germinate. The requirement for exposure to heat for spore activation is shared by many coprophilous fungi. Heat activation for coprophilous fungi may be accomplished during passage through the alimentary canal and by the high heat generated by microbial activity in the deposited dung. The effects of heat may be exerted on some proteins, denaturing them, and also on the lipids in the cellular membranes, which undergo changes in some properties such as permeability.

2. *Exposure to cold temperatures.* The majority of the fungi whose spores are stimulated to germinate by exposure to cold temperatures are the basidiomycetes and especially plant parasites such as the smuts belonging to the genus *Tilletia*. *Tilletia* spores may require exposure to temperatures as low as 5 to 10°C for an indefinite period of time to germinate. The requirement for cold temperatures provides a timing mechanism whereby the spores of the parasites remain dormant over the winter and germinate only when their hosts renew their growth.

3. *Light.* Some spores are activated by light, and the majority of these are plant parasites. A large number of the parasites are rust and smut species; however, not all rust and smut species require exposure to light to germinate, and only spore stages may have a light requirement (e.g., the urediniospores of *Cronartium ribicola* require light while the teliospores do not). As in the case of cold activation, a requirement for exposure to light for spore germination may correlate the activities of the parasite and host.

    In contrast, germination of some spores can be inhibited by light. This has been reported for conidia of *Verticillium agaricinum* (Osman and Valadon, 1981) and for urediniospores of species of *Puccinia*. Although germination of urediniospores of *P. arachidis* is inhibited by light, growth of the germ tube after germination is not affected. The effects of light exposure are not permanent, and the spores will germinate normally if returned to the dark (Subrahmanyam et al., 1988).

4. *Chemicals.* Under laboratory conditions, spore germination may be triggered by the addition of a diverse range of chemical agents, including inorganic ions, solvents (alcohols, ethers, chloroform, acetone, furans, pyrroles, and thiophenes), aliphatic esters, amino acids, carbohydrates, lipids, and vitamins (Fig. 11–21). Spores are often not specific for any one chemical, and various chemicals or chemical combinations may be effective.

    An interesting situation exists with the commercial mushroom *Agaricus brunnescens*. The basidiospores germinate only if they are present in large numbers or if its own actively growing mycelium is present. Mycelium of other fungi and the yeast *Saccharomyces cerevisiae* can

Figure 11–21 Respiratory changes induced after activation of ascospores of *Neurospora crassa* by furfural. [From A. S. Sussman, 1953, Am. J. Botany 40:401–404.]

also stimulate spore germination. The effect can be duplicated if the spores are exposed to the volatiles from a dilute solution of isovaleric acid, a long-chain fatty acid normally present in the volatiles emitted by *Saccharomyces cerevisiae* (Elliott, 1985). According to Rast and Stäubel (1970), the incorporation of isovaleric acid activates the tricarboxylic acid cycle, triggering germination.

MECHANISMS REGULATING CONSTITUTIVE DORMANCY

Activated spores often show an abrupt increase in the rate of respiration and other physiological changes before the morphological changes of germination occur. Triggering mechanisms in some manner initiate changes leading to a chain of physiological events, often directly or indirectly involving respiration. If the physiological nature of the triggering event is known, the scientist can gain some insight into the factors that maintain the spore in its dormant state. Experimental evidence derived from various fungi reveals that spore dormancy may be controlled by a variety of factors, perhaps working in combination. These include compartmentation, permeability, and metabolism.

Dormancy may be the direct cause of compartmentation in the cell—that is, endogenous reserves are isolated from the enzymes that would degrade them by a cellular membrane. In *Neurospora crassa,* the enzyme trehalase is bound to the inner spore wall and separated by the plasmalemma from its substrate, trehalose, a sugar alcohol. Activation with heat treatment or furfural (Fig. 11–21) alters the permeability of the plasmalemma, allowing the trehalase to diffuse into the cytoplasm, where it can hydrolyze the trehalose. Hydrolysis of trehalose makes it possible for it to be used as a substrate for respiration (Hecker and Sussman, 1973). In *Myrothecium verrucaria* and *Trichoderma reesei,* trehalose and amino acids and other endogenous reserves are sequestered in the vacuole. Activating treatments cause an increase in the permeability of the vacuolar membrane, permitting the substrates to diffuse outward and to be utilized as substrates for respiration. Effective activating agents for the latter fungi include sublethal freezing or heating and azide treatment (Mandels, 1981).

Dormancy can also be caused directly by metabolic factors such as the level of enzyme activity. This has been demonstrated in the heat activation of sporangiospores of *Phycomyces blakesleeanus* (Fig. 11–22). In *P. blakesleeanus* heat treat-

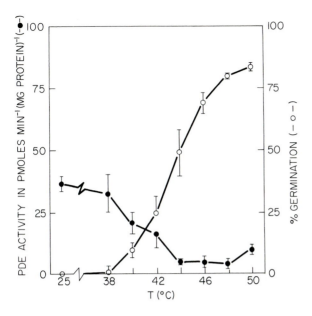

Figure 11–22 Cyclic AMP phosphodiesterase activity (●) and percentage germination (○) of *Phycomyces* spores heated for 5 minutes at different temperatures. [From A. J. Van Laere, 1986, Exptl. Mycol. 10:52–59.]

ment inactivates phosphodiesterase, an enzyme that normally catalyzes the breakdown of cyclic AMP. As a result, the cyclic AMP level increases markedly even before heat treatment is completed, but decreases soon thereafter. The cyclic AMP activates some enzymes by phosphorylation, initiating some of the physiological changes that occur during germination. One of these changes is the activation of trehalase, which then can act on the trehalose reserves (Van Mulders and Van Laere, 1984; Van Laere, 1986). The rise in cyclic AMP is often the first metabolic change that can be detected as a result of treatment by acetate or pyrosulphite in *P. blakesleeanus* (Van Mulders and Van Laere, 1984) as well as in *Pilobolus longipes*, which can be activated by glucose or glucose analogs. In *P. longipes* it is suggested that the uptake of glucose may depolarize the plasmalemma, leading to an increase in cyclic AMP (Bourret, 1984). Although the initial stimulus leading to the accumulation of cyclic AMP may differ from that involved in heat activation, the rise in cyclic AMP itself appears to be the metabolic event that actually causes the shift from dormancy to germination by enzyme activation.

A second factor that is involved in maintaining the dormancy of *Phycomyces blakesleeanus* sporangiospores is apparently related to a metabolic block. Uptake of pyruvate by the mitochondria is limited, preventing subsequent steps in respiration. The activating heat treatment enhances the ability of the mitochondria to take up pyruvate, perhaps by causing a conformational change in the carrier molecule (Van Laere et al., 1982).

Exogenous Dormancy

Spores, either with or without constitutive dormancy, may have a dormancy imposed upon them by environmental conditions unfavorable for vegetative growth. Exogenous dormancy may be imposed by inhibitors in the soil or marine environment or by inhibitors produced by plants. Inhibition in soil is discussed in Chapter 12. Temperatures lower than those permitting germination or vegetative growth will often impose a dormancy on the spore, while the spore will usually die at temperatures higher than that permitting vegetative growth (Sussman and Halvorson, 1966). The majority of the spores require a high relative humidity (usually 80 to 100%) and will become dormant if the relative humidity is too low.

Generally, dormancy will be broken and the spores activated when the restraining environmental condition is alleviated. Unlike constitutive dormancy, these factors allowing the germination of the spores are required for continued vegetative growth and are not triggering mechanisms.

Like constitutive dormancy, exogenous dormancy may play an ecologically important role in the life of the fungus. It may serve as a timing mechanism in the case of overwintering spores and, in general, allows the spores to germinate only when conditions are such that their vegetative growth will be supported.

SPORE GERMINATION

The irreversible chain of events transforming the self-contained spore (sometimes dormant) to its vegetative form constitutes germination. These changes include both morphological and physiological events. Spore germination is usually marked by a swelling of the spore and a number of cytological changes. These typically include swelling and then division of the nucleus, an increase in the quantity of endoplasmic reticulum and number of ribosomes and mitochondria, disappearance of lipid bodies, and sometimes vacuole formation. If the spore is to give rise to a mycelium (and not to zoospores), one or more germ tubes are formed. A germ tube may appear from any point on the surface of the spore or from a predetermined pore, marked initially by a thin wall. The formation of the germ tube is often first visible when the outer wall of the spore cracks and the germ tube protrudes through it as a papilla. The wall of the germ tube usually appears to be continuous with the inner wall of the spore (if an inner wall is present), which is often extensible and increases in size through the synthesis of additional wall materials. Migration of a nucleus and other cellular organelles into the germ tube takes place, and subsequently the germ tube grows by apical extension to give rise to a mycelium.

Many physiological changes take place upon germination, all more or less involved with a general increase in metabolic rate that converts the dormant spore with a low metabolic rate to an actively growing vegetative state. These changes may include the synthesis of macromolecules (such as DNA, RNA, protein, carbohydrates, and lipids); an excretion of metabolites, enzymes, or vitamins; an increase in dry weight and volume; absorption of water; and activation of enzyme systems (Fig. 11–23). Generally, these processes occur with great rapidity, as is the case for protein synthesis, apparently due in part to the presence

Figure 11–23 Metabolic events occurring during germination of *Neurospora* ascospores. Lipids are utilized as an endogenous energy source supporting fermentative respiration during dormancy. Activation is accompanied by the ability of trehalase to hydrolyze the sugar, trehalose, which together with lipids serves as a substrate for respiration. Germination is accompanied by the oxidative respiration of lipids. [From A. S. Sussman, 1961, Spores, Vol. II, Burgess, Minneapolis, Minn.]

of preformed messenger RNA, which has been detected in some spores and becomes active as spore germination is initiated. This phenomenon occurs in *Rhizopus stolonifer* sporangiospores, which are estimated to contain 6.6×10^5 preformed RNA molecules of the type that can function as mRNA, and translation into proteins occurs immediately after the spores are placed in a suitable medium (Van Etten and Freer, 1978). One of the most remarkable changes, however, is that which occurs in respiration. The dormant spore characteristically respires at a low rate, apparently utilizing the Embden-Meyerhof (EM) pathway in most cases. When germination begins, there is a great increase in the respiratory rate, marked by an increased oxygen intake and carbon dioxide output and a shift from anaerobic respiration to the aerobic respiration characteristic of fungal mycelia (Fig. 11–21). For *Neurospora tetrasperma*, the respiratory rate may increase 20 to 30 times upon germination (Sussman, 1953). Some spores require an exogenous source of carbon or nitrogen during germination, while others do not require exogenous sources of nutrients but utilize their reserve materials (such as lipids, carbohydrates, or protein) as a substrate for respiration.

REFERENCES CITED

ALASODURA, S. O. 1963. Fruiting in *Sphaerobolus* in reference to light. Ann. Botany (London) 27:125–145.

BAKER, G. E. 1966. Inadvertent distribution of fungi. Can. J. Microbiol. 12:109–112.

BOURRET, J. A. 1984. Glucose transport by germinating *Pilobolus longipes* spores. Exptl. Mycol. 9:48–55.

BRODIE, H. J. 1956. The structure and function of the funiculus of the Nidulariaceae. Svensk Botan. Tidskr. 50:142–162.

——, and P. H. GREGORY. 1953. The action of wind in the dispersal of spores from cup-shaped plant structures. Can. J. Botany 31:402–410.

BULLER, A. H. R. 1909. Researches on fungi. Longman, New York, Vol. 1, pp. 1–287.

——. 1924. Researches on fungi. Longman, New York, Vol. 3, pp. 1–611.

——. 1933. Researches on fungi. Longman, New York, Vol. 5, pp. 1–416.

——. 1934. Researches on fungi. Longman, New York, Vol. 6, pp. 1–513.

CRAIGIE, J. H. 1927. Discovery of the function of the pycnia of the rust fungi. Nature 120:765–767.

DIXON, P. A. 1961. Spore dispersal in *Chaetomium globosum* Kunze. Nature 191:1418–1419.

DOWDING, E. S. 1955. *Endogone* in Canadian rodents. Mycologia 47:51–57.

EDGERTON, C. W. 1910. The bean anthracnose. Louisiana State Univ. Agr. Expt. Sta. Bull. No. 119, pp. 1–55.

ELLIOTT, T. J. 1985. The general biology of the mushroom. *In:* Flegg, P. B., D. M. Spencer, and D. A. Wood, Eds. The biology and technology of the cultivated mushroom. Wiley, New York, pp. 9–22.

FOUDIN, A. S., P. B. BUSH, V. MACKO, J. K. PORTER, J. D. ROBBINS, and W. K. WYNN. 1977. Isolation and characterization of three self-inhibitors of germination from aeciospores of *Cronartium fusiforme*. Exptl. Mycol. 1:128–137.

GREGORY, P. H. 1949. The operation of the puff-ball mechanism of *Lycoperdon perlatum* by raindrops shown by ultra-high-speed schlieren cinematography. Trans. Brit. Mycol. Soc. 32:11–15.

——. 1952. Spore content of the atmosphere near the ground. Nature 170:475–477.

——. 1973. Microbiology of the atmosphere. 2nd ed. Leonard Hill, Glasgow, Scotland, 377 pp.

——, E. J. GUTHRIE, and M. E. BRUCE. 1959. Experiments on splash dispersal of fungus spores. J. Gen. Microbiol. 29:328–354.

HAARD, R. T., and C. L. KRAMER. 1970. Periodicity of spore discharge in the Hymenomycetes. Mycologia 62:1145–1169.

HEALD, F. D., and R. A. STUDHALTER. 1914. Birds as carriers of the chestnut-blight fungus. J. Agr. Res. 2:405–422.

HECKER, L. I., and A. S. SUSSMAN. 1973. Localization of trehalase in the ascospores of *Neurospora:* relation to ascospore dormancy and germination. J. Bacteriol. 115:592–599.

INGOLD, C. T. 1933. Spore discharge in the ascomycetes. I. Pyrenomycetes. New Phytologist 32:175–196.

——. 1934. The spore discharge mechanism in *Basidiobolus ranarum.* New Phytologist 33:274–277.

——. 1956. The spore deposit of *Daldinia.* Trans. Brit. Mycol. Soc. 39:378–380.

——. 1971. Fungal spores—their liberation and dispersal. Clarendon Press, Oxford, 302 pp.

——. 1975. Convergent evolution in aquatic fungi: the tetraradiate spore. Biol. J. Linn. Soc. 7:1–25.

——, and V. J. COX. 1955. Periodicity of spore discharge in *Daldinia.* Ann. Botany (London) II 19:201–209.

——, and V. J. DRING. 1957. An analysis of spore discharge in *Sordaria.* Ann. Botany (London) II 21:465–477.

——, and B. MARSHALL. 1963. Further observations on light and spore discharge in certain Pyrenomycetes. Ann. Botany (London) II 27:481–491.

——, and M. NAWAZ. 1967. Sporophore development in *Sphaerobolus:* effect of blue and red light. Ann. Botany (London) 31:791–802.

——, and B. A. OSO. 1968. Increased distance of discharge due to puffing in *Ascobolus.* Trans. Brit. Mycol. Soc. 51:592–594.

IQBAL, S. H., and J. WEBSTER. 1973. The trapping of aquatic hyphomycete spores by air bubbles. Trans. Brit. Mycol. Soc. 60:37–48.

LEACH, C. M. 1976. An electrostatic theory to explain violent spore liberation by *Drechslera turcica* and other fungi. Mycologia 68:63–86.

MACKO, V., R. C. STAPLES, Z. YANIV, and R. R. GRANADOS. 1976. Self-inhibitors of fungal spore germination. *In:* Weber, D. J., and W. M. Hess, Eds. The fungal spore—form and function. Wiley, New York, pp. 73–98.

MANDELS, G. R. 1981. Compartmentation of metabolic systems in the regulation of dormancy in fungus spores. Exptl. Mycol. 5:278–291.

NAWAZ, M. 1967. Phototropism in *Sphaerobolus.* Biologia 13:5–14.

OSMAN, M., and L. R. G. VALADON. 1981. Effect of light (especially near-U.V.) on spore germination and ultrastructure of *Verticillium agaricinum.* Trans. Brit. Mycol. Soc. 77:107–109.

RAST, D., and E. J. STÄUBEL. 1970. On the mode of isovaleric acid in stimulating the germination of *Agaricus bisporus* spores. New Phytologist 69:557–566.

ROBINSON, P. M., D. PARK, and M. K. GARRETT. 1968. Sporostatic products of fungi. Trans. Brit. Mycol. Soc. 51:113–124.

SUBRAHMANYAM, P., P. M. REDDY, and D. MCDONALD. 1988. Photosensitivity of urediniospore germination in *Puccinia arachidis.* Trans. Brit. Mycol. Soc. 90:229–232.

SUSSMAN, A. S. 1953. The effect of furfural upon the germination and respiration of ascospores of *Neurospora tetrasperma.* Am. J. Botany 40:401–404.

——. A. S. 1966. Dormancy and spore germination. *In:* Ainsworth, G. C., and A. S. Sussman, Eds. The fungi—an advanced treatise. Academic Press, New York, Vol. 2, pp. 733–764.

——, and H. O. HALVORSON. 1966. Spores—their dormancy and germination. Harper & Row, New York, 354 pp.

TALBOT, P. H. B. 1952. Dispersal of fungus spores by small animals inhabiting wood and bark. Trans. Brit. Mycol. Soc. 35:123–128.

VAN ETTEN, J. L., and S. N. FREER. 1978. Polyadenylate-containing RNA in dormant and germinating sporangiospores of *Rhizopus stolonifer.* Exptl. Mycol. 2:301–312.

VAN LAERE, A. J. 1986. Cyclic AMP, phosphodiesterase, and spore activation in *Phycomyces blakesleeanus.* Exptl. Mycol. 10:52–59.

——, R. R. VAN DEN BOSCH, and A. R. CARLIER. 1982. Pyruvate metabolism by mitochondria from dormant and activated *Phycomyces blakesleeanus* spores. J. Gen. Microbiol. 128:1537–1545.

VAN MULDERS, R. M., and A. J. VAN LAERE. 1984. Cyclic AMP, trehalase and germination of *Phycomyces blakesleeanus* spores. J. Gen. Microbiol. 130:541–547.

WEBSTER, J., R. A. DAVEY, G. A. DULLER, and C. T. INGOLD. 1984-a. Ballistospore discharge in *Intersonilia perplexans*. Trans. Brit. Mycol. Soc. 82:13–29.

——, ——, and INGOLD, C. T. 1984-b. Origin of the liquid in Buller's drop. Trans. Brit. Mycol. Soc. 83:524–527.

WOODWARD, R. C. 1927. Studies on *Podosphaera leucotricha* (Ell. & Ev.) Salm. I. The mode of perennation. Trans. Brit. Mycol. Soc. 12:173–204.

YOON, K. S., and D. J. MCLAUGHLIN. 1986. Basidiosporogenesis in *Boletus rubinellus*. II. Late spore development. Mycologia 78:185–197.

ZOBERI, M. H. 1973. Influence of water on spore release in *Cookeina sulcipes*. Mycologia 65:155–160.

Additional Recommended References

GREGORY, P. H. 1966. Dispersal. *In:* Ainsworth, G. C., and A. S. Sussman, Eds. The fungi: an advanced treatise. Academic Press, New York, Vol. 2, pp. 709–732.

INGOLD, C. T., and B. A. OSO. 1969. Light and spore discharge in Ascobolus. Ann. Botany (London) 33:463–471.

KRAMER, C. L., and S. M. PADY. 1970. Ascospore discharge in *Hypoxylon*. Mycologia 62:1170–1186.

MEREDITH, D. S. 1965. Violent spore release in *Helminthosporium turicum*. Phytopathology 55:1099–1102.

WALKER, L. B., and E. N. ANDERSEN. 1925. Relation of glycogen to spore-ejection. Mycologia 17:154–159.

WALKEY, D. G. A., and R. HARVEY. 1966. Spore discharge rhythms in Pyrenomycetes. I. A survey of the periodicity of spore discharge in Pyrenomycetes. Trans. Brit. Mycol. Soc. 49:583–592.

——, and ——. 1967. Spore discharge rhythms in Pyrenomycetes. II. Endogenous and exogenous rhythms of spore discharge. Trans. Brit. Mycol. Soc. 50:229–249.

——, and ——. 1968. Spore discharge rhythms in Pyrenomycetes. IV. The influence of climatic factors. Trans. Brit. Mycol. Soc. 51:779–786.

WEBER, D. J., and W. M. HESS, Eds. 1976. The fungal spore: form and function. Wiley, New York, 895 pp.

WEBSTER, J. 1959. Experiments with spores of aquatic Hyphomycetes. I. Sedimentation and impaction on smooth surfaces. Ann. Botany. (London) II 23:595–611.

——, and R. A. DAVEY. 1984. Sigmoid conidial shape in aquatic fungi. Trans. Brit. Mycol. Soc. 83:43–52.

ZOBERI, M. H. 1961. Take-off of mould spores in relation to wind speed and humidity. Ann. Botany (London) II 25:53–64.

PART III ECOLOGY AND UTILIZATION BY HUMANS

12 Fungi as Saprophytes

Like other saprophytes, saprophytic fungi feed on nonliving organic matter. The *substratum*, or food source, may be the bodies of dead animals or plants; separate parts of the plants or animals such as leaves, fallen branches, or hair; waste products from plant or animal sources; some synthetic compounds such as plastics; and any soluble products diffusing from the above.

Saprophytes outnumber parasites in nature, and it is likely that they constitute the predominant flora in most habitats. Freshwater fungi occur in ponds, lakes, rivers, streams, and bogs. The plant material, such as leaves or twigs that has settled to the bottom, can provide a particularly rich assortment of fungi, and many aquatic fungi can be recovered from the water itself by baiting techniques. A number of fungi can be recovered from the ocean, where they are particularly common as saprophytes on decaying wood. Terrestrial habitats that support saprophytic fungi are diverse. These include the surfaces of animals and plants where some fungi can grow saprophytically. Although animal surfaces are generally an inhospitable environment for saprophytic fungi, many more occur on the aerial portions of plants. Yeasts and some filamentous fungi on the flower can feed on the sugary secretions, while those on the leaf can use nutrients carried in rain, dust, or in fluid that is secreted from the hydrathodes in guttation. This fluid may contain amino acids, carbohydrates, and minerals. Rain and dew can also leach organic materials (particularly carbohydrates and organic and amino acids) from the leaves. Saprophytic fungi are particularly numerous in the vicinity of the roots, the *rhizosphere*, where organic nutrients are abundant. During growth, the root tip is protected by a mucilage cap and root cap cells that are sloughed off into the soil. Epidermal and cortical cells are also sloughed off, and all of these cells provide cellulose and other cellular constituents. The mucilage consists of acidic polysaccharides. Additional organic matter is made available in exudates which leak out of the cells. The exudate can contain minerals and low-molecular-weight organic compounds (amino acids, organic acids, simple sugars, nucleotides, and enzymes). The soil is a particularly important habitat for saprophytic organisms, in part because of the abundance of plant roots, which can provide nutrients while living as well as after they have died.

DECOMPOSITION

When plants or animals die or lose any of their parts, there is a large corpus, or body, consisting of a mixture of organic compounds. The individual organic compounds that comprise the substratum are the actual materials used as a food source. These include the great variety of carbohydrates, proteins, lipids, and nucleic acids that are produced by organisms generally. Any individual compound used as a nutrient is known as the *substrate*. It is by the individual decomposition of these compounds that the larger mass is eliminated, or *decayed*. Any species of microorganism is capable of decomposing only certain components in the plant or animal residue, and a number of species are required to decompose a substratum completely. Fungi, bacteria, and protozoa often are involved in decomposing a single substratum, the relative importance of each varying with the prevailing conditions.

To the microorganism, the process of degradation is simply a means of obtaining nutrients by digestion. Bacteria, actinomycetes, and fungi secrete digestive enzymes into the environment which degrade a specific molecule into its simpler components, making it go into solution. The food substances then enter the metabolic pathways, such as those described in Chapter 9. Although some compounds are broken down to respiratory carbon dioxide or released as metabolites, much of the substratum is incorporated into the protoplasm of the decay-causing saprophyte and will be liberated only when the microorganism itself dies and is decayed by other microorganisms.

Decomposition is complete when the organic compounds are returned to the environment in their inorganic or mineral form. For carbon, that form is carbon dioxide; for nitrogen, it is ammonia; and for phosphorus, it is phosphate.

Methods for Study of Decomposition

Determining the rates at which a substratum is being decomposed, how these rates are affected by environmental conditions or nutrients, and the course of decomposition of the individual substrates is very important in the study of saprophytic fungi. Disappearance of the substratum may be followed by calculating the loss in total weight or analyzing the metabolic or waste materials produced as decomposition occurs. For example, the rate of carbon dioxide evolution can be used as an index of carbohydrate decomposition (Fig. 12–1), while ammonia evolution may simultaneously serve as an index of decomposition of nitrogenous substances. Chemical analyses of the composition of the substratum at different stages in the decomposition process can be useful in determining the fate of individual substrates.

The Decomposition Process

The individual substrates are attacked at different rates. Readily soluble sugars and amino acids are apparently absorbed directly and metabolized within the microorganism and are among the first components to disappear. Simple carbohydrates (sugars, starches, hemicelluloses, etc.) and some of the proteins are decomposed rapidly by a great variety of microorganisms. Cellulose, fats, and oils are decomposed more slowly and usually by specific microorganisms; lignins, waxes, and tannins are the most resistant to decomposition and are at-

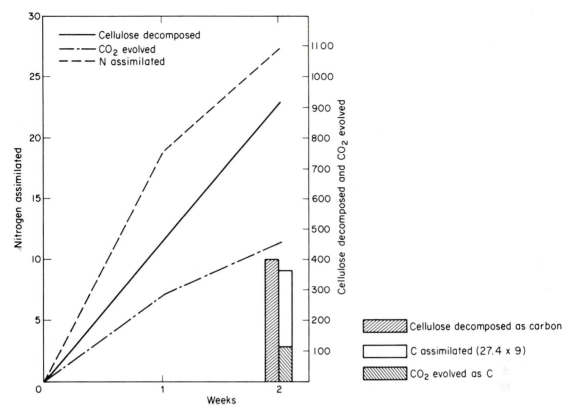

Figure 12–1 Relation between cellulose decomposition by microorganisms and assimilation of nitrogen or its transformation into protoplasm of the microorganism. [From S. A. Waksman, 1952, Soil microbiology, John Wiley & Sons, Inc., New York.]

tacked by comparatively few fungus species. It may take years for some leaves and stems to decay; a pine needle may undergo decay for 8 to 9 years before it can no longer be recognized (Kendrick, 1959).

Maximum decomposition will occur only as long as there is an abundant supply of nitrogenous and carbon compounds and other essential nutrients (especially phosphorus) present in the substratum or soil (Waksman, 1952). The ratio of carbon to nitrogen is especially important, as it primarily determines the amount of microbial growth that can occur and, therefore, indirectly the amount of decomposition that can take place (Fig. 12–1). Nitrogen is needed by microorganisms to synthesize protoplasm and, in the absence of adequate nitrogen, fungi will be considerably impaired in their ability to decompose available carbohydrates. If the nitrogen supply is in excess of the amount required for protein synthesis (generally, more than 1.5 to 1.7% of the substratum), the excess is liberated as ammonia. If the carbon supply (and therefore the energy source) is inadequate, as much as 50 to 80% of the total nitrogen may be released as ammonia. Under favorable conditions, fungi will assimilate an average of 30 to 40% of the carbon of the substratum that has been decomposed.

Young, green plants or animal residues high in nitrogen and phosphorus will undergo rapid and almost complete decay without added nutrients, particularly under aerobic conditions (Fig. 12–2). In contrast, materials low in nitrogen (such as straw, stubble, and forest litter) decompose slowly and incompletely, perhaps leaving 50 to 60% of their original weight after 3 to 10 months of decomposition. Agriculturalists often add fertilizers containing nitrogen and phosphorus to the soil to increase the decomposition of fibrous or woody plant remains consisting primarily of cellulose or lignin, such as straw or stubble.

Figure 12–2 Course of decomposition of an alfalfa plant. *A* indicates aerobic, *AN* indicates anaerobic. [From S. A. Waksman, 1952, Soil microbiology, John Wiley & Sons, Inc., New York.]

It is often difficult to determine the chemical steps involved in decomposition in nature, where substrata are present as complex mixtures of compounds and environmental conditions are quite variable. It is difficult to duplicate these conditions in the laboratory, but an approach may be made by utilizing media which contain purified components and inoculating these with pure or mixed cultures of microorganisms. In addition, studies of enzyme systems may be utilized. Information obtained in this manner is generally equivalent to that presented in Chapter 8, but it is not necessarily representative of activities of fungi in soil. A fungus may be able to degrade a substance (cellulose, for example) in culture, but this does not indicate that it can establish itself and feed on this complex compound in nature under intense competitive conditions. The chemical pathway of decomposition may also be modified according to the particular organisms that are involved and by the prevailing environmental conditions. Although at least some fungi are capable of degradng most substrata in nature, fungi are especially active in the decomposition of celluloses, hemicelluloses, and lignins.

CELLULOSE DECOMPOSITION

Cellulose, a glucose polymer, occurs widely in nature in plant cell walls. Actinomycetes, bacteria, fungi, protozoa, and some insects are cellulose decomposers but, of these, the fungi are particularly active as cellulose decomposers. The cellulose-decomposing fungi include members of the Ascomycotina, Basidiomycotina, and Deuteromycotina as well as some chytrids that may be trapped by using filter paper as a bait. Some members of the Holobasidiomycetidae are the most active in the decomposition of cellulose, and the majority of the members of this group are probably able to decompose cellulose.

The first enzymes involved in cellulose decomposition are the cellulases. The cellulases may occur as a complex, and the particular combination of enzymes in "cellulase" may be different from organism to organism. Cellulose is first cleaved into long, linear chains and then into disaccharide units (cellobiose) by enzymes in the cellulase complex. Cellobiose is then hydrolyzed to glucose by cellobiase. The overall scheme of the decomposition is as follows:

Decomposition of glucose depends on the nature of the organisms and the environmental conditions, as they may favor the growth of one group of organisms over another and may also determine the speed of decomposition. Among the environmental conditions, oxygen supply is especially important. If aerobic conditions prevail, glucose may be converted to organic acids or carbon dioxide and water by filamentous fungi. If, on the other hand, anaerobic conditions prevail, yeasts or bacteria may produce fermentative end products.

HEMICELLULOSE DECOMPOSITION

The hemicelluloses occur abundantly in plant walls interspersed through cellulose and lignin. Next to cellulose, hemicelluloses are the most abundant compound in the plant cell wall. The hemicelluloses are a heterogeneous group of polysaccharides and, upon hydrolysis, yield hexoses, pentoses, and frequently uronic acids. Decomposition of the hemicelluloses involves hydrolysis of the polymer to smaller units by specific enzymes. The actual enzymes involved and the biochemical steps depend on the particular hemicellulose which is being degraded.

Hemicelluloses tend to disappear rather rapidly during the early stages of decomposition and are decomposed by aerobic and anaerobic bacteria, actinomycetes and fungi. Fungi are more active in the early stages of decomposition, while the actinomycetes are more active in the later stages. The hemicellulose-decomposing fungi belong to all major groups, and the total number of species which can decompose hemicellulose is far greater than those capable of decomposing cellulose. Generally, the microorganisms that degrade hemicellulose do not exhibit a high degree of substrate specificity and often may use organic acids and simple sugars.

LIGNIN DECOMPOSITION

Lignin is decomposed almost entirely by fungi and especially by basidiomycetes, such as members of the Aphyllophorales and Agaricales. Lignin occurs abundantly in plant material, where it is bound with polysaccharides in the secondary cell wall and also occurs in the intercellular lamella. The lignin content of the plant increases both as the plant ages and with an increasing abundance of woody tissues. There are many types of lignins, with different types occurring in different species or even coexisting in the same plant. The lignin molecule is a highly complex polymer with the nature and arrangement of the monomers subject to variation. Although the structure of the lignin molecule is incompletely known, it is believed that the basic unit of the polymer is a phenylpropane structure consisting of a three-carbon side chain on a benzene ring:

$$\langle \text{benzene ring} \rangle - C - C - C$$

Although the biochemical details of lignin decomposition are not known, it has been suggested that the lignin-decomposing basidiomycetes secrete extracellular enzymes which release the aromatic units from the polymer. Further breakdown of the aromatic units probably can be carried out by a wide range of fungi other than the primary lignin decomposers.

Lignin and Humus Formation. Organic matter in the soil often is found in two principal fractions. The surface litter layer, prevalent in forests, consists primarily of recently fallen or deposited organic materials of which the origin may still be recognized. Beneath the litter is a dark humus layer in which the organic material is in later stages of decomposition, and the origin of the organic matter is no longer recognizable. In addition to the partially decomposed organic matter, the humus layer contains *humic* acid, a compound similar to lignin. The structure of humic acid is incompletely known, although numerous phenolic compounds may be chemically isolated from it which are generally similar to those in lignin. Humic acid is produced mainly under aerobic conditions and involves polymerization reactions which are poorly understood (Flaig et al., 1975). Decomposition of the original organic material and humic acid synthesis take place concurrently.

Humus is important in maintaining soil fertility (the ability of the soil to support plants). Humus has larger quantities of nitrogen and carbon than is found in the bodies of organisms. Through the comparatively slow but sustained decomposition of humus, these elements and others are constantly available to plants as nutrients. Humus also provides a physically favorable medium for plant growth as it is possible for the humus to modify the pH of the soil, to absorb substances which are injurious to plants, and to retain water and air. A high humus content is invariably associated with a high degree of soil fertility.

Fungi play a role in maintaining the humus level, and therefore, the soil fertility. Fungi decompose organic matter which is used in the formation of the humic acid molecule. The lignin molecule and the humic acid molecule bear a large number of structural similarities, suggesting that the decomposition of lignin may provide the monomers required for humic acid synthesis.

The extent to which fungi play a role in the polymerization of these monomers in nature is unknown although such polymerization may occur in cultures (Flaig et al., 1975). Fungi can also degrade humus, thereby releasing nutrients trapped within it.

ROLE OF SAPROPHYTES IN NATURE

The majority of the substrata that decay-causing organisms attack represent the "garbage of nature" that litters the habitats of living creatures. The sheer abundance of this debris could reach astronomical proportions. The weight of leaves formed by a single tree may be calculated to be 2 tons over a 10-year period. In a hardwood forest, an estimated 1 to 2 tons of leaves and branches are dropped per acre annually; in a tropical rain forest the figure rises to 60 tons of debris (Burges, 1965). Farmers usually harvest only a small amount of a crop, such as the grain or fruit alone, leaving the remainder of the crop plant to decay in the field. We discard huge quantities of waste paper and other household garbage annually. Without decay of this debris, the earth would become covered with plant and animal bodies, probably excluding most living organisms from their natural habitats.

The real significance of decay, however, lies in the return of those chemical compounds locked in the organisms to nature's cycles. There is a finite amount of materials available for the construction of living organisms, and life would eventually cease if chemicals were indefinitely bound in dead organisms and unavailable to living creatures. Carbon compounds are of special significance in this connection.

In nature, carbon regularly cycles through its inorganic form (carbon dioxide) and organic forms that are required by all living organisms (Fig. 12–3). Autotrophic organisms (chemosynthetic bacteria, photosynthetic bacteria, and green plants) are the only organisms that are capable of converting inorganic carbon (carbon dioxide) into organic compounds. All other life is ultimately dependent upon these autotrophic organisms as a source of organic compounds. Especially significant are the photosynthetic green plants, which serve as the base of the food chain by directly supporting herbivorous animals and, in turn, supporting carnivorous animals.

The principal source of carbon dioxide required by plants during photosynthesis is the atmosphere, which contains carbon dioxide in the comparatively low concentration of 0.03% or about 2.1 billion tons. Land plants fix annually about 80,000 million (8×10^{10}) tons of carbon dioxide. The amount of carbon dioxide in the atmosphere would only be sufficient to last a few decades unless replenished (Kononova, 1966).

Carbon dioxide may be restored to the atmosphere as an end product of cellular respiration of plants and animals; from the decomposition of limestone and other rocks; as a result of volcanic eruptions; or from the combustion of coal, petroleum, and gas or other materials. The major source, however, is from respiration of microorganisms, which provides approximately 80% of the carbon dioxide required by land plants. Soil microorganisms have been estimated to yield 63.9×10^9 tons of carbon dioxide annually, and fungi contribute approximately 13% of this amount (Kononova, 1966).

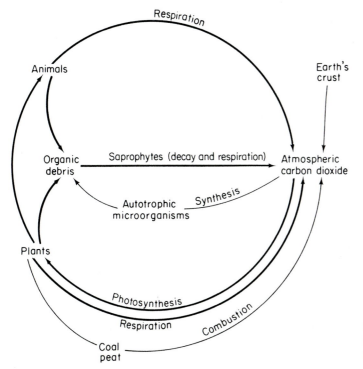

Figure 12–3 The carbon cycle in nature. Principal pathways are indicated by darker lines.

Other materials, including nitrogen, sulfur, phosphorus, potassium, calcium, and iron, are required by photosynthetic plants and are absorbed in organic forms and converted into organic compounds. Like carbon, these chemicals are cycled through inorganic and organic forms.

Saprophytic Fungi in the Soil

From the preceding section, it is evident that nature's cycles are largely initiated through the autotrophic green plants. The plants assimilate inorganic chemicals and convert them to organic compounds used by other organisms; eventually, the organic compounds are returned to the inorganic form by decay-causing microorganisms. These cycles take place in lakes, rivers, oceans, and on land. Fungi are involved in causing decay in all of these environments, but the greatest amount of decay by fungi occurs in the soil.

Soil is a complex medium, consisting of inorganic minerals and organic debris which serve as boundaries to channels and pores containing air and water. Substrata for decay-causing fungi are abundant in soil. Plant roots slough off their outer tissues, excrete organic materials (principally carbohydrates, amino acids, vitamins, and organic acids), and drop their leaves and dead branches upon the soil. Green manure crops such as clover and unharvested remains of crops are plowed under. Animals migrate through and over the soil and leave their excrement, outer sloughed tissues, shed hair and feathers, and their bodies when death overcomes them. This abundance of organic material provides nourishment either directly or indirectly for an extensive and varied microflora and microfauna. A conservative estimate of the numbers of microorganisms in 1 gram of fertile soil is (Nicholas, 1965)

True bacteria	10^6 to 10^9
Actinomycetes	10^5 to 10^6
Protozoa	10^4 to 10^5
Algae	10^1 to 10^3
Fungi	10^4 to 10^5

This estimate does not include larger soil animals such as nematodes, earthworms, mites, ants, or other worms and arthropods.

METHODS FOR STUDY OF SOIL FUNGI

Investigators of soil fungi often need to determine the extent of mycelial activity in the soil, the species present, and the relative distribution of the fungi. Such determinations are necessary for the study of effects of soil treatments, such as the addition of fertilizer, seasonal changes, different soil types, or close association with roots. Direct visual observation of fungi in the soil is hindered by the difficulty of making microscopical observations in an opaque medium and the irregularity with which spores or reproductive structures may be found. Many highly specialized methods have been developed for the study of soil fungi, but generally each method is highly selective for certain types of fungi and several methods must be used in combination to obtain an approximate picture of soil fungus activities at any given time (Fig. 12–4). Some frequently used techniques follow.

Rossi–Cholodny Contact Slide. The Rossi–Cholodny contact slide method is useful for examining the soil microflora in its natural distribution within the soil. A glass microscope slide is buried in the soil and allowed to remain in place for several days or weeks. Microorganisms coming into contact with the slide adhere to it, and after its removal from the soil the slide may be

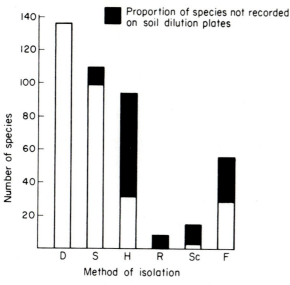

Figure 12–4 A comparison of fungi obtained from wheat-field soil by different isolation techniques. Numbers of fungi obtained and the proportion of species are different from those isolated on soil-dilution plates. [From J. H. Warcup, 1965, Ecology of soil-borne plant pathogens:—prelude to biological control. Reprinted by permission of the Regents of the University of California.]

D = Soil dilution plates   R = Isolation from rhizomorphs
S = Warcup soil plates   Sc = Isolation from sclerotia
H = Hyphal isolation   F = Isolation from fructifications

stained and examined microscopically. The fungi observed are those that came into contact with the slide through the entire period it was buried. Changes in the fungal population may be determined by burying a number of slides, which are removed at intervals during the test period. This technique is especially useful for determining the abundance of mycelia in different soils at different times. Owing to the general lack of sporulating structures formed, it is not useful for the identification of soil fungi.

Dilution Plates. The dilution plate method can be used to recover soil fungi as living colonies which are suitable for identification. Typically, a soil sample is diluted to a concentration such as 1:10,000, an aliquot is placed in a petri dish, and a cooled nutrient medium added. Usually, the medium is acidified, or an agent such as streptomyocin or rose bengal is added to retard bacterial growth. In a few days, each viable spore or hyphal fragment will produce a fungal colony. Although this method can be useful in recovering living fungi from the soil and in determining relative numbers of fungi, its chief disadvantage is that the number of fungal colonies isolated will reflect the extent of sporulation that occurs in the soil rather than the extent of actual mycelial activity. Members of the Deuteromycotina are recovered most frequently in soil dilution plates, whereas the basidiomycetes and nonsporulating fungi (Mycelia Sterilia) are rarely recovered, although they occur widely in the soil.

Warcup Soil Plate. The previous method often fails to recover nonsporulating fungi in the Mycelia Sterilia (such as *Rhizoctonia solani*) because mycelial fragments cling to the soil particles which are discarded. The standard dilution methods may be modified by placing a small amount of soil in the empty petri dish, crushing it with a spatula, and then adding cooled medium. Generally, the method is similar to the previous one and also favors the heavily sporulating species. It has the advantage of being somewhat simpler to use.

Direct Isolation. The previous techniques generally fail to recover basidiomycetes, although a casual glance at the many basidiocarps would indicate that basidiomycete mycelia do exist in the soil. Warcup (1959) found it possible to obtain basidiomycete mycelium in culture through germination of their sclerotia or from fragments of rhizomorphs.

Fungi are most abundant in the upper 4 inches of soil and are chiefly limited to the upper 12 inches. Their numbers and types decrease with increasing depth. The top 6 inches of arable land that contains 60,000 pounds per acre of dry organic matter may be estimated to contain 300 pounds dry weight of fungus mycelium (Nicholas, 1965).

Most types of fungi from chytrids to agarics, from saprophytes to parasites of plants and animals (including man), and from predators of soil animals to symbiotic mycorrhiza-forming fungi may be detected at one time or another in the soil. Representatives of all fungal groups occur in the soil. Those most frequently isolated are members of the Deuteromycotina, often *Aspergillus*, *Penicillium*, and *Trichoderma* (Fig. 6–1).

Only some of the many fungi occurring in the soil lead active saprophytic lives there and, for these fungi, the soil is their usual habitat. Other fungi are transient organisms that invade the soil on the plant or animal hosts that they parasitize. The transient fungi may not be able to survive in the absence of their hosts or may be able to lead a saprophytic existence of a limited duration. Often the transient fungi are capable of temporarily colonizing substrata in the soil but are unable to compete successfully with the native fungi. Numerous fungi are represented in soil as dormant structures; these include sporangiospores, conidia, oospores, and ascospores and probably basidiospores, chlamydospores, and sclerotia. For many of these fungi, occurrence in soil is incidental; they do not play a significant role in saprophytic activities in the soil. Active soil fungi exist as similar dormant structures, however, when food is lacking.

The dormant condition is more or less typical for fungi in the soil, and there is comparatively little fungus activity in the absence of an available food source. A suitable substratum is often as small as a single cellulose fiber that supports its own rather typical microflora. Many such substrata, each constituting a microhabitat, occur in the soil in a three-dimensional array or mosaic of separate ecological niches. Between the individual ecological niches are spaces where fungi are comparatively inactive or dormant (Garrett, 1955). Fungus structures in the less active areas include spores, chlamydospores, sclerotia, rhizomorphs, and mycelia (often dead or inactive). When a suitable substratum becomes available, dormant fungus structures in the immediate vicinity of the new substratum burst into activity; colonize, exploit, and exhaust the substratum; and then become dormant again (Garrett, 1951; Warcup, 1967). The mycelium may lyse when nutrients are no longer available, making survival dependent on the formation of dormant spores (Lockwood, 1964).

Fungistasis.   In the absence of a suitable substratum, fungal mycelia and spores are generally maintained in a dormant condition. This dormancy is maintained by a widespread fungistatic effect of the soil *(soil fungistasis)*. Fungistasis is found in practically all soils (except in some deep subsoils, highly acid soils, or soils exposed to cold temperatures). Fungistasis may be reversed by the addition of readily decomposable organic materials such as glucose or by the growth of roots into the region and subsequent secretion of organic compounds by the roots. Roots may secrete a wide variety of organic compounds, including amino acids, soluble sugars, and vitamins. With such a reversal of fungistasis, fungus spores germinate and the fungal mycelium resumes active growth.

Fungistasis is invariably associated with the presence and activity of soil microorganisms. Soil sterilization reverses fungistasis, and subsequently fungistasis may be restored by the addition of a small amount of unsterilized soil. Those soil types in which fungistasis is lacking are those without a substantial microbial population. Fungistasis is generally assumed to be a result of the general microbial activity in the soil and not necessarily a product of any specific group of microorganisms.

There is little knowledge or agreement concerning the mechanism involved in fungistasis. One possible mechanism is that the living microorganisms compete for nutrients required by fungus spores and the germination of the fungus spores is inhibited until nutrients become available from a newly introduced substratum or from roots (Lockwood, 1964). Alternative suggestions are that fungistasis results from the accumulation of staling substances produced by soil microorganisms (Jackson, 1965; Park, 1960) or that specific inhibitors are produced by microorganisms (Park, 1967; Watson and Ford, 1972). Volatile compounds such as acetaldehyde, ethanol, and ethylene are produced in culture by soil fungi and will inhibit spore germination, which leads some investigators to suggest that the inhibitors are volatile compounds which may be produced by fungi as well as by bacteria (e.g., Gray, 1976). Watson and Ford (1972) propose that fungistasis is maintained by volatile inhibitors that are primarily of microbial origin but may also depend on nonbiotic factors (perhaps the pH or the iron or calcium carbonate content of the soil), and that this inhibition may be released by stimulators, which are primarily biotic in origin and are mostly nutrients but may also include volatiles emitted by plant roots.

Conditions Affecting Growth in Soil. Factors affecting the growth of mycelia in the soil include temperature, moisture, carbon dioxide, oxygen, and acidity.

Soil moisture is a fairly complex environmental parameter and involves free water as well as relative humidity. Levels of soil moisture indirectly affect soil aeration, movement of solutes, and diffusion of nutrients or inhibitors (Griffin, 1969). Fungi are generally able to tolerate a wide range of moisture conditions in the soil, but if the moisture level becomes extremely low, both the available free water and the relative humidity are reduced, which results in little or no fungal activity. Fungal activities tend to be inhibited generally by relative humidity levels less than 75% and favored by high relative humidities near 100% (Griffin, 1963). If the free-water level becomes very high (to the extent that aeration is hindered), the number of active mycelia is also reduced. This effect of reduction in aeration is greatest for strongly aerobic fungi but may be negligible for many species, some of which may be abundant in waterlogged soils. A similar reduction in aeration may be caused by extreme compaction of the soil.

Soil fungi are relatively insensitive to reductions in oxygen, and there is typically little change in the growth rate until the oxygen concentration is lowered to about 4% in the gas phase. After this low level is reached, fungal species will respond differently to a further lowering (Griffin, 1972).

The carbon dioxide concentration in the soil atmosphere usually is between 0.002 and 0.02 atmosphere, but values of 0.1 atmosphere or more may occur at greater depths or in soil with a high water content (Griffin, 1972). High carbon dioxide concentrations are associated with lowered oxygen partial pressure. A study of fungi originating from different levels within the soil usually reveals that those from the surface levels of the soil are intolerant of carbon dioxide while those isolated from depths where carbon dioxide can accumulate are more tolerant of high carbon dioxide levels.

Temperatures may fluctuate greatly in the soil. Soil temperature is affected by moisture content, slope, color, and the nature of the vegetation which it supports. Typically, there is daily fluctuation of temperature, the lowest temperatures occurring at dawn and the highest temperatures an hour or two after noon. Daily fluctuations of as much as 35°C may be noted in the temperate zone. Seasonal variations may be even more extreme. Soil fungi are generally tolerant of wide fluctuations in temperature. This tolerance is probably chiefly a position effect; that is, those fungi killed by extreme temperatures at one level (as by freezing at the surface) are replaced by other fungi surviving at a different level where temperatures are less extreme (Raney, 1965). Fungi are apparently active in the humus layers even during the winter months, although their activity may decrease to approximately 43% of the summer level (Waid, 1960). Fungi may respond to the effects of temperature and relative humidity in combination: for example, they may survive longer in extremely hot ($56°C$) soils if the relative humidity is low (85 to 90%) rather than high (100%) (Chen and Griffin, 1966).

Fungi as a group can tolerate a wide pH range, but some are confined to either alkaline or acid soils, while other species can tolerate a wide pH range. Soil pH affects the availability of nutrients, but it is not known whether this effect is significant for soil fungi. Altering the soil pH by addition of lime will in turn alter fungal activity.

Antagonism between Soil Microorganisms. A soil fungus is in close proximity with other soil microorganisms and its ability to live in the soil is partially determined by its ecological interrelationships with other forms of soil life. These interrelationships are often *antagonistic* in nature. Antagonisms result when one or more of the organisms involved in the interrelationship are harmed or have their activities curtailed. In the case of soil fungi, antagonism may take the form of exploitation, antibiosis, and competition.

Exploitation. A soil fungus may be directly attacked by parasitic or predatory organisms. Bacteria may cause lysis of the mycelium and are often responsible for the early disappearance of fungus mycelia in the soil. Some soil fungi are capable of parasitizing other fungi in culture (see Chapter 14), but the extent to which parasitism occurs in nature is unknown. In addition, various members of the soil fauna feed on fungi in the soil.

Antibiosis. Fungi modify the environment by excreting metabolic wastes or by-products (staling products) These metabolic by-products are often of a potentially dangerous composition and, in high concentrations, may be toxic to the fungus that produced them or to any other soil microorganisms that might be sensitive to them (Fig. 12–5). This phenomenon is known generally as *antibiosis*. The metabolic products that can cause antibiosis are varied and include (1) respiratory carbon dioxide, alcohol, acids, and other simple compounds that have a nonspecific effect; and (2) the antibiotics, higher-molecular-weight by-products (often a weak acid), some of which have specific toxic effects on susceptible microorganisms. Antibiosis may make an environment generally toxic and therefore antagonistic to some fungi present. The degree of antagonism experienced is largely a matter of tolerance of the individual organisms and the ability to withstand toxic conditions favors the survival of a microorganism. Antibiosis is not necessarily a product of specific antibiotic production, although in some instances this is the case. Antibiosis differs from fungistasis in that the effects of antibiosis are localized and usually specific, whereas those of fungistasis are widespread and general.

Antibiotics. Antibiotics are produced by bacteria, actinomycetes, and fungi and may be toxic to members of any group. Production of antibiotics is widespread among the fungi, occurring in comparatively few of the lower fungi but in many members of the Ascomycotina, Basidiomycotina, and Deuteromycotina. The majority of the fungi producing antibiotics are soil fungi, which often gives these fungi an ecological advantage. Antibiotics may be isolated in small quantities from soil samples but may be inactivated by alterations in pH, absorption on clay or humus, or degradation by microorganisms (Gray, 1976). The role of antibiotics in competition will be discussed more completely in the next section.

Antibiotic-producing fungi which are effective against bacteria may be isolated from the soil. An agar plate is inoculated with bacteria which are allowed to develop for 12 to 24 hours; then small particles of soil are added to the plate. A fungus producing antibiotics (effective against the test bacteria) may be detected when the bacteria surrounding the growing hyphae become lysed, producing a

Figure 12–5 Antibiosis between *Trichoderma lignorum* (left) and *Mucor heterogamous* (right). *T. lignorum* produces antibiotic(s) which inhibit the growth of *M. heterogamous,* resulting in a clear zone with little growth where the fungi meet. [Courtesy Plant Pathology Department, Cornell University.]

clear zone around the hyphae. Portions of the hyphae may be cut out and cultured individually.

Most antibiotics are toxic to a broad spectrum of microorganisms, and the antibiotic producer may not be immune to its effects but may have a high tolerance. Many antibiotics are toxic to only certain organisms or types of organisms. Types of toxicity include (1) a specific toxicity to gram-positive bacteria, (2) semi-specific toxicity to acid-fast bacteria, affecting gram-positive bacteria at higher doses, and (3) a specific toxicity to fungi.

Resistance to antibiotics by microorganisms may take several paths: (1) An organism may produce an enzyme that will destroy the inhibitor; (2) production of metabolites may antagonize or neutralize the effects of the antibiotic; (3) possession of biosynthetic pathways that are not inhibited by the antibiotic; and (4) possession of barriers may prevent the movement of antibiotics into the cell. The last is believed to be especially significant in fungi. Commercial production of antibiotics is discussed in Chapter 16.

Competition. When some commodity of the environment is in short supply and cannot meet the demands of all organisms present, the organisms will attempt to obtain enough of that commodity to support themselves. This rivalry between organisms for a commodity is termed *competition.* Soil fungi most commonly must compete for a food source, and competition is especially keen in substratum colonization. This interplay may become antagonistic if a species is eliminated or curtailed, but the rivalry in itself is not necessarily antagonistic.

The physiological ability of a fungus to compete successfully for a substratum may be termed *competitive saprophytic ability* (Garrett, 1981). Competition for a substratum is indirect, usually taking the form of differential growth or ability to withstand prevailing conditions in the soil. Characteristics that may give a fungus a favorable competitive ability are (1) a high growth rate and rapid germination of spores in the presence of a substratum, (2) a high degree of metabolic efficiency in production of enzymes and utilization of substrata, (3) ability to produce substances (including antibiotics) that are toxic to other soil microorganisms, and (4) a high degree of tolerance to antibiosis. A successful competitor may have any combination of these traits, but probably it is endowed with at least one. However, no single factor has been conclusively demonstrated to be the sole cause for differential survival in the soil. Rapid germination of spores, a rapid growth rate, and an ability to quickly consume the available nutrients are comparatively obvious when one considers that the fungus that can get to the food source first and consume the most will get more of the food. Although the production of toxic byproducts is an incidental excretion of waste products by the producer, it can give that organism an advantage by keeping others away. Likewise, a resistance to toxic substances simply means that an organism is not likely to be deprived of the food by producers of the toxic substances.

Although antibiotic production may greatly benefit a given fungus, this is not always the case. In nature there is interaction between a large number of organisms and often a delicate system of checks and balances that determine the extent of activity of any given organism. The potential benefit of antibiotic production by a given fungus, *A*, may be modified in nature by (1) the relative numbers of organisms sensitive to *A*'s antibiotic; (2) the relative numbers of organisms able to inhibit *A* and whether or not they are sensitive to *A*'s antibiotic; and (3) the proportion of fungi eliminated by *A* that would otherwise have eliminated *A*'s antagonists (Park, 1960) (see Fig. 12–6).

Figure 12–6 Hypothetical situation where antibiotic production is a disadvantage to the producer, *A*. Left: antibiotic producer *A* is absent and antibiotic producer *B* eliminates sensitive competitors *C* and *D*; center; antiobiotic producer *A* arrives, eliminates sensitive *B*, and now must compete with both nonsensitive *C* and *D*; right: nonantibiotic-producing *E*, in the same situation as *A* in the center, must compete only with *B*.

SUBSTRATUM COLONIZATION AND SUCCESSION

Studies of substratum colonization and succession were pioneered by Garrett (1951, 1981). The concepts and terms presented below are drawn from his work. Upon the renewal of activity, the fungus is able to penetrate and decompose any of the nutrients within the substratum that it is capable of using. While the nutrients that it can utilize are available, the fungus grows actively; but it will become dormant when those particular nutrients are exhausted. The substratum may still have many nutrients which other fungi can utilize, and these fungi may then colonize the substratum and exploit it of those nutrients that they can use. This *succession* of fungi will continue until the substratum is completely decomposed or exhausted (Fig. 12-7). In principle, succession of fungi on a substratum is similar to that of higher plants on land because in each case environmental conditions are changed by the organisms. Unlike higher plant succession, succession by fungi depletes the environment and makes it less satisfactory for growth by other organisms, whereas higher plant succession constantly improves the environment for growth by other plants.

Succession of fungi occurs commonly in nature and may be observed on plant parts which remain temporarily above ground (standing dead trees or herbaceous stems), are on the surface of the ground (logs, nut hulls, leaves), or are initially present within the soil (roots) (Hudson, 1968). Fungi are especially important in the breakdown of plant remains and, therefore, succession of fungi on plant substrata has special significance. A simultaneous succession of bacteria and soil fauna occurs on the substratum alongside of the fungi. Fungal succession of plants usually begins even before death occurs. Unlike animals, plants do not ordinarily die a sudden death except in the case of accidents such as floods or fires, but individual roots or branches may die and leaves are shed. Rather than a sudden death, the plant typically undergoes a period of gradual decline, perhaps due to attack by a highly specialized plant parasite or to unfavorable growing conditions. In these instances, the plant is moribund or undergoing a period of gradual death. The tissues are highly susceptible to invasion by fungi that are comparatively unspecialized as parasites. Virtually no senescent plant tissue is free of parasites, as they occur in dying roots and even in autumn leaves before their fall, even though only isolated parts of the plant may

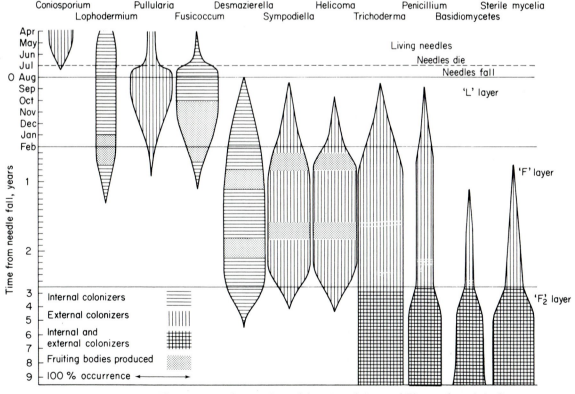

Figure 12-7 Succession of fungi on leaves of *Pinus sylvestris* in litter, showing the relative proportion of each genus and the temporal and spatial relationships. The sugar fungi are lacking in this succession. [From W. B. Kendrick and A. Burges, 1962, Nova Hedwigia 4:313–344.]

be moribund. With death of the entire plant or plant part, the parasite may continue to live saprophytically. In almost all cases, the parasites are the initial or pioneer invader of the plant tissue.

Either overlapping with the parasites or following them next in the succession are the *sugar fungi,* which suddenly invade the substratum after death of the plant is complete, and which occur in or on soil. The sugar fungi are predominantly members of the Mucorales and are capable of metabolizing the available sugars and simple carbon compounds. These invading fungi compete not only for the carbon source but also for the available oxygen and other nutrients such as nitrogen, phosphorus, and potassium. These fungi are able to dominate the environment because their spores germinate rapidly in the presence of a substratum and their hyphae have a rapid growth rate, both resulting in a sudden burst of activity. Eventually this group eliminates itself from the substratum because it is sensitive to the accumulation of its metabolic by-products, especially carbon dioxide. Formation of dormant structures and quiescence follows. Evidence from *Pythium mamillatum* indicates that these fungi may be restricted to the early stages of colonization before toxic metabolites accumulate. *P. mamillatum* actively colonizes virgin substrata, but colonization is reduced if the substratum has been precolonized (Barton, 1960).

The next stage of the succession consists of colonization by *cellulose-decomposing fungi* which either overlap with the sugar fungi or follow them. Cellulose-decomposing fungi are predominantly members of the Basidiomycotina, but they include chytrids and members of the Deuteromycotina and Ascomycotina. These fungi have the disadvantage of having to compete for a substratum that is somewhat depleted of readily available simple sugars and of nutrients such as nitrogen. The cellulose decomposers hydrolyze the cellulose molecule to cellobiose and then to glucose, which they can assimilate. The hydrolysis of cellulose occurs externally, and the simpler forms are free in the vicinity of the mycelium before their intake and assimilation. Many *secondary sugar fungi,* unable to degrade cellulose, exist in the vicinity of cellulose-decomposing fungi and feed upon the simpler carbohydrates alongside of the cellulose decomposers. The secondary sugar fungi and cellulose decomposers physiologically could have been the first to colonize the fresh substratum but could not compete with the rapid colonization characteristic of the primary sugar fungi. The fungi in this stage of succession decline with the exhaustion of available food.

The predominant material left after the decline of the cellulose-decomposing fungi is lignin. Lignin is the principal constituent of humus (comprising 30 to 60% of it, in contrast to the 10 to 30% in the original plant). Lignin tends to accumulate as there are comparatively few fungi which are capable of strongly decomposing it; all of them are higher basidiomycetes (Fig. 12–8). There is comparatively little competition at this stage in succession as the substratum is practically devoid of readily obtainable food that other fungi can use, thus giving the lignin fungi an advantage. The lignin fungi may combat the shortage of food by forming an extensive mycelial system through which food may be transported from one portion to another or one food base to another. Unlike the ephemeral sugar fungi that become dormant until food is available, the lignin decomposers (and some cellulose decomposers) form either mycelial strands or rhizomorphs that travel actively from one substratum to another. Evidence that a lignin-decomposing fungus can establish itself earlier only with difficulty comes from *Agaricus brunnescens.* Commercially, *A. brunnescens* is grown only on compost that is in a late stage of decomposition; the proportion of lignin is increased and the readily decomposable materials (sugars, hemicelluloses, etc.) are decreased, thus making the compost unfavorable for would-be competitors (Fig. 12–8).

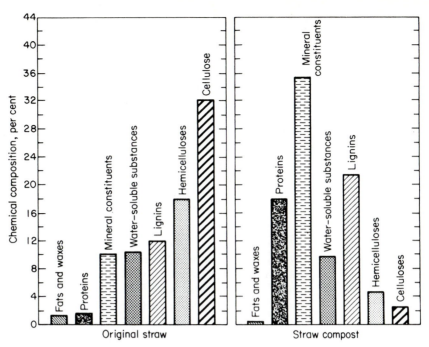

Figure 12–8 Comparative chemical composition of oat straw and of a compost prepared from it. [From S. A. Waksman, 1952, Soil microbiology, John Wiley & Sons, Inc., New York.]

DETERIORATION OF MATERIALS BY FUNGI

Many of the materials used by people to make products are of plant or animal origin. These include wood and wood products, paper, cotton and woolen fabrics, leathers, and food products. Any of the plant or animal parts used to make these goods are commonly encountered by fungi in nature, and any of these products may support the growth of saprophytic fungi if environmental conditions are favorable. In addition, other manufactured products may not be directly produced from plant or animal parts but may nevertheless support growth of saprophytic fungi. Products of this type include certain plastics and paints.

The growth of fungi on these products is a substratum-saprophyte relationship, with fungi attacking goods that can provide nourishment as long as favorable environmental conditions prevail. Generally, fungus attack is favored by high humidity or abundant moisture and warm temperatures. For many goods, the usually low humidity and moisture alone is adequate to prevent fungal growth.

We have established that growth of saprophytic fungi leads to decomposition in nature and that decomposition of natural materials is necessary to remove waste products and to recycle the components of these materials. In the same manner, decay of human goods which often find their way to the garbage dump is essential for the removal of the products of civilization. Fungi and other microorganisms play a role in the decomposition of paper, plastics, and other materials used by humans. However, if saprophytic fungi become established on goods that have not been discarded, *deterioration* results. Deterioration may be considered as a reduction in the useful properties of these goods due to their partial decomposition.

The role of fungi in deteriorating goods has long been familiar. It is not unusual to find mildew or molds (usually, a member of the Deuteromycotina) on shoes or books that have been left in a damp place. During World War II, many troops were in the tropics, where high humidity and temperatures favored the luxurious growth of fungi on a variety of goods. Fungi damaged or destroyed canvas tents, leather goods such as shoes and cases, clothing, packing boxes, electrical equipment (Fig. 12–9), rubber and plastic products, and optical equipment. The "kerosene fungus," *Cladosporium*, can even grow in storage and aircraft tanks containing kerosene-type fuels, where it causes corrosion of the aluminum (Hendey, 1964).

It is evident that fungi may cause deterioration of a wide variety of materials. Damage by fungi is commonly a problem on wood and wood products, paper, fabric, and foods. These are considered in greater detail below.

Wood and Wood Products

Forests constitute one of our most important natural resources, contributing wood for innumerable uses by humans. Of the types of materials affected by fungi, the greatest amount of damage occurs to wood in the form of standing timber, rough lumber, or finished wood products, including paper. Deterioration is caused by abrasion, termites, marine borers in salt water, but fungi are the most serious.

Fungi often attack the heartwood of standing, living trees. These fungi are saprophytes in that they attack and decay the nonliving heartwood of the tree, which may not interfere with the functions of the living sapwood. Such infected trees are structurally weakened and may meet an early death because of the heartwood destruction. In any event, heartwood is usually rendered unfit for lumber. Fungi attacking the heartwood are basidiomycetes, usually Aphyllophorales; they may be divided into (1) the "brown rot" fungi that remove the structural carbohydrates (cellulose and hemicellulose) and leave the lignin relatively unchanged, and (2) the "white rot" fungi that primarily attack the lignin but also remove a small amount of the structural carbohydrates, leading to a lightening of the wood.

Felled timber, in the form of logs or sawed lumber, is seasoned by air-drying or kiln-drying to rid it of excess moisture. Logs and lumber are subject to attack by a large number of fungi, especially during the early stages of seasoning by air-drying while the wood has a high moisture content. Air-drying is

Figure 12–9 Fungal growth (arrows) on electrical equipment. [Courtesy U.S. Army Frankford Arsenal, Philadelphia, Pa.]

usually accomplished by stacking the wood outdoors. Care must be taken to stack it in such a way that air may freely circulate throughout and to avoid as much contact with the ground as possible. If moisture accumulates, either within the stack or next to the ground, favorable conditions are created for the growth of fungi. Fungi growing under these conditions are often deuteromycetes and ascomycetes that live chiefly on cell contents and often cause a discoloration (staining) of the wood. The stain may be yellow, brown, or blue in color and the color may be a result either of the mycelium of the fungus within the wood or of pigmented by-products. Similar fungi may produce superficial (moldlike) colonies on the wood without staining and may simply be wiped away. Wood invaded by either of these superficial types of fungi may be used for most purposes, excluding those where the unsightly stains would be undesirable (furniture, etc.) or where the remaining mycelium and spores could contaminate food. In addition, the wood may be invaded by decay-causing basidiomycetes that decompose the wood, making it unsound for structural purposes. Generally the fungi that attack the wood are similar in their physiology to those causing heartrot, but they are different organisms. The sapwood is particularly vulnerable to decay in these stages (Fig. 12–10).

Vulnerability to fungus attack decreases when the moisture content of the wood becomes lower than the point at which free water within the cells is gone, usually between 25 and 30% moisture content. With complete drying to 20% or less moisture, the wood is ordinarily immune to fungus attack. Exceptions to the above are (1) invasion by dry rot fungi, such as *Serpula lacrymans*, which translocate water from a moist substratum to its mycelium, causing decay of dry

Figure 12–10 Rot of an oak log caused by *Trametes versicolor.* Note the basidiocarps on the surface of the log and the rotting sapwood in the region penetrated by the mycelium. [Courtesy Plant Pathology Department, Cornell University.]

Deterioration of Materials by Fungi

timbers in buildings; and (2) wood that is immersed in water or mud where anaerobic conditions prevail, as in piling for wharves, which prevents fungal growth.

For many wood products that would normally be maintained under dry conditions (such as packing crates, furniture, and wood for indoor construction) no further treatment beyond adequate seasoning is necessary to prevent destruction by fungi. Development of any existent mycelium is arrested as long as the moisture content remains under 20%, although it may become active if sufficient moisture becomes available. Wood intended for many uses where the product will be exposed to high humidity and moisture (especially next to the ground) requires additional treatment with a chemical preservative to prevent decay. Treatment may consist of applying coal-tar creosote under pressure to wood intended for use in contact with the ground or water (as telephone poles, fence posts, railroad ties, foundation piles, etc.). Creosote impregnation is generally considered to be the most effective treatment for prevention of decay, but it cannot be used where it might interfere with the application of paint or where the odor or appearance would be objectionable. For wood to be used for such things as boats and timber for construction, an effective treatment consists of applying a mixture of 5% pentachlorophenol and 95% oil carrier to the wood, usually by immersing the wood into the chemical for a short period of time or occasionally under pressure. Ordinary paints, varnishes, oils, and stains used as a wood finish have little, if any, value in protecting the wood from decay as moisture can penetrate through the surface (Fig. 12–11).

Paper

Raw materials and paper are vulnerable to fungus attack at every stage of paper manufacture. Pulpwood (90% of the raw material for paper) must be handled carefully during outdoor storage at the paper plant to avoid fungus attack. Decay of the pulpwood lowers the yield of pulp and destroys fibers which may lead to a 75% reduction in strength of the finished product if it is used in the paper (Gascoigne and Gascoigne, 1960). During manufacture, pulpwood is converted into pulp by mechanical grinding or through chemical treatments with sulfite, soda, or sulfate that separate the cellulose fibers. Pulp is vulnerable to

Figure 12–11 Fungal damage to paint on wood. Left: typical appearance of damage to paint; right: germinating spores of *Pullularia pullulans* on a latex paint film. *P. pullulans* is the fungus most often associated with paint damage. × 55. [Courtesy Nuodex Division of Tenneco Chemicals, Inc.]

fungus attack during storage by the usual mold fungi and wood-decaying fungi encountered in lumber. The most severe problem is the formation of slime in a watery suspension of pulp by bacteria and, to a lesser extent, by fungi. The slime, probably a product of hemicellulose digestion, damages paper made from the pulp by discoloring it or by producing shiny, translucent spots on it (Fig. 12–12). Fungi frequently isolated from slimes include *Penicillium, Cladosporium, Trichoderma* (Fig. 6–1), and *Mucor.* Additional damage during paper manufacture occurs from mildewing and discoloration of paper stocks.

Once the paper is finished, it provides a suitable substratum for cellulolytic fungi that may cause yellow, brown, or black discoloration and spotting through mildewing or destruction of the paper. Paper usually escapes attack by these organisms as it is not ordinarily used or maintained under very humid conditions, unlike some lumber. Numerous species of fungi have been isolated from paper, the majority of them members of the Deuteromycotina. *Aspergillus* and *Penicillium* are frequently encountered. Paper to be used under adverse conditions may be protected from fungi by incorporating a small amount of fungicide into the paper during manufacture.

Textiles

Textiles and cord made of linen, cotton, manilla, or jute consist almost entirely of cellulose and are subject to the attack by cellulolytic fungi, especially outdoors or under conditions of high humidity. Cellulolytic fungi attacking textiles in-

Figure 12–12 Spotting of finished paper resulting from slime carried along during processing. [Courtesy Institute of Paper Chemistry, Appleton, Wis.]

clude *Aspergillus* spp., *Chaetomium globosum, Cladosporium herbarum,* and *Humicola.* The marine fungus, *Zalerion,* a member of the Deuteromycotina, is a serious assailant on twines and ropes used in seawater. Principal damage to textiles results from musty odors, spotting and discoloration, loss in water repellency, loss in strength, and decrease in flexibility. Staining, both from pigment production and presence of fungus structures, is often the most objectionable effect. Fabric which is expected to be exposed to adverse environmental conditions (tents, cord, sandbags, netting, etc.) where it would be subject to fungus attack may be treated during manufacture with a fungicide, such as an aqueous solution of copper napthenate.

Woolens may be attacked by fungi, but comparatively few fungi attack wools and the damage is usually minor in comparison to fungus damage on cottons. *Microsporum* and *Trichophyton* and other wool decomposers are close allies of those fungi that cause skin diseases (see Chapter 13).

Food Spoilage and Mycotoxins

Every type of food is subject to infestation by fungi of one type or another. Food may be simply made unpalatable by a discoloration (as in rice), by production of off-flavors (as mustiness in coffee), or by the simple distaste of eating food with an obvious fungus growth. Food value is usually lessened, especially that of fats, carbohydrates, and proteins. Most often, food is made unfit for consumption by outright spoilage or through production of toxins. Losses during storage and marketing of food crops are due to mechanical damage, aging, and decay. Fungi are the agents most often responsible for decay.

Fruits and vegetables suffer from fungus infestation during storage, transport, and marketing. The fungi attacking fruits and vegetables may be field parasites, in addition to those fungi which occur predominantly in storage. Common examples of storage fungi are *Penicillium digitatum* on citrus fruits and *Rhizopus stolonifer* on sweet potatoes and other vegetables.

Microorganisms found in dairy products are usually bacteria, although fungi also occur and on occasion may cause deterioration of the product. Fungi in milk, butter, and cheese are primarily yeast and deuteromycetes *(Geotrichum candidum, Penicillium, Alternaria,* and *Cladosporium)* (Figs. 6–1 and 6–2). The predominant problem caused by these fungi is the development of a rancid or off-flavor.

Dried fruits are especially likely to support growth of yeasts on their surface where a sugary substance often accumulates. Infestation by yeasts leads commonly to souring of dried prunes both in storage and in the package and in the souring of dates after packaging.

Perhaps the most familiar example of fungi on food is that of *Rhizopus stolonifer,* the "bread mold," on bread. Numerous other fungi occur on bread, quickly making it unfit for consumption if the prevailing conditions are favorable for growth of fungi. The baker attempts to keep the bakery as free as possible of contaminating spores and to produce the bread under conditions as nearly aseptic as possible. In addition, preservatives are added to retard fungal growth.

MYCOTOXINS IN FOOD

Grains and forage plants may become infected by fungi as they stand and dry in the field or after they are harvested and stored. Infestation is most commonly by species of *Aspergillus* and *Penicillium,* which do not ordinarily infect the plants while living but are soil saprophytes. Fungal growth is encouraged if the grains are allowed to remain in the field during rainy weather or to overwinter in the field or, in the case of stored grains, if the grain is inadequately dried before storage, if it is damaged, or if the moisture or temperature levels are too high during storage. Nearly all the fungi on grains primarily attack the embryos, often completely destroying 14 to 15% of the embryos while not attacking the endosperm. These fungi have undesirable effects on the grains; including a decrease in the percentage of germination; discoloration of the embryos which appear as black specks in milled flour and may result in off-flavors; or the production of toxins.

The toxins may cause severe illness or death in animals or humans that eat the moldy grains, hay, or straw. For example, species of *Penicillium* and *Aspergillus* may invade standing or fallen corn in the late summer or early autumn, especially if the weather is rainy. This moldy corn is responsible for moldy corn toxicosis in pigs and cattle that forage on it in the field or consume animal feed prepared with moldy corn. Poisoning due to moldy corn toxicosis results in hemorrhaging, liver damage, and frequently death. A second example is the poisoning of horses that are confined to a stall and fed damp straw on which the saprophyte *Dendrodochium toxicum* has grown. The poison acts principally on the central nervous and cardiovascular systems and results swiftly in death (Brook and White, 1966).

The best-known example of poisoning of humans by moldy grains results in a condition known as alimentary toxic aleukia, which results in abnormalities of the blood system (Joffe, 1971). This condition has been observed sporadically in rural populations in Siberia and was especially widespread from 1941 to 1947 when epidemics resulted in poisoning of up to 10% of the population in some areas. Poisoning occurs when millet is allowed to overwinter in the field before it is harvested. The mild winters accompanied by numerous cycles of freezing and thawing favor growth of fungi that are resistant to cold and are even able to grow at $-10°C$. Toxin production is favored by temperatures between 1 and $4°C$. Although various fungi may be involved, the principal causal agent is *Fusarium sporotrichoides,* which produces several toxins. Milling of the grain removes the outer infected portions of the grain and most of the toxins, and therefore poisoning is virtually unknown in the urban parts of Russia, where refined grain products are used. About 1.5 kilograms of contaminated grain eaten over a period of about 6 weeks induces pathological changes in the blood (Feuell, 1969). Symptoms of the poisoning are somewhat variable, but typically include fever, headache, diarrhea, nausea, and vomiting during the first few days. As the condition progresses, there is a destruction of the blood-forming elements of the bone marrow. Erythrocytes, platelets, hemoglobin, and leucocytes decrease and other abnormalities in the blood occur. There is a pronounced tendency for hemorrhaging to occur, and bleeding may occur from the nose, throat, gums or internally in organs such as the lungs, liver, or brain. Hemorrhagic rashes occur on the skin as a result of friction, and bleeding ulcers appear on the mucous membranes. Swelling of the lymph glands occurs, and may be so pronounced that death occurs by strangulation. This condition is frequently fatal, but recovery can result if the symptoms are recognized early and the patient does not consume additional contaminated millet. Control also involves educating the public about the necessity for handling grain properly and sampling grain to determine if it is contaminated (Forgacs and Carll, 1962).

Aflatoxins. The toxins produced by fungi on grains include a subgroup known as the aflatoxins, all lactones with a similar molecular structure (Fig. 12–13) and capable of causing extensive pathological changes in the liver. The aflatoxins are produced by *Aspergillus flavus* and by some other closely related species of *Aspergillus* on stored wheat, corn, rice, barley, bran, flour, soybeans, and peanuts. These fungi occur commonly in the soil as saprophytes and do not generally seem to invade seeds prior to harvest.

The aflatoxins were discovered in the early 1960s when veterinarians, pathologists, microbiologists, nutritionists, and chemists teamed together to study

(a) Rubratoxin B

(b) Sporofusariogenin glycoside

$(C_5H_9O_4)(C_6H_{10}O_5)_2 - O -$

(c) Aflatoxin B$_1$

Figure 12–13 (a) Principal toxin causing moldy corn toxicosis produced by *Penicillium rubrum;* (b) one of the toxins causing alimentary toxic aleukia; (c) most commonly produced aflatoxin.

an apparently new disease, turkey X disease. In 1960, more than 100,000 turkeys died within a few months within a radius of less than 100 miles of London but widespread death of turkeys did not occur elsewhere. The disease was characterized by a loss of appetite, lethargy, and a weakness of the wings. The turkey died within a week and at death assumed a characteristic position with the head drawn back and the legs extended backward. Post-mortem studies showed that the liver had hemorrhaged and had necrotic lesions, and frequently the kidneys were swollen. The scientists were unable to demonstrate that a pathogenic microorganism was responsible for the disease and concluded that the turkeys were being poisoned. It was finally established that all birds affected had been fed with feed produced at one particular mill in London. Peanut meal from that mill caused similar symptoms in poultry which were experimentally fed with it. A toxin was extracted from the meal and was found to be active in the purified form. The toxin was named *aflatoxin* as it was determined that it had been produced by *Aspergillus flavus.* Meanwhile in the United States a similar situation was developing. Trout in hatcheries had been characteristically fed with fresh animal by-products from slaughter houses, but their diet was changed to a dry food, including grain derivatives. After two years, these fish developed tumors of the liver which has subsequently been attributed to aflatoxins in the feed (Halver, 1969).

Similar sporadic outbreaks of poisoning due to aflatoxins (*aflatoxicosis*) have been observed in farm animals and have been attributed to aflatoxins in their feed. Studies with aflatoxins show that a wide variety of farm animals are susceptible to them but differ in their reactions. Young animals are more susceptible than older ones. For example, calves are more susceptible than mature cattle. Day-old ducklings are among the most susceptible of the animals, while mature sheep are perhaps the most resistant. The first clinical signs of aflatoxicosis are a loss of appetite and weight. A few days before death, the animals lose muscular coordination, appear to be dull, and become inactive. Pathological changes occur in the liver and possibly in other organs and vary somewhat from species to species.

Not all animals die when they consume food containing aflatoxins, and it is important to know whether human food produced from the farm animals contains aflatoxins. The aflatoxins are excreted in the urine and are not retained in the tissues that are normally used as meat or incorporated in the eggs. One aflatoxin, B_1, is metabolized and converted into a closely related toxin or toxins, the M toxins, found in milk. The M toxins are stable when the milk is dried, and because they are bound to the casein fraction, they are incorporated into cheese if cheese is made from the milk (Allcroft, 1969). Rats fed primarily on milk containing the M toxin subsequently developed necrosis of the liver (Kraybill and Shapiro, 1969).

A high correlation can be shown between the consumption of food contaminated with aflatoxin and a high incidence of liver cancer in certain populations (Fig. 12–14). However it is not known to what extent the aflatoxin is responsible for causing the liver cancer or operates in conjunction with other predisposing factors such as malnutrition or parasitism. There are certain populations in parts of Africa south of the Sahara, Southeast Asia, Japan, and southern India where there is a high incidence of primary liver cancer and also a high utilization of grain products that may become contaminated by *Aspergillus flavus*. For example, a survey was conducted in Ethiopia, where it is the custom to store grain from one harvest to another, a period including the rainy season which encourages fungal growth. Gruels made from grain are eaten by nursing mothers, and these gruels and other grain products are an important part of the diet generally. Analysis of several samples of grains and grain products from a local market showed that *Aspergillus flavus* and aflatoxin B_1 were normally present (Kraybill and Shapiro, 1969). The highest rate of liver cancer occurs among the Bantu population in Mozambique, where primary liver cancer occurs at a rate approximately 500 times that in the United States. These people consume large quantities of corn meal. By drawing samples from food prepared for normal meals, it was found that 9.3% of the food sampled contained aflatoxin. The average daily intake of aflatoxin for each adult was calculated as 15 milligrams, the highest aflatoxin dosage known to occur in the human diet (Van Rensburg et al., 1974).

A practical problem is the production of animal or human foods that are free of aflatoxins. Contamination of peanuts is an especially important problem, as they are widely used for human and animal feed and aflatoxin production in them is commonplace. Therefore we will concentrate on peanuts to illustrate how aflatoxin levels in food may be controlled. The fungi are virtually absent in unblemished, immature peanuts just removed from the soil but invade the peanuts in the postharvest period while they are still moist and continue to do so until the moisture level drops to 10% or below. Growth of *Aspergillus flavus* on stored peanuts is encouraged by damage to the peanuts, maturity of the peanuts, high relative humidity (over 85%), and high temperatures (optimum 36 to

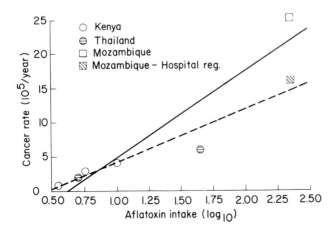

Figure 12–14 Relationship between the level of aflatoxin intake of population and the primary liver cancer rate, obtained by pooling international data. Solid line: the cancer rate for Mozambique was calculated by equating the incidence rate for goldminers to that for the general population. Broken line: Hospital registrations were used to calculate the cancer rate for Mozambique. [Figure and legend from S. J. Van Rensburg, J. J. Van Der Watt, I. F. H. Purchase, L. P. Coutinho, and R. Markham, 1974, S. African Med. J. 48:2508a–2508d.]

38°C) during storage (Diener and Davis, 1969). Therefore, control partially depends on harvesting immature peanuts in such a manner that they are not damaged, culling out damaged pods, drying the peanuts quickly after harvest, and maintaining them at low temperatures and low relative humidity whenever they are stored. *A. flavus* is more likely to invade defective kernels that are broken, shriveled, or discolored than those in good condition. After harvest, batches of peanuts can be assayed for their possible contamination level by sorting out the peanuts which are undersized or not in good condition. Weighing is used to determine what percentage these defective peanuts represent of the total batch, and samples are examined with a microscope to determine whether or not *A. flavus* is visible. If either *A. flavus* is visible or the percentage of damaged peanuts is high, the peanuts are used to make oil, as the aflatoxins are not retained in the oil but are left behind in the pressed cake, which is then used for fertilizers. Assuming that the batch of peanuts has passed this initial inspection, it can be further processed for use as food in the form of salted nuts, confections, peanut butter, or animal food. The peanuts are subsequently shelled, and samples removed at the time of shelling and analyzed chemically for the presence of aflatoxins. If aflatoxins are absent in this test, the peanuts are processed further. At each stage, defective or broken kernels are culled out. By following these procedures, it is possible to produce a final product such as peanut butter in which aflatoxins cannot be detected even though present in low levels in the unprocessed batch of peanuts (Kensler and Natoli, 1969).

REFERENCES CITED

ALLCROFT, R. 1969. Aflatoxicosis in farm animals. *In:* Goldblatt, L. A., Ed. Aflatoxin: scientific background, control, and implications. Academic Press, New York, pp. 237–264.

BARTON, R. 1960. Antagonism amongst some sugar fungi. *In:* Parkinson, D., and J. S. Waid, Eds. The ecology of soil fungi: an international symposium. Liverpool University Press, Liverpool, England, pp. 160–167.

BROOK, P. J., and E. P. WHITE. 1966. Fungus toxins affecting mammals. Ann. Rev. Phytopathol. 4:171–194.

BURGES, A. 1965. The soil microflora—its nature and biology. *In:* Baker, K. F., and W. C. Syder, Eds. Ecology of soil-borne plant pathogens: prelude to biological control. University of California Press, Berkeley, Calif., pp. 21–32.

CHEN, A. W., and D. M. GRIFFIN. 1966. Soil physical factors and the ecology of fungi. VI. Interaction between temperature and soil moisture. Trans. Brit. Mycol. Soc. 49:551–561.

DIENER, U. L., and N. D. DAVIS. 1969. Aflatoxin formation by *Aspergillus flavus*. *In:* Goldblatt, L. A., Ed. Aflatoxin: scientific background, control, and implications. Academic Press, New York, pp. 13–54.

FEUELL, A. J. 1969. Types of mycotoxins in foods and feeds. *In:* Goldblatt, L. A., Ed. Aflatoxin: scientific background, control, and implications. Academic Press, New York, pp. 187–222.

FLAIG, W., H. BEUTELSPACHER, and E. RIETZ. 1975. Chemical composition and physical properties of humic substances. *In:* Gieseking, J. E., Ed. Soil components: organic components. Springer-Verlag, New York, Vol. 1, pp. 1–211.

FORGACS, J., and W. T. CARLL. 1962. Mycotoxicoses. Advan. Vet. Sci. 7:274–382.

GARRETT, S. D. 1951. Ecological groups of soil fungi: A survey of substrate relationships. New Phytologist 50:159–166.

——. 1955. Microbial ecology of the soil. Trans. Brit. Mycol. Soc. 38:1–9.

——. 1981. Soil fungi and soil fertility, 2nd ed. Pergamon Press, Oxford, 150 pp.

GASCOIGNE, J. A., and M. M. GASCOIGNE. 1960. Biological degradation of cellulose. Butterworth, London, 264 pp.

GRAY, T. R. G. 1976. Survival of vegetative microbes in soil. Symp. Soc. Gen. Microbiol. 26:327–364.

GRIFFIN, D. M. 1963. Soil physical factors and the ecology of fungi, III. Activity of fungi in relatively dry soil. Trans. Brit. Mycol. Soc. 46:373–377.

——. 1969. Soil water in the ecology of fungi. Ann. Rev. Phytopathol. 7:289–310.

——. 1972. Ecology of soil fungi. Syracuse University Press, Syracuse, N.Y., 193 pp.

HALVER, J. E. 1969. Aflatoxicosis and trout hepatoma. *In:* Goldblatt, L. A., Ed. Aflatoxin: scientific background, control, and implications. Academic Press, New York, pp. 265–306.

HENDEY, N. I. 1964. Some observations on *Cladosporium resinae* as a fuel contaminant and its possible role in the corrosion of aluminum alloy fuel tanks. Trans. Brit. Mycol. Soc. 47:467–475.

HUDSON, H. J. 1968. The ecology of fungi on plant remains above the soil. New Phytologist 67:837–874.

JACKSON, R. M. 1965. Antibiosis and fungistasis of soil microorganisms. *In:* Baker, K. F., and W. C. Snyder, Eds. Ecology of soil-borne plant pathogens: prelude to biological control. University of California Press, Berkeley, Calif., pp. 363–373.

JOFFE, A. Z. 1971. Alimentary toxic aleukia. *In:* Kadis, S., A. Ciegler, and S. J. Ajl, Eds. Microbial toxins. Academic Press, New York, Vol. 7, pp. 139–189.

KENDRICK, W. B. 1959. The time factor in the decomposition of coniferous leaf litter. Can. J. Botany 37:907–912.

KENSLER, C. J., and D. J. NATOLI. 1969. Processing to ensure wholesome products. *In:* Goldblatt, L.A., Ed. Aflatoxin: scientific background, control, and implications. Academic Press, New York, pp. 334–359.

KONONOVA, M. M. 1966. Soil organic matter: its nature, its role in soil formation and in soil fertility. 2nd Engl. ed., translated by Nowakowski, Z., and A. C. D. Newman. Pergamon Press, Elmsford, N.Y., 544 pp.

KRAYBILL, H. F., and R. E. SHAPIRO. 1969. Implications of fungal toxicity to human health. *In:* Goldblatt, L. A., Ed. Aflatoxin: scientific background, control, and implications. Academic Press, New York, pp. 401–441.

LOCKWOOD, J. L. 1964. Soil fungistasis. Ann. Rev. Phytopathol. 2:341–362.

NICHOLAS, D. J. D. 1965. Influence of the rhizosphere on the mineral nutrition of the plant. *In:* Baker, D. F., and W. C. Snyder, Eds. Ecology of soil-borne plant pathogens: prelude to biological control. University of California Press, Berkeley, Calif., pp. 210–217.

PARK, D. 1960. Antagonism—the background to soil fungi. *In:* Parkinson, D., and J. S. Waid, Eds. The ecology of soil fungi: an international symposium. Liverpool University Press, Liverpool, England, pp. 148–159.

——. 1967. The importance of antibiotics and inhibiting substances. *In:* Burges, A., and F. Raw, Eds. Soil biology. Academic Press, New York, pp. 435–447.

RANEY, W. A. 1965. Physical factors of the soil as they affect soil microorganisms. *In:* Baker, K. F., and W. C. Snyder, Eds. Ecology of soil-borne plant pathogens: prelude to biological control. University of California Press, Berkeley, Calif., pp. 115–119.

VAN RENSBURG, S. J., J. J. VAN DER WATT, I. F. H. PURCHASE, L. P. COUTINHO, and R. MARKHAM. 1974. Primary liver cancer rate and aflatoxin intake in a high cancer area. S. African Med. J. 48:2508a–2508d.

WAID, J. S. 1960. The growth of fungi in soil. *In:* Parkinson, D., and J. S. Waid, Eds. The ecology of soil fungi: an international symposium. Liverpool University Press, Liverpool, England, pp. 55–75.

WAKSMAN, S. A. 1952. Soil microbiology. Wiley, New York, 356 pp.

WARCUP, J. H. 1959. Studies on basidiomycetes in soil. Trans. Brit. Mycol. Soc. 42:45–52.

——. 1967. Fungi in soil. *In:* Burges, A., and F. Raw, Eds. Soil biology. Academic Press, New York, 532 pp.

WATSON, A. G., and E. J. FORD. 1972. Soil fungistasis: a reappraisal. Ann. Rev. Phytopathol. 10:327–348.

Additional Recommended References

ALEXANDER, M. 1977. Introduction to soil microbiology, 2nd ed. Wiley, New York, 467 pp.

BROADBENT, D. 1966. Antibiotics produced by fungi. Botan. Rev. 32:219–242.

BURGES, A., and F. RAW, Eds. 1967. Soil biology. Academic Press, New York, 532 pp.

CHESTERS, C. G. C., and R. H. THORNTON. 1956. A comparison of techniques for isolating soil fungi. Trans. Brit. Mycol. Soc. 39:301–313.

CHRISTENSEN, C. M., and H. H. KAUFMANN. 1965. Deterioration of stored grains by fungi. Ann. Rev. Phytopathol. 3:69–84.

COOKE, R. C., and A. D. M. RAYNER. 1984. Ecology of saprophytic fungi. Longman Group, Harlow, Essex, England, 415 pp.

GREATHOUSE, G. A., and C. J. WESSEL, Eds. 1954. Deterioration of materials: causes and preventive techniques. Reinhold, New York, 835 pp.

HURST, H. M., and N. A. BURGES. 1967. Lignin and humic acids. *In:* McLaren, A. D., and G. H. Peterson, Eds. Soil biochemistry. Marcel Dekker, New York, pp. 260–286.

——, and ——, and P. LATTER. 1962. Some aspects of the biochemistry of humic acid decomposition by fungi. Phytochemistry 1:227–231.

JOHNSON, T. W., JR. 1968. Saprobic marine fungi. *In:* Ainsworth, G. C., and A. S. Sussman, Eds. The fungi: an advanced treatise. Academic Press, New York, Vol. 3, pp. 95–104.

JONES, E. B. G., Ed. 1976. Recent advances in aquatic mycology. Wiley, New York, 749 pp.

KIRK, T. K. 1971. Effects of microorganisms on lignin. Ann. Rev. Phytopathol. 9:185–210.

PARK, D. 1968. The ecology of terrestrial fungi. *In:* Ainsworth, G. C., and A. S. Sussman, Eds. The fungi: an advanced treatise. Academic Press, New York, Vol. 3, pp. 5–39.

PARKINSON, D., and J. S. WAID, Eds. 1960. The ecology of soil fungi: an international symposium. Liverpool University Press, Liverpool, England, 342 pp.

ROBINSON, P. M., D. PARK, and M. K. GARRETT. 1968. Sporostatic products of fungi. Trans. Brit. Mycol. Soc. 51:113–124.

SIU, R. G. H., and E. T. REESE. 1953. Decomposition of cellulose by microorganisms. Botan. Rev. 19:377–416.

SPARROW, F. K., JR. 1968. Ecology of freshwater fungi. *In:* Ainsworth, G. C., and A. S. Sussman, Eds. The fungi: an advanced treatise. Academic Press, New York, Vol. 3, pp. 41–93.

WAKSMAN, S. A., and E. S. HORNING. 1943. Distribution of antagonistic fungi in nature and their antibiotic action. Mycologia 35:47–65.

WARCUP, J. H. 1951. The ecology of soil fungi. Trans. Brit. Mycol. Soc. 34:376–399.

WHITE, W. L., R. T. DARBY, G. M. STECHERT, and K. SANDERSON. 1948. Assay of cellulolytic activity of molds isolated from fabrics and related items exposed in the tropics. Mycologia 40:34–84.

——, G. R. MANDELS, and R. G. SIU. 1950. Fungi in relation to degradation of woolen fabrics. Mycologia 42:199–223.

WOGAN, G. N., Ed. 1965. Mycotoxins in foodstuffs. MIT Press, Cambridge, Mass., 291 pp.

13 Fungi as Predators and Parasites

At least some fungi are capable of exploiting virtually every living organism as a source of food. Exploiting fungi are either predators or parasites. *Predators* actively trap other organisms, which they then kill and consume for food. *Parasites* invade a living plant or animal, feeding and multiplying within it at the host's expense but not contributing to the welfare of the host.

FUNGI AS PREDATORS

Several fungi obtain their livelihood by capturing animals (amoebae, rotifers, other protozoa, and nematodes) in a specialized trap and then feeding on them after their death. These are the *predacious* fungi; they are common in soil, dung, rotting wood, aquatic habitats, and especially mosses.

Predacious fungi may be obtained in culture by pouring cornmeal agar into petri dishes, allowing it to solidify, and then adding a small amount of substratum such as leaves, rotting wood, or moss plants with a small amount of soil clinging to their roots. Nematodes crawl from the debris, and the predacious fungi may often be found after an incubation period of 1 week to 2 months at room temperature (Duddington, 1955).

Predacious fungi constitute the order Zoopagales entirely and are also numerous in the family Moniliaceae of the Hyphomycetes. These two groups represent the majority of the predacious fungi, although predacious fungi occur in diverse taxonomic groups. All the members of the Zoopagales are predacious fungi. As discussed in Chapter 3, these fungi are obligate predators on amoebae and other protozoa and may occur either exogenously or endogenously.

421

Predacious Hyphomycetes

Predacious members of the Moniliaceae are predominantly captors of nematodes. Unlike the Zoopagales, these fungi have a comparatively broad or unrestricted host range and are not obligate predators, meaning that they may live saprophytically in the absence of prey. The predacious Hyphomycetes are predominantly exogenous, capturing their prey through a large variety of mechanisms.

ADHESIVE ORGANS

A comparatively simple mechanism for the capture of nematodes is the formation of short lateral hyphal branches that curl and anastomose with similar branches or with the parent mycelium, forming single loops or a network of loops. Nematodes are captured in the sticky secretion of these loops or by entanglement in the network. An example of this type of mechanism may be found in *Arthrobotrys oligospora* (Fig. 13–1). *A. oligospora* may be isolated by placing horse

(a)

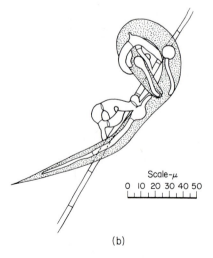

(b)

Scale-μ

0 10 20 30 40 50

Figure 13–1 *Arthrobotrys oligospora:* (a) network of loops (bars in the micrographs = 50 micrometers); (b) captured nematode with endogenous hyphae. [Micrographs courtesy B. Nordbring-Hertz; drawing from C. Drechsler, 1937, Mycologia 29:447–552.]

manure in a glass chamber and then culturing some of the milky fluid which rises on the sides of the container. This fluid is rich in both predacious fungi and nematodes. The presence of nematodes stimulates the formation of traps in *A. oligospora*, which at first are single small loops formed by lateral branches that recurve and anastomose with the parent hypha. Additional loops are formed until a network results. All loops are at right angles to each other, so that some loops may be flat upon the substratum while others are perpendicular to the substratum, making it extremely difficult for a nematode to escape the network. Copious amounts of an adhesive viscous material are secreted from the loops when contact is made with a nematode, enabling the fungus to trap large nematodes that are 500 to 600 micrometers in length. After a nematode is trapped by the loops, the dying nematode's integument is perforated by the fungus and a large globular organ practically equal to the width of the nematode's cavity is formed within the nematode. A number of hyphae arise from the globular organ and fill the cavity of the now deceased nematode, depleting the contents of the carcass. Death occurs within about 2 hours after capture.

A subspherical knob borne on a short lateral branch is the organ of capture in certain members of the genera *Dactylaria* and *Dactylella* (Fig. 13–2). In *Dactylella ellipsospora*, the stalks bearing globose knobs are formed at regular intervals along the hyphae and are usually vertical so that the knob is held slightly above the surface of the substratum. These knobs may be formed in the absence of nematodes but are more prevalent in their presence. When a nematode comes into contact with a sticky knob, it is caught and struggles to get away, coming

Figure 13–2 (a) Knobbed hypha and nematode parasitized by *Dactylella ellipsospora*; (b–e) *Dactylella lysipaga*: (b) living nematode encircled by eight rings, none of them having yet penetrated the animal; (c) hyphae with rings and knobs; (d) infected nematode; (e) conidiophore with conidia. [From C. Drechsler, 1935, Mycologia 27:447–552.]

into contact with more knobs and being held more firmly. When the nematode has become quiescent, an outgrowth from the knob penetrates its body and forms a globose bulb from which trophic hyphae that fill its body cavity are formed.

RINGS

Some predacious fungi form curved cells which join to form a closed ring. A nematode is trapped when it tries to pass through a ring in its wanderings and wedges itself in the ring when trying to squeeze through. Often the ring is broken off from the parent mycelium, and an active nematode may accumulate several rings before its activity is impaired. In *Dactylella lysipaga* and *D. leptospora*, fungi of this type, penetration of the nematode is accomplished by an outgrowth from the ring (even though isolated from the parent mycelium). This outgrowth is a globose knob which gives rise to trophic hyphae within the nematode.

Numerous predacious fungi, including *Arthrobotrys dactyloides* (Fig. 13–3), form rings similar in structure to the above but differing in that they are not passive. When a nematode pushes its head through a ring, the cells of the ring suddenly enlarge in diameter and shorten in length and thus firmly ensnare the nematode in the constriction. In *A. dactyloides*, constriction is so severe that the

Figure 13–3 *Arthrobotrys dactyloides:* (a) three-celled constricting rings; (b) captured nematode invaded by fungus; (c) conidia and conidiophore. [From C. Drechsler, 1937, Mycologia 29:447–552.]

nematode's internal organs are frequently severed at the point of strangulation, resulting in the paralysis and death of the nematode. Hyphae penetrate the nematode, ending in curious knobs. Mechanical stimulation of the rings by the nematodes is apparently responsible for ring closure, as it was found that contact with a micromanipulator would cause ring closure in *Dactylaria brochophaga*, in which the ring increased to three times its size and closed in less than 0.1 second. Ring closure in *D. brochophaga* is a result of swelling of the vacuoles (Comandon and de Fonbrune, 1938).

TRAP INDUCTION

In the absence of nematodes, the fungal trapping organs are sometimes entirely absent or present in only small numbers. The addition to a vegetative culture of a small amount of sterile water in which nematodes have lived is sufficient to stimulate trap formation, indicating that one or more diffusible substances produced by the nematode are responsible for initiating this response. A variety of biological materials will also induce a similar response: they include human blood serum, extracts from earthworms, yeast extract, and the amino acids valine, leucine, and isoleucine. Traps may be induced to form in *Arthrobotrys dactyloides* by water or nutrient deficiencies in culture, which leads to the suggestion that in nature both the presence of the prey and the generally unfavorable conditions lead to trap formation (Balan and Lechevalier, 1972).

Biological Control

Some studies have been made to determine whether it is feasible to utilize predacious fungi as biological agents to control plant-parasitic nematodes causing disease of crop plants. Addition of fungal spores alone to the soil does not increase the population of predacious fungi that occurs naturally. It is believed that agricultural soils ordinarily have a high population of indigenous nematode trapping fungi, and that as these are poor competitors, they trap nematodes under conditions adverse for their survival. However, if large quantities of green manure (fresh plant material that can be readily decomposed) or other organic matter is added to the soil along with the fungal spores, there may be a decrease in the nematode population. The green manure enhances the buildup of bacteria and free-living nematodes, which in turn encourages growth of the predacious fungi and trap formation, resulting in increased trapping of plant-parasitic nematodes (Barron, 1977).

FUNGI AS PARASITES

Parasitic fungi are those that invade a living plant or animal (the host), feeding and multiplying within it at the host's expense while not contributing to the welfare of the host. Some parasites may cause virtually no damage to the organisms on which they are feeding. The majority of parasitic fungi, however, are *pathogens* and cause an abnormal physiology in the host or a destruction of the host that develops concurrently with the parasitic relationship. This abnormal physiology culminates in *disease*, marked by detectable changes in function or morphology, *symptoms*.

Some parasitic fungi lead predominantly saprophytic lives (often in the soil) and become parasites only when they accidentally gain entrance into a potential victim, which is often weakened by injury or disease. These are the *facultative parasites*. The facultative parasites as a group are not specialized as parasites and often are so virulent as parasites that they cause the rapid death of their hosts. The early death of the host deprives the parasite of a source of food and may lead to starvation of the parasites.

In contrast to the facultative parasites, some fungi are *obligate parasites* and grow exclusively in living tissues. These fungi are highly specialized as parasites and establish a delicately balanced physiological relationship with their host. The host may show only mild symptoms and live indefinitely, although it is parasitized. In this manner, the life of the host is preserved and the parasite is ensured of a food supply for a long period of time.

Fungi as Parasites of Animals

Virtually no animal is free from the possibility of incurring an invasion by a parasitic fungus, thus serving as its host and source of nutrients. Fungus diseases of animals vary from those that cause only a slight, superficial irritation that gives the host little, if any, discomfort (such as the Laboulbeniales on insects) to those that are quickly lethal.

ARTHROPOD PARASITES

A large number of fungi are parasitic on insects, other small arthropods such as mites and spiders, and larger arthropods such as crayfish and lobsters. Insect parasitic fungi include some chytrids, almost all members of the Entomophthorales, a few yeasts, numerous members of the Ascomycotina and Deuteromycotina, and the basidiomycete *Uredinella*. Crayfish and lobsters are attacked by members of the Oomycetes and *Fusarium*. Specific examples of insect diseases will be discussed.

Entomophthora Infections. Most insects are susceptible to infection caused by one of the more than 40 species of *Entomophthora*. An *Entomophthora* disease which is familiar to many laypersons is that of houseflies caused by *E. muscae* (Fig. 13–4). Flies killed by this fungus may commonly be found attached to window panes, surrounded by a halo of sporangia forcibly discharged from sporangiophores which emerge between the abdominal segments of the flies.

Infection of the host by the parasite and subsequent development of the parasite are more or less similar for all species of *Entomophthora* (MacLeod, 1963). A sporangium comes into contact with the host; a germ tube formed from the sporangium then penetrates the integument of the insect, especially in the thinner intersegmental areas or joints of the appendages, and enters the body cavity. Inside the host, the fungus usually grows by budding and fission until the body cavity is almost completely filled with short, thick hyphal bodies containing a fatty protoplasm (Fig. 13–5). In most instances, these parasites invade the entire head, thorax, and abdomen and may even be found within the tarsi. Finally all internal structures of the host are destroyed, leaving only the fungus-filled exoskeleton. During the early stages of the infection, there may be no noticeable effects of the disease, but later the insect becomes restless. The females may become sterile, which leads to a reduction in the insect population. Just before death, usually 5 to 8 days after infection, the host may crawl upward on blades of grass, settle on the underside of branches or leaves, or seek other elevated positions. Death characteristically occurs in the afternoon between 3 and 7 P.M. Upon death, the host is attached to its final substratum by rhizoids produced by the fungus which simultaneously forms sporangiophores and sporangia. The sporangiophores usually emerge through the intersegmental areas, forming cushion-like rings surrounding portions of the insect. The sporangiophores form overnight and discharge their sporangia the next morning while dew is still present. Sporangium formation may be interrupted, with sexual reproduction following, which leads to the formation of zygospores within the body cavity (Fig. 13–5).

(a)

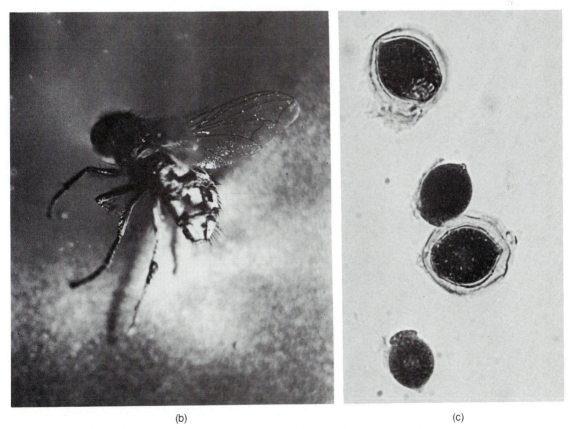

(b)   (c)

Figure 13–4 *Entomophthora muscae* on insect hosts: (a) note the cushionlike ring of sporangiophores which protrude between the segments of the insect host, × 4; (b) white mass of sporangia is distributed around the dead fly, × 5; (c) discharged sporangia. [Courtesy J. Weiser, 1969, An atlas of insect disease, Academia, Prague.]

(a)

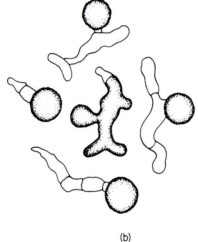

(b)

Figure 13–5 *Entomophthora sphaerosperma:* (a) hyphal bodies from a live, adult insect; (b) stages in formation of zygospores (resting spores). Both × 400. [From A. G. Dustan, 1927, J. Econ. Entomol. 20:68–75.]

Massospora Infections.   The genus *Massospora* is also in the Entomophthorales but, unlike *Entomophthora*, consists of relatively few species. The best known member of this group is *Massospora cicadina* (Fig. 13–6), a parasite on the 17-year locust (MacLeod, 1963). It is possible that infection of the locust occurs underground at some time during its long dormancy. Infection is confined to the softer tissues in the posterior portion of the locust's body and, in some instances, is confined to male insects. Unlike *Entomophthora* species, this parasite forms oval sporangia in clusters within cavities in the abdomen that are not forcibly discharged. During the later stages of infection, posterior portions of the abdomen drop away successively. The exposed segment becomes filled with sporangia that burst through the confining chitinous ring as a creamy mass. As the mass of sporangia dries, it drops away, eventually followed by the entire abdominal segment. The insects will continue to fly around when left with only a head and thorax. Dark brown resting spores (zygospores) may be formed within the abdomen after the decline of the sporangial stage.

Cordyceps Infections.   The largest single genus of insect parasites in the ascomycetes is the genus *Cordyceps* (Sphaeriales), which has about 200 species. Insects infected by *Cordyceps* are chiefly in the orders Hemiptera, Diptera, Lepidoptera, Hymenoptera, and Coleoptera. The larvae are most commonly infected. A few species attack spiders.

Species of *Cordyceps* were the first insect-attacking fungi to be recognized, owing to their conspicuous, colored clublike stroma that arises from the diseased

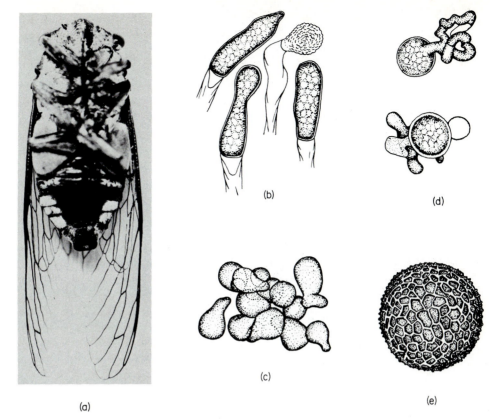

Figure 13-6 *Massospora cicadina;* (a) diseased insect with fungus localized in the posterior segments, × 8; (b) stages in formation of sporangia on sporangiophores, × 568; (c) hyphal bodies, × 268; (d) young
resting spores with hyphal bodies attached, × 268; (e) mature resting
spore, × 568. [(a) Courtesy J. Weiser, 1969, An atlas of insect disease,
Academia, Prague, (b–e) from A. T. Speare, 1921, Mycologia 13:72–82.]

insect (Fig. 13–7). These infected insects were originally described as ''vegetable
wasps,'' ''plant worms,'' or ''trees'' growing from insects (McEwen, 1963). In
China, caterpillars infected with *Cordyceps* are considered to be both a tonic and
a delicacy and are prepared in a broth. As they are expensive, only the middle
classes or well-to-do can afford this delicacy (Hoffman, 1947).

Cordyceps militaris is perhaps the best known member of this group and
will serve as an example of the life history of these fungi (McEwen, 1963). The
discharged ascospore fragments transversely into a number of smaller fragments, which may then land on the moist skin of a larva. The spore fragment
becomes rounded, and a germ tube is put forth which directly penetrates the
integument presumably by hydrolyzing the chitin. Within the larva, the rather
thick hyphae ramify between the muscles and fatty tissue throughout the body
cavity. The hyphae break down into a number of cylindrical segments, and they
pass into the blood, where they elongate up to several times their original size.
The segments divide repeatedly by transverse wall formation and by budding
to form additional single-celled bodies. These cells are dispersed throughout the
larvae until they fill the body cavity, meanwhile replacing or digesting the blood
and softer tissues. At this point, the larva becomes soft and relaxed and dies.
After death of the insect, the entire body cavity (except the alimentary canal)
becomes filled with the budded cells, and the corpse expands to its former size
and turgidity. In a few days, most of the insect tissues within the exoskeleton

Figure 13-7 Left: ascocarps of the insect parasite, *Cordyceps militaris* on pupae of its host, approximately × 0.4; right: cordycepin. [Photograph courtesy Plant Pathology Department, Cornell University.]

are replaced by fungus tissues, and the entire mass becomes hardened. The insect tissues are extremely resistant to decay due to the production of an antibiotic, *cordycepin* (Fig. 13-7), by the parasite (Madelin, 1966). The sexual stage of the fungus, a club-shaped yellow or red fleshy stroma, arises from the stromatized larva. Perithecia are formed in the swollen apical portion of the stroma.

Muscardine Diseases. "Muscardin" in French means a bonbon, and the term "muscardine" as adapted from it by French biologists is a descriptive term for a number of fungus diseases which transform the insects into a white mummy, somewhat reminiscent of a bonbon or sugared almond. The fungus responsible for causing a large number of the so-called muscardine diseases is *Beauveria bassiana*, a member of the Deuteromycotina. *B. bassiana* may cause disease in the silkworm, European corn borer, coddling moth larvae, and other insects (Fig. 13-8). This fungus is also weakly pathogenic on rodents and humans (MacLeod, 1954; Steinhaus, 1949).

Muscardine diseases caused by *B. bassiana* are similar and the well-known infection of silkworm larvae serves as an example (Beauverie, 1940; Steinhaus, 1949). In silkworms, conidia of *B. bassiana* are transmitted from one individual to another through direct contact. The conidia germinate by means of a germ tube that directly penetrates the intact integument. Although direct penetration through the integument is the preferred route in the insects studied, Gabriel (1959) showed that conidia placed in the alimentary canal by means of a micropipette will in some instances germinate and penetrate through the lining of the alimentary canal. Presumably penetration of the silkworm occurs through a combination of mechanical pressure and enzyme secretion, as in the case of the corn borer (Lefebvre, 1934). The mycelial filament advances through the silkworm larvae and digests chitin in its path, destroying the chitin and causing a molt. As the filament penetrates through the cuticle and into the body cavity of the insect, the number of blood cells increases in the vicinity of the mycelium.

Figure 13–8 *Beauveria bassiana* on dead beetle. × 2. [Courtesy J. Weiser, 1969, An atlas of insect disease, Academia, Prague.]

The hypodermal tissue is destroyed, and the fungus multiplies in the blood in the form of short filaments or hyphal bodies that are disseminated to all parts of the insect's body, eventually destroying all internal organs.

As the disease progresses, the volume of blood and number of blood cells decrease, while the parasite shows a corresponding increase. Certain changes also take place in the composition of the blood: normal acidity changes to near neutrality, certain amino acids disappear (especially after death), and oxalic acid crystals are formed. A toxin is also formed in the blood and, if injected into healthy larvae, will cause a swelling and stiffening of the body. In later stages of infection, blood circulation slows and then stops, and the consistency of the blood becomes pasty. The actual killing of the insect may be partly attributed to the increased viscosity of the blood (Madelin, 1963).

After death, the insect's body becomes progressively more hardened and reddish. Hyphae of *B. bassiana* emerge within 24 to 48 hours from the corpse to the exterior where they produce conidia under humid conditions. Meanwhile there is also the formation of a crystalline inflorescence, perhaps magnesium and ammonium oxalates.

Muscardine of silkworms may present a serious economic threat to growers of silkworms. This disease has been particularly severe in France and Italy, where infested silkworm nurseries are quarantined and workers from such a nursery are not allowed to visit healthy ones. Contaminations of healthy individuals with conidia of *B. bassiana* comes primarily from contact with diseased individuals. A silkworm is easily contaminated if conidia are present in the rearing cages or rooms, and the longevity of the conidia (up to 5 years in dry air) contributes greatly to this problem. If the disease becomes apparent during the rearing season, the infected larvae may be separated by hand or, more conveniently, all larvae are separated from food by a barrier of wire mesh. The healthy silkworms will be able to crawl through the barrier to obtain food, while the diseased ones will not be able to do so and can be culled out. The diseased larvae and litter should then be burned. Between rearing seasons, the surroundings may be disinfected with a fungicide.

Fungus Parasites as Biological Control Agents. Naturally occurring infections may be responsible for limiting insect populations. For example, the citrus mealybug is not a serious pest in Florida because it is held in check by *Entomophthora fumosa*, which is favored by the rainy, warm growing season. By contrast, the citrus mealybug is a serious pest in California, where the rainy season coincides with the cooler months, therefore not favoring growth of *E. fumosa* and allowing the insect pest to multiply in great numbers (Steinhaus, 1949). If a population of an insect pest is to be substantially reduced by a fungus in the field, there must generally be a high population density of the host, high levels of the fungal spores to serve as inoculum, and favorable climatic conditions. Intermittent rains that maintain a high relative humidity together with a water film are generally favorable. The humidity maintains the fungal inoculum in a viable condition, while the water film favors germination. Temperature requirements vary, but the optimum generally falls between 20 and 30°C (Ferron, 1978; Hagen and Van den Bosch, 1968; Mackauer and Way, 1976; Lisansky and Hall, 1983).

Considerable effort has been expended to learn how to utilize infection of insects by pathogenic microorganisms to control insect pests. An advantage of biological control is that many microorganisms are host specific and could be utilized to control specific insect pests while not injuring beneficial species. In contrast, insecticides kill both useful and harmful insects and also leave chemical residues on plant parts that will be used for food. Many research programs are being actively pursued in countries around the world (Ferron, 1978). In general, there has been some measure of success, especially with viral and bacterial parasites. Success with fungal parasites has been somewhat sporadic.

The use of a parasitic fungus to control insects may involve (1) manipulation of the environment by agricultural practices to enhance the rate of naturally occurring infection, for example, spraying crops with water to increase the humidity, or (2) the introduction of a suitable insect parasite into a population of insects. Most emphasis has been placed on the second method. Pioneer research of this type was carried out by Dustan (1927), who reared diseased insects in cages and then released them in orchards to infect healthy insect pests. Partial control of the green apple bug and the European apple sucker was obtained. In more recent studies, strains of pathogenic fungi are selected in the laboratory, grown in culture to obtain large numbers of spores, and then artificially disseminated in the field. Appropriate strains must be selected that are virulent for the insect pests and are host specific so that they not only will be restricted to the insect pests but also will not become pathogenic on vertebrates. At present, there is little evidence that any substantial danger exists from a lack of specificity for insect hosts. Once a suitable strain is isolated and cultured, it must be released in the field under appropriate conditions. This is frequently difficult, as success often depends upon density of the insect host population, humidity, or rainfall (Ferron, 1978). For example, although species of *Entomophthora* infect aphids and reduce their populations naturally, it is difficult to obtain a higher kill rate by using introduced spores because the aphids do not generally occur in dense enough populations to enhance the activity of the fungus (Hagen and Van den Bosch, 1968).

Field studies indicate that certain insects may be controlled by pathogenic fungi under experimental conditions. For example, up to 60% of a population of mosquito larvae hatching from disease-free eggs can become infected by *Coelomomyces* in nature, which suggests that the artificial spread of *Coelomomyces* by humans may be useful in mosquito control (Chapman, 1974). In the Soviet Union, suspensions of conidia of *Beauveria bassiana* have been used to control the Colorado beetle and coddling moth. Utilization of *Metarhizium anisopliae* has been effective in reducing population of the coconut palm rhinoceros beetle in Tonga and of leafhoppers and froghoppers in pastures and on sugar cane in Brazil. A preparation containing conidia of *M. anisopliae* is now available to farmers in Brazil, who are beginning to use biological control in their everyday practice. In general, more progress is required before insect diseases are used as a matter of course in agriculture to control insect pests (Ferron, 1978).

Birds and mammals, including humans and their domesticated animals, are susceptible to attack by numerous fungi that cause disease. A disease of animals is frequently termed a *mycosis* and the name for the disease is generally derived by adding *-osis* or *-mycosis* to the name of the fungal agent. For example, the disease sporotrichosis, especially common among florists, gardeners, and berry pickers, is caused by *Sporotrichum schenckii*, a fungus which occurs on plants and enters the human body at the site of an injury, perhaps caused by the penetration of a splinter. Coccidioidomycosis, a lung disease of rodents, other animals, and humans, is caused by *Coccidioides immitis*. Some diseases, such as athlete's foot or ringworm, are simply known by their common descriptive name.

Mycotic diseases of higher animals and humans are ubiquitous, and it is probably rare that an individual escapes mild infection by a fungus throughout life. Common diseases such as ringworm of the scalp often occur in epidemic proportions among schoolchildren or among farm animals. Many individuals have had mild diseases from which they have recovered spontaneously and may learn that they had such a disease only when they are screened by appropriate tests (see further discussion on page 435). Some of the diseases are especially likely to occur among people in certain occupations. For example, histoplasmosis occurs frequently among poultry farmers and explorers of caves because the causal fungus occurs in soil and in the eliminated feces of chickens, other birds, and bats and the spores can be inhaled from the air. Despite the common occurrence of fungus diseases, there are comparatively few deaths among humans that are attributed directly to these diseases. In 1973, only 530 deaths in the United States were attributed to fungal diseases (Emmons et al., 1977).

The Fungi. The fungi causing diseases of higher animals belong to different fungal groups, ranging from a few possible representatives in the Chytridiales to some members of the Agaricales. Diseases of cattle are caused by members of the Mucorales (*Mucor, Absidia,* and *Rhizopus*), which may invade the central nervous system or cause abortion. The majority of the diseases are caused by conidial stages of members of the Ascomycotina or by members of the Deuteromycotina. In the Ascomycotina, the pathogenic fungi include yeasts, Plectomycetes, and Pyrenomycetes. Members of the Basidiomycotina known to be parasitic on animals include *Ustilago maydis* (causal agent of corn smut), which was isolated from an epithelial infection of a farm laborer (Fischer, 1965), *Ustilago hypodytes,* which causes skin lesions among sugarcane workers and basket makers (Bereston, 1974), and the agarics *Schizophyllum commune* and *Coprinus cinereus* (Rippon, 1978).

Host ranges are comparatively broad in this group, and a parasitic fungus typically can infect a number of animal species. Most of the diseases that attack humans also occur on domesticated and wild animals. *Entomophthora coronata,* a member of the Entomophthorales that causes insect disease, also causes a disease in horses and humans characterized by nasal polyps, subcutaneous granulomas of the face, and infection of the sinuses. A degree of host specificity may be found in certain instances. For example, coccidioidomycosis occurs most frequently among dark-skinned males (Negro, Mexican, Filipino) between the ages of 30 and 50 years, while light-skinned females enjoy the greatest immunity to this disease.

Most fungi that may cause diseases of the higher animals are facultative parasites and these fungi may be isolated with great regularity from the soil. They are opportunists and if entrance into an animal through a wound or the respiratory tract is gained, they may become parasitic. *Coccidioides immitis, Aspergillus fumigatus,* and *Rhizopus* are common saprophytes that are widely distributed in the environment but may cause a fatal disease. Although spores from many hundreds of species of saprophytic fungi must be breathed in by humans and animals, come into contact superficially with the skin and nails, or be introduced in wounds, a very small number of species are apparently capable of causing infection. Only about 60 species of fungi may be regularly isolated from human subjects, and at least double that number of species have been isolated in rare instances (Emmons, 1960). With the exception of the dermatophytes, the soil serves as the source of inoculum for pathogenic fungi. Animal-to-animal or person-to-person contact resulting in the spread of a pathogen is rare. Obligate parasitism is rare or absent among the fungi that cause diseases of higher animals, and no fungus is known to be an obligate parasite in humans.

Fungi causing disease of animals fall into three major groups: (1) those fungi which cause systemic infections deep within the tissues (involving the vital organs or nervous system) and which are often fatal; (2) those fungi (the dermatophytes) which cause superficial and usually mild infections of the skin, nails, or subcutaneous tissues; and (3) those fungi causing infections intermediate in severity between the two extremes above.

Systemic Diseases. The fungi that cause systemic disease exist in nature as sporulating saprophytes and often gain entrance to the body through inhalation of spores (direct transmission from individual to individual rarely occurs). If spores are inhaled, the lungs are the organs which are first infected. Symptoms are similar to those of tuberculosis and include nodule or cavity formation within the lungs or inflammation of the lungs. Alternatively, the spores may be introduced into the body in a superficial wound caused by a splinter or other object. In this latter case, open pitlike ulcers are formed in superficial tissues. The parasite is disseminated to other parts of the body in the blood and lymph, and additional infections occur in other internal organs or in the nervous system. Although some infections are comparatively mild and may not be noticed, an alarming percentage are chronic or fatal.

Once the fungus has made contact with the body, it may or may not be able to invade the tissues as a parasite. Some of the fungi that cause systemic infections are especially successful as parasites. They have the ability to adapt to the higher temperatures of the host's body and to the lower oxidation-reduction potential of the tissues. The systemic fungi are also able to overcome the host's defense mechanisms by a combination of a rapid growth rate and relative insensitivity to these defense mechanisms. Unlike many pathogenic fungi, the most successful systemic fungi do not depend on a debilitated host or injuries to become established. One of the unique features that the systemic fungi share with a few other pathogenic fungi is that they are *dimorphic* (Rippon, 1980).

Dimorphic fungi may occur in two distinct morphological forms. One of these is typical of the saprophytic phase and the second is typical of the parasitic phase. While parasitic within tissues, these fungi form unicellular bodies which resemble yeasts or are thick-walled spores. If these fungi are cultured on artificial media at moderate temperatures (about 25°C), colonies consisting of mycelium and conidia are formed. At higher temperatures (about 37°C), the cell form typical of the parasitic phase is produced (Fig. 13–9). If a parasitic fungus that ordinarily would become dimorphic and be disseminated in a warm-blooded animal is injected into a cold-blooded animal such as a frog or lizard, which is then maintained at temperatures of approximately 25 to 27°C, the fungus remains in the mycelial stage and is incapable of producing a systemic infection. Unlike the dimorphic fungi that cause systemic infections, those fungi that attack superficial tissues are not dimorphic but occur within the lesion in their mycelial form. If a fungus that ordinarily causes a superficial infection can be induced in culture to form a unicellular form, the unicells may be able to cause a systemic infection if injected into an animal. Such evidence indicates that the

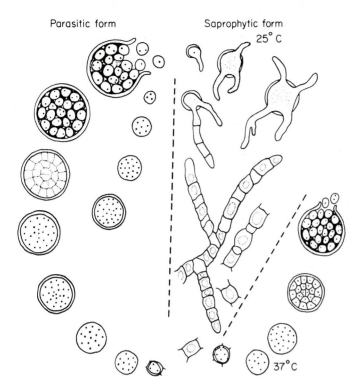

Parasitic form   Saprophytic form
25° C

37° C

Figure 13–9 Life cycle and dimorphism of *Coccidioides immitis*, showing both parasitic forms and saprophytic forms at 25 and 37°C. [From J. W. Wilson and O. A. Plunkett, 1965, The fungous diseases of man. Reprinted by permission of the Regents of the University of California.]

ability of a fungus to cause systemic infection is correlated with its ability to become dimorphic and form unicells.

An example of a systemic disease is coccidioidomycosis, which is also known as San Joaquin Valley fever and is caused by *Coccidioides immitis*. This dimorphic fungus produces mycelium and arthrospores in the soil. The arthrospores are inhaled and become spherical within a few hours or days. The enlarged arthrospore functions as a sporangium when mature and cleaves internally into sporangiospores. The sporangiospores are released when the outer wall ruptures, and after they have been disseminated within the body and have become mature, each sporangiospore may function as a sporangium and produce additional sporangiospores by cleaving internally. Although sexual stages are unknown, this fungus resembles the Zygomycetes because its mode of sporangiospore cleavage is similar to that occurring in members of the Zygomycetes (Emmons et al., 1977).

Coccidioidomycosis occurs in rodents, dogs, cattle, horses, and sheep as well as in humans. It occurs principally in the southwestern parts of the United States, where dry dusty conditions prevail and the fungal spores are readily blown about. In that region, virtually all human residents have had the disease at one time or another; the infection may have been so mild that its existence was determined only by a hypersensitivity to an antigen injected intradermally (Emmons, 1960). Mild infections may go unnoticed or produce symptoms similar to those of pneumonia or tuberculosis. In severe cases, the fungus may be disseminated throughout the body in the blood stream, which leads to the formation of lesions of the skin, subcutaneous tissues, bones, joints, internal organs, and brain. Symptoms include fever, chills, prostration, progressive emaciation, and usually death in those cases where dissemination has occurred. In domestic animals, the infection by *Coccidioides immitis* is generally not severe and causes no apparent illness. Past or present infection is often not detected unless the animal is tested with an antigen or lesions in the lung or lymph nodes are detected after slaughter (Blood and Henderson, 1960). In a sample of sheep tested in California, 52% yielded a positive skin test for coccidioidomycosis, while up to 20% of the slaughtered cattle examined in Arizona had the characteristic lesions (Maddy, 1960).

Usually, antibodies and a subsequent immunity result from an infection by a systemic fungus. This immunity helps the individual combat the original infection and prevents further infections by the same species. An immunity developed during the course of the disease may be adequate to cure the disease spontaneously without the benefit of additional treatment.

Treatment of systemic diseases includes bed rest and supplemental nutrients. These enable natural immunity to develop and cure the disease. Certain antifungal drugs may be used with varying degrees of success, and the effectiveness of a particular drug varies with the particular disease. The antibiotic amphotericin B is frequently helpful. In some instances, an infected body part may be amputated or a lesion surgically removed if the infection is localized.

Intermediate Diseases. Some parasitic fungi occur on the skin, subcutaneous tissue, lungs, or mucous membranes and may extend to a considerable depth within the tissue. Unlike the systemic diseases, the infection is not regularly distributed to other parts of the body. Infection may be acquired by inhalation of spores, by introduction of spores into a wound, or possibly by superficial contact with the fungus or by ingestion of spores. Unlike the systemic diseases, a specific immunity against these diseases is difficult to demonstrate. Examples of the intermediate diseases include candidiosis and aspergillosis.

Candidiosis (also known as "thrush") is caused by various species of *Candida*, most commonly by *Candida albicans* (Fig. 13–10). This fungus is a member of the Deuteromycotina and predominantly forms yeastlike cells, but there is some formation of mycelium. *C. albicans* is a normal saprophytic inhabitant of the intestinal tract or vagina but quickly becomes parasitic under debilitating circumstances such as malnutrition, unsanitary conditions, the presence of other diseases, or prolonged antibiotic therapy. Candidiosis may affect the skin around the fingernails or between the fingers of people who keep their hands wet for prolonged periods of time (e.g., dishwashers) or may affect the lungs. However, candidiosis is primarily a disease of the mucous membranes, on which soft, gray-white lesions will form (Fig. 13–10). In humans, candidiosis may occur as a vaginal infection, often appearing during pregnancy, when the glycogen content of the mucous membranes in the vagina increases, favoring the growth of the fungus. During birth, the infant is infected by fungal cells from the vagina and often develops candidiosis in the mouth, where lesions occur on gums, tonsils, tongue, or mucous membranes. In some cases, gasteroenteritis will develop in humans and may occur when infants swallow some of the *C. albicans* cells from their mouth. The fungus may be disseminated in the body in a small percentage of the cases, and may be lethal, especially in patients who are suffering from other diseases. Candidiosis also occurs in other mammals and birds (Small, 1971). It is common in caged pet birds such as canaries and parakeets when cages are not maintained in a sanitary condition and debilitating factors such as vitamin B deficiencies are present. The disease is most commonly localized in the crop of the bird. Baby pigs may develop candidiosis if they are reared on an artificial diet (Blood and Henderson, 1960). The piglets vomit frequently, become emaciated by 2 weeks of age, and have a more or less continuous white lining which covers the back of the tongue and esophagus. Candidiosis has also been observed in cattle in a feedlot, where they had diarrhea and symptoms similar to pneumonia, and resulted in the death of some of the cattle.

Treatment of candidiosis varies according to the severity and location of the infection. Systemic infection may be treated by injections of the antibiotic amphotericin B. For less severe localized infections, a second antibiotic, nystatin,

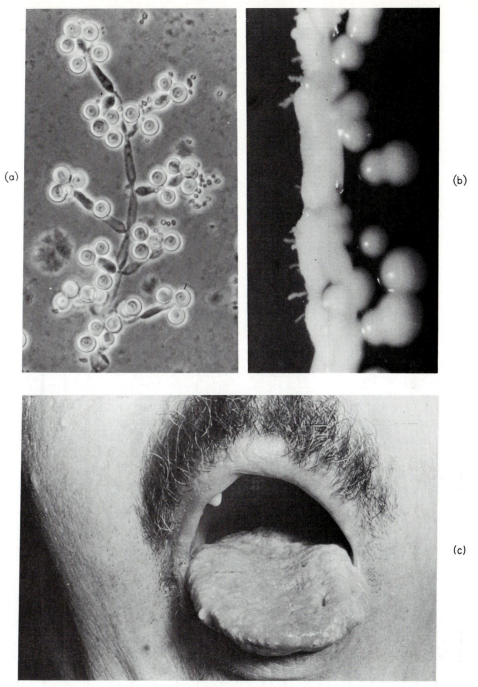

Figure 13–10 *Candida albicans* and the disease it causes, candidiosis: (a) conidia and conidiophores of *C. albicans*, × 600; (b) yeastlike colonies of *C. albicans*, × 5; (c) candidiosis of the tongue. [Courtesy P. M. D. Martin.]

may be administered orally or incorporated into topical creams or ointments for application to the infected area.

Aspergillosis, caused by *Aspergillus fumigatus*, is a disease that is common among newly hatched fowl, often occurring in epidemic proportions. The lungs of infected chicks become covered by a felt of mycelium and spores to the extent that breathing is impossible. This fungus grows abundantly on decomposing vegetation, and compost piles often serve as an important source of inoculum. Although primarily a parasite of fowl, *A. fumigatus* may cause abortion in cattle and sheep, corneal ulcer of the eye if introduced with an injury, and a disease of the honeybee (Emmons, 1960). Normal humans are immune to infection by *A. fumigatus;* but if infection does occur, symptoms are often necrosis and cavity formation in the lung.

Superficial Diseases.   Superficial diseases are caused by fungi that attack the skin or its appendages (nails, feathers, and hair). These fungi are known as *dermatophytes* and are members of the Deuteromycotina or conidial stages of the Ascomycotina. They are almost entirely confined to superficial tissues, surviving in deeper tissues only in rare instances. Diseases of this type include various ringworms, athlete's foot, and epidermophytosis (Fig. 13–11).

Dermatophytic diseases are found in all parts of the world. The relative importance of any given disease varies from region to region, influenced by the local climatic conditions, nutrition, hygiene, or local clothing habits that tend to make perspiration accumulate in a particular part of the body. These fungi are widely distributed as soil saprophytes that may cause a disease upon contact of a potential host with the soil; as infective agents on dogs, cats, horses, cattle, sheep, and monkeys that will cause infection of humans if given the opportunity; or as infective agents that occur preferentially on humans. They are passed from person to person upon contact. The last type may be responsible for epidemics among humans. Spores and mycelia are abundant in the environment in skin, hairs, and feathers shed from diseased individuals. Repeated contact with dermatophytes is inevitable, and infections may be severe under certain conditions. For example, during the Vietnam War, some American soldiers developed severe dermatophytic infections on the body and in the groin area. These infections were favored by the hot, damp conditions as these particular soldiers frequently walked in paddies and swamps and were immersed in water to the waist or higher. Most people have at least one dermatophyte infection

(a)   (b)

Figure 13–11 Dermatophytic disease, epidermophytosis: (a) symptoms on the hand; (b) the fungus, *Epidermophyton,* within the tissue of the hand prepared by biopsy, × 480. [Courtesy P. M. D. Martin.]

during their lifetime and sometimes repeated infections. For example, athlete's foot is of extremely common occurrence, sometimes clinically diagnosed in more than 50% of the individuals of a given sample. Despite the ubiquity of the parasites and the occasional occurrence of an epidemic, the actual number of dermatophytic infections remains comparatively low, considering the abundance of inoculum of many types.

The low incidence of infection by contacted dermatophytes may be attributed in part to the fact that a high proportion of humans are apparently immune to attack by dermatophytes. In other instances, infection never takes place because the inoculum is washed away with soap and water and favorable circumstances for infection are not present. Circumstances favoring infection by dermatophytes include (1) frequent wetting of a body part, such as the hands of a dishwasher; (2) natural occurrence of a moist, oily skin, which is much less resistant than a dry skin, which continually sheds its outer cells; (3) close-clipping of hair, as commonly done at the nape of the neck in men so that the highly susceptible hair follicles which otherwise would be protected by a covering of resistant, inert hair are exposed; (4) wearing of tightly fitting, poorly ventilated shoes which do not allow perspiration to evaporate; and (5) obesity, which leads to folds and crevices in the flesh where moisture accumulates.

Any dermatophyte may infect the skin, hair, or nails although usually a preference is shown for one of these. Symptomatology is primarily a function of the body part infected and, to a lesser extent, of the fungus causing the disease. For this reason, dermatophytoses are usually named for the body part infected and not for the fungal agent. Typical symptoms may include inflammation or redness of the infected part, scaling, brittleness and fissures of the nails, and loss of hair in the part infected.

As an example of a dermatophytic disease, we briefly examine ringworm. Ringworm is a dermatophytic disease that is important in humans as well as in dogs, cats, horses, and cattle. Among the animals, this disease may impair the growth of calves or damage the coat of fur-bearing animals or the leather produced from infected animals. The disease may be transmitted from one animal species to another or to humans by direct contact or by contact with contaminated bedding, harnesses, grooming aids, or feeding utensils. In rural areas, it has been estimated that 80% of the ringworm on humans results from infection by animals (Blood and Henderson, 1960). Ringworm is caused by species of *Trichophyton* and *Microsporum* that invade the keratinized layers of the skin, the hair fibers, or both. Autolysis of the hair occurs and it breaks off. There is also the production of a characteristic, circular dry, scabby lesion. The fungus invades the tissue and causes exudation to take place, and the exudate combines with the shedding epithelial material to produce the scab. Once a scab is established, the fungus cannot survive underneath it near the center because anaerobic conditions are established, and therefore the fungus dies at the center but survives at the periphery. The fungus then continues to grow centrifugally, extending the circular lesion that gives this disease its name. The infection is favored by a slightly alkaline pH. In humans, children are more readily infected because the pH of their skin is about 6.5, but the pH drops to about 4.0 at puberty when fatty acids are secreted. Both the lower pH and the fungistatic action of the fatty acids contribute to a lower incidence of ringworm in adults. Spontaneous cure may occur in both animals and humans. Ringworm may be treated by removing the scabby lesions by scraping or brushing and then applying an antifungal agent. Systemic treatment with griseofulvin is useful in both veterinary and human medicine.

The majority of the dermatophytic diseases may be effectively treated by the oral administration of griseofulvin (Fig. 13–12), an antifungal antibiotic prepared from *Penicillium griseofulvum.* Griseofulvin is incorporated into the epidermal cells in high enough concentrations to inhibit the growth of the dermato-

Figure 13-12 Griseofulvin.

phyte, although it does not kill the fungus. Topical application of fungicidal materials is often ineffective because the medication cannot penetrate through the infected tissue. However, localized infections which are not extensive may be treated by the application of ointments or solutions containing antifungal agents such as selenium sulfide or sodium thiosulfate.

Fungi as Parasites of Plants

Diseases of plants may be caused by a number of organisms, including fungi, bacteria, algae, nematodes, viruses, and mites. The number of plant parasitic fungi is greater than the number of all other parasites combined.

All plant species are susceptible to attack by fungi, and examples of plant diseases may be found among the algae, which frequently suffer from chytrid infections in epidemic proportions; among mosses infected by the basidiomycete *Eocronartium muscicola*; among ferns, which are parasitized by a number of rusts; and among the higher vascular plants, which may have diseases such as rusts, smuts, powdery mildews, and others. The most ancient examples of fungi are plant-inhabiting fungi fossilized together with their vascular plant hosts (Fig. 13-13). It is undetermined whether these fungi were parasites or symbionts.

The most familiar fungi are those that have caused disease of food crops and have caused hunger or monetary loss since the dawn of agriculture. There are numerous passages in the Bible referring to "blasts," "mildews," and "blights" which often led to famine. As an example (I Kings 8:36, 37, 39),

> Then hear Thou in heaven, and forgive the sin of Thy servants. . . . If there be in the land famine, if there be pestilence, blasting, mildew, locust . . . or if there be caterpillar; if their enemy besiege them in the land of their cities; whatsoever plague, whatsoever sickness there be. . . then hear Thou in heaven Thy dwelling place, and forgive. . . .

From about 700 B.C. into the beginning of the Christian era, the Romans held annual rituals to placate Robigus, the god whom they believed to be re-

Epidermis

Cortex

Figure 13-13 Mycelium of a fossil fungus within the cortex and epidermis of roots of a plant from the Carboniferous period. [From T. G. B. Osborn, 1909, Ann. Botany 23:603-661.]

Chap. 13 / Fungi as Predators and Parasites

sponsible for causing plant rust diseases when he was displeased. Starvation due to plant diseases may be found in even more recent times. From 1845 to 1860, 1 million people died in Ireland as a result of illness and starvation after their potato crops were ravaged by late blight of potatoes (caused by *Phytophthora infestans*). An additional $1\frac{1}{2}$ million people (about one-fourth of the population) emigrated to other countries.

At the present time, plant disease epidemics may still wipe out an entire crop if conditions favorable for disease develop. The yield of cereal crops (which supply 80% of human food) may be reduced by 80% during a single epidemic of the stem rust disease. In an advanced country, where agriculture is sufficiently diversified and there is a sound economy, such an epidemic may not result in starvation, as it would have in previous centuries, but even lesser epidemics would spell starvation and disaster for an underdeveloped country.

Although sudden loss of an entire crop from epidemics may pose a serious threat, the majority of losses result from smaller outbreaks of disease which often occur with greater regularity. These losses are due to the reduction in crop yield and also to the expense of utilizing control measures as a regular agricultural practice. In the United States the average annual loss has been estimated to be in the billions of dollars for all types of plant disease, while the average loss in yield from each crop as a result of disease is 10%.

Since 1974, a combination of factors has reduced the world's grain reserves to extremely low levels. Factors contributing to this shortage are the increasing populations in developing countries; the rising consumption of beef, which results in an increased demand for feed grains; shortages and high costs of pesticides, fertilizers, and fossil fuels; adverse weather conditions; and finally, the cumulative effects of a number of plant disease and pest problems (Apple, 1977). The unrelenting pressures on the world's food supplies mean that productivity from crop plants must be maximized and plant diseases controlled.

FUNGI THAT CAUSE DISEASE

Fungi that cause plant disease may be found in all major groups of fungi, ranging from the morphologically simple *Plasmodiophora brassicae*, which causes clubroot of cabbage, to the complex members of the Agaricaceae such as *Armillaria mellea* (the ''honey fungus''), which causes root rots of standing trees. The groups with the largest number of plant pathogens are the Deuteromycotina and Ascomycotina. Such a diverse group of fungi has little in common other than the ability to parasitize living plants. Plant parasitic fungi range from those that are facultative parasites and lead predominantly saprophytic lives in the soil to those that are obligate parasites of plants.

Pathogenicity. The plant parasites vary greatly in their *pathogenicity*, the ability to cause disease in different host plants. Some fungi, especially the facultative parasites, have comparatively wide host ranges. An example is *Verticillium albo-atrum* which has been isolated from more than 70 host genera. Many parasites will infect several host species that are closely related; we may cite *Venturia inaequalis* (causal agent of apple scab) which attacks species in the host genus *Malus*. Some parasites such as the Erysiphales are so specialized that they are capable of attacking only one host species. Host specificity is carried to its ultimate by some races of plant parasites which can attack only certain genetic races or lines of a single host species, as the rusts do. A high degree of host specificity is typical among the obligate parasites.

It is known for many host-parasite combinations that both resistance of the host and the alternative state, susceptibility, and the pathogenicity of the fungus and its alternative state, nonpathogenicity, are genetically controlled. One of the better understood relationships is the *gene-for-gene* mechanism which occurs between the flax rust *Melampsora lini* and its host. For each gene controlling

Fungus race no.	Fungus race genotype			Reaction[a] of plants possessing genes							
				llmmnn	L	M^a	N^1	LM^a	LN^1	M^aN^1	LM^aN^1
108	$a_L a_L$	$A_M{}^a$	$A_N{}^1$	S	S	R	R	R	R	R	R
123	A_L	$a_M{}^a a_M{}^a$	$A_N{}^1$	S	R	S	R	R	R	R	R
52	A_L	$A_M{}^a$	$a_N{}^1 a_N{}^1$	S	R	R	S	R	R	R	R
156	$a_L a_L$	$A_M{}^a$	$a_N{}^1 a_N{}^1$	S	S	R	S	R	S	R	R
192	A_L	$a_M{}^a a_M{}^a$	$a_N{}^1 a_N{}^1$	S	R	S	S	R	R	S	R
154	$a_L a_L$	$a_M{}^a a_M{}^a$	$A_N{}^1$	S	S	S	R	S	R	R	R

[a]S = susceptible; R = resistant

Figure 13–14 Host reaction to different combinations of three sets of host–parasite complementary genes in the gene-for-gene mechanism of *Melampsora lini* on flax. [From H. H. Flor, 1955, Phytopathology 45:680–685.]

pathogenicity in *M. lini,* there is a corresponding gene in the flax controlling resistance. Resistance of a specific host race to a specific pathogen race occurs only where the complementary genes of both the host and parasite are dominant (Fig. 13-14). A similar gene-for-gene mechanism exists between several plant hosts and their pathogens, including several smut, rust, and powdery mildew fungi (Flor, 1971), as well as the apple scab fungus *Venturia inaequalis* (Keitt et al., 1959).

New physiologic varieties, races, or biotypes of pathogenic fungi may arise through mutation or hybridization of sexually reproducing fungi (as, for example, in the rusts, Chapter 5), and new genetic lines of nonsexually reproducing pathogens may arise through mutation, heterokaryosis, and the parasexual cycle, previously discussed in Chapter 7. Race formation provides for a flexibility of genotypes, and the resultant phenotype may be endowed with greater pathogenicity, allowing the establishment of a fungus on a new host in nature.

THE PLANT DISEASES

Penetration. Probably the first step in establishing the disease relationship is making the required physical contact between the pathogen and host. Plants can sometimes attract plant pathogenic fungi by producing chemicals that induce positive chemotaxis of fungal zoospores toward the root, or positive tropic growth of mycelial fungi.

Parasitic fungi may enter the plant through (1) wounds, (2) natural openings, or (3) direct penetration. A fungus may be capable of utilizing only one means of entrance or any combination. Wounds are created when the plant is damaged, often during cultivation, by insects or animals, or by growth when roots penetrate the cortex. Natural openings such as lenticels or stomates provide a means of entrance for some fungi (Fig. 13–15). A third major means of entrance is by direct penetration of the intact plant epidermis. Fungi that penetrate the epidermis typically form an enlarged, flattened *appressorium* that adheres firmly to the cell wall. Penetration is effected by a slender outgrowth from this, the sytletlike *infection peg,* which forces its way into the cell mechanically, probably aided by partial enzymatic digestion of the cuticle and cell wall. The fungi that enter through the stomates or directly through the epidermis are sensitive to contact stimuli from the plant's surface as well as to chemical stimuli. They respond with a number of directional growth responses that allow them to enter the plant. These are directional response of germ tubes, adherence and directional growth of the hyphae, appressorium formation, and directional emergence of infection pegs (Wynn, 1981).

Generally, younger plants are more susceptible to invasion through natural

Figure 13–15 Mycelium of *Thielaviopsis basicola* entering a leaf through a stomate. [Courtesy Plant Pathology Department, Cornell University.]

openings and direct penetration than older plants, which may have developed additional cutin, cork, or lignin, all barriers to the parasite, or a toughened epidermal cell wall. A plant may be resistant to a parasite if the fungus cannot penetrate the plant because of a structural barrier in the plant; for example, *Cercospora beticola* enters sugar beet leaves only through open stomata. Young leaves in which stomatal movement is not active escape infection; so do old leaves which are not invaded because stomatal movement is slight. Mature leaves with actively functioning stomata are seriously infected (Pool and McKay, 1916).

Establishment within the Host. The parasite may or may not be able to live within the host after having successfully penetrated it. A prime requisite is that the fungus be able to utilize the immediately available food supply, which requires the presence of appropriate enzymes. These may include a wide variety of hydrolytic enzymes capable of cleaving pectin, cellulose, lipid, and protein into their monomers (a root-rotting fungus that enters the leaves of a susceptible plant by accident may not survive because it cannot utilize the nutrients immediately available in the leaves). A second requisite is that the fungus be able to survive and penetrate normally through the tissues of the plant in order to maintain a source of food (e.g., the host must not be resistant). The fungus may not be able to do this if (1) the host protoplasm provides an unfavorable environment, (2) a hypersensitive reaction by the host occurs, or (3) mechanical barriers to the growth of the fungus are formed by the host.

The protoplasmic factors that make a cell incompatible for fungal growth constitute the primary line of defense for the plant, and can vary from one situation to another. In many cases, the plant can be incompatible because of the variety of potentially toxic compounds that it normally manufactures. These often include phenolic compounds, tannin, catechol, and chorogenic acid. An example of this may be found in colored onions resistant to *Colletotrichum circinans* (the causal agent of onion smudge), owing to the presence of protocatechuic acid, which causes bursting of the spores or young hyphae of the pathogen. This compound is lacking in nonpigmented susceptible varieties (Walker et al., 1929). In other instances, the host will produce toxic metabolites as a result of wounding, toxic chemicals, or invasion by a pathogen. These toxic compounds are known as *phytoalexins,* and each host plant produces specific compounds as a result of injury which differ from the phytoalexins of other host plants. Once induced to form, the phytoalexins can help curtail further invasion by the pathogen causing them to form, or by other potential pathogens at a later time. Still other instances of protoplasmic incompatibility may result from unfavorable pH, osmotic concentration, or nutrient inavailability. An example of this last situation is found in *Venturia inaequalis,* which has specific nutritional requirements that must be met by the host. A resistant host may lack biotin, and a nonpathogenic fungus strain that has a biotin requirement may become pathogenic if an external source of biotin is available (Keitt et al., 1959).

Hypersensitive plants are those that are so sensitive to infection by the parasite that the invaded cells die quickly, isolating the parasite in a few killed cells where it can no longer live. A rapid browning and collapse of the host cells accompanies this reaction. Although a few cells are killed, the plant remains healthy. A hypersensitive host may produce large quantities of phytoalexins in the cells that die, and these exert some control against the living parasite when they diffuse into the adjacent living host cells. The hypersensitive reaction typically occurs for incompatible pathogen and host combinations controlled by the gene-for-gene mechanism and is likely to occur when the first haustorium penetrates the plant cell or when several haustoria are formed (Bushnell, 1972).

Mechanical barriers simply are cellular components that make it impossible for the pathogen to penetrate further into the host tissue. Many diseased plants have localized hemispherical thickenings that often appear to be part of the walls but differ in their composition and may be callose deposits. Such thickenings are formed by the host cell at the site of penetration by the fungus and may actually encase the fungus in a sheath. Although it is difficult to prove, it has been suggested that these sheaths may function as a mechanical barrier (Aist, 1976). Other mechanical barriers include (1) the formation of wound cork or a gum deposit in response to infection, (2) the natural occurrence of cell walls that are difficult to penetrate (often owing to secondary thickenings), and (3) the natural occurrence of middle lamellae that are often unusually resistant to enzymatic degradation.

Infection. Assuming that the parasite is able to establish itself within the host, we may consider the means by which it develops within the tissues and the means by which it causes damage to the host.

One group of parasites (those causing soft rots, a watery disintegration of tissues) kills tissues in advance of their spread by secreting pectic enzymes that destroy the middle lamellae and the pectins of the cell walls, causing the cells to separate and hence the death of the tissue. Death is associated with rapid permeability changes resulting from plasmalemma damage. Although the exact mechanism of plasmalemma damage is not known, it may be related to the swelling due to osmotic pressure exerted in the absence of the restraining walls and permitting electrolyte leakage to occur (Mount, 1978). It is possible that nonenzymatic toxins are also involved in killing the cells. The fungi then live saprophytically on the dead cell material. Rotting of sweet potatoes by *Rhizopus stolonifer* occurs by this mechanism.

A second major group of pathogens consists of those that feed directly from the living plant cells, either with or without eventually killing them. They must either penetrate the cells or live in close conjunction with them. These fungi may be (1) intercellular with haustoria (Fig. 13–16), (2) intercellular with no means of penetrating the cells, (3) intracellular, or (4) any intergradation of these. Pathological damage to the host may be a disintegration of the tissue or interference with normal growth, reproduction, or physiological processes (as water transport, or respiration).

Tissue Disintegration. Tissue and cell disintegration may occur by the destruction of the middle lamellae by pectic enzymes or of the cell walls by cellulolytic enzymes or both. This is equivalent to that type of tissue disintegration occurring in advance of the fungus, as found in the soft rots (above). The extent of wall breakdown or softening may vary, and the effect, unlike that found in

Figure 13–16 Haustorium of the rust fungus, *Cronartium quercuum* f. sp. *fusiforme*, penetrating a cell of its plant host, the northern red oak. [From S. R. Khan and J. W. Kimbrough, 1982, Mycologia 74:809–819, The New York Botanical Garden.]

the soft rots, may simply result in a weakening of the tissue rather than its maceration.

A second type of cell damage is that in which the protoplast is attacked directly. Although comparatively little may be known about the manner in which the protoplast is attacked for a particular disease, it is possible that the primary damage may be (1) a destruction or alteration of the semipermeable membrane, with a subsequent loss of water and metabolites from the cell and an unrestricted movement of molecules into the cell; or (2) a direct action of a proteolytic enzyme or nonenzymatic toxin on the protoplast, disturbing or inhibiting normal metabolic processes. Toxins are produced by a wide variety of pathogenic fungi, and the damaging effects may be experimentally exerted by toxins in the absence of the fungi which produced them. Many toxins are known that affect the plasmalemma; for example, the toxin *victorin* produced by *Helminthosporium victoriae* causes the host cells to become leaky and lose large quantities of electrolytes and other materials, while uptake and accumulation of mineral salts and other materials are inhibited. Eventually, the oat cells infected by *H. victoriae* undergo false plasmolysis and die. Although many toxins from other fungi have similar effects, unique effects such as the increase in the uptake of nitrate may be induced by other toxins (Wheeler, 1968). The fungus benefits from these morbid activities because it is able to use the damaged cell contents for growth.

Tissue disintegration impairs the normal physiological processes of the plant part. For example, a leaf may no longer be able to produce food or a root may no longer be able to absorb water. Also it results in the death of a plant part, such as areas of the leaves or stems, otherwise termed a *necrosis*. Necrotic symptoms include *dry rots*, shrinkage and collapse of the tissue with no release of fluids (found in dry rot of potato caused by *Fusarium coeruleum*); *canker*, a wound or localized lesion that is often sunk beneath the surface of the stem and is surrounded by healthy tissue (found in chestnut blight caused by *Cryphonectria parasitica*); *damping-off*, in which there is a watery disintegration of the cortical tissues of young seedlings at the point of emergence from soil (found in most species of seed plants and caused by various fungi, including several species of *Pythium*); *root rots*, the rotting of cortical and phloem tissue of roots and basal

Fungi as Parasites

Figure 13–17 Effects of *Fusarium eumartii* on leaves, stems, and tubers of potato. Symptoms include an interveinal yellowing and shriveling of the leaflets, vascular discoloration of the stems and tubers, and a brown flecking and necrosis of the stem pith. [Courtesy Plant Pathology Department, Cornell University.]

stem portions (caused by various species of *Fusarium* and *Rhizoctonia* (Fig. 13–17); *leaf spots,* localized and limited lesions of collapsed cells on the leaf (found in apple scab caused by *Venturia inaequalis*) (Fig. 13–18); and *blight,* a generalized and rapid browning of the tissue that results in death (found in raspberry cane blight caused by *Leptosphaeria coniothyrium)* (Fig. 13–19).

Interference with Normal Growth. The parasite may not cause an outright destruction of the cells and tissues of the host but rather qualitative or quantitative changes from the normal development pattern. Development in plants is controlled by auxins and by inhibitory substances or systems that control the synthesis and response of the cells to the growth-promoting auxins. Some metabolites formed by pathogenic fungi are similar in their action to those regulatory substances occurring naturally in plants. Other metabolites are capable of eliciting unusual morphogenic responses from the plant, unlike those given by the auxins and regulatory mechanisms ordinarily present. A third possible mechanism is that the pathogenic fungus may not release a morphogenic metabolite but may interfere with the regulatory system of the host that would ordinarily limit or control development.

A symptom developing from an interference with normal growth is typically a *hypertrophy,* an overdevelopment of a plant part. Examples include the curling of the leaf due to excess tissue formation and excess anthocyanin development in peaches infected by the peach leaf curl pathogen, *Taphrina deformans;*

Figure 13–18 Spots on leaves and fruit caused by the apple scab pathogen, *Venturia inaequalis*. [Courtesy Plant Pathology Department, Cornell Univeristy.]

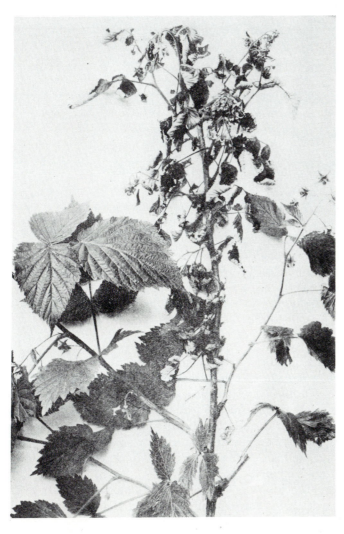

Figure 13–19 Raspberry cane blight caused by *Leptosphaeria coniothyrium*. Note the widespread death of the plant parts. [Courtesy U.S. Department of Agriculture.]

adventitious root formation in tomato plants with the *Fusarium* disease; and galls (enlarged tumor-like masses of tissue) such as those formed in cabbage by the clubroot organism, *Plasmodiophora brassicae* or by the corn smut organism, *Ustilago maydis* (Fig. 13–20). Generally, the abnormal development in itself is not harmful to the plant, but it invites malfunction of other physiological processes, such as water movement or reproduction, and, especially, diverts and uses food which would otherwise be available to support normal growth.

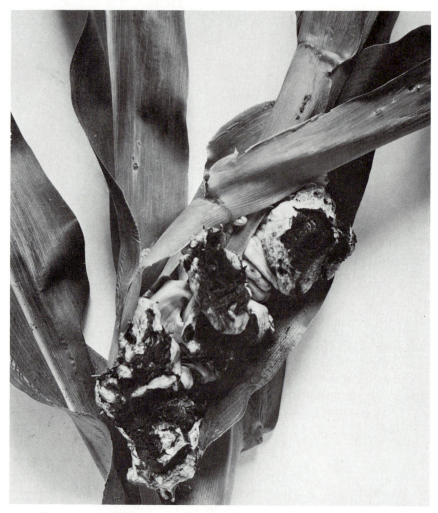

Figure 13–20 Galls on ear of corn resulting from the abnormal enlargement of the kernels in the corn smut disease caused by *Ustilago maydis*. The black teliospores are visible through the ruptured host tissue. [Courtesy U.S. Department of Agriculture.]

One of the best studied examples of growth interference by a pathogen is that of the bakanae or foolish seedling disease of rice caused by the ascomycete *Gibberella fujikuroi* (= *Fusarium moniliforme*). In rice, the leaves yellow and curl inward and often growth of the affected seedlings is arrested. The yield is often reduced. In some instances, the affected plants grow more rapidly than healthy ones and are conspicuous because of their greater size, etiolated appearance, and early flowering. The biologically active compounds produced by the pathogen were found to be gibberellic acids (Fig. 13–21). When introduced into test plants, these gibberellic acids were similar in their biological activity to growth-regulating compounds that occur naturally in higher plants (Braun, 1959).

Figure 13–21 Gibberellic acid.

Interference with Physiological Processes. Many plant parasitic fungi interfere with the normal processes of the plant, such as nutrition, uptake and movement of water and minerals through the plant, and respiration. Tissue may or may not be disintegrated in these cases. Disturbance in one physiological process may alter the course of another physiological process, so that ultimately these disturbances may have far-reaching effects in the plant.

A parasite may interfere with the nutritional processes of the plant (1) by impeding photosynthesis, (2) by preventing or impeding translocation of photosynthates to nongreen portions of the plant (perhaps by damage to the phloem), or (3) by causing a redistribution of nutrients or by assimilating them, thus preventing their normal utilization by the plant. In reference to the last point, those nutrients most often utilized by the parasite are carbohydrates and are probably intermediate compounds in photosynthesis or glycolysis. These include glyceraldehyde, phosphoglyceric acid, dihydroxyacetone, oxaloacetic acid, or ribulose diphosphate. Symptoms resulting from an interference with nutrition often include an underdevelopment or dwarfing of the plant parts involved, resulting ultimately in a reduced yield as well as premature ripening or senescence. Others lead to a degeneration of the plant; for example, both rust and powdery mildew diseases impair photosynthesis and the normal distribution of photosynthates. In these diseases, nutrient "sinks," areas where nutrients accumulate in unusually high concentrations at the expense of surrounding tissue, first develop in the vicinity of the haustoria of the parasite. The cells in the vicinity of the haustoria may sometimes enlarge and their life span is temporarily prolonged. Eventually, there is a reduction of chlorophyll in all parts of the leaf and all parts become yellowed (Bushnell, 1972; Huber, 1978; Kosuge, 1978).

Water and mineral absorption may be impaired by a number of conditions, including root damage or disease, a generalized disintegration of tissue through which water must pass, and gum formation in the xylem region. Some toxins produced by pathogenic fungi cause the stomates to remain permanently open, resulting in loss of large quantities of water through excessive transpiration. Pathogens such as *Venturia inaequalis* and many of the rust fungi develop in a subcuticular position, and the localized rupturing of the impermeable cuticle results in water loss by transpiration at the site of the rupture. Losses in water are generally first expressed as a turgor loss of the cells most directly affected, followed by a wilting or drooping of leaves and stem tips, and finally by a reduction in growth. Many plant diseases indirectly cause a disruption of the water flow, but the wilt diseases, such as *Fusarium* wilt of tomato (caused by *Fusarium lycopersici*) and Dutch elm disease (caused by *Ceratocystis ulmi*), are primarily diseases of the vascular system. The wilt disease fungi cause a local necrosis in the xylem. Pectic enzymes produced by the parasite act on the middle lamellae that are exposed in the pits, releasing pectic acid and other products of hydrolysis, which are then converted into calcium gels and upon further hydrolysis into gums. The fungi may also produce mucilaginous polysaccharides that form a viscous mass. Although the exact fate of these mucilaginous polysaccharides is not known, it has been suggested that they become dislodged from the hyphae and occlude the vessels or form a suspension or solution in the water in the xylem, increasing its viscosity and resistance to flow. The water flow may also be reduced to some extent by the hyphae if large quantities are present in a heavily infected cell, but the hyphae themselves are usually not considered to be a factor in the decreased water flow. *Tyloses* may form as a result of the infection: these are protoplasmic, balloonlike protrusions extending through the pits from adjacent parenchyma cells into the vessels. Tyloses may become so large and numerous that they become packed to form a tissuelike mass completely blocking the vessels (Talboys, 1978). In addition to the actual blockages in water flow, flow of minerals within the plant is severely disrupted. Movement of radioactive phosphorus can be reduced by 96 to 98% in the xylem of tomato plants afflicted by *Fusarium* wilt (Huber, 1978). Toxic metabolites such as ethylene and phenol are also released. The phenols become intermixed with the gums, and upon oxidation the phenols become red-brown, causing the characteristic discoloration of diseased tissues. The ethylene is carried upward to the leaves, where it causes them to droop. As the disease progresses, the leaves become yellow, die, and then drop off. The actual severity of the wilt diseases may vary greatly. If only a few cells or a small section of the vascular tissue is infected, the effects may be minimal and involve only those cells to which water transport is decreased. In contrast, an extensive infection resulting in massive blockage will result in the death of the plant (Talboys, 1978).

The respiratory rate of diseased tissues generally increases, possibly depleting the plant of available energy reserves (Fig. 13–22). A rise in temperature usually accompanies the increase in respiration. The increase in respiration is apparently due to loss of control at one or more points in the metabolic pathway (perhaps the uncoupling of oxidative phosphorylation) (Kosuge, 1978). Many ATP-utilizing reactions (primarily synthetic, defense reactions) of the host tissue are increased in response to infections. Such reactions include cork formation, increased protoplasmic streaming, hypertrophic cell growth, increased synthesis of compounds, and accumulation and mobilization of compounds to the site of infection. It has been suggested that the respiratory increase is a prerequisite for the initiation and continuation of the resistant reactions of the plants (Uritani and Akazawa, 1959).

ENVIRONMENT

Plants lack the complex regulatory systems found in higher animals, and the growth of the plant is more closely related to the prevailing environmental conditions than is the growth of an animal. The extent of plant disease, unlike

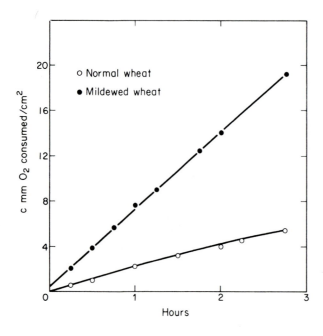

Figure 13–22 Increased respiration in tissues of wheat infected by the mildew pathogen, *Erysiphe graminis*, over that in normal tissues. Oxygen uptake is used as an index for the rate of respiration. [From P. J. Allen and D. R. Goddard, 1938, Am J. Botany 25:613–621.]

Chap. 13 / Fungi as Predators and Parasites

animal disease, is therefore more directly controlled by external environmental conditions. As a plant disease is the product of interaction between parasite and host, any environmental factor is potentially capable of affecting the parasite, the host, and the manner in which they interact. Both the parasite and host have individual reactions to each environmental factor; for example, a parasitic fungus has a minimum, maximum, and optimum temperature for growth (see Chapter 8) and this range may be quite different from that for the host plant. A fungus has an ideal temperature for infection, while the plant has a temperature range in which it is best able to withstand infection. Therefore, a plant disease develops at its maximum in that narrow range where the environmental conditions are favorable to the fungus but unfavorable to the host plant.

Environmental conditions exerting a profound influence on plant diseases include pH of the soil, available nutrients in the soil, soil moisture, aeration of soil, rainfall, atmospheric humidity, temperature range and fluctuation, and light intensity. Each parasitic fungus–host combination reacts uniquely to environmental conditions, and those conditions favoring one disease would be inhibitory to another disease. The relative importance of the various environmental factors varies from disease to disease, but temperature and moisture variations are those which are most often the critical factors.

CONTROL

In the control of plant disease it is of comparatively little value to attempt to save an individual plant (unless it is a valuable ornamental plant such as an elm tree). Instead, attention is directed toward preventing the spread or additional outbreaks of the disease. Control of plant disease may take many forms.

One type of control measure involves the exclusion of a parasitic fungus from an area where the fungus does not occur. Exclusion takes the form of quarantine, and plants that might bear the parasite are not transferred into a region or country which is free of that pest. A second means of control is the eradication of a susceptible host species from an area, thus indirectly eliminating the parasite. This type of control is useful for heteroecious rust fungi which require two hosts to complete their life cycle. The host which is not of economic importance is eliminated; for example, eradication of the barberry may be used to control stem rust of wheat.

Frequently, disease control is achieved by cultural manipulation or protection of the plant. These may include controlling the environment, sanitation, crop rotation, and application of fungicides to growing plants. As indicated above, environmental factors may influence the course of disease. Agriculturalists may control specific plant diseases by modifying certain aspects of the environment so that they are favorable for plant growth but are unfavorable for the development of diseases. For example, the moisture level may be varied by irrigating, either by use of sprinklers or by flooding, while temperature may be varied by shading the soil or altering the time at which the crops are planted. Sanitation procedures include the destruction of diseased plant parts and sterilization of equipment that might carry the parasite and facilitate its spread from field to field. Crop rotation is the cyclical planting of unrelated plants in a field (e.g., soybeans, potatoes, and rye may be alternated). Crop rotation prevents the buildup of parasitic fungi by depriving them of a susceptible crop and causing their starvation. Chemical fungicides (such as copper, sulfur, and dithiocarbamates) may be applied externally to a plant as a spray or dust. The fungicides kill fungi by inhibiting chitin synthesis or by causing disturbances in the plasmalemma, respiration, sterol metabolism, nuclear division, or synthesis of DNA, RNA, and protein. A single fungicide may impair any one or several of these functions.

Disease forecasting is becoming increasingly important in the control of disease. Records are kept of fluctuating biological conditions (e.g., spore release) and of the climate and are correlated with known factors favoring a particular disease. Computers may be used to store and correlate the data and also to simulate models of plant growth and disease. Local and regional forecasting centers then advise the agriculturalists if an unusual outbreak of a disease is predicted. In turn, appropriate preventive measures such as additional sanitation measures, application of a fungicide, or a change of the time of planting may be made in the field.

The control measures outlined above are expensive to apply and are often ineffective, especially in the midst of an epidemic. The ideal control method is to plant crops that are immune to the parasites and will not become diseased. Development of immune varieties involves extensive plant breeding programs. These breeding programs are complicated by the fact that often large numbers of physiological varieties, races, and biotypes of a fungus species exist which differ in their pathogenicity. This makes it comparatively difficult to develop a host which is resistant to all forms of the fungus. Once a resistant variety is developed, there remains the probability that new genetic lines of the parasitic fungus will arise and that new pathogenic gene combinations will be established through natural selection. Formation of new pathogenic fungus types makes it mandatory to introduce new genetic lines of crop plants frequently to achieve control of the disease.

REFERENCES CITED

AIST, J. R. 1976. Papillae and related wound plugs of plant cells. Ann. Rev. Phytopathol. 14:145–163.

APPLE, J. L. 1977. The theory of disease management. *In:* Horsfall, J. G., and E. B. Cowling, Eds. Plant disease: an advanced treatise. Academic Press, New York, Vol. 1, pp. 79–101.

BALAN, J., and H. A. LECHEVALIER. 1972. The predacious fungus *Arthrobotrys dactyloides:* induction of trap formation. Mycologia 64:919–922.

BARRON, G. L. 1977. The nematode-destroying fungi. Canadian Publications, Guelph, Ontario, 140 pp.

BEAUVERIE, J. 1914. Les Muscardines—le genre *Beauveria* Vuillemin. Rev. Gen. Botan 26:81–105, 157–173.

BERESTON, E. S. 1974. Occupational fungus infections. *In:* Robinson, H. M., Ed. The diagnosis and treatment of fungal infections. Charles C Thomas, Springfield, Ill., pp. 99–111.

BLOOD, D. C., and J. A. HENDERSON. 1960. Veterinary medicine. Williams & Wilkins, Baltimore, 1008 pp.

BRAUN, A. C. 1959. Growth is affected. *In:* Horsfall, J. C., and A. E. Dimond, Eds. Plant pathology: an advanced treatise. Academic Press, New York, Vol. 1, pp. 189–248.

BUSHNELL, W. R. 1972. Physiology of fungal haustoria. Ann. Rev. Phytopathol. 10:151–176.

CHAPMAN, H. C. 1974. Biological control of mosquito larvae. Ann. Rev. Entomol. 19:33–59.

COMANDON, J., and P. DE FONBRUNE. 1938. Recherches expérimentales sur les champignons prédateurs de nematodes du sol. Les pièges garrotteurs. Compt. Rend. Soc. Biol. Paris 129:620–622.

DUDDINGTON, C. L. 1955. Fungi that attack microscopic animals. Botan. Rev. 21:377–439.

DUSTAN, A. G. 1927. The artificial culture and dissemination of *Entomophthora sphaerosperma* Fres. a fungous parasite for the control of the European apple sucker (*Psyllia mali* Schmid B). J. Econ. Entomol. 20:68–75.

EMMONS, C. W. 1960. The Jekyll–Hydes of mycology. Mycologia 52:669–680.

——, C. H. BINFORD, J. P. UTZ, and K. J. KWON-CHUNG. 1977. Medical mycology, 3rd ed. Lea & Febiger, Philadelphia, 592 pp.

FERRON, P. 1978. Biological control of insect pests by entomogenous fungi. Ann. Rev. Entomol. 23:409–442.

FISCHER, G. W. 1965. The romance of the smut fungi. Mycologia 57:331–342.

FLOR, H. H. 1971. Current status of the gene-for-gene concept. Ann. Rev. Phytopathol. 9:275–296.

GABRIEL, B. P. 1959. Fungus infection of insects via the alimentary canal. J. Insect. Pathol. 1:319–330.

HAGEN, K. S., and VAN DEN BOSCH, R. 1968. Impact of pathogens, parasites and predators on aphids. Ann. Rev. Entomol. 13:325–384.

HOFFMAN, W. E. 1947. Insects as human food. Proc. Entomol. Soc. Wash. 49:233–237.

HUBER, D. M. 1978. Disturbed mineral nutrition. In: Horsfall, J. G., and E. B. Cowling, Eds. Plant disease: an advanced treatise. Academic Press, New York, Vol. 3, pp. 163–181.

KEITT, G. W., D. M. BOONE, and J. R. SHAY. 1959. Genetic and nutritional controls of host-parasite interactions in apple scab. In: Holton, C. S., Ed. Plant pathology: problems and progress 1908–1958. University of Wisconsin Press, Madison, Wis., pp. 157–167.

KOSUGE, T. 1978. The capture and use of energy by diseased plants. In: Horsfall, J. G., and E. B. Cowling, Eds. Plant disease: an advanced treatise. Academic Press, New York, Vol. 3, pp. 85–116.

LEFEBVRE, C. L. 1934. Penetration and development of the fungus, Beauveria bassiana, in the tissues of the corn borer. Ann. Botany (London) 48:441–452.

LISANSKY, S. G., and R. A. HALL. 1983. Fungal control of insects. In: Smith, J. E., D. R. Berry, and B. Kristiansen, Eds. The filamentous fungi: fungal technology. Edward Arnold, London, Vol. 4, pp. 327–345.

MACKAUER, M., and M. J. WAY. 1976. Myzus persicae Sulz. an aphid of world importance. In: Delucchi, V. L., Ed. Studies in biological control. Cambridge University Press, Cambridge. Intern. Biol. Programme 9:51–119.

MACLEOD, D. M. 1954. Investigations on the genera Beauveria Vuill. and Tritiachium Limber. Can. J. Botany. 32:818–890.

——. 1963. Entomophthorales infections. In: Steinhaus, E. A., Ed. Insect pathology: an advanced treatise. Academic Press, New York, Vol. 2, pp. 189–231.

MADDY, K. T. 1960. Coccidiomycosis. Advan. Vet. Sci. 6:251–286.

MADELIN, M. F. 1963. Diseases caused by hyphomycetous fungi. In: Steinhaus, E. A., Ed. Insect pathology: an advanced treatise. Academic Press, New York, Vol. 2, pp. 233–271.

——. 1966. Fungal parasites of insects. Ann. Rev. Entomol. 11:423–448.

MCEWEN, F. L. 1963. Cordyceps infections. In: Steinhaus, E. A., Ed. Insect pathology: an advanced treatise. Academic Press, New York, Vol. 2, pp. 273–290.

MOUNT, M. S. 1978. Tissue is disintegrated. In: Horsfall, J. G., and E. B. Cowling, Eds. Plant disease: an advanced treatise. Academic Press, New York, Vol. 3, pp. 279–297.

POOL, V. W., and M. B. McKAY. 1916. Relation of stomatal movement to infection by Cercospora beticola. J. Agr. Res. 5:1011–1037.

RIPPON, J. W. 1978. Mycosis (pathogenesis and epidemiology). In: Vinken, P. P. J., and G. W. Bruyn, Eds. Handbook of clinical neurology. North-Holland Publishing Co., Amsterdam, Vol. 35, pp. 371–381.

——. 1980. Dimorphism in pathogenic fungi. CRC critical reviews in microbiology. CRC Press, Boca Raton, Fla., pp. 49–97.

SMALL, E. 1971. Candidiasis (moniliasis). In: Kirk, R. W., Ed. Current veterinary therapy. IV. Small animal practice. W.B. Saunders, Philadelphia, pp. 288–289.

STEINHAUS, E. A. 1949. Principles of insect pathology. McGraw-Hill, New York, 757 pp.

TALBOYS, P. W. 1978. Dysfunction of the water system. In: Horsfall, J. G., and E. B. Cowling, Eds. Plant disease: an advanced treatise. Academic Press, New York, Vol. 3, pp. 141–162.

URITANI, I., and T. AKAZAWA. 1959. Alteration of the respiratory pattern in infected plants. In: Horsfall, J. G., and A. E. Dimond, Eds. Plant pathology: an advanced treatise. Academic Press, New York, Vol. 1, pp. 349–390.

WALKER, J. C., K. P. LINK, and H. R. ANGELL. 1929. Chemical aspects of disease resistance in the onion. Proc. Natl. Acad. Sci. USA 15:845–850.

WHEELER, B. E. J. 1968. Fungal parasites of plants. *In:* Ainsworth, G. C., and A. S. Sussman, Eds. The fungi: an advanced treatise. Academic Press, New York, Vol. 3, pp. 179–210.

WYNN, W. K. 1981. Tropic and taxic responses of pathogens to plants. Ann. Rev. Phytopathol. 19:237–255.

Additional Recommended References

AINSWORTH, G. C. 1968. Fungal parasites of vertebrates. *In:* Ainsworth, G. C., and A. S. Sussman, Eds. The fungi: an advanced treatise. Academic Press, New York, Vol. 3, pp. 211–226.

ANDERSON, J. M., A. D. M. RAYNER, and D. W. H. WALTON, Eds. 1984. Invertebrate-microbial interactions. Cambridge University Press, Cambridge, 349 pp.

BAKER, E. D., and L. P. CADMAN. 1963. Candidiasis in pigs in northwestern Wisconsin. J. Am. Vet. Med. Assoc. 142:763–767.

BATEMAN, D. F., and R. L. MILLAR. 1966. Pectic enzymes in tissue disintegration. Ann. Rev. Phytopathol. 4:119–146.

BECKMAN, C. H. 1964. Host responses to vascular infection. Ann. Rev. Phytopathol. 2:231–252.

BERGER, R. D. 1977. Application of epidemiological principles to achieve plant disease control. Ann. Rev. Phytopathol. 15:165–183.

CAMERON, J. W. M. 1963. Factors affecting the use of microbial pathogens in insect control. Ann. Rev. Microbiol. 8:265–286.

CAMPBELL, R. 1981. Plant microbiology. Edward Arnold, London, 191 pp.

COUCH, J. N., and C. J. UMPHLETT. 1963. *Coelomyces* infections. *In:* Steinhaus, E., Ed. Insect pathology: an advanced treatise. Academic Press, New York, Vol. 2, pp. 149–188.

CRUICKSHANK, I. A. M. 1963. Phytoalexins. Ann. Rev. Phytopathol. 1:351–374.

DIMOND, A. E. 1970. Biophysics and biochemistry of the vascular wilt syndrome. 1970. Ann. Rev. Phytopathol. 8:310–322.

DRAKE, T. E., and H. I. MAIBACH. 1974. Cutaneous candidiasis. *In:* Robinson, H. M., Ed. The diagnosis and treatment of fungal infections. Charles C Thomas, Springfield, Ill., pp. 5–28.

DUDDINGTON, C. L., and C. H. E. WYBORN. 1972. Recent research on the nematophagous Hyphomycetes. Botan. Rev. 38:545–565.

EMMONS, C. W. 1951. The isolation from soil of fungi which cause disease in man. Trans. N.Y. Acad. Sci. Ser. II 14:51–54.

———. 1955. Mycoses of animals. Advan. Vet. Sci. 2:47–63.

GRISEBACH, H. 1983. Phytoalexins. *In:* Bennett, J. W., and A. Ciegler, Eds. Secondary metabolism and differentiation in fungi. Marcel Dekker, New York, Vol. 5, pp. 377–428.

INGHAM, J. L. 1972. Phytoalexins and other natural products as factors in disease resistance. Botan. Rev. 38:343–424.

KLIGMAN, A. M., and E. D. DeLAMATER. 1950. The immunology of the human mycoses. Ann. Rev. Microbiol. 4:283–312.

LYR, H. 1977. Mechanism of action of fungicides. *In:* Horsfall, J. G., and E. B. Cowling, Eds. Plant disease: an advanced treatise. Academic Press, New York, Vol. 1, pp. 239–261.

MADELIN, M. F. 1968. Fungal parasites of invertebrates. 1. Entomogenous fungi. *In:* Ainsworth, G. C., and A. S. Sussman, Eds. The fungi: an advanced treatise. Academic Press, New York, Vol. 3, pp. 227–238.

ORLOB, G. B. 1971. History of plant pathology in the Middle Ages. Ann. Rev. Phytopathol. 9:1–20.

PADDOCK, W. C. 1967. Phytopathology in a hungry world. Ann. Rev. Phytopathol. 5:375–389.

SHRUM, R. D. 1978. Forecasting of epidemics. *In:* Horsfall, J. G., and E. B. Cowling, Eds. Plant disease: an advanced treatise. Academic Press, New York, Vol. 2, pp. 223–238.

STEINHAUS, E. A. 1957. Microbial diseases of insects. Ann. Rev. Microbiol. 11:165–182.

SZANISZLO, P. J., and J. L. HARRIS, Eds. 1985. Fungal dimorphism: with emphasis on fungi pathogenic for humans. Plenum Press, New York, 395 pp.

VEY, A., and J. FARGUES. 1977. Histological and ultrastructural studies of *Beauveria bassiana* infection in *Leptinotarsa decemlineta* larvae during ecdysis. J. Invertebr. Pathol. 30:207–215.

WEBSTER, R. K. 1974. Recent advances in the genetics of plant pathogenic fungi. Ann. Rev. Phytopathol. 12:129–137.

WILLIAMS, P. H. 1979. How fungi induce disease. *In:* Horsfall, J. G., and E. B. Cowling, Eds. Plant disease: an advanced treatise. Academic Press, New York, Vol. 4, pp. 163–179.

WOOD, R. K. S. 1960. Pectic and cellulolytic enzymes in plant disease. Ann. Rev. Plant Physiol. 11:299–322.

WRIGHT, E. T. 1974. Coccidiomycosis: mycology, pathology, immunology. *In:* Robinson, H. E. M., Ed. The diagnosis and treatment of fungal infections. Charles C Thomas, Springfield, Ill., pp. 341–353.

ZENTMYER, G. A., and J. G. BALD. 1977. Management of the environment. *In:* Horsfall, J. G., and E. B. Cowling, Eds. Plant disease: an advanced treatise. Academic Press, New York, Vol. 1, pp. 121–144.

ZIMMERMAN, M. H., and J. McDONOUGH. 1978. Dysfunction in the flow of food. *In:* Horsfall, J. G., and E. B. Cowling, Eds. Plant disease: an advanced treatise. Academic Press, New York, Vol. 3, pp. 117–140.

14

Fungi as Hosts

In Chapter 13 we learned that the fungi are extremely active as parasites of a wide variety of organisms. The fungi themselves can be involved in the parasitic relationship as hosts. They can be parasitized by other fungi or by viruses.

FUNGI AS PARASITES OF FUNGI

Fungi frequently grow upon other fungi in nature and perhaps occur exclusively in this partnership. Examples include the Pyrenomycete *Cordyceps*, which occurs on ascocarps of *Elaphomyces*, a member of the subterranean Tuberales; the Pyrenomycete *Hypomyces*, which may occur on agarics; and *Boletus parasiticus*, which arises from the base of a puffball (Figs. 14–1 and 14–2). The relationship of these fungi to each other and many similar associations remain uncertain, as positive demonstrations of a parasitic relationship are largely lacking. Association does not imply a parasitic relationship; for example, Fitzpatrick (1915) studied the association of the agaric *Claudopus subdepluens* on *Coltricia perennis* (Fig. 14–3) and found that the mycelium of *C. subdepluens* could be traced through the tissues of the polypore and through the stipe, and it probably extended into a common substratum from which both fungi were deriving their nourishment. In addition, no physical connections (feeding structures such as haustoria) or damage to the host could be detected.

A true parasite–host relationship exists between many fungi in which the parasite relies directly on the host for its maintenance at the expense of the host. There are two fundamentally different groups of these parasites. A treatment of these two groups follows.

Figure 14–1 *Cordyceps capitata*, a parasite on *Elaphomyces granulatus*, a subterranean fungus. [Courtesy Plant Pathology Department, Cornell University.]

Biotrophic Parasites

The majority of the fungal parasites, the *biotrophic* parasites, develop a delicate physiological balance with their host and feed from the living host cell. These parasites vary greatly in the manner in which they parasitize their host, ranging from those that contact the host without penetrating it to those that develop

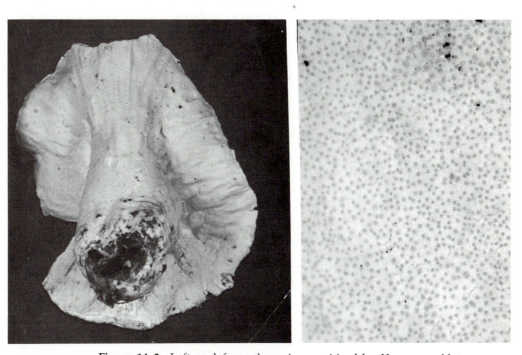

Figure 14–2 Left, a deformed agaric parasitized by *Hypomyces*. Note the absence of gills. Right, an enlarged view of the diseased agaric showing perithecia of *Hypomyces*, which appear as black dots. [Courtesy Plant Pathology Department, Cornell University.]

Figure 14–3 *Claudopus subdepluens* growing on, but apparently not parasitizing, *Coltricia* sp. [Courtesy Plant Pathology Department, Cornell University.]

within the host. Effects on the host may be rather insignificant or may eventually kill the host. A high degree of host specificity exists within this group, and a parasite may be restricted to a single host species or to a few closely related species. Many of these fungi are obligate parasites. These parasites include members of the chytrids, Mucorales, and Deuteromycotina.

More than 30 chytrid species are known to parasitize other chytrids and filamentous oomycetes. The parasitic chytrids penetrate and develop within or upon their hosts and finally destroy the host protoplasm as maturity is reached. Many of the chytrid parasites belong to the genus *Rozella*, which causes a marked enlargement of the host, a septation of the host cell, or both. In an ultrastructural study, it has been demonstrated that *Rozella polyphagi* exists as an unwalled cell within its host cell, where it feeds by phagocytosis (Powell, 1984). The host for *R. polyphagi* is another chytrid, *Polyphagus euglenae,* which in turn parasitizes an alga, a species of *Euglena*. A parasite that occurs on a second parasite in this manner is called an *epiparasite*. Other chytrids can exist as epiparasites. A three-membered host–parasite relationship exists in which *Chytridium parasiticum* is parasitic on *Chytridium suburceolatum,* which parasitizes another chytrid, *Rhizidium richmondense* (Willoughby, 1956).

Many members of the Mucorales are parasites, the majority occurring on other Mucorales. These fungi penetrate the host and typically produce a haustorium with a number of slender, short branches (Fig. 14–4). *Piptocephalis* is an example of this type of parasite. Under laboratory conditions, parasitism by *Piptocephalis virginiana* causes no significant weight loss in the host (Berry, 1959). No species of *Piptocephalis* is known to grow indefinitely in the absence of a host and will develop normally only on a host. The degree of parasitism by *Piptocephalis* is correlated with the quantity of soluble nitrogen in the host mycelium, and is favored by a high nitrogen-to-carbon ratio and certain amino acids (Barnett and Binder, 1973). *Piptocephalis* can penetrate both young and mature hyphae, but is apparently incapable of establishing a nutritional relationship with the mature hyphae. The young hyphae have only a single wall layer, while the mature hyphae have two wall layers, probably contributing greatly to the host re-

Figure 14–4 Absorptive organs of biotrophic fungal parasites: (a) haustoria of *Piptocephalis;* (b) clasping, contact cells of *Gonatobotrys simplex* on its host, *Mucor* (stippled); (c) haustorium of *Dispira simplex* in swollen host cell. [(a) From O. Brefeld, 1872, Botan. Untersuch. Schimmelpilze 1:1–64; (b) from J. W. Whaley and H. L. Barnett, 1963, Mycologia 55:199–210; (c) from M. Brunk and H. L. Barnett, 1966, Mycologia 58:518–523.]

sistance. Continued wall development appears to result in the development of a sheath around the haustoria (Manochoa and Campbell, 1983). A few members of the Mucorales cause an enlarged gall to form at the point of contact with the host cell. This gall consists of both an enlarged or branched ending of the parasite hypha and an outgrowth of the host hypha.

Some members of the Deuteromycotina are parasites on other members of the Deuteromycotina, Oomycetes, Mucorales, or Ascomycotina. These parasites do not penetrate the host, but instead they form somewhat specialized hyphae which contact the host. The contacting hyphae may have flattened tips that adhere firmly to the host, globose or fingerlike projections that partially clasp the host hyphae, or other modifications (Fig. 14–4). The cytoplasm of the host and parasite may be in contact either through a large pore that forms in the wall at the interface of the two fungi or through plasmodesmata transversing that interface (Hoch, 1978). The biotrophic parasites may alter the morphology of the host or cause a slight reduction in the growth rate. *Calcarisporium parasiticum* and *Gonatobotrys simplex* are representatives of this group, and are discussed below.

PHYSIOLOGICAL ASPECTS OF THE CONTACT PARASITES

If a parasite is to establish itself upon a host, its spores must be able to germinate in the presence of the host and then make contact with the host. Although the spores of some parasites are able to germinate in water, those of other parasites require exogenous substances to germinate. These exogenous substances may be provided by the host or by natural products in the substratum (e.g., nutrients or excretions of the host fungus). Contact with the host may be a result of tropic response. Germ tubes of *Calcarisporium parasiticum* and *Gonatobotrys simplex* secrete stimulants which cause host hyphae within a distance of about 40 micrometers to branch or to change directions and grow directly toward the parasite (Barnett and Lilly, 1958; Whaley and Barnett, 1963).

Next, a successful parasite must be able to establish a nutritional relationship with the host. This is usually preceded by penetration of the host or by formation of cytoplasmic contact. Host fungi differ in susceptibility to a parasite. Resistance of a potential host to a parasite may be due to the parasite's inability to establish a nutritional relationship with the host or inability to penetrate the host walls or to make cytoplasmic contact. Resistance to infection by *Calcarisporium parasiticum* may serve as an example. This organism parasitizes species of *Physalospora:* a large pore is formed in the parasite and host walls at the point of contact and the cytoplasm becomes confluent (Hoch, 1977). Resistance to infection may be due either to the inability of *C. parasiticum* to establish this cytoplasmic connection or to a deficiency within the host of a specific growth factor required by the parasite. To elucidate this situation further, *C. parasiticum* cannot be grown in the absence of its host unless an extract of the host mycelium is added to the culture. Ordinarily, *C. parasiticum* can parasitize *Physalospora obtusa* vigorously while it is unable to parasitize *P. rhodina*. Water extracts of both *P. obtusa* and *P. rhodina* are able to support growth of the parasite, indicating (1) that the required growth factor is retained within the mycelium (not secreted into the environment) and (2) that *C. parasiticum* cannot parasitize *P. rhodina* owing to an inability to establish the initial contact (Barnett and Lilly, 1958). The growth factor is believed to be a specific, vitaminlike compound. This growth factor was found in extracts from numerous fungi from the Ascomycotina, Basidiomycotina, and Deuteromycotina (in which immunity would be due to an inability by the parasite to obtain the growth factor), but it was not found in some fungi examined (in which case immunity would be due to a lack of a growth factor required by the parasite). Parasites other than *C. parasiticum* require the same growth factor.

After the parasite is established, nutrition of the host may influence the degree of parasitism, although the host is not noticeably affected. Parasites differ in their response to different nutritional conditions of the host. For one group of parasites, a high ratio of carbon to nitrogen in the host's medium may decrease growth of the parasite. In these instances, the internal composition of the host is varied and becomes less favorable for growth of the parasite. Perhaps the high carbon concentration leads to a depletion of the nitrogen reserve in the host, leading to nitrogen starvation of the parasite (Shigo et al., 1961). In contrast, another group of parasites may not be favored by these same conditions.

Necrotrophic Parasites

Unlike the biotrophic parasites, the *necrotrophic* parasites destroy their fungal hosts (apparently by means of enzymes or other toxic substances) and then absorb the nutrients from the dead host cells. Although these fungi are nutritionally dependent on their host fungi, they do not have a requirement for any specific nutrient. The host range of these parasites is rather broad, which indicates that their toxic substances are probably nonspecific. The majority of these fungi are capable of growing saprophytically on common laboratory media as well as on dead fungus structures. These fungi include several members of the Deuteromycotina and several wood-rotting basidiomycetes.

Rhizoctonia solani serves as an example of this type of parasitism (Butler, 1957). This fungus is a common plant parasite that may become parasitic on members of the Mucorales, Peronosporales, and Deuteromycotina. There are two basic types of infection. The first is that in which the parasite coils about the host but never penetrates the host. The second is that in which the parasite penetrates the host and establishes internal hyphae, sometimes forming external

Figure 14–5 The mycoparasite *Arthrobotrys oligospora* encircles a second fungus, *Rhizoctonia solani*. Bar = 5 micrometers. [Courtesy Y. Persson and R. Campbell.]

coils in addition (similar to those in Fig. 14–5). *R. solani* may be induced to form coils about fibers of glass wool or to penetrate fibers of cotton in the absence of a host. The type of infection (with or without penetration) depends on the host species.

Coiling of parasite hyphae without penetration occurs when *Pythium* is the host. As *R. solani* overgrows the *Pythium* colony, tight coils are formed around the host hyphae. The protoplasm of the encircled hyphae either coagulates into oil-like globules or becomes progressively less dense and disappears. Finally the walls of the host collapse.

Penetration of the host occurs when *Gilbertella persicaria* is parasitized. Short, lateral infection hyphae produced by *R. solani* penetrate the host. After penetration, the parasite grows rapidly within the host where it branches and enlarges to fill the host mycelium. Some of the internal hyphae of *R. solani* may penetrate the host's wall and grow outward again. Oil-like globules may appear in the host's cytoplasm, and the host's cytoplasm is diminished (Fig. 14–6).

A given host may resist the invasion by *R. solani*. This resistance may be by (1) disintegration or lysis of the internal hyphae of the parasite by the host, or (2) formation of walls separating parasitized portions of the host mycelium from those portions which are yet unparasitized, thus preventing the spread of the parasite within the host (Fig. 14–7). Maximum infection of hosts by *R. solani* occurs under those environmental conditions which favor growth of both the host and parasite, but apparently no factor plays a disproportionate role in determining susceptibility.

SPOROPHAGY

A specialized case of necrotrophic parasitism can be found in *sporophagy*, a situation where the hyphae of one species attack and kill the spores of a second species. The hyphae of the parasite do not attack hyphae of the host species. Sporophagy been observed only in species of the hymenomycetes, and is particularly prevalent among the fungi that decay wood and litter. The susceptible spores apparently exude a substance that attracts hyphae of another species. The hyphae grow toward the spores, wind around them, enclose the spores in dense branches, and finally penetrate the host spores and kill them (Fries and Swedjemark, 1985).

(a)   (b)

Figure 14–6 Early (left) and later (right) infection stage of *Rhizoctonia solani* within a sporangiophore of *Gilbertella persicaria*. Note the hyphal network of the parasite, the diminished host cytoplasm, and oil-like drops in the later stage. [Courtesy E. E. Butler.]

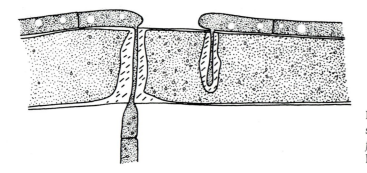

Figure 14–7 Infection hyphae of *Rhizoctonia solani* walled off by its host, *Rhizopus stolonifer*. Approximately × 600. [From E. E. Butler, 1957, Mycologia 49:354–373.]

MYCOVIRUSES

Viruses that infect fungi are called *mycoviruses*. They apparently are very common in the fungi, as there are numerous reports that mycoviruses or particles resembling them have been seen in fungal cells examined by electron microscopy. Judging from the frequency with which such particles are observed, it has been estimated that mycoviruses occur in about 5000 fungal species in nature (Hollings, 1978). Comparatively few mycoviruses have been isolated and characterized, and still fewer have been experimentally transmitted to test fungi and demonstrated to be infectious. Nevertheless, the numerous observations that have been made of the mycoviruses and mycoviruslike particles indicate that they are widespread among the fungi and occur in all major groups (Fig. 14–8).

Figure 14–8 A virus-infected basidiocarp of *Agaricus brunnescens* at the left is deformed, has a grayish color and sparse mycelium. An uninfected basidocarp is at the right for comparison. [Courtesy P. T. Atkey, AFRC Institute of Horticultural Research.]

This is demonstrated in a survey of 50 isolates of *Rhizoctonia solani* from the field, in which 49 were found to have RNA resembling that of mycoviruses (Zanzinger et al., 1984).

A typical mycovirus particle, the *virion*, is isometric in shape with a diameter between 25 to 48 nanometers. Mycoviruses are occasionally rod-shaped or have some other form (Fig. 14–9). Usually, the core contains a single molecule of double-stranded (ds) RNA but infrequently contains dsDNA. The core is surrounded by a protein capsid having a high molecular weight. A unique feature of the mycoviruses is that the genome is segmented with parts occurring within different virions; typically from two to five virions may be required to contain the complete genome. If the genome is to be fully analyzed, all virions containing parts of it must be recovered from the cell. This is unlike the dsRNA viruses occurring in plants and animals, where the entire genome is represented within a single virion.

Young growing hyphal tips are generally free of mycoviruses, while virus particles can be detected by electron microscopy elsewhere. In the younger cells, the mycoviruses are predominantly free in the cytoplasm, while they also occur in membrane-bound vesicles and vacuoles in older cells. Once established, a mycoviral infection tends to be maintained in the mycelium indefinitely—one of the early strains of *Penicillium chrysogenum* developed for commercial production of penicillin became infected by a mycovirus, and now all commercial strains are similarly infected because they were derived from the infected strain. Mycoviruses are regularly transmitted in the cytoplasmic contact occurring when hyphal fusions occur. They can also be carried into the developing fungal spores but apparently cannot infect through intact cell walls. Experimentally, it is relatively difficult to quantify the extent of transmission of mycoviruses that occurs naturally or to isolate, purify, and transmit them to cause infection of test strains, proving that they are infectious. These constraints have greatly hampered research with the mycoviruses.

Mycoviruses replicate within the host cell, producing new virions. This process involves the production of new dsRNA molecules as well as their enclosure in newly synthesized polypeptides that form the capsid. There is comparatively little detailed information about replication, as this mechanism has been studied in only a few mycoviruses and no general pattern emerges. The most extensive work has been carried out with the killer mycovirus of *Saccharomyces cerevisiae*, which is described further below (Tipper and Bostian, 1983; Williams and Leibowitz, 1987). RNA polymerase is required for transcription and the production of new RNA strands, and is produced by the mycovirus. One strand of

(a)

(b)

Figure 14–9 Viruses isolated from infected mushrooms such as that illustrated in Fig. 14–18: (a) virus having spherical particles 29 nanometers in diameter; (b) virus having rod-shaped particles. Bar = 100 nanometers. [Courtesy P. T. Atkey, AFRC Institute of Horticultural Research.]

the dsRNA is transcribed, forming a complementary single RNA strand (sometimes called the + strand). Once formed, the ssRNA is released from the virion into the host's cytoplasm, where it functions as messenger RNA and specific segments direct the synthesis of capsid protein. Further events occurring in the killer mycovirus replication have not been established conclusively but appear to follow the general pattern established for the replication of reoviruses that occur in vertebrates. In the reoviruses, after protein synthesis, the messenger-

RNA is used as a template for the production of a complementary RNA strand (the − strand), and the dsRNA is then reconstituted and incorporated in a complete virion. The events described for replication of the killer virus are different from those occurring in the replication of mycoviruses of *Penicillium stoloniferum* (Buck, 1975) and *Aspergillus foetidus* (Ratti and Buck, 1979). In the latter two fungi, the dsRNA serves as a template for the production of ssRNA, which is released into the cytoplasm to function as mRNA. Then the dsRNA serves as a template for the *semiconservative* production of new dsRNA, in which one strand will be retained from the old molecule and will be paired with a newly transcribed strand. In contrast, the replication of the killer mycovirus is *conservative* because the template RNA is not retained as part of the new molecule.

The mycoviruses are unusual because they generally have no apparent effect on their fungal hosts and appear to be avirulent, because otherwise healthy cells can contain large numbers of virus particles. Yeast cells, for example, usually harbor about 100 to 300 virions in each cell with no apparent ill effects (Bobek et al., 1983). One reason for the avirulence may be the timing of viral replication. In a study of virus-infected *Penicillium stoloniferum* it was found that synthesis of dsRNA and virions occurred mostly after the exponential growth phase of the fungal host had been completed and synthesis of its macromolecules had declined (Still et al., 1975). In some cases, however, the mycoviruses cause a readily apparent disease and the infected cell lines grow slowly or develop abnormally. Lysis of infected cells occurs infrequently. Viral infection is a major problem in commercial production of the edible mushroom *Agaricus brunnescens* because it causes degeneration of the mycelium and development of malformed basidiocarps, resulting in a reduction of yield (Fig. 14–8).

It is possible that mycoviruses might be involved in *hypovirulence,* a phenomenon marked by the reduced virulence of plant pathogenic fungi, making them less dangerous for their plant hosts. Hypovirulence occurs in *Cryphonectria parasitica,* the fungus responsible for chestnut blight, which caused the almost total elimination of the chestnut in North America. If a hypovirulent strain of *C. parasitica* attacks a chestnut tree, the tree not only remains comparatively healthy but also resists attack by more virulent strains if both are present in the same canker. It was determined that this trait was transmitted from the hypovirulent strain to the virulent strain by hyphal anastomosis and transmission of dsRNA through cytoplasmic contact. As a result of this transmission, a normally virulent strain becomes hypovirulent. Although this dsRNA has many characteristics of viral dsRNA and is probably viral in origin, virions have not been seen in the hyphae or recovered from the hypovirulent strains, and no capsids are formed around the dsRNA. Instead, dsRNA may be found in vesicles formed by the host. Naturally occurring hypovirulent strains have effectively controlled chestnut blight in Italy, and the artificial spread of such strains in France has been effective in reducing the incidence of this disease. Attempts at using this method of biological control in the United States have not yet been successful, however, although this is still being pursued (Anagnostakis, 1987; Van Alfen et al., 1975; Van Alfen, 1982; Van Alfen and Hansen, 1983).

The "killer" systems in the yeast *Saccharomyces cerevisiae* and the smut *Ustilago maydis* are correlated with viral infections. Phenotypically, the killer systems in each of these species can be divided into three groups. The killer strains produce extracellular proteinaceous toxins that can be lethal to sensitive strains. Still other strains are neutral and neither produce toxins nor are sensitive to toxins from the killer strains. The killer systems are under complex genetic control involving both nuclear and cytoplasmic genes. In both species, the killer strains are immune to their own toxins. This autoimmunity is conferred by the dsRNA resulting from the viral infection and some segments of the dsRNA code for toxin production (Koltin, 1983; Wickner, 1981).

REFERENCES CITED

ANAGNOSTAKIS, S. L. 1987. Chestnut blight: the classical problem of an introduced pathogen. Mycologia 79:23–37.

BARNETT, H. L., and F. L. BINDER. 1973. The fungal–host parasite relationship. Ann. Rev. Phytopathol. 11:273–292.

———, and V. G. LILLY. 1958. Parasitism of *Calcarisporium parasiticum* on species of *Physalospora* and related fungi. West Virginia Univ. Agr. Expt. Sta. Bull. 420T:1–37.

BERRY, C. R. 1959. Factors affecting parasitism of *Piptocephalis virginiana* on other Mucorales. Mycologia 49:374–386.

———, and H. L. BARNETT. 1957. Mode of parasitism and host range of *Piptocephalis virginiana*. Mycologia 49:374–386.

BOBEK, L. A., J. A. BRUENN, L. J. FIELD, J. D. REILLY, and K. W. GROSS. 1983. Cloning of cDNAs homologous to a yeast viral dsRNA. *In:* Compans, R. W., and D. H. L. Bishop, Eds. Double-stranded RNA viruses. Elsevier, New York, pp. 451–456.

BUCK, K. W. 1975. Replication of double-stranded RNA in particles of *Penicillium stoloniferum* virus S. Nucleic Acids Res. 2:1889–1902.

BUTLER, E. E. 1957. *Rhizoctonia solani* as a parasite of fungi. Mycologia 49:354–373.

FITZPATRICK, H. M. 1915. A parasitic species of *Claudopus*. Mycologia 7:331–342.

FRIES, N., and G. SWEDJEMARK. 1985. Sporophagy in Hymenomycetes. Exptl. Mycol. 9:74–79.

HOCH, H. C. 1977. Mycoparasitic relationships. III. Parasitism of *Physalospora obtusa* by *Calcarisporium parasiticum*. Can J. Botany 55:198–207.

———. 1978. Mycoparasitic relationships. IV. *Stephanoma phaeospora* parasitic on a species of *Fusarium*. Mycologia 70:370–379.

HOLLINGS, M. 1978. Mycoviruses: viruses that infect fungi. Advan. Virus Res. 22:1–53.

KOLTIN, Y. 1983. Exclusion mechanisms among the *Ustilago* dsRNA viruses. *In:* Compans, R. W., and D. H. L. Bishop, Eds. Double-stranded RNA viruses. Elsevier, New York, pp. 495–501.

MANOCHOA, M. S., and C. D. CAMPBELL. 1983. Host–parasite relations in a mycoparasite. VIII. Age-related host resistance in *Choanephora cucurbitarum*. Mycologia 75:588–596.

POWELL, M. J. 1984. Fine structure of the unwalled thallus of *Rozella polyphagi* in its host *Polyphagus euglenae*. Mycologia 76:1039–1048.

RATTI, G., and K. W. BUCK. 1979. Transcription of double-stranded RNA in virons of *Aspergillus foetidus* virus S. J. Gen. Virol. 42:59–72.

SHIGO, A. L., C. D. ANDERSON, and H. L. BARNETT. 1961. Effects of concentration of host nutrients on parasitism of *Piptocephalis xenophila* and *P. virginiana*. Phytopathology 51:616–620.

STILL, P. E., R. W. DETROY, and C. W. HESSELTINE. 1975. *Penicillium stoloniferum* virus: altered replication in ultraviolet-derived mutants. J. Gen. Virol. 27:275–281.

TIPPER, D. J., and K. A. BOSTIAN. 1983. Transcripts of the dsRNA virus-like plasmids of killer yeast. *In:* Compans, R. W., and D. H. L. Bishop, Eds. Double-stranded RNA viruses. Elsevier, New York, pp. 485–493.

WHALEY, J. W., and H. L. BARNETT. 1963. Parasitism and nutrition of *Gonatobotrys simplex*. Mycologia 55:199–210.

WICKNER, R. B. 1981. Killer systems in *Saccharomyces cerevisiae*. *In:* Strathern, J. N., E. W. Jones, and J. R. Broach, Eds. The molecular biology of the yeast *Saccharomyces*: life cycle and inheritance. Cold Spring Harbor Laboratory, Cold Spring Harbor, N.Y., pp. 415–444.

WILLIAMS, T. L., and M. J. LEIBOWITZ. 1987. Conservative mechanism of the *in vitro* transcription of killer virus of yeast. Virology 158:231–234.

WILLOUGHBY, L. G. 1956. Studies on soil chytrids. I. *Rhizidium richmondense* sp. nov. and its parasites. Trans. Brit. Mycol. Soc. 39:125–141.

VAN ALFEN, N. K. 1982. Biology and potential for disease control of hypovirulence of *Endothia parasitica*. Ann. Rev. Phytopathol. 20:349–362.

———, and D. R. HANSEN. 1983. Hypovirulence: a unique phenotypic expression of dsRNA in a pathogenic fungus. *In:* Compans, R. W., and D. H. L. Bishop, Eds. Double-stranded RNA viruses. Elsevier, New York, pp. 464–467.

——, R. A. JAYNES, S. L. ANAGNOSTAKIS, and P. R. DAY. 1975. Chestnut blight: biological control by transmissible hypovirulence in *Endothia parasitica*. Science 189:890–891.

ZANZINGER, D. H., B. P. BANDY, and S. M. TASVANTZIS. 1984. High frequency of finding double-stranded RNA in naturally occurring isolates of *Rhizoctonia solani*. J. Gen. Virol. 65:1601–1606.

Additional Recommended References

BARNETT, H. L. 1963. The nature of mycoparasitism by fungi. Ann. Rev. Microbiol. 17:1–14.

BOOSALIS, M. G. 1964. Hyperparasitism. Ann. Rev. Phytopathol. 2:263–376.

BUCK, K. W. 1980. Viruses and killer factors of fungi. *In:* Gooday, G. W., D. Lloyd, and A. P. J. Trinci, Eds. The eukaryotic microbial Cell. Cambridge University Press, Cambridge, pp. 329–375.

GHABRIAL, S. A. 1980. Effects of fungal viruses on their hosts. Ann. Rev. Phytopathol. 18:441–461.

GRIFFITH N. T., and H. L. BARNETT. 1967. Mycoparasitism by basidiomycetes in culture. Mycologia 59:149–154.

KUYKENDALL, W. R., D. F. HINDAL, and H. L. BARNETT. 1983. Parasitism and nutrition of the contact mycoparasite *Acladium tenellum*. Mycologia 75:656–665.

LEIBOWITZ, M. J., D. J. THIELE, and E. M. HANNING. 1983. Structure of separated strands of double-stranded RNA from the killer virus of yeast: a translational model. *In:* Compans, R. W., and D. H. L. Bishop, Eds. Double-stranded RNA viruses. Elsevier, New York, pp. 458–462.

LEMKE, P. A., Ed. 1979. Viruses and plasmids in fungi. Marcel Dekker, New York, Series on mycology, Vol. 1, 653 pp.

——, and C. H. NASH. 1974. Fungal viruses. Bacteriol. Rev. 38:29–56.

MANOCHA, M. S., and R. GOLESORKHI. 1979. Host–parasite relations in a mycoparasite. V. Electron microscopy of *Piptocephalis virginiana* infection in compatible and incompatible hosts. Mycologia 71:565–576.

SAKSENA, K. N. 1977. Mycoviruses. *In:* Maramorosch, K., Ed. The atlas of insect and plant viruses including mycoplasmaviruses and viroids. Academic Press, New York, pp. 421–430.

SLIFKIN, M. E. 1963. Parasitism of *Olpidiopsis incrassata* on members of the Saprolegniaceae. II. Effect of pH and host nutrition. Mycologia 55:172–182.

15 Fungi as Symbionts

Some fungi regularly form partnerships with other organisms, sometimes to the exclusion of an independent existence. Such a regular association is termed *symbiosis*, and the partners are termed *symbionts*. Symbiosis includes all relationships in which there is a regular association, and this may include parasitism. As we learned from Chapters 13 and 14, a parasitic relationship harms one partner. Alternatively, both partners may gain from the symbiotic relationship, or the partners may share food but may be neither harmed nor benefited from the relationship. In this chapter we explore the relationships of some fungal symbionts that cause little or no harm to their hosts. In all cases, the fungus benefits nutritionally from the relationship, but the benefits (if they do indeed occur) for the nonfungus member are varied.

FUNGI AS PARTNERS OF PLANTS

Fungi are found in dual associations with plants ranging from the algae (in which case the dual plant is a *lichen*) to the bryophytes or fern prothalli and finally to an association with roots of vascular plants (termed a *mycorrhiza*). It is generally unknown to what extent these analogous relationships are similar in their biology. Attention has been focused primarily on the economically important higher plants. We discuss the two better known types, lichens and mycorrhizae.

Lichens

An association between a fungus and an alga that develops into a unique morphological form that is distinct from either partner is termed a *lichen*. Excluded from this concept are associations in which an alga and fungus enter into limited interactions that do not form a new morphological entity, such as an alga over-growing a mushroom or limited parasitization of algal cells by fungi.

Lichen associations may be preserved by nonsexual reproduction, and their morphology is so constant that they were thought to be genetically autonomous plants until the Swiss botanist Simon Schwendener described their dual nature in 1868. Even at the present, the more than 16,000 lichen "species" may be classified in their own form class, the Lichenes, by using characteristics peculiar to the lichen association. As the lichens are a biological group and not a genetically related group, some taxonomists are correctly giving scientific names to both the fungus and algal partners and then integrating these into their proper respective positions in the mycological and phycological classifications.

Lichens are able to survive under severe conditions that cannot be tolerated by the majority of plants. They may be found in virtually every environment from the desert to cold sandy beaches, from bare rock to fertile soil or living leaves, and from the tropics to the arctic regions. Lichens often are the dominant vegetation of some sites, such as mountains above the tree line or in northern locations such as Greenland, Iceland, or Alaska. Lichens are conspicuously absent from the flora in and surrounding large cities because of their inability to withstand air pollution. They are especially sensitive to sulfur dioxide and fluorine, which are common pollutants.

Lichens have been important to humans in various ways. In the arctic region and to a lesser extent in the antarctic, lichens are harvested as fodder for reindeer. Various people have used lichens as food. For example, people in Iceland have prepared soups, desserts, and breads from lichens. It is thought that the manna referred to in the Bible may have been *Lecanora esculenta*, a lichen that grows in the mountains of Israel and may be blown loose into the lowlands. Desert tribes still eat this lichen, which is dried and mixed with dry meal to form a flour. An important way of using lichens has been to prepare blue, red, brown, or yellow dyes for cloth. Lichens have also been used as a source of medicines, poisons, cosmetics, perfumes, and essential oils by various peoples (Perez-Llano, 1948). The indicator pigment used to prepare litmus paper is derived from a lichen.

BIOLOGY OF THE SYMBIONT

Ordinarily the dominant member of the lichen symbiont is the fungus, which controls the morphological form that the lichen will assume (that is, lichens formed with a single fungus species but with different algal species may have the same form). In some lichens the situation is reversed and the alga determines the final form. The thallus may be leaflike and flat on the substratum (*foliose*), crustaceous and flat on the substratum (*crustose*), or upright and branched or pendulous (*fruticose*) (Figs. 15–1 and 15–3). The lichen consists of tissues that are generally similar to those which have been described in Chapters 1, 4, and 5. The most characteristic tissue is one consisting of loosely interwoven, branched hyphae that form a netlike structure. The algal cells may be evenly distributed among the hyphae or, most commonly, may occur in a thin layer. Anatomically, the most complex lichens are some of the foliose lichens, which are typically organized into the following layers (from top to bottom): (1) an epidermislike upper cortex of pseudoparenchymatous tissue, (2) a usually thin algal layer in which the algae cells are penetrated by fungal haustoria, (3) a medullary layer of loosely interwoven hyphae, and (4) a lower cortex. Crustose thalli are not divided into as many layers, as they lack a lower cortex, and some exceptionally simple ones may consist only of a single homogeneous layer containing the fungal hyphae and algal cells (Fig. 15–2). Fruticose forms consist of a medulla surrounded by a pseudocortex. The algal cells are distributed either evenly or in patches in the medulla (Fig. 15–3). Pores that allow the exchange of air may be present on the undersurface. Other modifications of the thallus

(a)

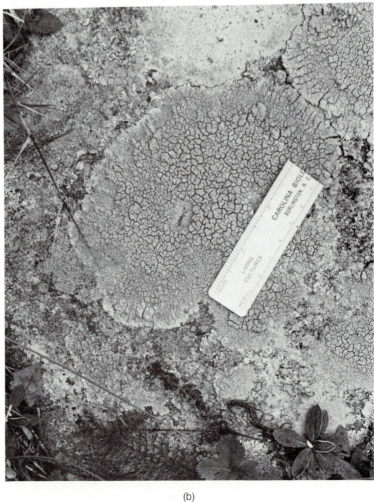

(b)

Figure 15–1 Foliose lichen, *Parmelia caperata*, approximately × 1; (b) crustose lichen, *Diploschistes scruposus*, approximately ½ × . [Courtesy M. E. Hale.]

Chap. 15 / Fungi as Symbionts

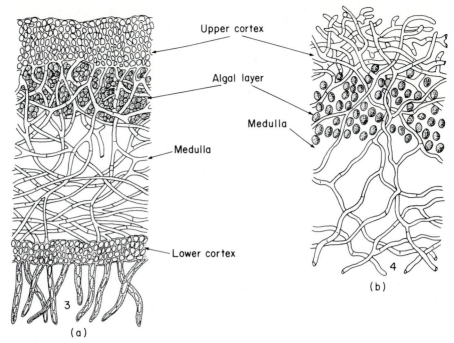

Figure 15–2 Longitudinal section through the vegetative thallus of a foliose lichen (a) and a crustose lichen (b). Compare with the fruticose lichen in Fig. 15–3. [From A. Schneider, 1897, A text-book of general lichenology, W. N. Clute & Co., Binghamton, N.Y.]

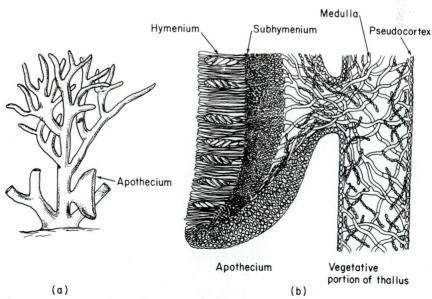

Figure 15–3 Fruticose lichen: (a) habit of the thallus; (b) section through a thallus showing a portion of an apothecium at the left and a vegetative portion at the right. [Adapted from A. Schneider, 1897, A text-book of general lichenology, W. N. Clute & Co., Binghamton, N.Y.]

may include loosely attached scales, columnar outgrowths *(isidia)*, pockets where the alga occurs alone, and *soredia* (discrete structures consisting of a few algal cells surrounded by hyphae).

Reproduction may occur by nonsexual or sexual means. Nonsexual reproduction occurs by fragmentation of the thallus and is especially likely to occur by separation of the columnar isidia, which appear to be an adaptation for nonsexual reproduction. Also the soredia may become detached and disseminated in a manner similar to spores and then develop into a new thallus. The fungus may also be propagated nonsexually by conidia formed in pycnidia. As the majority of the lichen fungi are in the Ascomycotina, the sexual reproductive structures are usually ascocarps (Figs. 15-3 and 15-4). Ascospores are formed in asci and are discharged. Although direct evidence is lacking, it is assumed that mycelium from germinating conidia or ascospores may come into contact with a suitable alga and reestablish the lichen relationship.

Growth of lichens is generally favored by periods of high humidity, cool temperatures, and low light intensities. Of these, moisture is the most critical factor because lichens apparently have no special mechanisms for uptake or conservation of water. In respect to water, they behave very much like an agar gel, rapidly absorbing water when it is available (sometimes up to 100 to 300% of its dry weight) and also rapidly losing water during a drought (to as low as 2 to 15% of its dry weight) (Smith, 1962). Both respiration and photosynthesis are favored by optimum water content of the thallus; they decrease with drying and sometimes with excess saturation. Moisture in the form of fog is especially favorable for lichen growth because it provides the necessary moisture while allowing photosynthesis to take place in the optimum low light intensity. Moisture availability is the critical factor in determining lichen distribution in nature.

Lichens grow extremely slowly, partially because they are exposed to optimal growing conditions for only a few hours a morning when there is still sufficient moisture from fog, dew, or recently melted snow to allow photosynthesis to occur at its optimum rate. Average annual increments for many lichens are less than 1 millimeter, and the greatest known increment is 4 centimeters. Lichens may reach maturity and produce ascocarps between 4 and 8 years of age. Longevity of lichens is often measurable in tens or hundreds of years. The age of some lichens in the alpine–arctic areas has been estimated to be between 1000 and 4500 years. This longevity can partially be attributed to the ability of lichens to survive long periods (up to 2 to 3 months) of drought unharmed.

Figure 15–4 Apothecia of the pixie-cup lichen, *Cladonia* sp. [From E. J. Moore and V. N. Rockcastle, 1963, Fungi, Cornell Science leaflet.]

Physiological studies of the composite lichen in culture are valuable but are difficult to carry out for extended periods of time because the lichen association tends to break down under the luxuriant conditions of moisture and nutrition inadvertently provided in culture. It is advantageous to study the physiology of the fungus and alga partners individually to better understand the lichen association. In order to carry out such studies, it is necessary to separate the components from each other and culture them independently.

The algal component may be isolated by first washing the lichen thallus in running tap water for 15 minutes, crushing the thallus, and removing the algal cells with attached hyphae with a sterile micropipette under a microscope. (This technique eliminates the possibility of culturing a transient contaminant.) These algal cells are then passed through four or five drops of sterile water and introduced onto slants of a solid inorganic mineral medium, which are incubated in the light. The fungus will not grow under such conditions. A simpler but less reliable method consists of peeling off the cortex of the lichen, fragmenting the remainder of the thallus, and then incubating a fragment on the illuminated medium. If this second method is employed, it is likely that foreign algal cells will be cultured; the algal partner should be distinguished by comparison with a microscopic mount of the lichen and possibly purified by subculturing (Ahmadjian, 1967-b).

The fungal partner may be isolated by using standard mycological techniques. Since the majority of lichen fungi are in the Ascomycotina, one may take advantage of spore discharge by preparing a petri dish with a 2% agar medium, soaking the lichen ascocarps in cold water for 15 minutes, and then attaching the inverted ascocarp to the underside of the petri dish lid with a dab of petroleum jelly. As the wet lichen dries, it will discharge large numbers of spores onto the agar. After germination, a block of the agar bearing mycelium is transferred to an enriched medium (such as malt-yeast agar) for further culturing (Ahmadjian; 1967-a).

THE COMPONENTS OF THE SYMBIONT

The Algal Partner.   At least 26 alga genera are members of the lichen association. These include 8 blue-green algae and 1 yellow-green alga; the remainder are green algae. *Nostoc* is the blue-green alga most frequently found in lichens, while the green alga *Trebouxia* is found in 75% or more of the lichens of the temperate zones (Ahmadjian, 1966-a). All the algal partners are probably able to exist independently in nature.

Numerous nutritional and physiological studies have been made of the algae (most often of *Trebouxia*) in pure culture. The majority of the isolates make improved growth on a medium with added carbohydrates or organic nitrogen, and some can grow heterotrophically in the dark with these added nutrients. The algae apparently do not require an exogenous source of vitamins and are extremely slow growers. Their slow rate of growth partially accounts for the slow growth rate of the lichen as a whole.

Lichens with *Nostoc* as the algal partner are able to fix atmospheric nitrogen. The isolated *Nostoc* not only excretes considerable amounts of nitrogenous compounds into the culture medium, but it also excretes polysaccharides and vitamins (biotin, thiamine, riboflavin, and nicotinic acid).

The Fungal Partner.   It is not known whether the fungal partners have an extensive independent existence in nature, although it is not unlikely that both a saprophytic mycelial stage and a nonlichenized sexual state might occur. Most of the lichen fungi are members of the Ascomycotina, while a few others are in the Basidiomycotina or Deuteromycotina. Taxonomically, the lichenized fungi are widely distributed in the Ascomycotina, and many of these fungi can form lichens. Lichenized and nonlichenized fungi may occur in the same order. For example, most of the members of the Order Dothideales (Loculoascomycetes), are nonlichenized but some may form lichens. In contrast, the Orders Caliciales and Lecanorales in the Discomycetes are comprised almost entirely of lichenized fungi. When isolated, the fungus retains the scientific name of the lichen.

Like the alga and the composite lichen, the fungi grow extremely slowly in culture and may achieve a size of only 1 to 2 millimeters within a year. All are either wholly or partially dependent on an exogenous source of thiamine and/or biotin and, to a lesser extent, the other vitamins.

RESYNTHESIS OF THE LICHEN FROM THE PARTNERS

Attempts to resynthesize the lichen from the isolated partners have met with limited success. These studies are of value primarily in indicating those conditions most favorable for resynthesis of the lichen, thereby giving additional clues about their biology. The lichen association can only be established if growing conditions are unfavorable for the independent growth of both partners. For example, resynthesis of the lichen thallus by *Endocarpon pusillum* was achieved only on soil that had been subjected to alternating periods of drying and wetting and also light and darkness. Low nutrient levels were also required. The most critical condition favoring resynthesis by *E. pusillum* was slow drying, a feature generally shared with other lichens that have been resynthesized (Ahmadjian, 1973). Fluctuating environmental conditions, particularly light and moisture, also favor the growth of the lichen once it is established (Pearson, 1970). A formed lichen will dissociate into its components if favorable conditions of moisture, light, and adequate nutrition are made available (this may occur in nature), or even if the environmental conditions are nonfluctuating. These observations indicate that the lichen association is a forced one that depends on adverse conditions for its maintenance.

NATURE OF THE INTERACTION

The physiological relationship of the partners in the lichen association is a matter of speculation. The most widely held view is that the photosynthetic alga provides organic material for both partners and that the fungus protects the alga from desiccation, high light intensities, and injury and provides it with water and minerals. This interpretation may be regarded as either (1) controlled parasitism, if the alga is exploited and is not dependent on the ''benefits'' that the association yields, or (2) a relationship in which the alga gains something that it would not ordinarily have if living in the free state. This broad interpretation has been only partially supported experimentally and, further, it cannot be assumed that all lichen relationships are similar in view of the many different algae and fungi that may be involved. We explore below some specific points that may aid in interpreting the nature of the interaction.

Possible Advantages for the Fungal Partner. It is well established that the fungus gains nutritionally from the alga in the lichen relationship (Fig. 15–5). In culture, the alga produces excess vitamins corresponding to the vitamin deficiencies in the cultured fungus, which suggests that the alga provides the fungus with necessary growth factors in the lichen. In addition, the excretion of nitrogenous compounds by some nitrogen-fixing algal partners may benefit the fungus in a similar fashion.

Carbon compounds manufactured by the algal partner during photosynthesis move into the medullary tissues of the lichen where they are absorbed by the fungus. This can be confirmed experimentally by incubating lichen discs in a medium containing isotopic carbon, ^{14}C. Clearly, the carbohydrates produced by the algae from photosynthesis are in excess of the amount that they require,

(a)

(b)

Figure 15–5 Haustoria of the fungus partner which penetrate the algal cells: (a) as seen with the light microscope; (b) as seen with the electron microscope. *Al W* indicates algal cell wall; *FW* indicates fungal cell wall; *N* indicates nucleus of fungal cell; *H* indicates haustorium; *Ch* indicates chloroplast of algal cell. The lichen in (b) is *Cladonia cristatella*. × 17,000. [(a) From A. Schneider, 1897, A text-book of general lichenology, W. N. Clute & Co., Binghamton, N.Y.; (b) courtesy R. T. Moore and J. H. McAlear, 1960, Mycologia 52:805–807.]

but they cannot divert this food surplus to reproduction or additional growth as they would if living in the free state.

The flow of nutrients from the alga to the fungus is substantial as the algal cells have been estimated to occupy only 3 to 10% of the total mass of the lichen by weight, but yet must supply carbohydrates to the entire lichen. As much as 70 to 80% of the carbon fixed by *Nostoc* or *Trebouxia* is probably passed to the fungus (Farrar, 1976). The flow of nutrients into the fungus is at its maximum when the thallus is saturated with moisture and is in the light. Conversely, fungal activity may be inhibited by low moisture levels in either the light or dark, and apparently little of the carbohydrate flows into the fungus under these conditions (Harris and Kershaw, 1971).

Possible Advantages for the Algal Partner. It is not so easy to assess the possible advantages of the lichen relationship for the algae. The algae may be at a disadvantage as they cannot grow and reproduce at their maximum rates because they are unable to utilize all the carbohydrates that they produce. The algae apparently retain all or most of the carbohydrates that they produce only when water levels are low (thereby inhibiting fungal activity) and light is present so that photosynthates can be produced. Such conditions may provide the principal periods during which the alga can store photosynthates and perhaps grow (Harris and Kershaw, 1971).

An interesting view has been given by Ahmadjian (1966-b). He noted that *Trebouxia*, the most common algal partner, is rarely found in the free-living state in nature. However, the most widely distributed free-living alga, *Pleurococcus vulgaris*, is never found as a member of a lichen association. This may be because *P. vulgaris* (and many other algae) cannot survive the initial parasitic contacts with a lichen fungus, while *Trebouxia* can withstand these encounters. *Trebouxia* has characteristics that place it at a competitive disadvantage in the free state (a low optimum light intensity, slow growth rate, and a preference for organic nitrogen) but allow it to thrive in the lichen association. The lichen association is favorable for this alga because it is removed from competition with free-living algae and is placed in its own ecological niche.

We may conclude that the lichen association is usually of doubtful value to the algal component. Although the alga may be able to inhabit new environments because of this association, it sacrifices its ability to grow and reproduce at its maximum rate and is not dependent on this relationship for its welfare. An exception is perhaps *Trebouxia* which is unable to compete effectively when free-living, and for this alga the lichen association is beneficial.

Mycorrhizae

Fungi can be found in association with the roots of almost all kinds of plants. Although the roots are infected by the fungi, the fungi cause little or no damage to the roots. The fungus–root combination is termed a *mycorrhiza*. The fungus and plant live together in a mutually beneficial symbiotic relationship under most conditions, and the fungi only rarely become parasites when the normal balance is overthrown.

Different types of mycorrhizae occur, which are distinguished by their morphology and to a certain extent, in their physiology. These include the *ectomycorrhizae*, characterized by an external sheath, and various types of *endomycorrhizae*, all lacking such a sheath. Three of these types are discussed further below.

ECTOMYCORRHIZAE

The ectomycorrhizae consist of a dense sheath of fungal cells surrounding the root. This external sheath can be more than 100 micrometers thick, and may be divided into more than one tissue layer. The fungus often penetrates between the cells of the epidermis and the first few cells of the cortical region, forming an intercellular hyphal network, but does not penetrate the cells themselves (Fig. 15–6). Characteristics of the sheath seem to be determined primarily by the fungus rather than by the plant, and morphological variations are used as an aid in identifying the fungal partner (Jackson and Mason, 1984). One of the most conspicuous differences is that of color; some fungi form jet black mycorrhizae while others may be yellow-brown or some other color. Sheaths may appear to be enveloped in a dense weft of hyphae or smooth, appearing to lack radiating hypha (in nature, gradations from one condition to the other occur). Rhizomorphs sometimes originate at the sheath and extend far into the soil.

The Ectomycorrhizal Plants. Ectomycorrhizae are typical of forest trees of the temperate region, where they occur more commonly on roots in the hu-

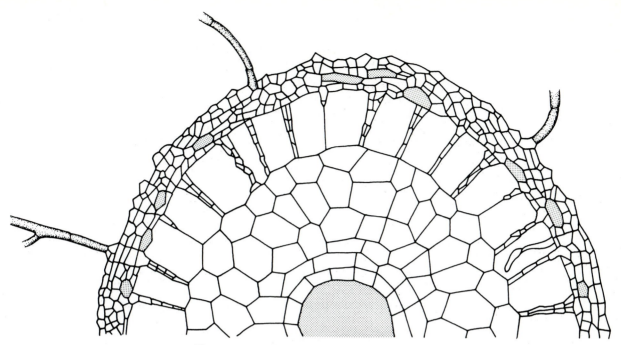

Figure 15–6 Cross section of an ectomycorrhiza, showing fungal sheath and intercellular hyphae. [From J. L. Harley, 1965, Ecology of soil-borne plant pathogens: prelude to biological control, University of California Press, Los Angeles. Reprinted by permission of the Regents of the University of California.]

mus layer than at greater depths. This association between the fungus and root is established on actively growing secondary roots. Some forest trees are obligate ectomycorrhizal formers and will fail to grow in the absence of mycorrhizal fungi: these include species of pine, oak, beech, and spruce. Other trees are facultative ectomycorrhizal formers: these may thrive in the absence of mycorrhizal fungi, although they may form ectomycorrhizae if the fungi are present. Facultative ectomycorrhizal trees include species of maple, juniper, willow, and elm, which are also often among the first invaders of wasteland, and are pioneers in forest succession. When moving into a wasteland, the facultative ectomycorrhizal plant species can survive in the absence of the established mycorrhizal fungus flora. As these plants grow, the mycorrhizal fungi eventually become established, and later the obligate ectomycorrhizal plant species may eventually become established (Meyer, 1973). It is sometimes desirable to establish trees in unfavorable locations such as barren wastelands or sites where there has been an extensive disruption resulting from construction of roadways. Either facultative ectomycorrhizal plant species or nursery-grown trees with established ectomycorrhizae may often be successfully planted in such sites.

The Ectomycorrhizal Fungi. Fungi that form ectomycorrhizae are usually members of the Basidiomycotina, especially members of the Agaricales and the Gasteromycetes. Some members of the Ascomycotina in the orders Eurotiales and Tuberales also form ectomycorrhizae. Sporocarps of certain fungi are frequently found under trees of a single species, and the assumption is often made that these are symbionts. Examples include the regular association of the agaric *Lactarius deliciosus* with *Pinus pinea* and also the bolete *Suillus granulatus* and the agaric *Russula emetica* with *Pinus pinaster*. The hypogeous genus *Rhizopogon* is also regularly associated with conifers. Some ectomycorrhizal fungi have a much broader host range and will apparently form the symbiotic relationship with several genera of host trees or sometimes live saprophytically. The actual number of ectomycorrhizal fungi is unknown, and many species apparently living entirely as saprophytes may actually be ectomycorrhizal.

In many instances, the degree to which the fungus is dependent on the ectomycorrhizal relationship remains obscure. Attempts to culture the fungus partner of ectomycorrhizae are sometimes futile, which implies the existence of intricate nutritional relationships with the host. Some fungi are apparently obligate ectomycorrhizal formers and are not known to sporulate in the absence of the mycorrhizal association; it is not known whether they exist saprophytically apart from their hosts. Melin (1953) found in his cultural studies of some ectomycorrhizal fungi that they cannot utilize a complex carbohydrate such as cellulose or lignin but require soluble carbohydrates. In addition, these fungi all require an exogenous source of thiamine, and some require other vitamins, growth factors in root exudates, or amino acids. Although inability to utilize complex carbohydrates is apparently of common occurrence among the ectomycorrhizal formers, it is not universal. In general, a dependency on soluble carbohydrates would make the ectomycorrhizal fungi poor competitive saprophytes in the soil and dependent on the sugars in the root as a food source. As explained further in the next section, the fungi do obtain sugars from the root.

Nature of the Interaction. The plant seeds germinate and give rise to an uninfected plant, and the ectomycorrhizae are established some weeks later. Only actively growing secondary roots are invaded. Germination of fungal spores and growth of mycelium in the vicinity of the root are stimulated by exudates from the root, which contain a variety of amino acids and vitamins. The fungus forms a weft of hyphae over the surface of the root, which gradually thickens into the sheath. Some hyphae invade the root by mechanically penetrating between the cells of the cortex (Harley and Smith, 1983). Further invasion of the fungus into the root seems to be inhibited by endogenous volatile compounds (Melin and Krupa, 1971). Typically, the root is maintained in a juvenile condition and fails to develop suberin, root hairs fail to develop, the root is extremely short and sometimes widened, and additional branching occurs (Fig. 15–7). Branching is usually dichotomous in the pines, whereas it is pinnate to irregular in other hosts. Anatomical studies of beech mycorrhizae show that the root continues to grow although the meristematic region is considerably reduced in length, owing to both a decrease in the division of meristematic cells and cell elongation. The surface area of the mycorrhizal roots is increased and their life prolonged.

Hormones produced by the fungus probably influence the development of the root, in part causing the distinctive morphological changes (Slankis, 1973). The ectomycorrhizal fungi are generally capable of secreting a variety of hormones in culture. These include auxins, cytokinins, and gibberellins which are homologous to those formed normally by the plant and which regulate cell division, growth, and other physiological processes, such as the mobilization and control of nutrient translocation.

After the fungus becomes established within the root, nutritional dependence on the plant can be demonstrated. Melin and Nilsson (1957) determined that carbohydrates move from the root into the fungus. They grew pine seedlings in an atmosphere containing ^{14}C-labeled carbon dioxide and showed that the ^{14}C fixed in photosynthesis could be detected in organic compounds that were translocated from the root into the fungal sheath. In ectomycorrhizae generally, the soluble carbohydrates that pass from the root into the fungal sheath are converted into insoluble storage polysaccharides (mannitol, trehalose, and glycogen) within the sheath. These polysaccharides are available to the fungus but not to the plant. These events help maintain a flow of carbohydrates to the

Figure 15–7 Roots of pine seedlings with (left) and without (right) the ectomycorrhizal fungus *Pisolithus tinctorius*. These are roots from the seedlings shown in Fig. 15–8. [Courtesy D. H. Marx, U.S. Department of Agriculture, Forest Service.]

roots by maintaining a gradient in the concentration of soluble carbohydrates in the root and by having a ''sink'' into which the carbohydrate moves.

Ectomycorrhizal fungi aid the plant in mineral uptake because they extend far into the soil and have the ability to accumulate ions in a very dilute solution. The ions are transported through the hyphae to the sheath, where most accumulate. For phosphates, as much as 90% of the phosphate may remain in the fungus sheath, where it is stored as polyphosphate granules while only 10% passes into the root. The quantity of phosphate entering the root increases when the external supply of phosphate becomes deficient. Under woodland conditions, quick uptake of ions by ectomycorrhizae is significant, as soluble ions are quickly flushed from the seasonal litter fall, passing immediately into the deeper mineral layers of the soil. These ectomycorrhizae can accumulate ions during the brief period when they are in luxurious supply and regulate their passage into the plant during the remainder of the year, when they are scarce. Further, mineral uptake is enhanced because of the increase in the surface area of ectomycorrhizal roots. The increased ability to accumulate ions and to pass them selectively to the host plant is advantageous to the plant, especially when it is grown in nutrient-deficient soils (Fig. 15–8). Probably all soils but the most fertile agricultural soils are somewhat deficient in nutrients, which makes ectomycorrhizal formation obligate for most plants in woodlands.

Ectomycorrhizae can also benefit the plant by increasing water absorption. Some ectomycorrhizal fungi can survive at very low soil moisture levels, and may even survive under drought conditions that would cause the wilting of plants (Harley and Smith, 1983). Duddridge et al. (1980) demonstrated that specialized hyphae, from 20 to 60 micrometers in diameter, formed within rhizomorphs when the transverse walls broke down. Water moved through these ''vessel'' hyphae into the roots and into the young leaves, where much of it was lost by transpiration. The fungi that form rhizomorphs may be particularly valuable as symbionts in dry habitats.

Figure 15–8 Enhanced growth occurs in the pine seedling that has established an ectomycorrhizal relationship with *Pisolithus tinctorius*. [Courtesy D. H. Marx, U.S. Department of Agriculture, Forest Service.]

VESICULAR–ARBUSCULAR MYCORRHIZAE

Unlike the sheath-forming ectomycorrhizal fungi, the vesicular–arbuscular (VA) fungi are endomycorrhizal and invade deeply into the root (Fig. 15–9). These fungi are abundantly represented in fossils from the Devonian and later periods and apparently coevolved with their hosts. Host plants include herbaceous plants, shrubs, and trees in both the temperate and tropical habitats. VA mycorrhizae are abundantly represented in herbaceous plants growing in cultivated soils and in grasslands. Many of the plants invaded by VA fungi are economically important: These include the legumes, grasses, tomatoes, apples, strawberries, peaches, and coffee. Overall, the VA mycorrhizae are more common than the other mycorrhizal types and occur almost universally in plants lacking other types, including some forest trees.

The VA fungi are all members of the Division Zygomycotina, belonging to the genus *Endogone* or closely related members of the Endogonaceae. The symbiotic association is obligate for these fungi, and they have not been successfully cultured apart from their hosts. However, not all members of *Endogone* are VA fungi and they live saprophytically.

Endogone forms sporocarps in the soil (Chapter 3) and produces an odor that attracts rodents, which dig up the sporocarps, eat them, and scatter the spores. The spores as well as the vegetative mycelium can serve as the inoculum for establishing the infection within the root. When Mount St. Helens erupted in 1980, widespread destruction of plants and larger animals occurred and large tracts of sterile soil resulted from the volcanic activity. Up to 200 meters of sterile rock material covered previously fertile soils, and both mycorrhizal plants and spores of VA fungi were absent in the volcanic debris. Gophers survived the eruption, and their tunneling activities proved to be significant in transporting viable VA spores up from the buried soil to the surface, where the spores would be in the vicinity of potential plant host roots. VA mycorrhizae were reestablished in the vicinity of gopher activity, a significant factor in reestablishing the ecosystem. Additionally, mammals such as elk appear to be important in long-

Figure 15–9 Cross section of VA mycorrhiza. Note the penetrating hypha (*PH*), branched arbuscules (*A*), globular vesicles (*v*), digestion of haustoria by plant (*D*), and external sporocarps (*S*) with spores. [Adapted from J. L. Harley, 1965, Ecology of soil-borne plant pathogens: a prelude to biological control, University of California Press, Los Angeles. Reprinted by permission of The Regents of the University of California.]

distance distribution of the VA mycorrhizae because grass seedlings already infected with the VA fungi could frequently be isolated from elk feces (Allen, 1987).

The VAM fungi grow as individual strands or loose wefts over the roots and extend into the soil. They penetrate host cells by forming cellulolytic enzymes that digest small portions of the cell walls. The fungus may be limited to the root hairs or other epidermal cells but also may penetrate cells of the cortex. Within the cells, the fungus produces enlarged or globular *vesicles* and clusters of dichotomously branched hyphal endings (the *arbuscles),* giving us the name for this mycorrhizal type (Fig. 15–9). The vesicles develop as swellings along the hyphae or at the tips, and their walls thicken and lipids are deposited in their cytoplasm as they mature. They probably function as storage organs and perhaps as survival structures under adverse conditions. The arbuscles develop when a hyphal tip penetrates a host cell and begins to branch dichotomously, providing a much greater area of contact between the fungus and the plant's cytoplasm. Relatively large quantities of phosphate (stored as polyphosphate granules) can be detected in the arbuscles, indicating that they are transporting phosphate from the soil to the root. Transfer of minerals from the fungus to the host and of carbohydrates from the host to the fungus probably occurs largely across the plasmalemma of the arbuscles. The arbuscles lyse when they become senescent, also releasing nutrients into the plant cell. Unlike the ectomycorrhizae, the morphology of the roots usually remains essentially unchanged, although sometimes they become darker, slightly thickened, or more brittle.

Many aspects of the physiology of the VA mycorrhizae are generally similar to that of the ectomycorrhizae. Like the ectomycorrhizae, their formation is favored by high light intensities and soils with low fertility. The VAM relationship benefits the plant by enhancing phosphate uptake and perhaps that of other minerals, such as potassium, iron, copper, calcium, and zinc. This ability to increase mineral uptake favors plant growth in poor soils. VA roots have a greater capacity to take up water and may withstand the stress of low levels of available moisture better than do nonmycorrhizal roots. For example, they recover from wilting better than nonmycorrhizal plants do and are better able to withstand the shock of being transplanted.

ORCHID MYCORRHIZAE

Almost all members of the Orchidaceae, the orchid family, enter into a symbiotic relationship with fungi that produce clamps, and are therefore members of the Division Basidiomycotina or are members of the genus *Rhizoctonia,* nonsporulating members of the Deuteromycotina with basidiomycetous affinities.

The orchids constitute a diverse group of temperate and tropical members, including both terrestrial and epiphytic forms. Orchids produce a very large number of minute seeds in each capsule, sometimes up to several million. The seeds contain a very small embryo and an infinitesimal food reserve that is depleted after the first few divisions of the embryo. After the food reserves in the seed are used and before the plant becomes completely autotrophic, virtually all orchids exist temporarily as symbionts. The length of the symbiotic period varies from a few months or years to their entire lifetime in the case of some non-chlorophyll-containing orchids. Prior to the symbiotic period, orchid seeds under natural conditions are invaded by hyphae of a basidiomycete or *Rhizoctonia* species that forms endomycorrhizae with the developing roots or other absorbing organs. The fungi are intercellular, forming coils of hyphae within the cells (Fig. 15–10).

Endomycorrhizal formation is obligate for the orchids under naturally occurring conditions. In the absence of fungal invasions, orchids fail to grow un-

Figure 15–10 Cross section of an orchid mycorrhiza showing fungal penetration and intracellular digestion (*D*). [From J. L. Harley, 1965, *Ecology of soil-borne plant pathogens: prelude to biological control,* University of California Press, Los Angeles. Reprinted by permission of The Regents of the University of California.]

less they are supplied with an external source of organic carbon compounds and sometimes vitamins. The fungi provide carbohydrates and possibly other accessory metabolites such as vitamins to the orchids, so in reality the orchids are parasitic on the fungi. Unlike those fungi that form ectomycorrhizae, the orchid fungi are able to utilize complex carbohydrates, often including lignin, pectins, and cellulose. The fungus hyphae digest organic materials in the surrounding environment (either the soil or supporting tree, in the case of the epiphytes), and assimilate these nutrients to support their own growth and to produce glucose, ribose, and other simple carbohydrates. The carbohydrates and other nutrients are translocated within the hyphae and finally released into the orchid host. This release is partially accomplished by the host's digestion of some of the intracellular coils, during which the coils swell and then disintegrate. In addition, it is likely that metabolites pass through the hyphal membranes and into the host cells without digestion of the fungus. Although the major direction of nutrient flow is from the fungus to the orchid, it is important to remember that the fungus is also deriving nutrients from the orchid. The fungi are apparently sometimes dependent upon the orchid as a source of amino acids or exogenous vitamins, such as thiamine or perhaps only one of the moieties of thiamine. However, experiments utilizing radioactive carbon have indicated that carbon is not transferred from the orchids to the fungus (Arditti, 1979).

The prime importance of the digestive process is apparently to control the degree of invasion by the fungus and to prevent it from parasitizing the seedling. Frequently invasion by a particularly virulent fungus completely parasitizes and kills the host plant. Conversely, the orchid seedling may so actively digest the fungus that an endomycorrhizal relationship is not established, which again leads to the death of the host. An additional resistance mechanism is the formation of at least three antifungal substances that are produced by some orchids in response to fungal invasion. These antifungal substances include *orchinol,* a high-molecular-weight phenolic compound, which inhibits fungal growth.

The fungi forming orchid endomycorrhizae are different from the ectomycorrhizal fungi in several important respects. The orchid fungi are widely distributed and have an extensive saprophytic existence apart from the orchids. Specificity for mycorrhizal formation with the orchids is low, and some of the fungi are virulent plant parasites. For example, some pathogenic strains of *Rhizoctonia solani* isolated from wheat, cauliflower, and tomato were capable of forming an endomycorrhiza with an orchid (Downie, 1957).

FUNGI AS INSECT SYMBIONTS

A number of fungi have symbiotic relationships with insects. Insect symbionts may be either *endosymbionts,* which are harbored within the insect, or *ectosymbionts,* which occur externally to the insect.

Endosymbionts

Insect endosymbionts are analogous to the intestinal flora of man. These insect endosymbionts include numerous species of bacteria and some yeasts (predominantly species of *Candida* but also *Torulopsis* and *Taphrina)* that occur routinely in the gut, fat bodies, Malpighian tubules, or other internal organs of some insect species. Endosymbionts are especially common in insects that live on a restricted diet such as wood, grains, or humus. These are rich in cellulose but poor in nitrogen and deficient in vitamins.

The endosymbionts may be harbored in specialized pouches of the gut or in modified cells (Fig. 15–11). The endosymbionts may be transmitted to offspring in a number of ways, including the following: (1) The offspring may suck or lick endosymbiont-containing droplets from the mother's body; (2) the eggs are smeared with material containing the bacteria or yeast in their passage through the ovipositor, and the young consume fragments of the shell upon hatching; or (3) the endosymbiont may be included within the egg and thereby included within the developing embryo.

Endosymbionts are provided with a relatively constant, favorable environment and are richly supplied with nutrients. From studies of nutritional requirements of symbiont-free insects, it has been determined that those endosymbionts studied produce vitamins (especially B vitamins) and some amino acids needed by the host to grow and develop normally and, in some instances, to remain alive. In some experiments, the symbiont-free insects throve if provided with amino acids alone but not with vitamins alone, which suggests that the dependence on amino acids as a source of nitrogen may be greater than the dependency on vitamins.

Ectosymbionts

SEPTOBASIDIUM

Septobasidium, a member of the Subclass Phragmobasidiomycetidae, is abundant in the tropics, semitropics, and southeastern United States (Couch, 1935). This fungus forms irregular, flattened, resupinate colonies that adhere closely to the bark or leaves of living trees. The colonies may be very inconspicuous (only a few millimeters in diameter) or may cover the entire lower side of the branches. The colonies may be dry and crustaceous or spongy and range in color from whitish through various shades of brown and black. Some species of *Septobasidium* are directly parasitic on the tree, causing a hypertrophy of the tis-

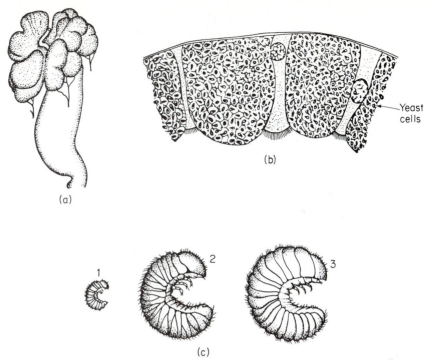

Figure 15–11 Endosymbiosis of a beetle by a yeast: (a) yeast-inhabited evaginations of the midgut of a larva; (b) wall of the evaginations; (c) effect of the symbiotic yeast on growth of larvae. Left: abnormally small larva from which the symbiont has been removed; center: larva from which the yeast has been removed but the larva was fed on a yeast-supplemented diet; right: normal larva, containing yeast symbionts. [From P. Buchner, 1965, Endosymbiosis of animals with plant microorganisms, John Wiley & Sons, Inc., New York.]

sues. The majority of the species, however, share a symbiotic relationship with scale insects.

The scale insects occupy chambers within the middle layer of the fungus colony. These chambers are only slightly larger than the insect, and there are numerous tunnels connecting chambers. Over this labyrinth, there is a dense fungus mat, on which the hymenium is formed (Fig. 15–12). Some of the insects are parasitized by *Septobasidium,* and the fungus forms several coiled or knotted haustoria within their hemocoels; numerous hyphae extend from the insect into the tissues of the fungus colony. These parasitized individuals occupy a chamber, where they are attached to the plant by their sucking apparatus. The plant sap therefore directly nourishes the insects and indirectly nourishes the fungus. A parasitized insect is much smaller than uninfected individuals and usually cannot reproduce. Many insects within the colony are not parasitized, often outnumbering the parasitized individuals. The fungus colony overwinters in this manner.

Growth of the colony is renewed in the spring, and the colony grows only a few millimeters annually. Some species may form a new layer of mycelium over the old one. With the renewal of colony growth over the insect, females give birth to nymphs while the fungus probasidia germinate to form basidia and basidiospores. The basidiospores then bud in a yeastlike manner. The young nymphs crawl about and, if they venture on the surface of the colony, will pick up the fungus buds which adhere to their surface. These buds germinate and penetrate the integument of the scale insect, thereby infecting these individuals. The nymphs settle down under the same mother colony, crawl out and settle

Figure 15–12 Transverse section through portion of *Septobasidium* colony and scale insect with sucking apparatus inserted into medullary rays of wood. *b*, basidium; *ys*, young insect; *sp*, basidiospores; *pb*, probasidium; *fs*, fruiting surface of fungus; *tl*, top layer of fungus; *bl*, bottom layer of fungus; *fm*, fungus mat; *ct*, thread connecting fungus mat with insect; *ss*, spindle-shaped hyphal threads in insect; *c*, coiled haustoria within insect; *st*, stylet of insect; *sh*, sheath secreted around stylet; *b*, bark of tree; *c*, cambium of tree; *mc*, medullary ray of tree. [From J. N. Couch, 1931, Quart. J. Microscop. Sci. II 74:383–437.]

under other fungal colonies, or crawl out and settle down on the bark where there is no fungal growth. These latter individuals initiate new colonies.

Species of *Septobasidium* that are parasitic on the insects are unable to grow independently of the insects in nature, and the amount of fungus growth that occurs is directly dependent on the number of parasitized insects. In addition to depending on the insects for food, the fungus is dependent on this relationship for spore distribution.

Although the scale insects can grow in the absence of the fungus in nature, free-living individuals occur only rarely. Life is too precarious for the free-living insects, as they are subjected to extremes of heat and cold from which the fungus colony protects them. They are also much more likely to be eaten by birds and to be parasitized by their most deadly enemy, those wasps which lay eggs within the scale insect's body. Although some insects are parasitized and may be killed by the fungus, the majority of the insects benefit from this protective relationship.

AMBROSIA FUNGI

Some scolytid beetles are inhabitants of wood, forming tunnels in the sapwood of diseased trees or felled logs. These tunnels consist of an entrance passageway from the outside, and they widen into a number of cavelike chambers or elongated galleries where eggs are laid and larvae develop. The tunnels, chambers, and galleries are lined with a cushionlike growth of *ambrosia* fungi (Fig. 15–13), which usually is the sole source of food for the larvae and is also important in the diet of the adults, who feed on the wood in addition. This symbiotic relationship between the beetle and fungus is highly specific, with only certain species of the beetle or fungus occurring together. Neither the ambrosia fungi nor the beetles are found without the other in nature.

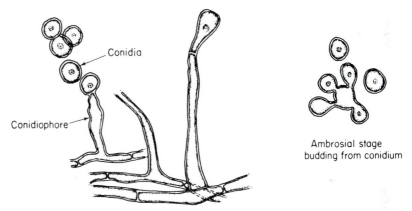

Figure 15–13 Ambrosia fungus, *Ambrosiella canadensis*. × 900. [From L. R. Batra, 1967, Mycologia 59:976–1017.]

The ambrosia fungi were originally seen as a saltlike crust lining beetle tunnels and were thought to be dried, exuded sap (the "ambrosia"). Later fungi were found in the tunnels but were thought to be incidental invaders that fed on the ambrosia and acquired the name of ambrosia fungi. The ambrosia fungi are filamentous members of the Hemiascomycetes (especially *Endomycopsis* and *Ascoidea*) and members of the Deuteromycotina (especially *Monilia* and *Ambrosiella*). These fungi form hyphae with conidiophores, conidia, and yeastlike sprout cells that bud from the conidia. The yeastlike cells constitute the ambrosial stage on which the beetles feed.

The beetles carry spores of the ambrosia fungus with them during their overwinter hibernations and during their migrations. The spores are stored in specialized tubes or pouches *(mycetangia)* that occur in different locations of the body among the insect species (Fig. 15–14). The mycetangia occur in the one sex of the beetle that initiates the tunnel-building; this is usually the female who prepares the tunnels for egg-laying. The mycetangia produce secretions that protect the thin-walled spores from desiccation and provide nutrients needed for the germination of the spores. When the beetles bore new tunnels in the spring, they prepare a mixture of feces and wood fragments which they smear on the tunnel walls as a bed for the fungal growth. The beds are inoculated with

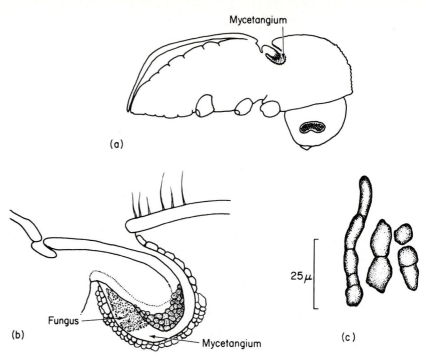

Figure 15–14 Mycetangium of *Anisandrus dispar:* (a) position of myce-tangium in insect, × 7.3; (b) enlargement of mycetangium showing glandular cells associated with it, and the enclosed fungus, × 180; (c) ambrosia fungus removed from the mycetangium. [From L. R. Batra, 1963, Trans. Kansas Acad. Sci. 66:213–236.]

ambrosia fungi through incidental contact with the mycetangia. A palisade layer of ambrosia fungus then develops, and the yeast-like ambrosial stage predominates. The ambrosial stage is dominant as long as the larvae and beetles are present, but when the beetles vacate the tunnel, the mycelial form dominates (as it does in pure culture). The beetles tend to maintain a pure culture of the ambrosia fungus and are able to do so when the tunnels are fresh. The means by which the beetles are able to suppress growth of foreign fungi is unknown, but perhaps the larvae and beetles have secretions that promote growth of the ambrosia fungi, which in turn are antagonistic to other fungi. Alternatively, the constant enlargement of the tunnels, coupled with the discardation of the debris, may eliminate the foreign fungi. Often secondary ambrosia fungi may appear in the tunnel, and the beetles may feed on them (the Hemiascomycete *Dipodascopsis*, discussed in Chapter 4, is of this type). Unlike the primary ambrosia fungus, these secondary fungi are not transported in mycetangia, and the beetle is not dependent on their presence. When the beetles vacate the tunnels, contaminating fungi of several types develop profusely.

The relationship between the fungus and beetles is mutually beneficial and is obligatory for both partners. Larvae are dependent on the ambrosia fungus as a source of food and vitamins, and even the adults may starve in the absence of ambrosial growth. There is evidence that at least some of the ambrosia beetles may have an absolute requirement for the ergosterol produced by the fungi (Kok, 1979).

Advantages of a fungus diet are that the fungus provides a high nitrogen, low residue food that can be obtained with a minimum of tunneling as the nutrients are translocated for some distance through the mycelium. Those beetles which feed instead on wood (a low nitrogen, high residue food) must actively tunnel through the wood to obtain enough food. The fungus benefits because it is protected from desiccation during transport in the mycetangia, the number of cells available for inoculation of wood are increased by the mycetangial secretions, it is disseminated directly to and placed with a suitable substratum, injury of the wood by the beetle facilitates rapid penetration by the mycelium, and urea and uric acids in the beetle's feces serve as an important source of nitrogen. Attempts to establish the fungus in the host plant in the absence of the beetle have been unsuccessful.

FUNGUS-GROWING INSECTS

Fungi may be cultivated by termites and ants, which tend their fungal gardens with great diligence. These insects depend on the fungi for their food (Figs. 15–15 and 15–16). The fungus-cultivating termites are especially prevalent in Af-

Figure 15–15 Excavation of termite nests used as brood chambers showing the spongelike combs that provide food for the insects. The combs appear light because of the abundant mycelium of the symbiotic fungus *Termitomyces striatus*. The mycelium gives rise to the aboveground basidiocarps after the termites desert the nest. [From P. A. Dixon, 1983, Trans. Brit. Mycol. Soc. 80:131–139.]

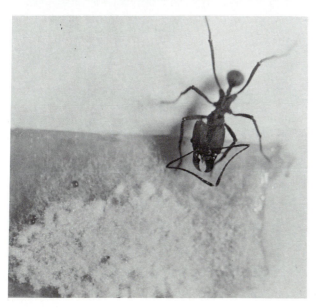

Figure 15–16 Ant with its cultivated fungus. [Courtesy N. A. Weber, 1966, Science 153:587–609.]

rica and southern Asia, where they usually construct conspicuous clay mounds that are up to 6 meters high and cover their nests and fungal gardens (Batra and Batra, 1979). These fungus-cultivating termites share many similarities with the ants, which are discussed further below.

Fungus-Growing Ants. The fungus-cultivating ants belong to a New World tribe of ants, the Attini, which are widely distributed on the American continent but are especially prevalent in the rain forests of South America. The ants and fungi are dependent on each other for their maintenance, and neither is found alone in nature (Weber, 1972-a,b, 1979).

The fungi are cultivated in well-circumscribed gardens. The gardens consist of a light gray to brown flocculent, spongy mass of leaves overgrown with the fungus and may be honeycombed with several tunnels or cells. The ants grow their gardens either in protected places on the ground such as under rocks or overhanging logs or, more usually, in chambers within nests (Fig. 15–17). The chambers may be craterlike excavations at the surface of the ground, or they may be irregular or domelike cavities in the ground that are connected to the exterior by tunnels. A single fungus garden may vary from 20 millimeters to 50 centimeters in diameter. As many as 2000 chambers may be found in a nest. The area of ground covered by the nest may be as large as 11 × 15 meters.

(a)   (b)

Figure 15–17 Two fungus gardens cultivated by ants. (a) A section of a nest made by *Trachymyrmex wheeleri* in Colombia. Compare the size of the nest with the forceps which are 113 millimeters long. (b) A section of a nest made by *Acromyrmex histrix* in Venezuela. There are two gardens in this nest. Compare the size of the nest with the knife, which is 25 centimeters long. [Courtesy of N. A. Weber; (a) 1966, Monogr. 1, Prog. Biol del Suelo, Actas Prim. Coloq. Latin. Biol. del Suelo; (b) from W. M. Wheeler, 1937, Mosaics and other anomalies among ants, Harvard University Press, Cambridge, Mass.]

Fungi associated with the ant gardens belong in one instance to the yeasts *(Tyridomyces formicarum)*, but otherwise they are mycelial fungi. While the identity of the fungus is unknown in most instances, basidiocarps of agarics in the genus *Lepiota* have been obtained by culturing the mycelium from several ant nests (Weber, 1979). The mycelia form clusters of inflated hyphae, which occur either in the ant nest or in pure culture in the absence of ants (Fig. 15–18). The protoplasm-rich inflations are the sole food of the larvae, although adult ants feed on both hyphae and hyphal inflations. The fungus is capable of actively degrading the cellulose in the substratum and of providing the ants with a nutritious diet. More than 50% of the dry weight of the fungus is available as soluble nutrients (including carbohydrates, amino acids, proteins and lipids) (Martin et al., 1969).

Figure 15–18 Fungi cultivated by ants. The enlarged hyphal endings may be as wide as 50 micrometers in diameter. [From N. A. Weber, 1966, Science 153:587–604.]

Some ants begin their fungus gardens by cutting leaves, while other genera scavenge for plant debris. The ants forage for plant material in the daytime and carry large cut pieces of leaves back to the nest. The ants maintain the cleanliness of the cut leaves by carrying the pieces high above the ground; also, a small ant may ride upon a section of a leaf carried by a larger ant, licking the leaf to cleanse it during its transport (Weber, 1972-a). When the ants reach the garden, they cut the leaf into small pieces (about 1 or 2 millimeters in diameter) and make these pieces pulpy by pinching them with their mandibles. They also lick the pieces repeatedly and then deposit a fecal droplet from the anus on it. Maceration of the leaves encourages the initial invasion of the tissue by the hyphae, and the application of fecal droplets facilitates this invasion because these droplets contain protein-degrading enzymes (the fungi themselves are deficient in these enzymes). The bits of plant material are packed into place and then the ant plants the garden with several tufts of mycelium carried in from an adjacent garden. Additional material may be placed in the nest, such as woody particles which may be dragged in, defecated on, and then inoculated with the fungus. Some ants also drag insect carcasses or insect excreta into the garden. The ants continue to deposit anal drops on the fungal garden even after it is established. These anal drops are an important source of nutrients for the fungus because they contain ammonia and a variety of amino acids. Continual deposition is required because the fungus is constantly utilizing the nutrients present within the anal drops (Martin and Martin, 1970). The ants also tend the garden by probing the fungus with their antennae to determine the condition of the fungus and by licking the hyphae. The ants can distinguish any invading fungi from those that they are cultivating and will remove hyphae of the unwanted species by pulling the hyphae out with their mandibles and placing the discarded hyphae on their refuse heap. Any unused substrata is carried out of the nest, and the ants may open or close the entrances of the tunnels to regulate the temperature. While the garden is maintained in good condition, the foreign bacteria and yeasts normally present exist at relatively low levels and are apparently simply suppressed in their growth by competition with the cultured fungus. Once the nest is abandoned, these bacteria, yeast, and other fungi completely overrun the nest as the gardens deteriorate. The ants abandon their nest if it is disturbed or if they migrate. Whenever the ants leave their nest, they carry portions of their garden along with them to serve as inoculum for the gardens that they will build in their new nest.

REFERENCES CITED

AHMADJIAN, V. 1966-a. Lichens. *In:* Henry, S. M., Ed. Symbiosis. Academic Press, New York, Vol. 1, pp. 35–97.

——. 1966-b. Cultural and physiological aspects of the lichen symbiosis. *In:* Jensen, W. A., and L. G. Kavaljian, Eds. Plant biology today: advances and challenges. Wadsworth, Belmont, Calif., pp. 148–163.

——. 1967-a. The lichen symbiosis. Blaisdell, Waltham, Mass., 152 pp.

——. 1967-b. A guide to the algae occurring as lichen symbionts: isolation, culture, cultural physiology, and identification. Phycologia 6:127–160.

——. 1973. Resynthesis of lichens. *In:* Ahmadjian, V., and M. E. Hale, Eds. The lichens. Academic Press, New York, pp. 565–579.

ALLEN, M. F. 1987. Re-establishment of mycorrhizas on Mount St. Helens: migration vectors. Trans. Brit. Mycol. Soc. 88:413–417.

ARDITTI, J. 1979. Aspects of the physiology of orchids. Advan. Botan. Res. 7:421–655.

BATRA, L. R., and S. W. T. BATRA. 1979. Termite–fungus mutualism. *In:* Batra, L. R., Ed. Insect–fungus symbiosis: nutrition, mutualism, and commensalism. Allanheld, Osmun, Totowa, N.J., pp. 117–163.

COUCH, J. N. 1935. *Septobasidium* in the United States. J. Elisha Mitchell Sci. Soc. 51:1–77.

DOWNIE, D. G. 1957. *Corticium solani*—an orchid endophyte. Nature 179:160.

DUDDRIDGE, J. A., A. MALIBARI, and D. J. READ. 1980. Structure and function of mycorrhizal rhizomorphs with special reference to their role in water transport. Nature 287:834–836.

FARRAR, J. F. 1976. The lichen as an ecosystem: observation and experiment. *In:* Brown, D. H., D. L. Hawksworth, and R. H. Bailey, Eds. Lichenology: progress and problems. Academic Press, New York, pp. 385–406.

HARLEY, J. L., and S. E. SMITH. 1983. Mycorrhizal symbiosis. Academic Press, New York, 483 pp.

HARRIS, G. P., and K. A. KERSHAW. 1971. Thallus growth and the distribution of stored metabolites in the phycobionts of the lichens *Parmelia sulcata* and *P. physodes*. Can. J. Botany 49:1367–1372.

JACKSON, R. M., and P. M. MASON. 1984. Mycorrhiza. Edward Arnold, London, 60 pp.

KOK, L. T. 1979. Lipids of ambrosia fungi and the life of mutualistic beetles. *In:* Batra, L. R., Ed. Insect–fungus symbiosis: nutrition, mutualism, and commensalism. Allanheld, Osmun, Totowa, N.J., pp. 32–52.

MARTIN, M. M., R. M. CARMAN, and J. G. MacCONNELL. 1969. Nutrients derived from the fungus cultured by the fungus-growing ant *Atta colombica tonsipes*. Ann. Entomol. Soc. Am. 62:11–13.

——, and J. S. MARTIN. 1970. The biochemical basis for the symbiosis between the ant, *Atta colombica tonsipes*, and its food fungus. J. Insect Physiol. 16:109–119.

MELIN, E. 1953. Physiology of mycorrhizal relations in plants. Ann. Rev. Plant Physiol. 4:325–346.

——, and S. Krupa. 1971. Studies on ectomycorrhizae of pine II. Growth inhibition of mycorrhizal fungi by volatile organic constituents of *Pinus silvestris* (Scots pine) roots. Physiol. Plantarum 15:337–340.

——, and H. Nilsson. 1950. Transfer of radioactive phosphorus to pine seedlings by means of mycorrhizal hyphae. Physiol. Plantarum 3:88–92.

——, and ——. 1957. Transport of C^{14}-labelled photosynthate to the fungal associate of pine mycorrhiza. Svensk Botan. Tidskr. 51:166–186.

Meyer, F. H. 1973. Distribution of ectomycorrhizae in native and man-made forests. *In:* Marks, G. C., and T. T. Kozlowski, Eds. Ectomycorrhizae: their ecology and physiology. Academic Press, New York, pp. 79–105.

Pearson, L. C. 1970. Varying environmental conditions in order to grow intact lichens under laboratory conditions. Am. J. Botany 57:659–664.

Perez-Llano, G. A. 1948. Economic uses of lichens. Econ. Botany 2:15–45.

Slankis, V. 1973. Hormonal relationships in mycorrhizal development. *In:* Marks, G. C., and T. T. Kozlowski, Eds. Ectomycorrhizae: their ecology and physiology. Academic Press, New York, pp. 231–298.

Smith, D. C. 1962. The biology of lichen thalli. Biol. Rev. Cambridge Phil. Soc. 37:537–570.

Weber, N. A. 1972-a. The fungus-culturing behavior of ants. Am. Zool. 12:577–587.

——. 1972-b. The attines: the fungus-culturing ants. Am. Scientist 60:448–456.

——. 1979. Fungus-culturing by ants. *In:* Batra, L. R., Ed. Insect–fungus symbiosis: nutrition, mutualism, and commensalism. Allanheld, Osmun, Totowa, N.J., pp. 77–116.

Additional Recommended References

Ahmadjian, V. 1965. Lichens. Ann. Rev. Microbiol. 19:1–20.

——, and M. E. Hale, Eds. 1973. The lichens. Academic Press, New York. 697 pp.

——, L. A. Russell, and K. C. Hildeth. 1980. Artificial establishment of lichens. I. Morphological interactions between the phycobionts of different lichens and the mycobionts *Cladonia cristatella* and *Lecanora chrysoleuca*. Mycologia 72:73–89.

Anderson, J. M., A. D. M. Rayner, and D. W. H. Walton, Eds. 1984. Invertebrate-microbial interactions. Cambridge University Press, Cambridge, 349 pp.

Batra, L. R. 1967. Ambrosia fungi: a taxonomic revision, and nutritional studies of some species. Mycologia 59:976–1017.

——, Ed. 1979. Insect–fungus symbiosis: nutrition, mutualism, and commensalism. Allenheld, Osmun, Totowa, N.J., 276 pp.

Batra, S. W. T., and L. R. Batra. 1967. The fungus gardens of insects. Sci. Am. 217:112–120.

Bliss, L. C., and E. B. Hadley. 1964. Photosynthesis and respiration of alpine lichens. Am. J. Botany 51:870–874.

Brown, D. H., D. L. Hawksworth, and R. H. Bailey, Eds. 1976. Lichenology: progress and problems. Academic Press, New York, 551 pp.

Francke-Grossman, H. 1967. Ectosymbiosis in wood-inhabiting insects. *In:* Henry, S. M., Ed. Symbiosis. Academic Press, New York, Vol. 2, pp. 141–205.

Graham, K. 1967. Fungal–insect mutualism in trees and timber. Ann. Rev. Entomol. 12:105–126.

Hale, M. E. 1959. Studies on lichen growth rate and succession. Bull. Torrey Botan. Club 86:126–129.

——. 1983. The biology of lichens, 3rd ed. Edward Arnold, London, 190 pp.

Koch, A. 1967. Insects and their endosymbionts. *In:* Henry, S. M., Ed. Symbiosis. Academic Press, New York, Vol. 2, pp. 1–106.

Marks, G. C., and T. T. Kozlowski, Eds. 1973. Ectomycorrhizae: their ecology and physiology. Academic Press, New York, 444 pp.

Meyer, F. H. 1974. Physiology of mycorrhiza. Ann. Rev. Plant Physiol. 25:567–586.

Powell, C. L., and D. J. Bagyaraj, Eds. 1984. VA mycorrhiza. CRC Press, Boca Raton, Fla., 234 pp.

Scott, G. D. 1956. Further investigations of some lichens for fixation of nitrogen. New Phytologist 55:111–116.

———. 1967. Studies of the lichen symbiosis. 3. The water relations of lichens on granite kopjes in central Africa. Lichenologist (Oxford) 3:368–385.

Skye, E. 1979. Lichens as biological indicators of air pollution. Ann. Rev. Phytopathol. 17:325–341.

Slankis, V. 1971. Formation of ectomycorrhizae of forest trees in relation to light, carbohydrates, and auxins. *In:* Hacskaylo, E., Ed. Mycorrhizae. U.S. Dep. Agr. Forest Serv. Misc. Publ. 1189, pp. 151–167.

———. 1974. Soil factors influencing formation of mycorrhizae. Ann. Rev. Phytopathol. 12:437–457.

Smith, A. H. 1971. Taxonomy of ectomycorrhizae-forming fungi. *In:* Hacskaylo, E., Ed. Mycorrhizae. U.S. Dep. Agr. Forest Serv. Misc. Publ. 1189, pp. 1–18.

Smith, D. C. 1963. Experimental studies of lichen physiology. *In:* Symbiotic associations. 13th Symp. Soc. Gen. Microbiol. 13:31–50.

———, and E. A. Drew. 1965. Studies in the physiology of lichens. V. Translocation from the algal layer to the medulla in *Peltigera polydactyla*. New Phytologist 64:195–200.

Tinker, P. B. H. 1975. Effects of vesicular–arbuscular mycorrhizas on higher plants. Symp. Soc. Exptl. Botany 29:325–349.

Trappe, J. M. 1962. Fungus associates of ectotrophic mycorrhizae. Botan. Rev. 28:538–606.

16 Fungi and Humans

Not only do fungi exploit other organisms (as saprophytes, parasites, or symbionts), but other organisms exploit fungi in numerous ways. The most usual of these is the use of fungi as food by various animals. It may be assumed that thousands of species consume fungal mycelium or the sporocarps of fungi. As we noted in Chapter 12, many soil animals consume fungi. Particularly avid mycophagists include the slugs and the red squirrel. Slugs will attack most fleshy Hymenomycete species, including *Amanita muscaria* and *A. phalloides*, which are poisonous to humans. The red squirrel collects large numbers of fungi in the late autumn and stores these fungi in tree trunks, in birds' nests, and on branches of trees whereupon the fungi dry out and serve as a supply of food for the winter (Buller, 1922). People are also avid mycophagists, but in addition they exploit the fungi in numerous ways which are not shared by other organisms. The remainder of the chapter deals with exploitation of fungi by humans.

Like other forms of life, fungi have long been objects of wonder and speculation. Some of the earliest recorded observations of the fungi are those of the Greeks and Romans, who were curious about the nature and origin of fungi. Pliny could not decide whether fungi were living or nonliving but nevertheless concluded that they were earthy concretions or imperfections having no vital connections with the earth (he was obviously unaware of the mycelium). A belief rather widely held by the Greeks (about 300 B.C.) was that truffles originate from thunder; that is, a generating fluid was contained in the thunder which, when mixed with heat, pierces the earth and forms the fungi (Buller, 1914).

The interest of some cultural groups in fungi may have been based on religious beliefs. For example, mushrooms are illustrated in some ancient Mayan hieroglyphics, and a possible interpretation is that these are sacred mushrooms being held in the hand of an animal deity. The Mayans also made stone carvings in the shape of mushrooms. These carvings are about 25 to 50 centimeters high and have some human features (many are pregnant females) or are effigies of

Figure 16–1 Mushroom stone featuring a human–animal effigy with claws on the base. The height is 29.5 centimeters and the mushroom pileus is 16.5 centimeters. [Courtesy B. Lowy.]

animals such as toads (Fig. 16–1). Although these are relics of a culture that existed 1000 to 4000 years ago, they only came to the attention of anthropologists in the early part of this century and are being recovered from excavations in Guatemala. These carvings may be relics of a culture in which mushrooms were consumed during religious ceremonies, during which some may have been used as seats (Lowy, 1971, 1972).

Fungi have been utilized by humans for a variety of purposes since ancient times. Some puffballs were recovered from British archeological sites that were approximately 2000 years old and are believed to have been used to stop the bleeding of wounds or as tinder (Watling and Seaward, 1976). Puffballs have long been utilized for these purposes, and as late as World War I were used to dress wounds when bandages were in short supply. Presently in Nigeria, fungi are extremely important to the Yoruba people, who have a culture rich in mycological folklore (Oso, 1975, 1977). These people may use various fungi as food; for medicinal purposes; as toothbrushes; in charms intended, for example, for use in making oneself disappear in the face of danger; or in a preparation that is rubbed on the gun, bow, or arrow along with incantations to make game drowsy and easy to kill.

FUNGI AS FOOD FOR HUMANS

One of the predominating interests that humans have had in fungi is their utilization as food. The Greeks and Romans were fond of truffles, mushrooms, and puffballs, all of which were such delicacies that they were the only foods that the wealthy insisted upon preparing themselves. The Roman Coelius Apicus gives us a recipe for the preparation of truffles (Buller, 1914):

. . . slice, boil, sprinkle with salt, and transfix with a twig; partly roast, and place in a cooking-vessel with oil, liquor, sweet boiled wine, unmixed wine, pepper and honey; while boiling, beat up with fine flour, take out the twigs and serve.

Although methods of preparation have changed with time, the view that fungi are delicacies has not.

Fungi collected for food include truffles, morels, mushrooms, puffballs, coral fungi, and nonwoody polypores. The truffles are probably the most highly esteemed of all and in France are collected with the aid of trained dogs or pigs that locate the scent of the underground ascocarps. Truffles are also commonly sold in the markets in North Africa and the Near East.

Cultivation of Fungi

Oriental people culture a variety of fungi for food. These include the jelly fungus *Auricularia* and various kinds of mushrooms, particularly the Shiitake mushroom *(Lentinus edodes)*, which has been cultured on inoculated logs in Japan for more than 2000 years. Truffles are cultivated in France, where truffle spores are introduced into the ground where oak trees are growing, and large numbers of ascocarps may be harvested later. Only one species, *Agaricus brunnescens*, is widely cultivated commercially in the United States and other Western countries, although there is increasing production of other kinds of mushrooms, such as the Shiitake. The world production of all types of mushrooms is more than 1 million tons, with *Agaricus brunnescens* accounting for about 70% of the total production (Farr, 1983).

Basidiocarps of *A. brunnescens* are prized primarily as a flavorful food but also provide many nutrients. They are higher in protein than many vegetables and fruits, as about 3 to 4% of their fresh weight is protein. Nutritionally, protein from mushrooms is intermediate in quality between that of vegetables and meats because it has lower quantities of some amino acids (particularly of methionine and cysteine) than meat but higher concentrations of lysine and tryptophan than most vegetable proteins. Mushrooms are lower in carbohydrates than are vegetables such as carrots or brussels sprouts, making them attractive for low-calorie diets. They are an excellent source of some vitamins (particularly riboflavin and nicotinic acid), the essential fatty acid linoleic acid, and minerals (particularly phosphorus and iron) (Manning, 1985).

The commercial mushroom, *A. brunnescens*, was originally cultivated on the open ground, but about 1810 a French gardener discovered that he could obtain year-round production by cultivating them in abandoned quarries near Paris. Since then, mushrooms have been cultivated in cellars, abandoned quarries and mines, and caves. They are most profitably cultivated in specially constructed windowless buildings in which shallow boxlike shelves are arranged in tiers, often with six or more shelves in a tier (Fig. 16–2). Such buildings are desirable because temperature and air quality can be controlled easily.

The substratum used to support growth of the mushrooms is compost, usually prepared from horse dung and straw but sometimes from other organic materials, such as hay, corncobs, or cotton and coconut wastes. A satisfactory artificial compost can be prepared from mixtures containing wheat straw and corncobs as carbohydrate sources, with ammonium nitrate or urea as a nitrogen source. Compost preparation requires two phases. The first phase is accomplished out-of-doors, where the thoroughly wetted composting materials are piled into large stacks. The material is mixed and turned every few days to maintain uniform decomposition and aerobic conditions, which encourage bacterial and fungal activity. Fungi involved in the early stages of composting include

Figure 16–2 Commercial mushroom farming. [Courtesy Butler County Mushroom Farm, Inc.]

those common as saprophytes on plants or on stored grains, and the sugar fungi (Chapter 12). After about 3 days, thermophilic fungi and bacteria predominate because much of the heat generated by respiration is retained within the organic matter and the temperature eventually rises to about 65°C. This first phase of composting is completed after 10 to 14 days, and depletes most readily available sugars, starches, hemicelluloses, polysaccharides, and proteins.

The second phase of composting is much more rigorously controlled and is accomplished indoors. The compost is packed into the shelf beds where the fungi will be grown, or treated in bulk and later transferred to beds. The air temperature is raised to at least 60°C with steam and maintained at this temperature for at least 3 hours, which kills microbial and insect pests. The thermophilic fungi and bacteria continue to be active, converting the ammonia into protein and reducing the pH. These activities are beneficial for the culture of *A. brunnescens* because both ammonia and the alkaline pH range are inhibitory for its growth and it also requires a source of insoluble nitrogen. At the end of this 5- to 8-day period, the compost has a high concentration of lignin, also favoring growth of *A. brunnescens* (Fermor et al., 1985; Hatch and Finger, 1979).

The compost is cooled to about 25°C and is transferred to the growing area. It is inoculated with *spawn* of *A. brunnescens*. Mushroom spawn is a pure culture of mycelium usually derived from germinated basidiospores and grown under laboratory conditions. Mushroom strains are selected to provide desirable characteristics such as basidiocarp size or tolerance of a wider range of environmental conditions, making cultivation easier. The mycelium is cultured in flasks or other containers on a medium of cereal grains, bran, or similar material. After the mycelium colonizes the compost for a period of about 18 days, a shallow *casing* layer is spread over the compost. The casing layer is 3 to 5 centimeters thick and consists of soil or a mixture such as sphagnum and chalk, which has favorable water-holding capacity. The mushroom beds must be moistened with a fine water spray perhaps 2 to 4 times weekly, and the casing layer is important because it absorbs water and helps prevent excessive drying of the underlying compost. The casing layer also has other attributes favoring basidiocarp development. The first mushrooms are ready to harvest about 17 to 21 days after casing. Subsequently, new crops of mushrooms appear about every 6 to 12 days for about 6 weeks. The mushrooms are usually harvested by hand.

A number of environmental conditions affect both mycelium growth and basidiocarp development and must be controlled. These include moisture, temperature, and carbon dioxide levels. Mushroom houses are maintained at a temperature of about 14 to 18°C because the optimum temperature for basidiocarp development is in that range. However, temperature tends to rise as result of respiration carried out by the mycelium and other microbes in the compost. Adequate ventilation is required to control the temperature as well as to remove excess levels of carbon dioxide, also resulting from respiration. The mycelium actually requires minute amounts of carbon dioxide for growth (the optimum concentrations are 0.1 to 0.5%), but concentrations greater than 0.2% inhibit basidiocarp formation while promoting mycelial growth. Mushroom growers actually use the carbon dioxide levels to synchronize crops for easier harvesting. Carbon dioxide is allowed to accumulate while the mycelium is growing to encourage it to grow rapidly and evenly through the casing layer. Then the aeration is increased, causing the carbon dioxide concentration to drop between 0.04 and 0.08%, promoting basidiocarp formation (Flegg and Wood, 1985).

Poisonous Fungi

People have been aware of mushroom poisoning for centuries, with some of the earliest records from Greece and Rome. In the fifth century B.C. cases of accidental mushroom poisoning were recorded. Not all cases of mushroom poisoning were accidental, however. The Roman emperor Claudius Caesar was murdered by his wife Agrippina (A.D. 54), who poisoned his dish of edible mushrooms, probably by addition of the poisonous mushroom *Amanita phalloides* (Wasson and Wasson, 1957). Also the prefect of Nero's guard and a number of other officers were poisoned at a banquet by dishes to which a poisonous fungus had been added. In what were probably the first attempts at classifying the fungi, the Greek medicine writer Dioscorides in the first century A.D. divided the fungi into edible and poisonous varieties. This system of classification was followed later by the herbalists (Buller, 1914).

The line between poisonous and nonpoisonous fungi is often hazy. Some fungi may cause minor gastrointestinal upsets due to individual sensitivities or allergies (just as common foods such as chocolate, eggs, or strawberries might upset certain people), to overindulgence, or to consumption of spoiled mushrooms. Numerous fungus species must be considered ''dangerous'' as they are poisonous to many individuals, although others can eat them with no ill effects; perhaps the most notorious example of this is the beefsteak morel, *Helvella esculenta,* which is considered a delicacy by many but which has also caused death in others. Some fungi such as the inky caps [*Coprinus* spp. (Fig. 5–43)] or *Clitocybe clavipes* may make the victim violently ill if consumed with alcohol but have no effect if eaten alone. In the United States, the death rate from mushroom poisoning is comparatively low, with approximately 50 fatalities annually (Block et al., 1955), but 90% of these deaths are attributed to consumption of species of *Amanita* (Fig. 16–3), particularly the white *Amanita phalloides* (the ''death angel'') and its close relatives *A. verna* and *A. virosa*. These last fungi are universally poisonous, and as little as 1 cubic centimeter will cause death.

It has been commonly thought that a poisonous species will blacken a sil-

Figure 16–3 *Amanita bisporigera,* one of the deadly white *Amanita* mushrooms. [Courtesy A. H. Smith.]

ver coin during cooking, that poisonous ones have a foul odor, or that poisonous fungi may be easily peeled while nonpoisonous ones cannot be. These tests are not reliable, and especially so when one considers that there is no clear distinction between ''poisonous'' and ''safe'' fungi because many species affect people differently. If one wants to collect fungi for consumption, the only safe way to proceed is to learn to identify species in the field, consume only those that are universally recognized as being safe, and eat at first only small quantities in the event that one is sensitive to that particular species.

CLINICAL ASPECTS OF FUNGUS POISONING

A large number of the fungi are poisonous and, upon analysis, the poisons are found to be biochemically diverse and sometimes unique for a fungal species (Benedict, 1972). Although a given fungal species may produce unique symptoms, the effects of some of these fungi are often generally similar. Partially to aid the physician who does not have either a specimen of the fungus or an accurate identification of it in cases of accidental poisonings, working classification schemes have been devised that are based on the similarities of symptoms resulting from poisonings. Although these schemes vary from author to author (e.g., Benedict, 1972; Lincoff and Mitchell, 1977), the three major groups discussed below are those generally recognized.

Fungi Affecting the Gastrointestinal Tract.   Fungi such as *Russula, Lactarius, Boletus, Chlorophyllum molybdites,* and *Entoloma lividum* and most of those to which individual sensitivity occurs belong to this group. Typical symptoms are nausea, vomiting, and diarrhea, which may vary in severity but terminate spontaneously within a day or so. There are practically no fatalities from this type of poisoning although a person who is aged or debilitated by disease may succumb to it. In this group of fungi, the toxic principles are poorly known, and further-more, it is not known whether partial decay of the basidiocarp or strain differ-ences may affect the toxicity (Lincoff and Mitchell, 1977).

Fungi Affecting the Nervous System. Fungi belonging in this category are those that cause sensitization to alcohol (*Coprinus* spp., etc.), *Amanita mus-caria*, and the hallucinogenic fungi. The inky cap, *Coprinus atramentarius*, con-tains a toxic derivative of the amino acid L- glutamine. This toxin has chelating properties similar to those of disulfiram (Antibuse); it binds molybdenum, blocks acetaldehyde dehydrogenase, and arrests ethanol metabolism at the acet-aldehyde stage. The symptoms appear about $\frac{1}{2}$ to 1 hour after drinking alcohol if the mushroom has been eaten within the preceding 4 or 5 days or along with the alcohol. Symptoms include flushing of the face and neck, protrusion and throbbing of the neck veins, a sensation of swelling and paresthesia of the hands and feet, and often chest pain. Nausea, sweating, and vomiting then follow. Recovery usually occurs spontaneously and completely within 2 to 4 hours (Lin-coff and Mitchell, 1977).

The red or yellow *Amanita muscaria* (Fig. 16–4) is responsible for a large number of mushroom poisonings. This species is popularly known as the fly-agaric because it was formerly used to kill flies (the pileus was brought indoors, and sprinkled with sugar to attract flies, which nibbled upon it). Toxic principles include the amines muscarine and muscaridine (Fig. 16–5). Muscarine may also occur in large quantities in other fungi, including various species of *Clitocybe* and *Inocybe*. After *A. muscaria* or another species containing muscarine is eaten, there is a latent period of 1 to 6 hours before symptoms begin. Symptoms in-clude salivation, nausea, vomiting, abdominal pains, thirst, and mucous and bloody stools. Respiration is at first rapid, but later becomes slow. At first, the symptoms are similar to those of acute alcoholic intoxication, as the victim stag-gers, may lose consciousness and become delirious, and occasionally suffers from hallucinations, manic conditions, and stupor. Convulsions may occur in severe cases. In fatal cases death results from respiratory arrest. Death results

Figure 16–4 *Amanita muscaria.* [Courtesy A. H. Smith.]

$(CH_3)_3\overset{+}{N}CH_2CHCH_2CHOHCHCH_3(Cl^-)$

$\qquad\qquad|\underline{\qquad}O\underline{\qquad}|$

Muscarine

$(CH_3)\overset{+}{N}CH_2CH_2CH_2CHOHCHOHCH_3(Cl^-)$

Muscaridine

Psilocin, R=H
Psilocybin, R= PO_3H_2

LSD

Amatoxins

		R_1	R_2	R_3	R_4
a	α – Amanitin	OH	OH	NH_2	OH
b	β – Amanitin	OH	OH	OH	OH
c	γ – Amanitin	H	OH	NH_2	OH
d	Amanin	OH	OH	OH	H
e	Amanullin	H	H	NH_2	OH

Figure 16–5 Structural formulas of some fungal poisons.

in probably less than 1% of the cases of poisoning by *A. muscaria,* but the rate may be as high as 6 to 12% in the case of poisoning by *Clitocybe dealbata* or *Inocybe fastigiata.* Most of these deaths occur in children who have cardiac or pulmonary disease (Lincoff and Mitchell, 1977).

Ingestion of *A. muscaria* or other agarics containing muscarine is not ordinarily fatal because the symptoms begin to develop rapidly, and the earliest symptoms often include vomiting and diarrhea, which prevent further absorption of the poison through the gastrointestinal tract. Poisoning from these fungi is treated by prompt administration of emetics, gastric lavage, cathartic, and enemas, all of which clear the gastrointestinal tract. Antispasmodic agents or sedatives may be administered in addition.

Hallucinogenic Fungi. A number of fungi affect the nervous system in such a manner that the individual perceives nonexistent sights and sounds, or has hallucinations. *Amanita muscaria* is one of these hallucinogenic fungi and is eaten in orgies by some Siberian tribes in the Kamchatka peninsula. A similar usage of other hallucinogenic fungi in religious rites occurred among peoples over a wide geographic area (from northern Europe, eastward to Siberia, southward to Borneo and New Guinea, and westward to Peru) (Wasson and Wasson, 1957).

The Mexican Indians have used hallucinogenic mushrooms, which they believe to be sacred, since before the time of Montezuma's coronation in 1502, the date of the first record (Wasson, 1961). (The hieroglyphics and mushroom stone carvings mentioned on page 495 may have been related to this practice.) Although the Mexican Indians have been using the hallucinogenic mushrooms in sacred rites for centuries, they guarded their secret so well that only a handful of people (predominantly missionaries) knew of their practice until a few decades ago. Hallucinogenic mushrooms are species of *Psilocybe* (Fig. 16–6) and are consumed during religious rites when a serious problem must be solved through divine revelation. The active ingredients of the hallucinogenic mushrooms are the indole derivatives psilocybin and psilocin (Fig. 16–5). Although psilocybin occurs in greater quantities than psilocin, both exert similar effects. Other closely related toxins may occur in species of *Psilocybe* as well as some other agarics, including Japan's "laughing mushroom" *(Panaeolus papilionaceus),* which causes the euphoric victim to laugh.

Members of Occidental cultures are currently indulging in hallucinogenic experiences induced by LSD *(d*-lysergic acid diethylamide). One of the natural sources of LSD is *ergot,* which is treated more fully on page 521. Hallucinogenic mushrooms, purified psilocybin, and LSD produce similar effects when consumed. They produce a euphoric, dreamlike state in which the body may feel heavy but there is a sensation of an extraordinary lightness or hovering. Perception of time and space are lost, and there are changes in the awareness of oneself. Visions of colorful patterns and objects may seem to fill all space and to

Figure 16–6 Hallucinogenic mushroom, *Psilocybe strichper.* [Courtesy A. H. Smith.]

dance past in an endless succession. All other senses (auditory, touch, taste, and smell) seem to be equally sharpened to produce a feeling of intense reality.

Fewer than 1% of the cases of poisoning by hallucinogenic fungi are fatal. Treatment includes emptying of the gastrointestinal tract as described for muscarine poisoning, administration of sedatives, and reassurance (Lincoff and Mitchell, 1977).

Fungi Causing Cell Destruction. Fungi having hemolytic poisons capable of destroying red blood cells include *Helvella esculenta*, species of *Galerina*, and *Amanita phalloides* and its close relatives *A. verna* and *A. virosa*. The *Amanitas* are of particular concern here, as they are responsible for the majority of deaths (more than 95%) following ingestion of mushrooms; and the mortality rate after consumption of these fungi is over 50% and may be as high as 90%. *A. phalloides* contains several toxins that are divided into two families, the phallotoxins and the amatoxins, which contain at least five and six members, respectively. These toxins are cyclopeptides with a sulfur-containing bridge coupling the indole nucleus of a tryptophan and a hydroxylated amino acid. The phallotoxins and amatoxins differ in biochemical details of their structure and in the mode of their action. The phallotoxins act quickly and a large dose may cause death in a mouse within 1 hour. The amatoxins are much slower in their action and require more than 15 hours to exert their lethal effects, but they are about 10 to 20 times more toxic than the phallotoxins. The amatoxins are among the most lethal poisons known, and a dose of only 5 to 10 milligrams is lethal for an adult human. Because the phallotoxins may not be absorbed from the digestive tract and because of the extreme toxicity of the amatoxins, it is believed that the amatoxins are primarily responsible for the deaths from *Amanita* poisoning. One of the amatoxins, amanitin, interferes with DNA and RNA transcription by inhibiting RNA polymerase (Lincoff and Mitchell, 1977; Wieland and Wieland, 1972).

After ingestion of *A. phalloides*, there is a latent period of 10 to 12 hours during which the fungus is most likely to be thoroughly digested and much of its poison absorbed. During this time, the toxins reach the kidneys and liver and begin to do irreversible damage to these organs, but the victim is without outward symptoms. Symptoms begin with severe abdominal pains, nausea, vomiting, and diarrhea. The patient may become restless, suffer from delirium and hallucinations, and eventually collapse. Fragility of the blood cells increases. After 3 to 4 days jaundice may set in, as well as renal disturbances and toxic hepatitis. The liver becomes enlarged, dark red, and very brittle owing to an excessive accumulation of blood from hemorrhaging. Death results from destruction of the liver cells and occurs about 2 to 4 days after ingestion of the poisonous agarics. In addition to the damage to the kidneys and liver, damage sometimes occurs to the heart, adrenals, and skeletal muscles.

Treatment for *A. phalloides* poisoning consists of immediate evacuation of the gastrointestinal tract during the latent period, administration of cardiac stimulants in the event of circulatory failure, administration of fluids such as a glucose solution, and hemodialysis in the event of kidney failure. Intravenous administration of thioctic acid in glucose may be a lifesaving treatment (Lincoff and Mitchell, 1977).

Fungi and Food Processing

In addition to being directly consumed by humans, fungi are used in the processing of many foods. These include the worldwide production of bread, wine, beer, and cheese. The use of fungi in the processing of food has reached its greatest diversity in the Orient, especially Japan, where different kinds of foods are produced.

The most widely consumed foods processed with the aid of fungi are bread and alcoholic beverages, which have been produced for centuries. The alcoholic beverages include beer, wine, whiskeys, and other fermented beverages, such as saké, made in Japan. The origins of bread and beverage making are lost in antiquity, but it is known that the Sumerians produced a beer consisting of moistened, fermented bread prior to 7000 B.C. The Egyptians also made this "bread beer," and regarded it as a holy gift from Osiris, the god of the dead. They were also the first to make leavened bread. Both bread making and brewing were later refined by the Greeks and Romans, and numerous changes in the manner of making beer have been made since these early times. For example, hops have been used as an ingredient of beer only since the Middle Ages (Stewart, 1974). Both bread and alcoholic beverages have been held in great esteem: bread is often referred to as the "staff of life," and wines have traditionally had a role in sacramental rites.

The production of both leavened bread and alcoholic beverages utilizes yeasts, which are members of the Hemiascomycetes and often strains of *Saccharomyces cerevisiae*. As noted in Chapter 9, the yeasts ferment glucose to yield ethyl alcohol and carbon dioxide:

$$C_6H_{12}O_6 \longrightarrow 2C_2H_5OH + 2CO_2$$

The baker takes advantage of the carbon dioxide gas production to make his bread porous and of a light texture, while the brewer and wine maker rely on the production of ethyl alcohol to make the characteristic beverages. The yeast cells produced during the fermentation process may be harvested and dried for use as a nutritional supplement for humans or in livestock feeds.

Bread. The baker utilizes compressed yeast cakes or active dry yeast. Commercial production of bakers' yeast begins with the selection of a suitable strain of *Saccharomyces cerevisiae*, which should have an aerobic growth habit, a high yield when grown in a medium containing molasses, stability during storage, and a superior performance in dough. The selected strain is propagated in large closed, sterile tanks containing about 35,000 to 190,000 liters of diluted molasses, essential minerals and vitamins, and a source of amino nitrogen. A propagation period of about 12 hours is required, and during this time the tanks are maintained at 30°C, the pH is controlled, and additional minerals and sterilized molasses are added to support growth of additional cells. It is particularly important to pump sterile air into the tanks during this period to maintain strongly aerobic conditions which repress alcohol formation and encourage formation of cellular components. When the propagation period is completed, the medium containing the yeast cells is centrifuged and a slurry containing about 18% solids is obtained. This slurry may be further compressed to approximately 30% solids and this is used by the baker or processed with starch for public sales. Alternatively, the yeast slurry may be dried and packaged under oxygen-free conditions as active dry yeast (Peppler, 1979).

In preparation of bread dough, a combination of yeast, water, flour, salt, and possibly other ingredients is made. The mixture is then kneaded to cause the flour proteins (*gluten*) to form a network that will give the dough its elasticity and enable it to hold the carbon dioxide produced by the yeast. Fermentation begins, and the yeast at first uses the hexose sugars present in the flour and later utilizes some of the maltose released by hydrolysis of starch. A fermentation period of about 2 to 4 hours is needed at a temperature of 27°C. The carbon dioxide bubbles are trapped in the gluten network and produce a large number of pores that make the bread rise. As the dough rises, the yeast cells become separated from the nutrients and the dough rises much more slowly. It is then kneaded to place the yeast cells and the nutrients in close proximity so that the rising process may continue. Immediately before baking, the alcohol content of the bread is as high as 0.5%, about one-sixth that of ale. This alcohol is driven off during baking and helps give the bread a pleasant aroma (Duddington, 1961).

Wine.   Wine is fermented fruit or vegetable juice and is most often made from grape juice. The characteristic and distinctive flavors of wine are partially determined by the variety of fruit used, the environmental conditions under which the fruits were grown, the parts of the fruit used, and the ability of the yeast to form flavored by-products (particularly esters) (Phaff and Amerine, 1978).

Traditionally, the yeasts used to ferment the juices were those occurring naturally on the surface of the fruit, but recently the trend has been to utilize laboratory cultures of strains of *Saccharomyces cerevisiae* var. *ellipsoideus*. These strains are chosen for their ability to properly ferment the variety of grapes used and for their ability to impart flavors by formation of by-products which are characteristic for a particular wine. The wine maker may grow the yeasts in the winery by first inoculating about 1 to 5 liters of fruit pulp *(must)* from the laboratory culture slant and then serially transferring the inoculum to successively larger volumes of fresh must until a quantity suitable for bulk fermentation is obtained. The grape juice is treated with bisulfite to kill undesirable yeast and bacteria. The purity and final vigor of the yeast cultures produced in this way is often uncertain. Alternatively, quantities of the selected yeast may be produced on a large scale, like the bakers' yeast, and then either compressed or active dry wine yeast may be added to the tanks of must or strained juice. This latter method eliminates the laborious and often uncertain stepwise preparation of starter cultures and the need for the costly must required for the starter cultures.

Fermentation is carried out in vats, usually of wood or stone, that vary in capacity from a small keg to one that holds 190,000 liters or more. The vat is covered to maintain anaerobic conditions that encourage fermentation and also discourage growth of acetic acid bacteria which could convert the wine into vinegar. The course of fermentation (particularly the nature of by-products) and the concentration of alcohol produced are influenced by the temperature, sugar concentration, acidity, and tannin content of the must and the amount of sulfite added. Of these factors, temperature is especially important as the yield of alcohol is higher at the lower temperatures and it also encourages the formation of pleasant flavors. At higher temperatures the alcohol resistance of the yeast is decreased. A favorable temperature for fermentation is in the vicinity of 10°C. If the sugar content is over 30%, the fermentation is inefficient because alcohol accumulates and arrests the activity of the yeasts before all the sugar is converted to alcohol. A low acid content favors production of acetaldehyde, glycerol, and volatile and fixed acids, which have a detrimental effect on the flavor, color, and stability of the wine. In addition, a low acid content gives a lower yield of aromatic principles, which give the wine its bouquet.

After the wine is partially fermented, the *racking* operation begins and continues throughout the remainder of the fermentation period. In racking, wine is siphoned off from the sediment, which consists of dead yeast cells and other materials which settle at the bottom of the vat and is transferred to a fresh vat. This procedure is repeated a number of times until the desired clarity of the wine is achieved. Before the last racking, a final clarification is performed by adding some substance (such as isinglass, casein, or bentonite) that carries additional sediment to the bottom. Fermentation continues until the alcohol content is about 12%; above this point the alcohol kills the yeast cells so that no more fermentation takes place. The wine is then stored in completely filled and sealed tanks for aging. Additional blending, fortification with additional alcohol, heat treatments, refrigeration, or filtration may be required before the wine is bottled.

Beer.   Beer is an alcoholic beverage prepared from fermented grains, usually barley (Westermann and Huige, 1979). In the initial steps of beer production, barley is *malted* through a controlled germination. The barley grains are washed and soaked for 2 to 3 days to stimulate development of the embryo; the water is then drained off and the seeds germinated. The embryo is allowed to grow until the plumule attains a length equal to approximately three-fourths of the kernel length and is then dried to halt growth. During the brief period of germination, the embryo produces a number of digestive enzymes which begin to hydrolyze food reserves in the seed. From the brewer's standpoint, the most important of these are the carbohydrases which attack starches and sugars, and of these the most important are especially the amylases, which convert the starch into dextrins, maltose, and a little glucose. The maltose is the principal substrate for yeast fermentation.

The next major step in beer production is known as *mashing*. The dried malt is ground and mixed with hot water. Most of the enzymatic conversion of starch to maltose by amylases takes place during mashing, which requires about 2 hours. The aqueous extract is separated from insoluble materials and husks and is then boiled with hops, the female inflorescence of the hop plant containing essential oils and resins which impart characteristic flavors to the beer.

The aqueous extract is then cooled to a temperature between 8 and 16°C (the temperature is partially determined by the type of beer being produced) and placed in fermentation vats. A selected strain of yeast is added. Traditionally *Saccharomyces carlsbergensis* is used to produce lager beer while *S. cerevisiae* is used to produce ales (in the United States, *S. carlsbergensis* may be used for both). During a fermentation period of about 5 to 14 days, the sugar is converted to alcohol. The yeasts settle to the bottom of the vat in a flocculent material in the production of the lager type of beer. In contrast, bubbles of carbon dioxide rise to the top of the vat and carry with them the yeast cells and dark flocculent materials from the liquid in the production of the ales. The resultant foamy scum must be removed periodically.

After the fermentation period is completed, the newly formed beer is allowed to rest for a few days to allow the yeast cells to settle out and is then drawn off into casks or large tanks for a maturation period of several weeks. The beer is clarified by addition of gelatinous materials in a manner similar to that in which wines are cleared and may also be filtered through diatomaceous earth. Antioxidants are added since beer changes flavor upon oxidation, and carbon dioxide is added either as the pure gas or by mixing in some freshly fermented beer. The beer is then bottled for sale.

Whiskey.   Whiskeys are made from fermented grains (corn, wheat, barley malt, or rye malt), which are mixed in varying proportions according to the type of whiskey being produced, and fermented by yeasts. The processes involved are similar to those involved in beer production (Fig. 16–7), the major exception is that the fermented grain broth is distilled in order to concentrate the alcohol.

CHEESE

Cheese is a food product made from milk and it contains butterfat, lactose, the proteins albumin and casein, and the watery whey. Casein will coagulate to form a curd when it is acted upon by an acid or by the enzymes pepsin or rennin. Rennin was formerly obtained from calves' stomachs, but there has been a widespread shortage of calves' rennin because few calves are slaughtered and the consumption of cheese has been increasing. Fungi may produce extracellular enzymes, including rennin, which can coagulate milk. Rennins from

Figure 16–7 Rectified cane spirit is passed through these nine charcoal filters at a very slow rate to remove any aromatic impurities so that the final product, vodka, will be completely colorless, tasteless, and odorless. [Courtesy Gilbey-Santhagens (Pty.) Ltd.]

fungi have become increasingly important in cheese manufacture (one fungus which is commercially important as a source of rennin is the plant pathogen *Cryphonectria parasitica*) (Sternberg, 1976). The coagulated curd is the newly formed cheese that separates from a watery fluid, the whey (whey is sometimes used as a cheap substrate for the production of food yeasts). The cheese may be eaten in this form when freshly prepared, such as cottage cheese. Alternatively, the cheese may be further drained of moisture, pressed, or cooked, as in the production of Edam, Gouda, and Cheddar cheeses. Some cheeses may be processed with a characteristic bacterial or fungal flora, which will impart distinctive flavors during the ripening period (Olson, 1979). Fungi are used to process cheeses of two principal types: (1) those of the Camembert type, which are soft in texture and are covered on the outside with a rind formed through the agency of the mold; and (2) those semihard cheeses of the Roquefort type which have streaks and pockets lined with a blue-green mold.

Camembert Cheese. Camembert cheese was first made in the community of Camembert in northwestern France. Milk is coagulated with rennin; the curds are placed in hoops and allowed to drain overnight and then turned over to drain evenly. After 2 or 3 days, each cheese is rolled in salt and sprinkled with spores of *Penicillium camemberti*, a white mold. The layer of salt extracts the whey from the curd and forms a shiny cover of brine which flows away. The surface of the cheese becomes hardened after this extraction and forms a rind. During these initial stages, the cheese consists of a sour curd surrounded by a salty rind, both of which inhibit the growth of bacteria and the fungus, *Geotrichum candidum*. The spores of *P. camemberti* germinate and mycelium penetrates the rind and eventually spores are produced on its surface. After *P. camemberti* has become well established, it secretes proteolytic enzymes into the curd which makes the curd become soft and buttery and develop a mild flavor. The acidity also disappears, allowing the establishment of *G. candidum* and bacteria during the later stages of ripening, thus accounting for the ammoniacal flavor. The ripening process requires about 4 weeks and is carried out in ripening rooms held at 10 to 16°C with a humidity of 86 to 88% (Thom, 1944).

Roquefort Cheese. Roquefort cheese originated in the sheep-raising region surrounding Roquefort in southern France. This cheese is made from sheep's milk, while a similar one, blue cheese, is produced from cow's milk in the United States. The milk is curdled with rennin at low temperatures, and the curd is broken into small masses and drained until it is rather dry and the pieces no longer merge completely when placed together. The curd is heaped loosely in cheese hoops so that cracks and channels remain. It is then inoculated with bread crumbs containing spores of *Penicillium roqueforti* which was grown on the bread. The curd is then maintained at a cool temperature 18 to 20°C and at a high humidity (80 to 90%). During the next 3 to 10 days, the ripening curd is periodically scraped, washed, and salted. After the salting period, the cheese is perforated with additional holes that provide adequate aeration for the fungus to grow along the many interfaces within the cheese. A 2-month ripening period is required and then the cheese is ready for sale. During its growth in the cheese, the fungus secretes protein and fat-digesting enzymes. The pungent flavors characteristic of Roquefort cheese result from the hydrolysis of fats and the liberation of high-molecular-weight fatty acids (capric, caproic, and caprylic) and accompanying methyl ketones and alcohols (Moskowitz, 1979).

ORIENTAL FOODS

Foods altered by bacterial or fungal growth have been part of the human diet since the beginning of recorded history and have been widely used in diverse cultures. The microbial growth often improves the texture, flavor, and nutritional value of the food while delaying spoilage. These fermented foods have often been valued in cultures where people do not have access to electricity or to commercially prepared foods because these foods add variety to the diets and have longer useful life than nonfermented foods. Such foods may be used as a beverage, a condiment, a vegetable, or a substitute for meat or bread. The use of these foods has been especially popular in Africa, Asia, and the Middle East. We shall consider the Oriental foods further.

In Japan and other Oriental countries, large quantities of diverse foods are prepared from soybeans, wheat, and rice through either the deliberate or accidental inoculation with members of the Mucorales, various yeasts, and *Aspergillus oryzae* (a member of the Deuteromycotina). These fungi may be used in combination with each other or in combination with bacteria to yield a specific food product. Foods of this type include tempeh, a solid food prepared with soybeans processed with species of *Rhizopus*; sufu, a Chinese cheese prepared from soybeans and *Actinomucor elegans* and *Mucor* spp.; miso, or soybean paste, prepared from rice and soybeans fermented by *Aspergillus oryzae* and *Saccharomyces rouxii*; and shoyu, or soya sauce, prepared from soybeans and wheat and processed with *Aspergillus oryzae* and other organisms. The production of tempeh and shoyu will be considered in more detail.

Tempeh. Tempeh is a food which is produced in great quantities in Indonesia, New Guinea, and Surinam. In tempeh preparation, soybeans are washed and soaked overnight in water. The following day the seed coats are removed and the dehulled soybeans are boiled for $\frac{1}{2}$ hour, drained, and cooled. When cool, the soybeans are inoculated with spores of a strain of *Rhizopus*, usually *R. oligosporus*. In Indonesia, this inoculation is accomplished by adding pieces of a previously made batch of tempeh or from the wrapper in which a former batch was held. Small amounts of inoculated soybeans (about the quantity that would fill a petri dish) are wrapped in leaves and tied with string. These soybeans are held at room temperature for about 24 hours and during this time the mycelium permeates the soybeans so that they are embedded in a whitish mycelium and the entire mass is held together. The newly prepared tempeh is consumed after this 24-hour period and is prepared fresh daily. The tempeh may be thinly sliced, dipped into salt water, and fried in vegetable oil; it may be roasted; or it may be cut into cubes and used as a meat substitute in soup.

Untreated soybeans do not soften well during cooking and are difficult to digest, but after treatment they are quite easy to digest and have a pleasant, newly acquired flavor. During the brief period of treatment, the fungus produces large quantities of proteolytic enzymes which break down the soybean proteins. The percentages of both soluble solids and soluble nitrogen increase during the period of fungal growth and thereby become more digestible upon consumption. There is also an increase in the riboflavin, niacin, and B_{12} content of the treated soybeans, although there is a decrease of thiamine (Hesseltine, 1965).

Shoyu. Shoyu, or soya sauce, is a brown liquid which is used for seasoning a large number of foods in China, Japan, the Philippines, Indonesia, Europe, and the United States; it is produced in all of these countries. Shoyu may be made through the acid hydrolysis of soybean protein with an inorganic acid (such as hydrochloric acid), by hydrolysis with fungus enzymes during a fermentation period, or through a combination of these methods. Although manufacture is easier using an inorganic acid hydrolysis, the product is inferior to that produced utilizing fungal fermentation.

Japan is the foremost producer of shoyu today, and 90% of the shoyu made there utilizes fungal processing of either defatted soybean meal or whole soybeans. Since 75% of all shoyu is made from defatted soybean meal, this mode of manufacture is discussed. Soybeans are washed, soaked for 15 hours at room temperature, and then autoclaved for 1 hour. Meanwhile wheat is cleaned, roasted, and crushed, or wheat bran is steamed and roasted. Wheat and soybeans are then mixed together in a ratio of 45% wheat to 55% soybean; the material is cooled and then inoculated with either *Aspergillus oryzae* or *A. soyae*. The molding grain mixture is placed in rooms with the temperature held at 25 to 35°C for 3 to 4 days with at least two turnings of the grain. At the end of this period, the moldy grain is placed in large tanks and an equal amount of brine is added. The incubating mixture (mash) remains in the tanks for 3 to 4 months if it is warmed or up to 1 year if it is not warmed. The mash is occasionally stirred with compressed air. Bacteria produce lactic acid during the first part of the fermentation period. Later during the fermentation period, yeasts produce ethyl alcohol, carbon dioxide, and acetic acid. Finally there is a period of aging and relatively little fermentative activity. The microorganisms which carry out the last two fermentations are not deliberately introduced but carried along randomly in the procedures used. At the end of the aging period, the mash is compressed as a cake, and the liquid is extracted as shoyu, which is pasteurized to destroy any microorganisms and enzymes present (Hesseltine, 1965).

INDUSTRIAL PRODUCTION AND UTILIZATION OF FUNGUS METABOLITES

In the preceding section, it was indicated that some fungus metabolites become involved in the processing of foods and change the flavor and composition of the food which served as a medium for the growth of the fungi. Many fungus metabolites are useful products alone and are separated from the substratum for further use in the pure form. Fungus products used commercially include organic acids, alcohol, antibiotics, pigments, vitamins, and enzymes. A more complete listing may be found in Table 16–1. Fungi are also used to convert steroids

TABLE 16–1

Commercially Important Fungus Metabolites

Metabolite	Produced by:	Use
Organic acids		
Citric acid	*Aspergillus* spp. especially *A. niger*	Manufacture of citrates; flavoring ingredient; effervescent salts; in ink; in dyeing; medically; in silvering mirrors
Gluconic acid	*Aspergillus niger*	Medically—increases solubility of calcium when combined as a salt; manufacture of toothpaste
Itaconic acid	*Aspergillus terreus*	Manufacture of plastics
Alcohols		
Ethyl alcohol	Yeasts	Solvent; raw material in manufacture of ether, esters, acetic acid, synthetic rubber
Glycerol	*Saccharomyces cerevisiae*	Explosives
Fats	*Penicillium* spp. Yeasts	Soap; manufacture of food (useful in times of food shortage)
Antibiotics		
Griseofulvin	*Penicillium griseofulvum*	Oral and topical antibiotic
Penicillin	*Penicillium chrysogenum*	Oral and parenteral antibiotic
Drug		
l-Ephedrine	Yeasts	Drug for treatment of asthma; ingredient for nose drops and inhalants
Vitamins, pigments, and growth factors		
B vitamins	Yeasts	Nutritional supplement and medical therapy
Riboflavin	*Ashbya gossypii* *Eremothecium ashbyii*	Nutritional supplement and medical therapy
Gibberellin	*Fusarium moniliforme*	Growth-promoting substance (vascular plants)
β-Carotene	*Blakeslea trispora*	Precursor for vitamin A; coloring agent for margarine, baked goods
Enzymes		
Amylase	*Aspergillus* spp.	Starch for hydrolysis for alcoholic fermentation; starch removal agent in food preparation; removal of starch sizing from fabrics; converting starch in sizing materials
Rennet	*Cryphonectria parasitica* *Mucor* spp.	Milk coagulation in cheese manufacture
Invertase	*Saccharomyces cerevisiae*	Production of artificial honey and invert sugar
Pectic enzymes	*Penicillium* spp.	Pretreatment of fresh fruit juices to remove turbidity; removal of pectins before concentrating juice
Protease	*Aspergillus* spp. *Penicillium roqueforti* *Agaricus brunnescens* *Cantharellus cibarius*	Food processing

from one form to another, and as a direct source of alkaloids that they may contain.

The metabolites are biosynthesized along diverse pathways and, as we might expect, there is no single set of environmental or nutritional conditions common to all commercial syntheses. In addition, a single fungus is usually capable of producing several metabolites. Attaining a metabolite of choice is accomplished by forcing the fungus to favor certain biosynthetic pathways over others. This may often be accomplished by adjusting the pH so that some enzymes act normally while others are inhibited, by adding antagonists that affect certain enzymes only, by adding nutrients or maintaining conditions that do not favor growth and thereby divert assimilated nutrients into metabolic waste products rather than into protoplasm, or by adding precursors that favor certain pathways. Modern industrial fermentation techniques make it possible to produce millions of tons of useful compounds annually. Often it is cheaper to produce compounds by fermentation techniques than to synthesize them chemically or extract them from biological sources.

Methods Involved in Industrial Mycology

GENETIC DEVELOPMENT OF STRAINS

Generally, a fungus that has just been isolated from nature is incapable of making large enough quantities of a metabolite to make commercial production feasible. The strain must be genetically improved so that larger quantities of the metabolite can be obtained. Genetic improvement of strains has primarily involved treatment of cells or spores with a mutagenic agent such as mustard gas, ultraviolet radiation, or X-radiation followed by selection of higher-yielding mutants from the next generation. These new, higher-yielding strains are subjected to mutagenic treatment, and again the higher-yielding mutants are selected. This mutation-selection cycle is repeated numerous times. During the development of the commercially important penicillin strains, 21 mutation-selection cycles were carried out over the course of more than 30 years in government, university, and industrial laboratories, resulting in a commercial strain that has a 10,000-fold greater yield than the original isolate (Aharonowitz and Cohen, 1981). Development of high-yielding strains by repeated cycles of mutagenesis and selection is laborious and time consuming, perhaps requiring years for strain improvement. However, this method is effective for enhancing the yield from some processes that are controlled by complex pathways and numerous genes.

The probability of increasing yields can sometimes be increased by genetic recombination. Commercially important yeasts such as *Saccharomyces cerevisiae* have sexual cycles and genetic recombination can result from breeding selected lines. Strain improvement of a deuteromycete can be accomplished by forcing nutritional mutants to form a heterokaryon and then selecting desirable recombinants resulting from the parasexual cycle. Recently a technique involving *protoplast fusion* has been utilized to force the union between two cells that would not normally undergo plasmogamy. These can be yeast cells or mycelial fungi of the same mating type or strains of a deuteromycete. For experimental purposes, fusion has even been forced between yeast cells and erythrocytes of a hen, but does not result in a viable product. To accomplish protoplast fusion, the cell walls are enzymatically removed and the naked protoplasts are treated with a solution of polyethylene glycol and calcium ions. This treatment is required to induce fusion, which involves membrane changes and formation of cytoplasmic bridges where cells contact each other, perhaps by promoting a reorganization of the lipid molecules. After the protoplasts fuse together, the common protoplast contains both genomes. Fungal protoplasts are capable of regenerating cell walls, and filamentous fungi revert to the mycelial form. The fusion products are at first heterokaryons, but fused yeast cells can complete the sexual cycle, resulting in haploid recombinants. Filamentous fungi most often remain as heterokaryons in which complementation occurs, modifying the genetic expression. Sometimes the parasexual cycle follows in the filamentous fungi, resulting in unstable diploids and haploid recombinants (Ferenczy, 1981). Protoplast fusion makes it possible to combine slow-growing but high-yielding mutant strains with a vigorously growing ancestral strain to yield new strains that have both high growth rates and a high yield of the desired metabolite. The new recombinant DNA technology also provides new possibilities for strain development (page 523).

PILOT STUDIES

A great deal of preliminary work is required to determine those conditions favoring the production of a desired metabolite before large-scale plant operations can take place. The first steps are selection of a suitable fungus strain and the determination of optimum aeration, temperature, pH, and nutritional requirements. Following this small-scale laboratory work, the next step is to expand to a large-scale laboratory simulation of the proposed plant method. The large-scale laboratory methods may or may not prove to be satisfactory as the optimum conditions established in the initial laboratory tests may not be those optimum in larger operations.

LARGE-SCALE PRODUCTION

Large-scale production of fungal metabolites is ordinarily accomplished in vats or tanks made of stainless steel or some other metal and having a capacity of several thousand liters. Open vats may be used if the fungus makes conditions unfavorable for contaminating microorganisms; for example, in alcohol production, the medium quickly becomes anaerobic and inhibits growth of aerobic contaminants. If contamination is a potential hazard, enclosed tanks are utilized because aseptic conditions can be maintained. Either aerobic or anaerobic conditions may be maintained in a closed tank. In tank cultures, a filamentous fungus forms a submerged mycelium. The rate of metabolism and yields of the product are high, as the mycelium is bathed in a circulating medium that make nutrients readily available to the fungus.

In planning a medium for industrial processes, the prime considerations are cost and effectiveness in the desired operation. A good medium is inexpensive and consists of a cheap carbon source (such as molasses, which contains glucose) and a cheap nitrogen source (such as ammonia). Vitamins or purified amino acids may be added to meet the specific requirements of a given organism. If the organism does not grow well on a medium of the type above, owing to complex nutritional requirements, cheap biological materials may be added to the medium or used exclusively. These biological additives may be corn steep liquor, malt, fish solubles, and yeast autolysate. Other additives are used to control pH or to direct the course of biosynthesis.

Some products are associated with an actively growing and healthy organism. For these products, the medium is composed so that maximum growth of the cells will take place, and the product is harvested when growth ceases. If the product is not associated with growth, the medium is composed so that growth is limited by trace metals, unfavorable pH, or unfavorable nitrogen but not by carbon. In this latter case, an excess of nutrients may be required in the medium after growth ceases to provide carbon and energy for the synthesis of the product.

The medium is usually sterilized only if it is likely that contaminating organisms will become established during the large-scale production procedure. Frequently the organism being cultured will make conditions unfavorable for contaminants and sterilization is not required. Semiaseptic conditions may be required, and only certain additives that are likely to be a major source of contamination are sterilized. If aseptic conditions are required, the large tank is usually steam-sterilized while empty, and the medium is steam-sterilized while it is being pumped into the tank. Chemical sterilants (such as ozone, ethylene oxide, or formaldehyde) may be used for materials that cannot be heat-sterilized.

Inoculum is prepared from stock cultures and basically involves culturing the organism in successively larger batches. Two or three buildup stages are cultured in the laboratory, followed by one to three stages in tanks of increasing size until enough cells are obtained to inoculate the final tank in which actual production will take place. The amount of inoculum may vary from 1 to 20% of the volume of the production tank. The object is to add enough healthy cells so that the growth phase and buildup of cells in the tank will occur rapidly, thereby reducing as much as possible the time required for production of the metabolite.

A number of factors may need to be controlled during large-scale tank culture. Aeration may be required, and if so, air that has been sterilized by passage through filters is piped into the tank or the medium is mechanically agitated. Alternatively, anaerobic conditions can be maintained if required. Actively metabolizing organisms release heat (as much as 2000 calories per gram of dry yeast), and in many processes, this rising temperature is controlled by cooling coils within the jacket of the tank or vat. The pH is ordinarily controlled by constant monitoring and the addition of the required buffers.

After the metabolite is produced, the cells are separated from the medium by filtration. The metabolite is then extracted from the remaining liquid by distillation or chemical processes. The commercial operation may involve the *batch* method, in which the tanks are filled with fresh medium and then emptied after the metabolite is produced, or the *continuous* method, in which there is a flow of medium into the tank and the uninterrupted production of the metabolite once the process has been successfully initiated.

We discuss in detail three industrial processes. Two of the most widely produced fungus metabolites are citric acid (an organic acid) and penicillin (an antibiotic). Both require highly aerobic conditions, and the latter requires strict asepsis. Ethyl alcohol serves as an example of a process that requires neither air nor aseptic conditions.

Citric Acid

The most important organic acid produced by fungi is citric acid (Fig. 16–8). It is used commercially as a flavoring ingredient in beverages and foods, especially in dry mixtures such as gelatins and soft drink powders or tablets, and as the principal acid in the preparation of soft drinks, desserts, jams, jellies, candies, wines, and frozen fruits. Citric acid is rapidly and almost completely metabolized in the human body and has wide pharmaceutical uses. These include its incorporation in effervescent products and as citrates in blood transfusion. Citric acid is also used in astringent lotions to adjust the pH, in hair rinses and hair-setting preparations, in electroplating, in leather tanning, and in reactivating old oil wells where the pores of the sand become clogged with iron.

Citric acid was originally produced from calcium citrate obtained from cull lemons. Methods for producing citric acid from fungus metabolism were introduced in the United States in 1923, and now practically all the citric acid is produced by fungi. In the United States and Europe in 1979, annual production of citric acid was more than 100 million kilograms (Lockwood, 1979).

Citric acid is produced commercially with selected strains of *Aspergillus niger*, which actively produce citric acid under certain conditions. The most important of these is utilization of a medium deficient in some essential metallic nutrient (especially manganese, iron, or zinc) and, generally, conditions which limit growth.

In citric acid production, *A. niger* is cultured as submerged mycelium in vats. The carbohydrate source may be refined or crude sucrose, cane syrup, or beet molasses. Beet molasses is the least expensive carbohydrate source, but it

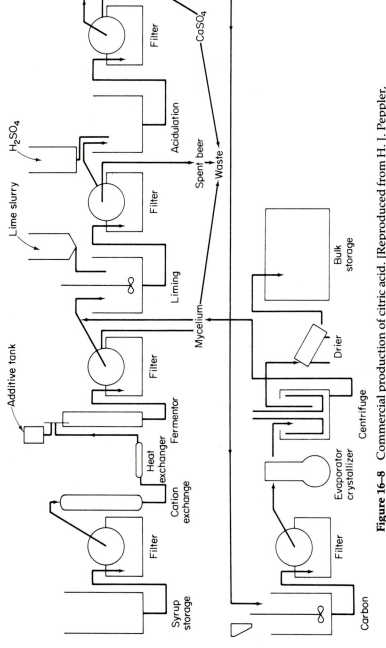

Figure 16-8 Commercial production of citric acid. [Reproduced from H. J. Peppler, Ed., Microbial technology, chapter by L. B. Lockwood and L. B. Schweiger, copyright 1967, by Reinhold Publishing Corp., New York, with permission.]

has a high metallic ion content which may cause low yields of citric acid to be produced. Beet molasses has been widely used in the pan techniques, although pure sucrose or highly refined syrups give higher yields in submerged cultures in vats. Metallic cations are removed from the carbohydrate source by the use of cation-exchange processes and absorbents or (in the case of beet molasses) by treatment with ferricyanide. For submerged culture, it is particularly important to make the medium deficient in manganese and iron. The carbohydrate is diluted to a concentration of about 20 to 25%—this high sugar concentration inhibits formation of acids other than citric acid. Production of citric acid requires a low pH, and hydrochloric acid is added to the medium to reduce the medium to the range of pH 2 to 5 when the spores in the inoculum germinate. In the vat, the pH is kept below 3.5. Ammonium salts or urea are added to provide nitrogen. For vat cultures, copper or organic ions may be added as antagonists to iron and aid in the control of mycelial growth which increases citric acid production. Highly aerobic conditions are required, and submerged cultures are aerated with sterile air and agitated. Also, during the incubation period, temperature must be maintained in the range of 25 to 30°C. The incubation period is 7 to 10 days in duration.

At the completion of the incubation period, the harvesting of the citric acid is started. Lime is added to the culture medium to precipitate any oxalic acid which formed, and the mycelium and calcium oxalate are filtered off. Additional filtration may be required to clarify the medium. A slurry of calcium hydroxide, $Ca(OH)_2$, is added to form calcium citrate, which precipitates out. The calcium citrate is filtered out and then treated with sulfuric acid which precipitates the calcium. The dilute citric acid solution is purified by treatment with carbon and is demineralized by successive passage through cation- and anion-exchange resins. This purified citric acid solution is then evaporated, leaving behind citric acid crystals, which are further purified by recrystallizations from water.

MECHANISM OF CITRIC ACID FORMATION

As indicated in Chapter 9, citric acid is one of the principal organic acids produced in the tricarboxylic acid cycle. Although many pathways and interconversions have been suggested for citric acid biosynthesis and accumulation under the conditions outlined, recent evidence indicates that the probable origin of citric acid is directly from the TCA cycle. When mycelium is actively growing, citric acid is produced as an intermediate in the TCA cycle and is further diverted to growth-promoting biosyntheses or energy release. For industrial purposes, culture conditions that are inhibitory to growth of the fungus are maintained, and the TCA cycle itself is inhibited or blocked by the enzymatic inhibition of the low pH or by specific enzyme inhibitors such as the copper ions. Biosyntheses leading to formation of citric acid are not inhibited, while those involving its conversion are inhibited, leading to the accumulation of citric acid as an "overflow" product.

During citric acid production, the activity of the condensing enzyme (operating in the condensation of acetyl-CoA and oxaloacetic acid to citric acid) is increased in activity, while the activities of isocitric dehydrogenase and aconitase disappear. Aconitase controls biosynthesis of isocitric acid from citric acid, and in turn isocitric dehydrogenase mediates in the hydrogen removal which yields oxalosuccinic acid from isocitric acid. It is strongly suggested that the inactivity of these latter enzymes is responsible for the accumulation of citric acid.

Alcohol

Production of ethyl alcohol as an end product of anaerobic fermentation is the oldest known industrial process utilizing fungi (in the context of beer and wine production). As is to be expected, processes similar to those used in making alcoholic beverages can be used to obtain pure ethyl alcohol for utilization as a solvent or for other industrial purposes. As late as 1930, fermentation of molasses accounted for 75% of the industrial alcohol produced. Molasses is a comparatively expensive medium and at times it has been in short supply . For many years (into the late 1960s), alcohol could be produced less expensively by chemical means from petroleum, and the mycological production of industrial alcohol declined to a point at which less than 25% of the industrial alcohol was made through yeast fermentation. There has been a renewed interest in the production of alcohol from cheap and readily available substrates since the early 1970s. The rising cost and decreasing supply of petrochemicals has again made fermentation processes an attractive alternative, especially as the ethyl alcohol may be used as an automotive fuel. For production of industrial alcohol, any substrate that can be fermented to alcohol may be used, and some countries or regions are utilizing waste products or surplus crops locally to make alcohol. For example, alcohol is being produced from cane sugar in India and Cuba, while cereal starch has been seriously considered in Australia and South Africa (Bu'Lock, 1979).

Much of the industrial alcohol is produced by fermentation of industrial wastes. Industrial wastes that can be fermented include *sulfite liquor,* the aqueous material left from the manufacture of cellulose or pulp from wood. This sulfite liquor contains sugars and other organic materials (amounting to more than half the original weight of the wood) and most of the inorganic chemicals used in processing the pulp; its utilization is sound both economically and from the standpoint of conservation. If this sulfite liquor were simply discarded in streams, aquatic wildlife would be in peril since the sulfite liquor may deplete the oxygen supply. For these reasons, pulp and paper companies often build a full-sized alcoholic fermentation plant to dispose of their wastes.

Production of industrial alcohol from sulfite liquor may be carried out using the Ekström process, which is as follows. The hot sulfite liquor is collected, neutralized with limestone, and stirred with steam. The sediment is allowed to settle out and then the liquor is cooled to about 30°C. Yeast and nutrients (such as ammonium sulfate or phosphoric acid) are added and the liquor is fermented in tanks. At the end of the fermentation period the medium contains about 1% alcohol by volume. It is degassed, or freed from the carbon dioxide, and the alcohol concentrated by distillation until 95% alcohol is recovered. A yield of about 11 liters of 95% alcohol may be expected from 1000 liters of waste sulfite liquor.

Penicillin

Production of antibiotics (antibacterial or antifungal compounds) is one of the largest and most important microbiological industries. In the United States in 1979, the value of the antibiotics produced from both bacteria and fungi was more than $1 billion, and more than 100,000 tons was produced worldwide annually at that time (Perlman, 1979). The role of antibiotics in ecology of fungi is discussed in Chapter 12.

Interest in utilization of antibiotics for therapy began in 1929 when Alexander Fleming found that a mold had contaminated his cultures of a pathogenic bacterium, *Staphylococcus aureus,* and had killed bacteria in its immediate vicinity leaving a clear ring of lysed bacteria. The mold was a *Penicillium* (later found to be *P. notatum).* Fleming named the unidentified lytic principle *penicillin* and realized the potential value of penicillin as a therapeutic agent. Little additional work was done as yields were low and the production of penicillin difficult (Fleming, 1929). It was not until World War II created large demands for effective therapeutic agents that the methodology for large-scale production and refinement of penicillin was developed.

Not only did penicillin engender interest in antibiotics and launch the antibiotic industry, but penicillin is still the most widely used antibiotic. It is the drug of choice when infection is caused by organisms susceptible to it. Penicillin is effective against gram-positive bacteria and also against some of the larger viruses and rickettsia. Penicillin may be administered orally, intravenously, or intramuscularly. If penicillin is given orally, the dosage must be at least five times as large as that given by injection because the penicillin is partially destroyed in the digestive tract and is incompletely absorbed.

Penicillin is a generic term applied to an entire group of antibiotics which are closely related in structure. Most penicillins are produced by species of *Penicillium* while others are semisynthetic. The naturally occurring penicillins have the same basic structure and differ in their R groups. The basic structure and the R group of the most commonly used penicillin (penicillin G) are given in Fig. 16–9.

Figure 16–9 Basic structure of penicillin. Penicillin G would have the radical shown as the R group.

PENICILLIN PRODUCTION

Penicillin is produced in submerged vat cultures using a selected strain of *P. chrysogenum* (Fig. 16–10). Composition of the medium is of prime importance in obtaining a good yield. A medium giving an unusually good yield of penicillin is corn steep liquor, a cheap bulk material which is a byproduct of cornstarch manufacture. Corn steep liquor is a concentrate of water soluble materials (principally nitrogenous compounds) from the corn grain. These nitrogenous compounds are of great importance in giving a good yield of penicillin. The principal carbohydrate in corn steep liquor is lactose, which is the most favorable carbon source for penicillin production as it is assimilated slowly, giving the fungus a

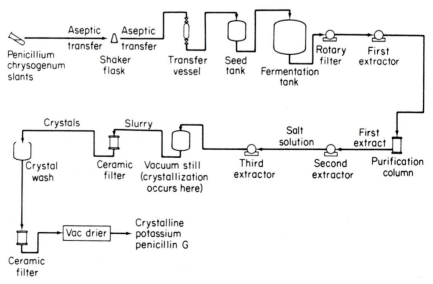

Figure 16–10 Commercial production of penicillin. [Reproduced from H. J. Peppler, Ed., *Microbial technology*, chapter by D. Perlman, 1967, Reinhold Publishing Corp., New York, with permission.]

Chap. 16 / Fungi and Humans

steady carbon source over a period of time and does not lead to the accumulation of organic acids. Precursor molecules (phenethylamine and phenylanine) needed for penicillin production are present in the corn steep liquor.

About 30,000 liters of the medium are placed in a tank, sterilized, and inoculated with an aqueous suspension of *P. chrysogenum* conidia. During the incubation period, the medium is aerated with sterile air and agitated. The tank is equipped with devices that allow the continuous addition of glucose syrups, sodium hydroxide, and sulfuric acid to maintain the pH between 6.8 and 7.4, cooling coils to maintain the temperature between 23 and 25°C, devices for the introduction of antifoam agents, and pumps for the addition of the acyl donor or precursor. The most commonly used precursor is phenylacetic acid which is required for penicillin G, but other precursors in the form of a salt, amide, or ester of the corresponding acid or amine may be added to yield penicillins bearing the desired acyl group.

At first, the mycelium grows actively and utilizes lactic acid and organic nitrogen compounds as sources of carbon. Ammonia is formed during the breakdown of these nitrogenous compounds. Upon exhaustion of the readily available nitrogenous compounds, the fungus then utilizes lactose as a carbon source and growth becomes substantially slower. Ammonia is now actively assimilated and the pH is lowered. Active penicillin production is associated with this phase of lactose and ammonia utilization. Addition of glucose increases the assimilation of ammonia, thereby increasing the yield of penicillin. Penicillin production ceases when the lactose is exhausted.

The duration of the incubation period is about 5 to 6 days. At the end of this time, the mycelium is filtered from the liquid medium which contains the penicillin. The filtrate, containing the penicillin, is mixed with a solvent (amyl or butyl acetate) and the resulting emulsion is centrifuged to extract the acetate solvent which now contains the penicillin. A phosphate buffer is added to the acetate, and the penicillin is extracted with the phosphate buffer by centrifugation. This last step may be repeated, using successively smaller quantities of liquid. Butanol is added to the aqueous mixture, and the potassium salt of penicillin is crystallized from the solution. Potassium penicillin may be further purified and used in this form or it may be converted to procaine penicillin (Perlman, 1979).

PENICILLIN BIOSYNTHESIS

Penicillin is produced along synthetic pathways not required for growth but superimposed upon those pathways required for maintenance. Key intermediates in the pathway are pyruvic acid, valine, and cysteine. As noted previously (Chapter 9), pyruvic acid is a compound required for the TCA cycle to take place. Penicillin production is at its maximum when the TCA cycle is not actively taking place or if a great deal of pyruvic acid is not being converted to acetyl-CoA and thence to fatty acids. The accumulating pyruvic acid can then be available for synthesis of L-valine, the amino acid which joins with cysteine or a derivative of cysteine (presumably originating from the TCA cycle) to form a dipeptide precursor of penicillin, L-cysteinyl-L-valine. This last intermediate is then combined with the acyl group from phenylacetic acid in the case of penicillin G or acyl groups from other specific precursors in the synthesis of the other penicillins. Many of the details concerning penicillin biosynthesis are not known.

Steroid Conversion

Steroids are a group of organic compounds which have the four-membered ring which appears in Fig. 16–11. Steroids are biologically active as hormones pro-

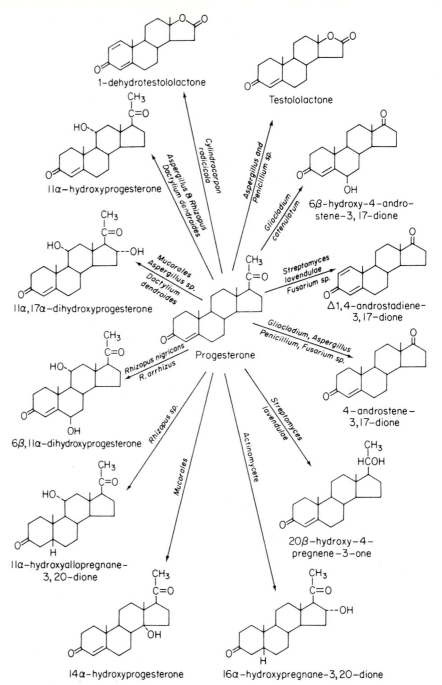

Figure 16-11 Some steroids produced from progesterone by biochemical alterations by several fungi, a Streptomycete (*Streptomyces lavendulae*) and an Actinomycete. [From S. C. Prescott and C. G. Dunn, Industrial microbiology, copyright 1959. Used with permission of McGraw-Hill Book Company.]

duced by the testes, ovaries, adrenal cortex, and placenta. The steroids differ in the nature of their side groups or side chains, and these differences in structure confer different biological properties on the steroids. Steroids are widely used medically as antiinflammatory agents, anesthetics, antifertility agents, or for the treatment of sterility. Steroids may be obtained directly from a natural source or they may be synthesized. The steroid nucleus can be produced chemically, and many of the transformations or additions of side chains may also be accomplished chemically. Some changes are impossible to accomplish by chemical means or may be very expensive and time consuming.

An example of a difficult conversion is the hydroxylation of carbon-11, necessary in the synthesis of cortisone and cortisol. It was discovered that *Rhizopus arrhizus* could accomplish the hydroxylation as a one-step process and produce 11- hydroxyprogesterone in low yields, while *R. stolonifer* could carry out the same conversion and give almost quantitative yields. After this pioneer discovery, it was found that many fungus and bacteria species can alter steroid compounds. Species of *Rhizopus* and *Aspergillus* are widely used for this purpose, but fungi capable of converting steroids are found in all the major fungal groups. Alterations that can be accomplished by fungi are hydrogenation, dehydrogenation, hydroxylation, epoxidation, side-chain cleavage, and the expansion of the five-membered steroid ring into a six-membered ring.

Microbiological transformation of steroids has been utilized greatly in the commercial production of steroid compounds, which are often synthesized using a combination of chemical and microbiological transformations. If a transformation can be readily accomplished chemically, it is done by the chemist. If such a reaction is chemically difficult, expensive, or otherwise impractical and if a microorganism can accomplish it readily and inexpensively, the transformation is done with the aid of a microorganism. In practice, chemical and microbiological transformations may be used stepwise and in any sequence to obtain the desired end product.

To carry out a microbiological transformation, the fungus or bacterium is grown in a flask or tank on a medium that will support good growth for a period of 17 to 48 hours with agitation and aeration. After adequate growth of the microorganism is attained, a measured quantity of the steroid precursor is dissolved in a solvent and added to the medium. The enzyme (or enzymes) produced by the microorganisms then effects the conversion of the steroid precursor. Optimum conditions of temperature, aeration, agitation, pH, and time differ with each conversion, and these factors must be carefully controlled. The usual period of transformation is 24 to 48 hours. The modified steroid is recovered from the medium by extraction with a solvent such as methylene chloride, ethylene chloride, or chloroform (Sebek and Perlman, 1979; Sebek, 1983).

Ergot, Ergotism, and Ergot Alkaloids

Ergot is the sclerotium of the Pyrenomycete, *Claviceps purpurea*, which infects rye and, less commonly, other grains. Overwintered sclerotia give rise to a fleshy perithecia-bearing stroma. The discharged ascospores infect developing rye ovules. Normal development of the grain is suppressed, and a hardened, purple-black sclerotium develops in place of the grain (Fig. 16–12).

If consumed in large quantities, as might happen if the sclerotia are included in milled flour, a disease known as *ergotism* develops. Cattle in the field may also be poisoned by the sclerotia if they eat infected pasture grasses. The cattle stagger, develop lameness, gangrene, and possible inflammation of the digestive tract (Brook and White, 1966).

Ergotism in humans is marked by vomiting, feelings of intense heat or cold, pain in the muscles of the calf, a yellow color in the face, lesions on the hands and feet, diarrhea, and an impairment of the mental functions. People with ergotism often become hysterical or hallucinate. Some have severe convulsions, while the limbs of others become gangrenous and drop off. Spontaneous abortion occurs in pregnant women. Severe cases of ergotism are fatal.

Epidemics of ergotism were common in Europe in the Middle Ages, when rye was consumed in large quantities. Ergotism changed the course of history in at least one event: Peter the Great in Russia was unable to gain control of some ports on the Black Sea in 1772 because his soldiers bought rye bread for

(a)

(b)

Figure 16–12 *Claviceps purpurea,* which causes ergot of grains; (a) fleshy ascocarps arising from sclerotized rye grains; (b) enlarged view of the apex of the ascocarp, showing perithecia embedded within the stroma. [Courtesy Plant Pathology Department, Cornell University.]

themselves and fed rye to their horses. Many of the men developed convulsions, while the horses developed the "blind staggers" (Carefoot and Sprott, 1967). It is also possible that the women accused of witchcraft because of their bizarre behavior in the colonial Massachusetts village of Salem had ergotism (Caporael, 1976). As recently as 1951, an epidemic of ergotism occurred in Pont-Saint-Esprit, France, resulting in the death of four people and the hospitalization of over 150 who were afflicted with hysteria. This epidemic was traced to contaminated flour.

The active principles in ergot that are responsible for ergotism are the alkaloids ergometrine, ergometrinine, ergotamine, and ergotaminine, which occur in the sclerotium. These alkaloids stimulate smooth muscle and selectively block the sympathetic nervous system; they have found modern medical use in stimulating the uterus to contract to initiate childbirth, and also to hasten the return of the uterus to its normal size after childbirth, in the treatment of certain peripheral circulatory disorders, and in the treatment of migraine headaches.

A modern mycological problem has been one of supplying the demands of the pharmaceutical industry for a source of ergot alkaloids. Until about 1950, efforts to produce the alkaloids in culture by conventional laboratory techniques had failed, and the only source of the alkaloids were the sclerotia obtained from naturally infected rye, or rye which was inoculated for this purpose. Inoculation of rye under field conditions is a risky affair as it depends on suitable environmental conditions for development of the sclerotia and, at best, only one crop of sclerotia can be obtained in a season. The search has continued for more reliable and productive methods of obtaining the ergot alkaloids. A process for their laboratory production was patented in France in 1965, but a large-scale production method has yet to be developed.

FUNGI AND RECOMBINANT DNA TECHNOLOGY

Recombinant DNA techniques involve the introduction of foreign DNA into an organism. The DNA may be derived from another organism or may be constructed *in vitro*. Unlike protoplast fusion, the entire genome from one cell is not transferred to another. Instead, only specific segments of DNA or genes are transferred by recombinant DNA techniques. Once successfully introduced into the host organism, the foreign DNA can encode for specific proteins, changing the phenotype of the host cell. These genetically engineered cells provide powerful tools that are valuable in many scientific disciplines, such as genetics and cell biology. Scientists can learn a great deal about the mechanism of gene expression when the gene is actually removed from its usual cellular environment and functions in a foreign cell. Information can be obtained about biological events occurring in the cell, such as DNA replication, repair, and recombination. The proteins that are encoded by a specific gene can be identified with immunological techniques. A relatively large amount of recombinant DNA can be obtained by cloning and can be isolated and analyzed biochemically to determine the nucleotide sequence and to correlate this sequence with the genetic effects.

Methodology

Fungi can be modified by transformation, the uptake of exogenous DNA by a fungal cell. Following its introduction into the cell, the foreign DNA may become integrated into the chromosome and capable of being inherited by subsequent generations. The foreign DNA also can express the foreign phenotype.

Reliable and sophisticated recombinant DNA techniques based on transformation systems have been widely used with the bacterium *Escherichia coli*, which undergoes transformation readily. *E. coli* is a prokaryotic organism and difficulties are often encountered when its transformation is attempted with genes from eukaryotic organisms. Although genes from yeasts and filamentous fungi can be expressed in *E. coli,* this is the exception rather than the rule. Molecular biologists require a eukaryotic organism into which DNA from eukaryotic organisms can be introduced and expressed and which can serve as a model experimental system for many eukaryotic organisms, including humans. Recently, reliable methods utilizing the yeast *Saccharomyces cerevisiae* have been developed which make transformation possible through the use of foreign DNA (Boguslawski, 1985). Each cell of *S. cerevisiae* has about 50 to 100 *plasmids,* self-replicating DNA molecules, that occur in the mitochrondria. These plasmids are circular, about 2 micrometers in diameter, and consist of double-stranded DNA containing several genes concerned with maintenance and reproduction of the plasmids. Methods utilizing these plasmids for recombinant DNA techniques are similar to those developed for *E. coli*. Plasmids can be removed from the cell, segments removed by restriction enzymes, and a new piece of foreign DNA inserted. The restructured circular plasmids can be used to transform other cells. Compared with *E. coli*, transformation efficiency in yeasts is low, as the cells are not naturally competent to take DNA. The yeast cells must be treated with an enzyme or ions to render them more permeable, and only certain strains can be transformed. Once the plasmid is absorbed by the yeast cell, it can replicate autonomously as a plasmid for a number of cell generations, producing a clone of cells with the recombinant genotype. Alternatively, the DNA may become integrated into one of the chromosomes (the course of events depends on the nature of the new DNA). As a result of cloning, a much larger quantity of modified DNA is present within the cells than was required for the initial transformation and sometimes can be recovered as plasmids. Due to the integration of the DNA into the chromosome of a yeast cell, it can be difficult to get large amounts of DNA directly from yeasts. This problem has been solved by modifying the yeast plasmid in such a manner that it can be cloned in either the yeast cell or in *E. coli*, which gives a much larger yield of DNA after cloning. The DNA can be extracted for further analysis, such as nucleotide sequencing; used to transform other cells; or subjected to mutagens *in vitro* and then reintroduced into a cell. As a result of these techniques, DNA from yeast mutants, filamentous fungi, bacteria, viruses, and mammals, including rabbits and humans, have been introduced into the yeast plasmid and used to transform yeast cells. After amplification by cloning in either yeast or bacteria cells, the recombinant DNA can be used to transform filamentous fungi.

Progress in transformation of filamentous fungi has lagged behind the advances achieved with bacteria and yeasts. Although a number of genes from filamentous fungi have been cloned in *E. coli* or yeasts (Fig. 16–13), it is comparatively difficult to obtain expression of these genes. Little progress has been made in transferring genes from one filamentous fungus to another. Part of the difficulty lies in finding suitable plasmids that can be isolated from the hyphae, modified, reintroduced into the hyphae, and recovered after cloning (Weiss et al., 1985). Successful transformation requires that the foreign DNA become integrated into the host's chromosome, an event which occurs with low frequency.

Fungi as Model Organisms

Because of their significant role as eukaryotic model organisms, the fungi are likely to continue contributing greatly to the development of molecular biology

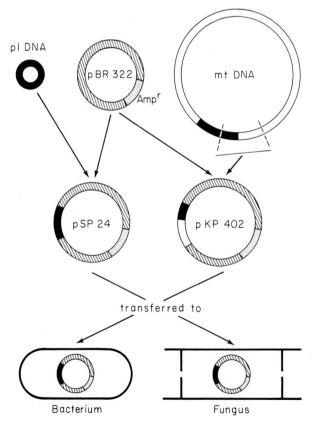

Figure 16–13 Construction of shuttle vectors with mitochondrial DNA of *Podospora anserina* being integrated in a bacterial plasmid pBR 322. In using either the *Podospora* mitochondrial plasmid (above to the left) or a part of the mitochondrial DNA containing an autonomously replicating sequence (above to the right), both hybrid vectors replicate and express equally in *E. coli* and in *Podospora*. [From K. Esser, U. Kuck, U. Stahl, and P. Tudzynski, 1984, *in:* D. H. Jennings, and A. D. M. Rayner, Eds., The ecology and physiology of the fungal mycelium, Cambridge University Press, Cambridge, pp. 343–352.]

in the future. They offer many unique advantages as model systems. They are readily grown in the laboratory on simple media, do not require much space, have rapid life cycles, and mutants can readily be obtained. The fungi are comparatively simple compared with other eukaryotes, and in particular, their genome is small in comparison. Much of the life cycle is spent in the haploid condition, which avoids the complication of gene masking by dominance, making gene mapping easier and facilitating isolation of mutants. Morphogenesis of reproductive structures can often be readily controlled, responding to appropriate nutritional and environmental conditions. The genetics of some organisms, such as the yeast *Saccharomyces cerevisiae* and the ascomycetes *Neurospora crassa* and *Podospora anserina,* are well known, enhancing their value as test organisms for molecular biologists and offering some distinct advantages. Yeasts such as *S. cerevisiae* are unicellular, which makes them easily manipulated. *S. cerevisiae* is safe to use in large-scale operations, can be adapted to continuous fermentations, grows easily on a variety of media, and has been characterized very well biochemically as well as genetically. It is possible to manipulate yeasts using standard microbiological methods. *N. crassa* and other ascomycetes are models of multinucleate sexually reproducing eukaryotes. Studies of meiosis in the ascus can provide important information about genetic recombination and its control because each ascus contains all the haploid nuclei resulting from a single meiotic division.

Practical Applications

Recombinant DNA experiments utilizing fungi are valuable in learning more about their molecular biology, as outlined above. But equally important, these methods provide a great potential for deriving greater benefit from commercially or ecologically important fungi. Some prospects for the future follow.

COMMERCIAL PRODUCTION OF FUNGAL PRODUCTS

The industrial production of compounds is an intricate process involving the response of the fungus to a variety of cultural conditions. Production of a particular compound is likely to be controlled by a large number of genes involving various metabolic events. Although genetic control of the production of a metabolite is complex, DNA recombinant methodology has the potential to contribute to the improvement of industrial fungus strains. Theoretically, these strains can be improved by eliminating undesirable characteristics, enhancing desirable characteristics, or even by introducing new characteristics into the fungus. Industrial microbiologists utilize genetic engineering in an effort to make an existing process more economical by modifying the physical or nutritional requirements of the fungi involved. In particular, it is desirable to develop fungal strains that can utilize cheaper and more readily available nutrient sources. One such source is wood, which consists mostly of cellulose. Comparatively few fungi produce the cellulases required to degrade cellulose to glucose, but glucose can be fermented to alcohol by yeasts or to citric acid by *Aspergillus niger,* or utilized by other filamentous fungi in producing metabolites. If the genes controlling cellulase production could be introduced into the commercially important fungi, wood could be used as a cheap carbon source because the organisms would degrade the wood as well as utilize the glucose to produce alcohol, citric acid, and other products. Development of new strains that produce more of an end product, such as an organic acid or antibiotic, or even strains that produce a new end product, is also a goal of genetic engineering. It should be possible to identify and clone the fungal genes that play a role in the synthesis of the end product and construct new strains, as is the case for alcohol production. Greatly improved alcohol-yielding strains may be developed after further studies of the cloned alcohol dehydrogenase genes and related loci are made, which will lead to an understanding of the signals that control alcohol metabolism (Bigelis, 1985).

Enzymes produced by microorganisms are becoming increasingly important in industrial processes, replacing many chemical procedures. Genetic engineering is potentially of value in enhancing enzyme production. One such application may be in the manufacture of plastics. Currently petrochemicals are used to produce alkene oxides which are then polymerized to form plastics. It has been proposed that enzymes from microorganisms can be used to synthesize these alkene oxides more economically and with less pollution. The three enzymes required for this synthesis can be obtained from the basidiomycete *Oudemansiella mucida*, the deuteromycete *Caldariomyces,* and the bacterium *Flavobacterium.* This process also yields economically important by-products in large quantities. Glucose is provided for the growth of *O. mucida* and is converted into fructose (used as a sweetener) and gluconic acid (added to dishwasher detergents to prevent spots). Recombinant DNA technology will undoubtedly be used to improve the yield of the three required enzymes and make this method of plastic production a reality (Eveleigh, 1981).

MEDICINE

Genetically engineered yeast cells can be used to express mammalian genes and to produce proteins of medical or agricultural importance. These include human interferons, growth factor, and hormones. Genes controlling pathogenesis in fungi causing disease of humans are being studied using recombinant DNA techniques. Factors regulating dimorphism in pathogenic fungi are also being investigated. Recombinant DNA techniques offer the possibility that the nature of the pathogenic fungi or the course of the disease could be altered.

PLANT PATHOLOGY

Genetic engineering may help us understand the relationship between a plant and its fungal pathogens at a molecular level. Studies with cutinase illustrate this. Many plant pathogenic fungi produce cutinase, an enzyme that can degrade the cutin on a plant's surface and facilitate penetration into the plant. A gene coding for cutinase production has been cloned and characterized from *Fusarium solani* (Soliday et al., 1984), which has the potential to provide the molecular basis for understanding this aspect of plant pathology better and perhaps to develop a strategy for plant protection (Mishra, 1985).

WASTE MANAGEMENT

Genetic engineering provides us with an opportunity to learn more about possible ecological applications. One unique property of many fungi is their ability to degrade cellulose and lignin, the major components of wood, in nature. Molecular cloning of genes that code for the production of enzymes responsible for digesting these compounds has met some success. The genes coding for cellulase production in *Trichoderma reesi* and other filamentous wood-rotting fungi have been used to transform yeast cells, which can then produce this enzyme. This method provides great promise for eventually being used to degrade accumulating waste materials, such as newspapers, manure, remains of crops, and by-products of manufacturing processes (Montenecourt and Eveleigh, 1985). Not only would the environment be improved by the removal of the waste, but it could be used as a substrate to produce cells that could be a source of protein for animals.

REFERENCES CITED

AHARONOWITZ, Y., and G. COHEN. 1981. The microbiological production of pharmaceuticals. *In:* Industrial microbiology and the advent of genetic engineering. W. H. Freeman, San Francisco, pp. 57–70.

BENEDICT, R. G. 1972. Mushroom toxins other than *Amanita. In:* Kadis, S., A. Ciegler, and S. J. Ajl, Eds. Microbial toxins. Academic Press, New York, Vol. 8, pp. 281–320.

BIGELIS, R. 1985. Primary metabolism and industrial fermentations. *In:* Bennett, J. W., and L. L. Lasure, Eds. Gene manipulations in fungi. Academic Press, New York, pp. 358–401.

BLOCK, S. S., R. L. STEPHENS, A. BARETO, and W. A. MURRILL. 1955. Chemical identification of the *Amanita* toxin in mushrooms. Science 121:505–506.

BOGUSLAWSKI, G. 1985. Yeast transformation. *In:* Bennett, J. W., and L. L. Lasure, Eds. Gene manipulations in fungi. Academic Press, New York, pp. 161–195.

BROOK, P. J., and E. P. WHITE. 1966. Fungus toxins affecting mammals. Ann. Rev. Phytopathol. 4:171–194.

BULLER, A. H. R. 1914. The fungus lore of the Greeks and Romans. Trans. Brit. Mycol. Soc. 5:21–66.

———. 1922. Researches on fungi. Longman, London. Vol. 2, 492 pp.

BU'LOCK, J. D. 1979. Industrial alcohol. *In:* Bull, A. T., D. C. Ellwood, and C. Ratledge, Eds. Microbial technology: current state, future prospects. Symp. Soc. Gen. Microbiol. 29:308–325.

CAPORAEL, L. R. 1976. Ergotism: the satan loosed in Salem? Science 192:21–26.

CAREFOOT, G. L., and E. R. SPROTT. 1967. Famine on the wind. Rand McNally, Skokie, Ill., 229 pp.

DUDDINGTON, C. L. 1961. Micro-organisms as allies: the industrial use of fungi and bacteria. Macmillan, New York, 256 pp.

EVELEIGH, D. E. 1981. The microbiological production of industrial chemicals. *In:* Industrial microbiology and the advent of genetic engineering. W. H. Freeman, San Francisco, pp. 71–82.

FARR, D. F. 1983. Mushroom industry: diversification with additional species in the United States. Mycologia 75:351–360.

FERENCZY, L. 1981. Microbial protoplast fusion. *In:* Glover, S. W., and D. A. Hopwood, Eds. Genetics as a tool in microbiology. Cambridge University, Cambridge, pp. 1–34.

FERMOR, T. R., P. E. RANDLE, and J. F. SMITH. 1985. Compost as a substrate and its preparation. *In:* Flegg, P. B., D. M. Spencer, and D. A. Wood, Eds. The biology and technology of the cultivated mushroom. Wiley, New York, pp. 81–109.

FLEGG, P. B., and D. A. WOOD. 1985. Growth and fruiting. *In:* Flegg, P. B., D. M. Spencer, and D. A. Wood, Eds. The biology and technology of the cultivated mushroom. Wiley, New York, pp. 141–177.

FLEMING, A. 1929. On the antibacterial action of cultures of a *Penicillium,* with special reference to their use in the isolation of *B. influenzae.* Brit. J. Exptl. Pathol. 10:226–236.

HATCH, R. T., and S. M. FINGER. 1979. Mushroom fermentation. *In:* Peppler, H. J., and D. Perlman, Eds. Microbial technology, 2nd ed. Academic Press, New York, Vol. 2, pp. 179–199.

HESSELTINE, C. W. 1965. A millennium of fungi, food, and fermentation. Mycologia 57:149–197.

LINCOFF, G., and D. H. MITCHELL. 1977. Toxic and hallucinogenic mushroom poisoning. Van Nostrand Reinhold, New York, 267 pp.

LOCKWOOD, L. B. 1979. Production of organic acids by fermentation. *In:* Peppler, H. J., and D. Perlman, Eds. Microbial technology, 2nd ed. Academic Press, New York, Vol. 1, pp. 241–280.

LOWY, B. 1971. New records of mushroom stones from Guatemala. Mycologia 63:983–993.

———. 1972. Mushroom symbolism in Maya codices. Mycologia 64:816–821.

Manning, K. 1985. Food value and composition. *In:* Flegg, P. B., D. M. Spencer, and D. A. Wood, Eds. The biology and technology of the cultivated mushroom. Wiley, New York, pp. 211–230.

Mishra, N. C. 1985. Gene transfer in fungi. Advan. Genet. 23:73–178.

Montenecourt, B. S., and D. E. Eveleigh. 1985. Fungal carbohydrases: amylases and cellulases. *In:* Bennett, J. W., and L. L. Lasure, Eds. Gene manipulations in fungi. Academic Press, New York, pp. 491–512.

Moskowitz, G. J. 1979. Inocula for blue-veined cheeses and blue cheese flavor. *In:* Peppler, H. J., and D. Perlman, Eds. Microbial technology, 2nd ed. Academic Press, New York, Vol 2, pp. 201–222.

Olson, N. F. 1979. Cheese. *In:* Peppler, H. J., and D. Perlman, Eds. Microbial technology, 2nd ed. Academic Press, New York, Vol 2, pp. 39–77.

Oso, B. A. 1975. Mushrooms and the Yoruba people of Nigeria. Mycologia 67:31–319.

———. 1977. Mushrooms in Yoruba mythology and medicinal practices. Econ. Botany 31:367–371.

Peppler, H. J. 1979. Production of yeasts and yeast products. *In:* Peppler, H. J., and D. Perlman, Eds. Microbial technology, 2nd ed. Academic Press, New York, Vol. 1, pp. 157–185.

Perlman, D. 1979. Microbial production of antibiotics. *In:* Peppler, H. J., and D. Perlman, Eds. Microbial technology, 2nd ed. Academic Press, New York, Vol. 1, pp. 241–280.

Phaff, H. J., and M. A. Amerine. 1979. Wine. *In:* Peppler, H. J., and D. Perlman, Eds. Microbial technology, 2nd ed. Academic Press, New York, Vol. 2, pp. 131–153.

Sebek, O. K. 1983. Fungal transformations as a useful method for the synthesis of organic compounds. Mycologia 75:383–394.

———, and D. Perlman. 1979. Microbial transformation of steroids and sterols. *In:* Peppler, H. J., and D. Perlman, Eds. Microbial technology, 2nd ed. Academic Press, New York, Vol. 1, pp. 483–496.

Soliday, C. L., W. H. Flurkey, T. W. Okita, and P. E. Kolattukudy. 1984. Cloning and structure determination of cDNA for cutinase, an enzyme involved in fungal penetration of plants. Proc. Natl. Acad. Sci. USA 81:3939–3943.

Sternberg, M. 1976. Microbial rennets. Advan. Appl. Microbiol. 20:135–157.

Stewart, G. G. 1974. Some thoughts on the microbiological aspects of brewing and other industries utilizing yeast. Advan. Appl. Microbiol. 17:233–264.

Thom, C. 1944. Molds in the cheese industry. J. N.Y. Botan. Garden 45:105–113.

Wasson, R. G. 1961. The hallucinogenic fungi of Mexico: an inquiry into the origins of the religious idea among primitive peoples. Botan. Museum Leaflets Harvard Univ. 19:137–162.

Wasson, V. P., and R. G. Wasson. 1957. Mushrooms, Russia, and history. Pantheon Books, New York, Vols. 1 and 2, 433 pp.

Watling, R., and M. R. D. Seaward. 1976. Some observations on puff-balls from British archeological sites. J. Archaeol. Sci. 3:165–172.

Weiss, R. L., D. Puetz, and J. Cybis. 1985. Expression of *Aspergillus* genes in *Neurospora*. *In:* Bennett, J. W., and L. L. Lasure, Eds. Gene manipulations in fungi. Academic Press, New York, pp. 279–292.

Westermann, D. H., and N. J. Huige. 1979. Beer brewing. *In:* Peppler, H. J., and D. Perlmann, Eds. Microbial technology, 2nd ed. Academic Press, New York, Vol. 2, pp. 1–37.

Wieland, T., and O. Wieland. 1972. The toxic peptides of *Amanita* species. *In:* Kadis, S., A. Ciegler, and S. J. Ajl, Eds. Microbial toxins. Academic Press, New York, Vol. 8, pp. 249–280.

Additional Recommended References

Backus, M. P., and J. F. Stauffer. 1955. The production and selection of a family of strains in *Penicillium chrysogenum*. Mycologia 47:429–463.

BATRA, L. R., and P. D. MILLNER. 1974. Some Asian fermented foods and beverages and associated fungi. Mycologia 66:942–950.

BUCK, R. W. 1961. Mushroom poisoning since 1924 in the United States. Mycologia 53:537–538.

CHANG, S. T. 1980. Mushrooms as human food. Bioscience 30:399–401.

COCHRAN, K. W., and M. W. COCHRAN. 1978. *Clitocybe clavipes:* Antibuse-like reaction to alcohol. Mycologia 70:1124–1126.

DUNN, C. G. 1958. The industrial utilization of yeasts. Econ. Botany 12:145–163.

GROVES, J. W. 1964. Poisoning by morels when taken with alcohol. Mycologia 56:779–780.

HOFFER, A., and H. OSMOND. 1967. The hallucinogens. Academic Press, New York, 626 pp.

HOPWOOD, D. A. 1981. The genetic programming of industrial microorganisms. *In:* Industrial microbiology and the advent of genetic engineering. W. H. Freeman, San Francisco, pp. 31–44.

KURTZMAN, C. P. 1983. Fungi: sources of food, fuel, and biochemicals. Mycologia 75:374–382.

LOWE, D. A., and R. P. ELANDER. 1983. Contribution of mycology to the antibiotic industry. Mycologia 75:361–373.

MAISCH, W. F., M. SOBOLOV, and A. J. PETRICOLA. 1979. Distilled beverages. *In:* Peppler, H. J., and D. Perlman, Eds. Microbial technology, 2nd ed. Academic Press, New York, Vol. 2, pp. 79–94.

PEPPLER, H. J. 1960. Yeast. *In:* Matz, S. A., Ed. Bakery technology and engineering. AVI, Westport, Conn., pp. 35–74.

PERBODY, J. F. 1979. Fungal protoplasts: isolation, reversion, and fusion. Ann. Rev. Microbiol. 33:21–39.

ROWLANDS, R. T. 1983. Industrial fungal genetics and strain selection. *In:* Smith, J. E., D. R. Berry, and B. Kristiansen, Eds. The filamentous fungi: fungal technology. Edward Arnold, London, Vol. 4, pp. 346–372.

WANG, H. L., and C. W. HESSELTINE. 1979. Mold-modified foods. *In:* Peppler, H. J., and D. Perlman, Eds. Microbial technology, 2nd ed. Academic Press, New York, Vol. 2, pp. 95–129.

WASSON, R. G. 1962. The hallucinogenic mushrooms of Mexico and psilocybin: a bibliography. Botan. Museum Leaflets Harvard Univ. 20:25–73.

WUEST, P. J. 1983. Resources needed to farm the 'champignon.' Mycologia 75:341–350.

YOUNG, F. M., and B. J. B. WOOD. 1974. Microbiology and biochemistry of soy sauce fermentation. Advan. Appl. Microbiol. 17:157–194.

Glossary

Acervulus pl.* **acervuli** (dim. of L.* *acervus* = heap). A subcuticular or subepidermal mass of closely clustered conidiophores and conidia which are not covered by fungal tissue.

Aeciospore (Gr.* *aikia* = injury, *sporos* = seed). A binucleate spore borne in an aecium.

Aecium, pl. **aecia** (Gr.* *aikia* = injury). The first sorus that is formed after plasmogamy and bears binucleate aeciospores, stage I (in the Uredinales).

Aerobe (Gr. *aēr* = air, *bios* = life). An organism that requires free oxygen for respiration.

Aerobic. With the qualities of an aerobe.

Agaric. A gill-bearing mushroom.

Alternation of generations. Succession of gamete- and spore-bearing thalli in the life cycle of an organism.

Anaerobic. Not requiring free, molecular oxygen for respiration.

Anamorph (Gr. *ana* = anew, *morphe* = form). The nonsexual (usually conidial) stage in the life cycle of a fungus (*cf.* telomorph).

Annulus, pl. **annuli** (L., ring). A membranous skirt surrounding the stipe of a Hymenomycete.

Antagonism. An ecological association between organisms in which one or more of the participants is harmed or has its activities limited.

Antheridium, pl. **antheridia** (dim. of Gr. *anthos* = flower). A male gametangium.

Antibiosis. (Gr. *anti* = against, *bios* = life). Production of antibiotic(s) that affects other organisms adversely.

Antibiotic. A compound of high molecular weight produced by a microorganism that has specific toxic effects on other microorganisms sensitive to it.

Apomictic (Gr. *apo* = lack, *mixis* = mingling). Developing by apomixis.

Apomixis. Development of a product that is normally the result of sexual reproduction without plasmogamy, karyogamy, or meiosis.

Apophysis (Gr. *apo* = lack, *physis* = growth). A swelling of the vegetative cell beneath the sporangium in some chytrids.

Apothecium, pl. **apothecia** (Gr. *apo* = lack, *theke* = sheath). An ascocarp in which the hymenium is not covered by fungal tissue at maturity.

Appressorium, pl. **appressoria** (L. *apprimere* = to press against). An enlargement on a hypha or germ tube which attaches itself to the host before penetration takes place.

Arbuscule. A branched ending of the fungal hypha formed by some mycorrhizal fungi and of unknown function.

Archicarp (Gr. *archi* = first, *karpos* = fruit). The female organ(s) and supporting cells (in the Ascomycotina).

Asocarp (Gr., *askos* = sac, *karpos* = fruit). The ascospore-bearing, multicellular sporocarp formed by a member of the Ascomycotina.

Ascogenous hyphae (Gr. *askos* = sac, *genesis* = being produced). Those hyphae, often arising from the archicarp, which give rise to asci.

Ascogonium, pl. **ascogonia** (Gr. *askos* = sac, *gone* = offspring). The cell or cells of the archicarp that function as the female gametangium and which usually receive the male nuclei and then give rise to the ascogenous hyphae.

Ascospore (Gr. *askos* = sac, *sporos* = seed). A spore borne in an ascus.

Ascostroma, pl. **ascostromata** (Gr. *askos* = sac, *stroma* = cover). A stroma bearing asci.

Ascus, pl. **asci** (Gr. *askos* = sac). A cell that is the site of meiosis and in which endogenous spores of sexual origin are formed.

Assimilation. The uptake of nutrients and their incorporation into the organism.

Autoecious (Gr. *autos* = self, *oikas* = house). Capable of completing a life cycle on one host (of Uredinales) (*cf.* heteroecious).

Autotrophic (Gr. *autos* = self, *trophe* = nourishment). Independent of an external source of organic food (*cf.* heterotrophic).

Azygospore (Gr. *azygos* = unmatched, *sporos* = seed). A body morphologically similar to a zygospore but which is of nonsexual origin (in the Zygomycetes).

Basidiocarp (Gr. *basidion* = small base, *karpos* = fruit). A sporocarp produced by a member of the Basidiomycotina and which bears basidiospores.

Basidiole (dim. of Gr. *basidion* = small base). A structure in a hymenium that is morphologically similar to a basidium without sterigmata and may be an immature basidium or a permanently sterile structure in the hymenium of a member of the Basidiomycotina (*cf.* cystidium).

Basidiospore (Gr. *basidion* = small base, *sporos* = seed). An exogenous sexual spore borne on a basidium.

Basidium, pl. **basidia** (Gr. *basidion* = small base). A cell in which karyogamy and meiosis take place and which bears exogenous spores of sexual origin.

Binding hyphae. Thick-walled, highly branched, aseptate hyphae in a basidiocarp that bind other hyphal types together (*cf.* generative, lactiferous, and skeletal hyphae).

Biotrophic parasite (Gr. *bios* = life, *trophe* = nourishment). A fungus parasitic on other fungi which does not kill its host but feeds on living cells.

Bipolar sexuality (L. *bi* = two, *polus* = pole). A type of sexual compatibility controlled by multiple-allelic genes at a single locus (*cf.* tetrapolar sexuality).

Bitunicate (L. *bi* = two, *tunica* = covering). Having a double ascus wall (*cf.* unitunicate).

Blastic development (Gr. *blastos* = bud, embryonic cell). Formation of a conidium by enlargement and maturation of a conidial initial that is distinct from the vegetative hyphae or conidiophore and enlarges before septation separates the conidium from the parent cell. In **enteroblastic** development (Gr., *enteron* = that within), some of the wall layers of the new conidium are not continuous with the wall layers of the parent cell. In **holoblastic** development (Gr. *holos* = whole or entire), all wall layers of the parent cell are continuous with those of the conidium (*cf.* thallic development).

Bolete (Gr. *bolos* = a throw). A member of the Agaricales that possesses tubes and not gills.

Bromatium, pl. **bromatia** (Gr. *bromatos* = food). An enlarged hyphal ending, induced and used as food by fungi-cultivating ants.

Budding. Production of a new cell from a small outgrowth or protrusion from the mother cells (of yeasts or spores) (*cf.* fission).

Capillitium, p. **capillitia** (L. *capillus* = hair). A mass of sterile fibers interspersed among spores within a sporocarp (in the Gasteromycetes).

Carpogenic cell (Gr. *karpos* = fruit, *genesis* = being produced). A cell that gives rise to the ascogonium and supporting cells in the Laboulbeniales.

Catahymenium (Gr. *kata* = under or lower, complete; *hymen* = membrane). A hymenium type initiated as a palisade of cystidia through which basidia eventually penetrate; nonthickening (*cf.* euhymenium).

Centrum (Gr. *kentron* = center). The contents of an asocarp locule, including the ascogenous system and sterile cells.

Chemotropism. See **tropism.**

Chlamydospore (Gr. *chlamydos* = cloak, *sporos* = seed). A thick-walled, nonsexual spore that represents a transformed hyphal cell.

Clamp (connection). A recurving outgrowth of a cell that, at cell division, acts as a bridge to allow passage of one of the products of nuclear division into the penultimate cell, thereby ensuring maintenance of the dikaryotic condition (of members of the Basidiomycotina).

Cleistohymenial (Gr. *kleistos* = closed, *hymen* = membrane). Of apothecia in which the hymenium is covered by excipular tissue during part of the development (*cf.* gymnohymenial).

Cleistothecium, pl. **cleistothecia** (Gr. *kleistos* = closed, *theke* = case). An ascocarp with the asci surrounded by fungal tissue and without regularly formed openings.

Colonization. The invasion of a substratum or habitat by a fungus.

Colony. A discrete mass of cells or hyphae developing from a single cell, hyphal fragment, or spore.

Columella, pl. **columellae** (L. *columen* = column). A sterile column within a spore-bearing structure, often an extension of a supporting stalk.

Compatible. Describes fungal sexual partners that are able to produce a fertile cross with each other.

Competition. A more or less active demand on the part of two organisms for some commodity (space, food, etc.) that is inadequate to provide for all organisms present.

Competitive saprophytic ability. Relative capability of successfully competing with other saprophytes.

Complementation. Compensation by a nucleus for deficiencies of other nuclear type(s) in a heterokaryon.

Conidiophore (Gr. *konis* = dust, *phoros* = bearing). A hypha, often specialized in structure, that bears one or more conidia.

Conidium, pl. **conidia** (Gr. *konis* = dust). A thin-walled, nonsexual spore that is borne both terminally and exogenously on a conidiophore and is deciduous at maturity.

Constitutive dormancy. See **dormancy.**

Coprophilous (Gr. *kopros* = dung, *philein* = to love). Living on dung.

Cortina (dim. of L. *cortis* = curtain). A veil extending from the margin of the pileus in some Hymenomycetes.

Crozier. A recurved hook at the tip of an ascogenous hypha that will become the ascus.

Crustose thallus (L. *crusta* = crust). A flat, crustlike thallus that adheres closely to the substratum (of lichens) (*cf.* foliose thallus, fruticose thallus).

Culture (L. *colere* = to cultivate). (1) To grow an organism or (2) the resulting growth. Usually on artificial medium for experimental or industrial use.

Cystidium, pl. **cystidia** (Gr. *kyst* = sac). A sterile cell occurring among the basidia and often projecting beyond the hymenium and differing morphologically from the basidium (*cf.* basidiole).

Decay. See **decomposition.**

Decomposition. Decay; separation of a material into its component parts, for example, breakdown of organic matter by microorganisms.

Dermatophyte (Gr. *dermatos* = skin, *phyton* = plant). A parasitic fungus that attacks and causes a disease of the skin or its appendages.

Dikaryon (L. *di* = two, Gr. *karyon* = nucleus). A cell or hypha in which haploid, genetically different nuclei of two types are closely associated in pairs following plasmogamy (*cf.* heterokaryon, monokaryon).

Dikaryotic. The condition of being a dikaryon.

Dimorphic (L. *di* = two, Gr. *morphe* = form). Producing two morphologically different forms, not necessarily required to complete the life cycle (of animal parasites that can exist as single-celled and mycelial forms).

Dimorphism. Existence of two morphologically different forms in a dimorphic organism.

Dioecious (L. *di* = two, Gr. *oikos* = house). Having male and female reproductive structures on separate thalli.

Diplanetic (L. *di* = two, Gr. *planos* = wandering). Having a succession of two morphologically different zoospore stages separated by a resting stage (in the Oomycetes) (*cf.* monoplanetic).

Diplanetism. The succession of two zoospore stages in a diplanetic organism.

Diploid (Gr. *diploos* = double). Having the 2N number of chromosomes within a single nucleus.

Diploidization. Karyogamy of haploid nuclei to form a diploid nucleus, especially in a heterokaryon.

Disease. A malfunctioning of an organism, caused by external agents, which results in abnormalities.

Dissimilation. Metabolic decomposition within an organism.

Dormancy. A period of rest or minimal metabolic activity, especially in spores. Constitutive dormancy is imposed by the organism, while exogenous dormancy is imposed by external conditions.

Ectomycorrhiza (Gr. *ecto* = on the outside, *mykes* = fungus, *rhiza* = root). A mycorrhizal type in which a sheath of fungal hyphae is formed on the outside of the root (*cf.* endomycorrhizae).

Endobiotic (Gr. *endon* = within, *bios* = life). Growing entirely within a host (*cf.* epibiotic, interbiotic).

Endogenous (Gr. *endon* = within, *genesis* = being produced). Produced or borne within a structure.

Endomycorrhiza (Gr. *endon* = within, *mykes* = fungus, *rhiza* = root). A mycorrhizal type in which the fungal hyphae penetrate the root and a sheath is lacking (*cf.* ectomycorrhiza).

Endospore (Gr. *endon* = within, *sporos* = seed). The innermost layer of a spore wall (*cf.* epispore, perispore).

Enteroblastic development (see **blastic development**).

Enterothallic development (see **thallic development**).

Epibiotic (Gr. *epi* = upon, *bios* = life). Growing on the outside of a host (*cf.* endobiotic, interbiotic).

Epispore (Gr. *epi* = upon, *sporos* = seed). The outermost layer of the spore wall, which is often ornamented (*cf.* endospore, perispore).

Epithecium (Gr. *epi* = upon, *theke* = sheath or cover). A layer of tissue that covers the asci in some Discomycetes.

Eucarpic (Gr. *eu* = normal, *karpos* = fruit). A condition in which only part of an assimilative thallus is converted into reproductive structure(s) (*cf.* holocarpic).

Euhymenium (Gr. *eu* = normal, *hymen* = membrane). A hymenium type initiated as a palisade of basidia; sometimes thickening (*cf.* catahymenium).

Excipulum, pl. **excipula** (L., *vessel*). In the Discomycetes, the tissue beneath the subhymenium and forming the outermost sterile margin of the apothecium.

Exogenous (L. *ex* = out of, Gr. *genesis* = being produced). Borne or produced on the outside of a structure.

Exogenous dormancy. See **dormancy.**

Exploitation. An ecological interrelationship in which one organism inflicts harm on another for its own benefit (parasitism or predation).

Facultative parasite (L. *facultas* = ability). An organism that normally lives as a saprophyte but under certain conditions can live as a parasite (*cf.* obligate parasite).

Fermentation (L. *fermentum* = yeast). Anaerobic respiration; anaerobic dissimilation of an organic compound.

Fertilization. Sexual union of both cytoplasm (plasmogamy) and nuclei (karyogamy).

Fertilization tube. An extension of the antheridium which penetrates the wall of the oogonium to reach the oosphere and through which the male protoplast passes (in the Oomycetes).

Fission. Cell division by a cleavage (splitting) of the cell into two parts (*cf.* budding).

Flagellum, pl. **flagella** (L. *flagellum* = whip). A hairlike structure that propels a swimming cell.

Flexuous hyphae (L. *flexuosus* = bent). Those hyphae in a pycnium that undergo plasmogamy with pycniospores (in the Uredinales).

Foliose thallus (L. *folium* = leaf). A flat, leaflike thallus that is spread over the substratum but is attached at localized points only (of lichens) (*cf.* crustose thallus, fruticose thallus).

Fruticose thallus (L. *fruticosus* = shrub). An upright or pendulous thallus (of lichens) (*cf.* crustose thallus, foliose thallus).

Fungicide. A chemical or physical agent that kills fungi (*cf.* fungistasis).

Fungistasis. Inhibition of fungal growth that is not lethal (*cf.* fungicide).

Fungus, pl. **fungi** (L., *fungus* or *mushroom*). A eukaryotic heterotrophic organism whose usually walled, threadlike cells absorb nutrients in the form of a watery solution.

Funiculus, pl. **funiculi** (L. *funis* = rope). A hyphal cord that attaches a peridiole to the peridium (in the Nidulariales).

Gametangium,   pl. **gametangia** (Gr. *gamete* = germ, *angeion* = case). A differentiated cell that produces discrete gametes or whose undifferentiated protoplast functions in the stead of discrete gametes.

Gamete   (Gr. *gamete* = germ). A differentiated reproductive cell that is capable of undergoing plasmogamy with another similar or dissimilar reproductive cell.

Generative cell.   In the Trichomycetes, a cell that bears a sporangiole.

Generative hyphae.   Nonspecialized, thin-walled hyphae in a basidiocarp (*cf.* lactiferous, binding, and skeletal hyphae).

Geotropism.   See **tropism.**

Germination   (L. *germen* = sprout). The emergence of a protoplast, usually vegetative, from a spore.

Gill.   A hymenium-covered, platelike appendage that hangs from the underneath surface of the basiodiocarp of some Hymenomycetes.

Gleba   (L., *clod*). Spore-bearing tissue that is enclosed by a peridium throughout development (in the Gasteromycetes).

Gymnocarpic development   (Gr. *gymnos* = naked, *karpos* = fruit). Development of basidiocarps with the hymenium exposed throughout development (*cf.* hemiangiocarpic, pseudoangiocarpic).

Gymnohymenial development   (Gr. *gymnos* = naked, *hymen* = membrane). Development of apothecia in which the hymenium is exposed throughout development (*cf.* cleistohymenial).

Haploid   (Gr. *haploos* = single). Having a single set (*N*), or the reduced number, of chromosomes in a single nucleus.

Haploidization.   In a heterokaryon, the transformation of a diploid nucleus to a haploid nucleus through a series of mitotic divisions in which aneuploid nuclei are formed.

Hapteron   (Gr. *haptein* = fasten upon). An organ of attachment in some lichens or in the Nidulariales. In the latter, the hapteron is a mass of highly adhesive hyphae at the base of the funiculus.

Haustorium,   pl. **haustoria** (L. *haustus* = drawing of water). A specialized hyphal extension, formed by some parasitic fungi, which penetrates the host cell and absorbs nutrients.

Hemiangiocarpic development   (Gr. *hemi* = partly, *angeion* = case, *karpos* = fruit.) Development of basidiocarps in which the hymenium is formed endogenously during the early stages of development, but which becomes exposed before maturity (*cf.* gymnocarpic, pseudoangiocarpic).

Heteroecious   (Gr. *heteros* = different from, *oikos* = house). Requiring more than one host species to complete a life cycle (of Uredinales) (*cf.* autoecious).

Heterogamy   (Gr. *heteros* = different from, *gamete* = germ). Plasmogamy between gametes which are morphologically different (*cf.* isogamy).

Heterokaryon   (Gr. *heteros* = different from, *karyon* = nucleus). A cell that contains genetically different nuclei or a thallus made up of such cells (*cf.* dikaryon, homokaryon).

Heterokaryosis.   See **heterokaryotic.**

Heterokaryotic.   The condition of being a heterokaryon.

Heterothallic   (Gr. *heteros* = different from, *thallos* = shoot). The condition of being self-sterile, requiring a partner for sexual reproduction (*cf.* homothallic).

Heterothallism.   Self-sterility (*cf.* homothallism).

Heterotrophic.   (Gr. *heteros* = different from, *trophe* = nourishment). Deriving nourishment from organic matter formed by other organisms, living as a saprophyte or parasite (*cf.* autotrophic).

Hilar appendix.   The hilum and associated structures at the base of the basidiospore.

Holobasidium (Gr. *holos* = entire, *basidion* = small base). A single-celled basidium (*cf.* phragmobasidium).

Holoblastic development. See **blastic development.**

Holocarpic (Gr. *holos* = entire, *karpos* = fruit). A condition in which the entire assimilative thallus converts into a reproductive structure (*cf.* eucarpic).

Homokaryon (Gr. *homos* = alike, *karyon* = nucleus). A cell which contains nuclei of one genetic constitution only or a thallus made up of such cells (*cf.* heterokaryon).

Homokaryotic. The condition of being a homokaryon.

Homothallic. (Gr. *homos* = alike, *thallos* = shoot). The condition of being self-fertile, able to reproduce sexually without a partner (*cf.* heterothallic).

Homothallism. Self-fertility (*cf.* heterothallism).

Host. A living organism that provides nourishment, and possibly protection, to a parasite that lives in or on it.

Humus (L., earth). The dark organic portion of soil consisting largely of partially decayed plant material.

Hymenium, pl. **hymenia** (Gr. *hymen* = membrane). A palisade-like layer of asci or basidia, including any sterile cells, such as basidioles, paraphyses, or cystidia.

Hymenophore (Gr. *hymen* = membrane, *phoros* = carrying). The trama immediately beneath the hymenium and which bears the hymenium in a basidiocarp.

Hypertrophy (Gr. *hyper* = over, *trophe* = nourishment). Abnormal overdevelopment of a tissue, cell, or components of a cell.

Hypha, pl. **hyphae** (Gr. *hyphē* = a web). Filamentous part of a fungus, usually septate and consisting of several cells in linear succession.

Hyphal body. A segment of a hypha that becomes separated and reproduces by budding or fission (in the Entomophthorales).

Hypogeous fungi (Gr. *hypos* = below, *ge* = earth). Fungi that form sporocarps within the soil.

Inoculate (L. *inoculare* = to engraft). To introduce microorganisms, or a substance containing microorganisms, within or upon an organism or substratum.

Inoculum. That portion used in inoculation.

Inoperculate (L. *in* = not, *operculum* = lid). Of asci or sporangia that open by means of a pore lacking an operculum (*cf.* operculate).

Interbiotic (L. *Inter* = between, Gr. *bios* = life). With part of the thallus within a host and part external to the host (of Chytridiales) (*cf.* endobiotic, epibiotic).

Isidium (Gr. *is* = tissue, *idios* = one's own). A protuberance of the lichen cortex that may be cylindrical, clavate, or other shapes.

Isogamy (Gr. *isos* = equal, *gamete* = germ). Plasmogamy between gametes that are morphologically similar (*cf.* heterogamy).

Karyogamy (Gr. *karyon* = nucleus, *gamos* = marriage). Fusion of two nuclei.

Kinetosome (Gr. *kinein* = to move, *soma* = body). A basal body lying within the cell that gives rise to the fibers within the flagellum.

Lactiferous hyphae (L. *lactis* = milk). Hyphae filled with a milky fluid (*cf.* binding, generative, and skeletal hyphae).

Lacunar development (L. *lacuna* = small pit or gap). Development of Gasteromycetes in which the cavities are formed by tearing or stretching of the glebal tissue.

Lamella, pl. **lamellae** (dim. of L. *lamina* = a thin plate). See **gill.**

Lichen (Gr. *leichen* = tree-moss). A thallus consisting of an alga and fungus intermixed and living in a symbiotic relationship.

Life cycle. The succession of stages following a particular phase of development or spore form, which culminates with the production of that same phase or spore form.

Loculate (dim. of L. *locus* = place). Containing locules.

Locule. A cavity in a stoma which contains asci and is not bounded by a wall originating from the archicarp or other sexual structures.

Lomasome (Gr. *loma* = border, *soma* = body). A body of various forms which occurs between the plasmalemma and cell wall.

Long-cycled. Having an alternation of generations and all five spore stages in the life cycle (of the Uredinales) (*cf.* short-cycled).

Lysis (Gr. *lyein* = to break up). Breakdown of a substance or cells by an enzyme or other agent.

Metabasidium (Gr. *meta* = change, *basidion* = small base). The cell in which meiosis occurs in members of the Basidiomycotina (*cf.* probasidium).

Mitotic crossing over. Exchange of chromosome segments during haploidization.

Mold. A downy fungal growth on a substratum, usually consisting of a Hyphomycete mycelium.

Monoecious (Gr. *monos* = single, *oikos* = house). Having male and female reproductive organs on a single thallus (*cf.* dioecious).

Monokaryon (Gr. *monos* = single, *karyon* = nucleus). A cell or hypha in which there are one or more haploid nuclei, all of a single genetic type (*cf.* dikaryon, heterokaryon).

Monokaryotic. With the qualities of a monokaryon.

Monoplanetic (Gr. *monos* = single, *planos* = wandering). Having zoospores of one type only (of Oomycetes) (*cf.* diplanetic).

Monoplanetism. Formation of one type of zoospores only (of Oomycetes) (*cf.* diplanetism).

Mushroom. A fleshy Hymenomycete with gills, popularly used to refer to an edible variety.

Mycelium, pl. **mycelia** (Gr. *mykes* = fungus). A mass of hyphae, often used to denote all hyphae comprising a thallus.

Mycetangium, pl. **mycetangia** (Gr. *mykes* = fungus, *angeion* = vessel). In some beetles, a specialized pouch where their fungal symbionts can be harbored.

Mycology (Gr. *mykes* = fungus, *logos* = discourse). The science dealing with fungi.

Mycorrhiza, pl. **mycorrhizae** (Gr. *mykes* = fungus, *rhiza* = root). A symbiotic relationship between a fungus and the root of a higher plant.

Necrosis (Gr. *nekros* = dead body). Death of cells or tissues, especially as a result of disease.

Necrotrophic parasite (Gr. *nekros* = dead body, *trophe* = nourishment). A fungus parasitic on other fungi which kills its host and feeds on the dead cells (*cf.* biotrophic parasite).

Nonsexual reproduction. Production of progeny without the formation of spores of sexual origin.

Nuclear cap. A body, containing a large quantity of RNA, at the side of the nucleus in a zoospore or gamete.

Obligate parasite. An organism that is incapable of living as a saprophyte and must live as a parasite (*cf.* facultative parasite).

Oidium, pl. **oidia** (Gr. *oidion* = small egg). A thin-walled cell released by the breakdown of a hypha into its component cells. An oidium may function as a nonsexual spore or as a fertilizing agent.

Oogamy (Gr. *oon* = egg). A type of heterogamy in which plasmogamy takes place between a large nonmotile egg and a small motile male gamete or cytoplasm from an antheridium.

Oogonium, pl. **oogonia** (Gr. *oon* = egg, *gone* = offspring). A female gametangium that contains one or more discrete gametes.

Oosphere (Gr. *oon* = egg, *sphaira* = sphere). A large naked sphere of protoplasm that functions as the egg or female gamete in the oogonium.

Oospore (Gr. *oon* = egg, *sporos* = seed). A thick-walled sexual spore that develops from the oosphere after plasmogamy or sometimes by parthenogenesis.

Operculate (L. *operculum* = lid). An ascus or sporangium bearing an operculum (*cf.* inoperculate).

Operculum. A flap or lidlike covering over the opening of an ascus or sporangium.

Ostiole (L. *ostium* = small opening). A tubular passage terminating in a pore that extends through the neck of a perithecium; an opening or pore allowing the release of spores from any sporocarp.

Papilla, pl. **papillae.** A nipple-like extension, for example, on a zoosporangium.

Paraphysis, pl. **paraphyses** (Gr. *para* = beside, *physis* = a growth). Sterile, elongated cells that may occur in the hymenium, intermixed with asci; they grow upward and have free apices.

Parasexual cycle (Gr. *para* = besides). A sequence involving heterokaryon formation, diploidization, and haploidization, often resulting in the formation of recombinant nuclei. Unlike the sexual cycle, the parasexual cycle may occur at any point or continuously throughout the life cycle.

Parasite (L. *parasitus* = one who eats at the table of another). An organism that derives its nourishment from another living organism, usually (but not always) causing harm to the host organism.

Parthenogenesis (Gr. *parthenos* = virgin, *genesis* = being produced). Reproduction involving the development of female gametes without fertilization taking place.

Partial veil. A veil in some Hymenomycetes extending from the stipe to the margin of the pileus.

Pathogen (Gr. *pathos* = disease, *genesis* = being produced). A parasite that causes disease in its host.

Pathogenicity. Ability to produce a disease; capacity to be a pathogen.

Peridiole (Gr. *peridion* = small pouch). In the Nidulariales, a segregant of the gleba, containing basidiospores and surrounded by a distinct wall, which functions as a unit of dispersal.

Peridium, pl. **peridia** (Gr. *peridion* = small pouch). The outermost membranous covering of a sporocarp.

Periphysis, pl. **periphyses** (Gr. *peri* = around, *physis* = a growth). Short, hair-like filaments that occur in the ostiole of some Pyrenomycetes.

Periplasm (Gr. *peri* = around, *plasma* = something formed). In some Oomycetes, the protoplasm that surrounds the oosphere within the oogonium.

Perispore (Gr. *peri* = around, *sporos* = seed). A membrane surrounding the spore walls which is not permanently attached (*cf.* exospore, endospore).

Perithecium, pl. **perithecia** (Gr. *peri* = around, *theke* = case). A closed ascocarp whose wall is developed from the archicarp or other sexual structures, which has a regular means of opening, such as by a pore or slit.

Phialide (Gr. *phiale* = drinking vessel). A terminal cell of a conidiophore within or upon which blastogenously produced conidia are formed without a change in length of the conidiophore and in a basipetal fashion.

Phototropism. See **tropism.**

Phragmobasidium (Gr. *phragmos* = partition, fence; *basidion* = small base). A basidium that is divided into more than one cell by transverse or longitudinal septa (*cf.* holobasidium).

Pileus, pl. **pilei** (L. *pileus* = cap). The expanded caplike portion of some basidiocarps or ascocarps that supports the hymenium.

Plasmodium (Gr. *plasma* = something formed). A naked, multinucleate thallus that moves in an amoeboid fashion (in the Plasmodiophoromycetes).

Plasmogamy (Gr. *plasma* = something formed, *gamos* = marriage). The fusion of cytoplasm of two protoplasts, often sexual in nature and followed by karyogamy.

Polypore (Gr. *polys* = many, *poros* = pore). A common name for a member of the Aphyllophorales that has pores; a bracket fungus.

Pore (Gr. *poros* = pore). A small opening; in the Hymenomycetes, the mouth of a tube.

Predator. An organism that captures other organisms, which are then killed and used as food.

Probasidium (Gr. *pro* = before, *basidion* = a small base). The cell in which karyogamy occurs in the basidiomycetes (*cf.* metabasidium).

Progamete (Gr. *pro* = before, *gamete* = germ). In the Zygomycetes, a swollen hyphal tip that will differentiate into the gametangium and the suspensor.

Protoperithecium, pl. **protoperithecia** (Gr. *protos* = first). Early haploid stage that will develop into a perithecium after plasmogamy.

Pseudoangiocarpic development (Gr. *pseudes* = false, *angeion* = case, *karpos* = fruit). Development of basidiocarps in which the immature hymenium is exposed early in development but becomes covered by the incurving pileus margin and partial veil (*cf.* gymnocarpic, hemiangiocarpic).

Pseudoparaphysis, pl. **pseudoparaphyses** (Gr. *pseudes* = false, *physis* = growth). A sterile thread that grows downward in the cavity of some ascocarps, usually becoming attached at the bottom.

Pseudoparenchyma (Gr. *pseudes* = false). A fungal tissue consisting of more or less isodiametric cells and resembling the parenchyma of higher plants.

Pseudoseptum (Gr. *pseudes* = false, L. *septum* = hedge). A sievelike septum with numerous pores.

Pseudothecium (Gr. *pseudes* = false, *theke* = sheath or cover). An ascostroma resembling a flask-shaped perithecium.

Pycnidium, pl. **pycnidia** (dim. of Gr. *pyknos* = dense). A closed sporocarp that contains a cavity bearing conidia.

Pycniospore (Gr. *pyknos* = dense). A spore (spermatium) borne in a pycnium in the Uredinales.

Pycnium, pl. **pycnia** (Gr. *pyknos* = dense). In the Uredinales, stage 0, consisting of the male fertilizing elements (pycniospores) and female elements, the flexuous hyphae.

Receptacle. A stalk bearing a fertile region with asci or basidia (in the Laboulbeniales and Phallales).

Resting spore. A thick-walled spore that germinates after a resting (dormant) period, often winter.

Rhizoid (Gr. *rhiza* = root). A rootlike anchoring hypha; in the Chytridiales, a filamentous portion of the thallus that extends into the substratum.

Rhizomorph (Gr. *rhiza* = root, *morphe* = form). A compact, macroscopic strand of fungus tissue that resembles a root.

Rhizoplast (Gr. *rhiza* = root, *plastos* = formed). A strand that extends from the nucleus to the blepharoplast in motile cells.

Saprophyte (Gr. *sapros* = decayed, *phyton* = plant). A fungus, bacterium, or plant that feeds on dead organic matter.

Sclerotium, pl. **sclerotia** (Gr. *skleros* = hard). A hardened, resistant or resting body, consisting of a mass of fungal tissue, either with or without host tissue, which may give rise to a mycelium, a sporocarp, or a stroma.

Septal pore cap. A domelike configuration consisting of endoplasmic reticulum, which occurs directly over the pore in a septum (in the basidiomycetes).

Septum, pl. **septa** (L., fence). A cross wall in a hypha or spore.

Sexual reproduction. Production of progeny after karyogamy and meiosis take place at a regular point in the life cycle. Plasmogamy may or may not occur and may or may not be accomplished by specialized organs.

Short-cycled. Having fewer than five spore stages in the life cycle (of the Uredinales) (*cf.* long-cycled).

Skeletal hyphae. Thick-walled, branched or unbranched, straight or slightly flexed hyphae within a basidiocarp (*cf.* binding, generative, and lactiferous hyphae).

Soredium, pl. **soredia** (Gr. *sorus* = heap). A discrete group of algal and fungal cells formed by a lichen that can be disseminated and initiate a new lichen thallus.

Sorus, pl. **sori** (Gr., heap). A cluster or mass of spores, as in the Plasmodiophorales, Uredinales, or Ustilaginales.

Spermatium, pl. **spermatia** (Gr. *sperma* = seed or germ). A nonmotile, uninucleate male gamete which usually becomes detached before plasmogamy.

Sporangiole (Gr. *spora* = seed, *angeion* = vessel, L. *olus* = diminutive suffix). A small sporangium containing only one or a few sporangiospores and without a columella.

Sporangiophore (Gr. *spora* = seed, *phoros* = carrying). A modified hypha that supports the sporangium.

Sporangiospore (Gr. *spora* = seed). A nonsexual spore produced within a sporangium.

Sporangium, pl. **sporangia** (Gr. *spora* = seed, *angeion* = vessel). A sac that bears endogenous nonsexual spores (sporangiospores).

Spore (Gr. *spora* = seed). A discrete sexual or nonsexual reproductive unit, usually enclosed by a rigid wall, which is capable of being disseminated. Unlike seeds of the higher plant, the spore does not contain an embryo.

Sporocarp (Gr. *spora* = seed, *karpos* = fruit). A general term for a multicellular body or organ that contains or bears spores (e.g., an ascocarp or basidiocarp).

Sporodochium, pl. **sporodochia** (Gr. *spora* = seed, *doche* = receptacle). A cluster of conidiophores arising from a stroma or mass of hyphae.

Sporophore (Gr. *spora* = seed, *phoros* = carrying). A spore-producing or spore-bearing structure such as a conidiophore, ascocarp, or basidiocarp.

Sterigma, pl. **sterigmata** (Gr., support). A spicule-like structure on a basidium which supports the basidiospore.

Sterile. (1) Tissues or structures that are not producing spores; (2) uncontaminated by microorganisms.

Stipe (L. *stipes* = stem or stalk). The stalklike portion of some large ascocarps or basidiocarps; any spore-bearing stalk.

Stolon. A horizontal hypha that becomes anchored in the substratum at points of contact (in the Mucorales).

Stroma, pl. **stromata** (Gr., cover). A compact mass of vegetative tissue, sometimes intermixed with host tissue, which often bears sporocarps either within or upon its surface.

Subhymenium, pl. **subhymenia** (L. *sub* = under, Gr. *hymen* = membrane). A differentiated tissue immediately beneath the hymenium which gives rise to the asci or basidia and sterile elements in the hymenium.

Substratum.   A medium that serves as a food source and/or support for fungi.

Succession.   Progressive changes in the fungal population on a substratum.

Suspensor.   A specialized hyphal tip that supports a gametangium and eventually a zygospore (in the Mucorales).

Symbiont   (Gr. *sym* = with, *bios* = life). A partner in a symbiotic relationship.

Symbiosis.   The relationship in which dissimilar organisms live together. This relationship may be harmful, beneficial, or neutral to one or both of the partners.

Symptom.   Detectable abnormalities in either physiology or morphology produced during the course of a disease.

Synnema,   pl. **synnemata** (Gr. *syn.* = together, *nema* = thread). A fascicle of conidiophores, usually upright.

Teliospore   (Gr. *telos* = completion, fulfillment, *sporos* = seed). A thick-walled resting spore which is the site of karyogamy and produces the basidium (in the Uredinales and Ustilaginales).

Telium,   pl. **telia** (Gr. *telos* = completion, fulfillment). The final sorus (stage III) produced in the life cycle of the Uredinales and resulting in teliospores.

Telomorph   (Gr. *telos* = completion, fulfillment, *morphe* = form). The sexual stage in the life cycle of a fungus (*cf.* anamorph).

Tetrapolar sexuality   (Gr. *tetra* = four, L. *polus* = pole). A type of sexual compatibility controlled by multiple-allelic genes at two loci borne on separate chromosomes (*cf.* bipolar sexuality).

Thallic development   (Gr. *thallos* = shoot). Formation of conidia by septation of a vegetative hypha or conidiophore into discrete conidial initials, which may then enlarge. In **enterothallic** development (Gr. *enteron* = that within), additional wall layers are laid down within walls of the parent cell and are not incorporated into the spore. In **holothallic** development (Gr. *holos* = whole or entire), all wall layers of the parent cell are incorporated into the new spore (*cf.* blastic development).

Thallus,   pl. **thalli** (Gr. *thallos* = young shoot). Any simple vegetative plant body that lacks roots, stems, and leaves.

Tinsel flagellum.   A flagellum that bears small filaments which extend from the main axis of the flagellum (*cf.* whiplash flagellum).

Tissue.   An aggregate of hyphae or of shortened, more or less rounded, hyphal cells that are attached to each other along all sides.

Trama   (L. *trama* = woof). The sterile tissue of a basidiocarp.

Trichogyne   (Gr. *trichos* = hair, *gyne* = woman). That portion of the archicarp that initially receives the male protoplasm and is often long and filamentous.

Trichophoric cell   (Gr. *trichos* = hair, *phoros* = carrying). The cell that bears the trichogyne in the Laboulbeniales.

Tropism   (Gr. *trope* = an attraction for). A bending or growth in response to a one-sided stimulus, either toward (positive) or away from (negative), frequently used as a suffix (for example: chemotropism, response to a chemical agent; geotropism, response to gravity; phototropism, response to light).

Unitunicate   (L. *unus* = one, *tunica* = covering). Having a single-layered ascus wall (*cf.* bitunicate).

Universal veil.   A veil enveloping the entire basidiocarp during early developmental stages (in some Hymenomycetes).

Urediniospore   (L. *urere* = to burn, Gr. *sporos* = seed). A binucleate repeating spore borne in the uredinium in the Uredinales.

Uredinium,   pl. **uredinia** (L. *urere* = to burn). The sorus that bears uredinio-spores in the Uredinales (stage II).

Vegetative.   A cell or structure that is not producing reproductive structures, usually in the assimilative state.

Vesicle   (L. *vesica* = bladder). A bladderlike sac or an evanescent bubble within which zoospores mature; any bubblelike cell or bubblelike membranous structure within the cell.

Volva   (L. *volvere* = to roll). A cuplike structure at the base of a basidiocarp (in the basidiomycetes).

Whiplash flagellum.   A smooth flagellum (*cf.* tinsel flagellum).

Woronin body.   A rounded organelle occurring near septa in at least some ascomycetes or deuteromycetes.

Yeast.   A unicellular member of the Endomycetales; sometimes used for a nonmotile unicellular stage, for example, of the dimorphic animal parasites.

Zoosporangium,   pl. **zoosporangia** (Gr. *zoon* = animal). A sporangium within which zoospores are produced (in the lower fungi).

Zoospore   (Gr. *zoon* = animal, *sporos* = seed). A nonsexual, motile spore that bears one or two flagella.

Zygospore   (Gr. *zygon* = yoke, *sporos* = seed). A thick-walled sexual resting spore resulting from the fusion of gametangia (in the zygomycetes).

Zygosporophore   (Gr. *zygon* = yoke, *phoros* = bearing). A specialized hyphal branch that first bears a gametangium and then the zygospore in the zygomycetes.

Zygote   (Gr. *zogon* = yoke). A diploid cell resulting from the union of two gametes.

Author Index

Horstra, K., 25
Houlahan, M. B., 296
Howard, K. I., 57–59, 73–74
Huang, L. H., 134, 136, 166
Huber, D. M., 449–50, 453
Hubert, E. E., 200
Hudson, H. J., 407, 419
Hudspeth, M. E. S., 272
Hughes, S. J., 160, 166, 236, 243
Huige, N. J., 507, 528
Hunter, K., 328, 330–31
Hurst, H. M., 420

I

Ingham, J. L., 454
Ingle, M. R., 343, 357
Ingold, C. T., 85, 243, 367–68, 371, 374, 378, 380–81, 390–91
Ingram, D. S., 55, 75
Innocenti, F. G., 348, 357
Inoue, Y., 349, 357
Iqbal, S. H., 381, 390
Ishikawa, T., 351, 359
Iten, W., 9, 21

J

Jackson, R. M., 403, 419, 476, 492
Jahnke, K.-D., 273
Jain, K., 343, 355
Jaynes, R. A., 465, 467
Jennings, D. H., 282, 303–5
Jensen, J. D., 130
Jerebzoff, S., 360
Jockusch, J., 337, 359
Joffe, A. Z., 415, 419
Johns, R. M., 76
Johnson, D., 188
Johnson, G. T., 348
Johnson, T., 273
Johnson, T. W., Jr., 33, 63, 74, 420
Jones, A. C., 348
Jones, E. B. G., 74–75, 96, 301, 304, 420
Jones, E. W., 21, 22, 332
Joshi, N. C., 186

K

Kamat, M. N., 169
Karling, J. S., 25, 33–34, 36, 40, 52–55, 74
Kaufmann, H. H., 420
Kavanagh, F., 291–92, 304
Keitt, G. W., 442, 444, 453
Kendrick, B., 235–36, 238, 243
Kendrick, W. B., 395, 407, 419
Kennedy, L. L., 187, 228–29
Kensler, C. J., 418–19
Kershaw, K. A., 475–76, 492
Kerwin, J. L., 351
Khan, S. R., 71, 74, 77, 83, 99, 445
Kidby, D. K., 323
Killian, K., 162, 166
Kimball, J., 206
Kimbrough, J. W., 18, 21, 99, 106, 166, 169, 445
King, D. S., 99
Kinghorn, J. R., 289, 304
Kirk, P. M., 243
Kirk, T. K., 420
Klebs, G., 340, 357

Klemm, E., 348, 357
Kligman, A. M., 454
Kobayashi, Y., 34, 49, 74
Koch, A., 493
Kochert, G., 360
Kohlmeyer, J., 269, 272
Kok, L. T., 488, 492
Kolattukudy, P. E., 526, 528
Kole, A. P., 25, 76
Koltin, Y., 255, 272, 465–66
Kononova, M. M., 399, 419
Korf, R. P., 106, 112, 144, 147–48, 166–67, 169, 267–68, 272
Korhonen, K., 260–61, 271
Kornberg, H. L., 315, 331
Kosuge, T., 449, 453
Kozlowski, T. T., 493
Kramer, C. L., 112, 119, 166, 169, 372, 389, 391
Kraybill, H. F., 417, 419
Kreger-Van Rij, N. J. W., 112, 166, 169, 240, 243
Krug, J. C., 132
Kubicek, C. P., 315, 331
Kuck, U., 529
Kuehn, H. H., 121–22, 166–67
Kuhad, R. C., 351, 357
Kühner, R., 174–75, 228
Kurtz, M. B., 305, 360
Kurtzman, C. P., 27, 529
Kusano, S., 40, 74
Kuykendall, W. R., 467
Kwok, S., 266, 272
Kwon-Chung, K. J., 120, 166, 184, 228, 433, 435, 452

L

Lambowitz, A. M., 251, 271
Latter, P., 420
Lawler, G. C., 341
Leach, C. M., 346, 349, 357, 378
Leak, L. V., 45
Leal, J. A., 343, 357
Leatham, G. F., 360
Lechevalier, H. A., 425, 452
Lefebvre, C. L., 430, 453
Lehotský, J., 350, 356
Leibowitz, M. J., 463, 466–67
Lemke, P. A., 467
Leonards, K. S., 6, 21
Leppik, E. E., 274
Lewis, J., 13
Lichtwardt, R. W., 92–93, 95–99, 267, 272
Lilly, V. G., 283, 291–92, 299, 304–5, 343–44, 356–57, 460, 466
Lin, C. C., 56, 73
Lincoff, G., 500–2, 504, 527
Lindenmyer, A., 317, 321
Linder, D. H., 124
Ling-Young, M., 82
Lingappa, B. T., 40–41, 74
Link, K. P., 443, 454
Lisansky, S. G., 432, 453
Lockwood, J. L., 402–3
Lockwood, L. B., 514, 527
Lodder, J., 169
Lösel, D. M., 329–31
Lovett, J. S., 45
Lowe, D. A., 529
Lowe, J. L., 229
Lowry, R. J., 5
Lowy, B., 229, 496, 527

Lu, B. C., 344, 348, 357
Luttrell, E. S., 127, 134–35, 162, 164, 166, 265, 267–68, 272
Lyr, H., 454
Lysek, G., 354, 357
Lythgoe, J., 45

M

MacConnell, J. G., 491–92
MacDonald, D. J., 340, 357
MacDonald, J. A., 263, 272
Machlis, J., 334, 356
Machlis, L., 334, 357–58
MacIntyre, D., 75
MacKauer, M., 432, 453
Macko, V., 384, 389–90
MacLeod, D. M., 426, 428, 430, 453
MacNeill, B. H., 274
Maddy, K. T., 436, 453
Madelin, M. F., 342–43, 356, 360, 430–31, 453–54
Mai, S. H., 136, 166
Maibach, H. I., 454
Maiello, J. M., 343, 358
Maisch, W. F., 529
Malibari, A., 479, 492
Malloch, D., 129, 166, 346, 359
Manachère, G., 348, 350, 358
Mandels, G. R., 305, 386, 390, 420
Maniotis, J., 129, 167
Manning, K., 497, 528
Manochoa, M. S., 459, 466–67
Marchant, R., 10, 19, 21, 76
Markham, P., 18, 21
Markham, R., 417, 419
Marks, G. C., 493
Marshall, B., 368, 390
Marshall, M. R., 344, 358
Martin, G. W., 229
Martin, J. S., 491–92
Martin, M. M., 491–92
Martin, P. M. D., 437–38
Martin, R. W., Jr., 28, 67, 74
Martin, W. W., 62
Mason, P. M., 476, 492
Mathew, K. T., 292, 304
Matile, P., 9, 21, 114–15, 167
McAlear, J. H., 17
McClary, D. O., 169
McCoy, R. E., 347, 358
McDonald, D., 385, 390
McDonald, W. L., 344, 358
McDonough, J., 455
McEwen, F. L., 429, 453
McGahen, J. W., 134, 167, 256, 273
McKay, M. B., 443, 453
McKeen, W. E., 60, 74
McKerracher, L. J., 11–12, 21
McLaughlin, D. J., 22, 230, 372, 391
McMorris, T. C., 335, 356, 358
McMurrough, I., 319, 331
McNabbola, S. S., 55, 75
Mehrotra, B. S., 82, 98
Melin, E., 478, 491–93
Meredith, D. S., 377, 391
Merkus, E., 106, 167
Meyer, F. H., 477, 493
Michalenko, G. O., 4
Middleton, J. T., 68–69, 74
Miles, P. G., 343, 355
Millar, R. L., 454
Miller, C. E., 28, 36, 38–39, 53, 67, 73–74, 343, 356

Subject Index

Phacidiales, Order:
  classification, 113
  evolution, 268
  morphology, 144–47
Phacidium nigrum, 145
Phallales, Order:
  classification, 177
  morphology, 217–21
  spore dispersal, 219
Phallogaster, 218
Phallogaster saccatus, 218
Phallotoxins, 501, 504
Phialide, 235, 238
Phoma apiicola, 297
Phosphorus requirement, 290
Photoreceptor, 347–48
Phototropism:
  adaptive value, 366, 383
  of asci, 368–70
  of *Pilobolus,* 365–66
  of *Sphaerobolus,* 375
Phragmobasidiomycetidae, Subclass:
  classification, 177
  in evolution, 270
  morphology, 191–94
Phragmobasidium, 190–93
Phycomyces, 79, 81, 294, 387
Phycomyces blakesleeanus, 248, 253, 386–87
Phycomyces microsporus, 82
Phyllachora, 129
Phyllosticta antirrhini, 343
Phylogeny of fungi (evolution):
  evidence, 264–65
  relationships, 265–71
Physalospora, 460
Physalospora obtusa, 460
Physalospora rhodina, 460
Physical factors (*see also* pH; Light; Moisture; Oxygen requirements; Temperature, effects on; Water):
  effects on growth, 295–303
  effects on sporulation, 345–50, 355
Phytoalexins, 443
Phytophthora, 67, 70, 292, 343–44
Phytophthora erythroseptica, 329
Phytophthora infestans, 67, 70, 441
Phytophthora megasperma, 261–63
Phytophthora parasitica, 70
Pileus, characteristics, 195
Pilobolus, 79, 81, 266–67, 352, 364–66, 383
Pilobolus kleinii, 364–66
Pilobolus longipes, 387
Piptocephalis, 458–59
Piptocephalis lepidula, 85
Piptocephalis virginiana, 458
Pisolithus tinctorius, 479–80
Pityrosporum ovale, 328
Plant diseases (*see* Parasitic fungi, on plants)
Plasmalemma (*see also* Ultrastructure):
  characteristics, 6, 9–10
  lipid content, 330
Plasmids:
  characteristics, 11
  in cytoplasmic inheritance, 251
  in genetic engineering, 523–34
Plasmodiophora brassicae, 53–55, 441, 448
Plasmodiophorales (*see* Plasmodiophoromycetes)
Plasmodiophoromycetes:
  classification, 3

evolution, 31–32, 265–66
  morphology, 52–55
Plasmodium, characteristics, 53
Plasmogamy (*see also* Reproduction, sexual):
  Ascomycotina, general characteristics, 103–4
  Basidiomycotina, general characteristics, 170–71, 174
  genetic control, 257
  hormonal control, 334–38
  in life cycle, 19–20
  role in speciation, 260
Platypeltella angustispora, 160
Plectomycetes, Class:
  classification, 113
  evolution, 267
  morphology, 119–22
Pleospora, 159, 346
Pleosporaceae, Family, 159
Pleospora herbarum, 162, 346
Pleospora-type of development, 164
Pleurotus, 204–5, 350
Podospora anserina, 250–51, 354, 524–25
Poisonous fungi:
  ergotism, 521–22
  hallucinogenic fungi, 502–4
  mushrooms, 499–504
  mycotoxins, 414–17
  poisons as metabolites, 327
Polyphagus euglenae, 38, 42–44
Polyploidy, 226–27, 262–64
Polyporaceae, Family, 198
Polyporus, 198, 201
Polysaccharides:
  decomposition, 396–98
  production, 319–20
  utilization for growth, 285–86
  utilization for sporulation, 342–43
Polystigma, 129
Polystigma rubrum, 103
Polystigmataceae, Family, 129
Potassium requirements, 290
Powdery mildew fungi, 123–25
Predacious fungi:
  as biological control agents, 425
  nematode trapping, 421–25
  protozoa trapping, 90–92
Probasidium, 175
Progametangia, 79
Protomycetales, Order, in classification, 113–14
Protoplast fusion, 512–13, 523
Protozoa:
  in soil, 400
  trapped by fungi, 90–92
Protozoan ancestors of fungi, 265, 267
Psathyrella, 204
Pseudoangiocarpic development, 209–10
Pseudohydnum gelatinosum, 192
Pseudomeliola, 129
Pseudoparaphyses, 106, 110
Pseudoparenchyma, 16
Pseudosepta, 16–17, 61
Pseudothecium, 160
Psilocin, 327
Psilocybe, 204
Psilocybe strichper, 503
Psilocybin, 502–3
Psychrophilic fungi, 297
Puccinia arachidis, 385
Puccinia graminis, 180–83, 254, 384
Puccinia graminis var. *secalis,* 178

Puccinia graminis var. *tritici,* 178
Puccinia malvacearum, 183
Puccinia podophyllia, 178
Puccinia sorghi, 178
Puffballs, 220–24
Pullularia, 407
Pullularia pullulans, 407, 412
Pycnidium, 147, 240–42
Pycnium and pycniospore:
  in evolution, 269
  morphology, 178–79
Pyrenomycetes, Class:
  evolution, 267–69
  morphology, 122–39
  spore release, 367–69
Pyridoxine requirement, 292
Pyronema, 150
Pyronema doemsticum, 104–5, 110, 140–42, 144, 150, 342–44, 348, 350
Pyronema omphalodes, 141
Pythiopsis, 65
Pythium, 67–70, 344, 445, 461
Pythium butleri, 291
Pythium debaryanum, 69
Pythium mamillatum, 408

Q

Quinone synthesis, 323

R

Receptacle, 219
Reproduction (*see also* Sporulation):
  nonsexual:
    in Ascomycotina, 107, 112
    in Basidiomycotina, 175, 179, 214
    characteristics, 20
    in Deuteromycotina, 233–39
    in Endomycetales, 115–16
    in Mastigomycota, 24–31, 35–38, 64–66
    in Zygomycotina, 82–86
  sexual:
    in Ascomycotina, 101–12
    in Basidiomycotina, general, 170–75
    characteristics, 19–20
    in Chytridiomycetes, 38–41
    genetic control, 256–57
    in lichens, 472
    in Oomycetes (oogamy), 56–64
    in Plasmodiophoromycetes, 55
    in Teliomycetes, 178–79
    in Zygomycotina, 79–82
Respiration (*see also* Fermentation; Glycolysis; Oxygen requirements):
  in carbon cycle, 399
  cytochromes, 316–17, 347
  factors influencing, 317–18
  oxygen requirement, 308–13, 316–17
  in plant disease, 450
  respiratory chains, 316–17
  in spore germination, 386, 389
  in sporulation, 318, 341–42, 349–50, 352
Resting spores, 39–40, 44, 54–55
Rhizidium braziliensis, 33–34
Rhizidium Nowakowskii, 33
Rhizidium richmondense, 458
Rhizidium-type development, 36, 44
Rhizoctonia, 242, 446, 460–61, 482
Rhizoctonia solani, 18, 401, 462–63, 484

Rhizoids, characteristics, 32
Rhizomorphs:
  characteristics, 190
  luminescence, 212–13
  in soil, 402, 479
Rhizomyces ctenophorus, 155
Rhizophlyctis, 38
Rhizophlyctis rosea, 28
Rhizophydium, 36, 265
Rhizophydium subangulosum, 27
Rhizoplast, 25–26, 28
Rhizopogon, 477
Rhizopogon luteolus, 215
Rhizopus, 79, 81, 86, 254, 312–13, 433, 509, 520–21
Rhizopus arrhizus, 341, 520
Rhizopus nigricans, 78, 520
Rhizopus oligosporus, 509
Rhizopus stolonifer, 79, 389, 414, 444, 462
Rhizosphere, 393
Rhodotorula, 308
Rhodotorula gracilis, 308
Rhythms:
  in spore discharge, 368–69, 378
  in sporulation, 352–55
  types, 353
Rhytisma, 144
Rhytisma andromedae, 145
Riboflavin requirement, 292, 296
Ribosomes (*see also* Ultrastructure):
  characteristics, 7–8
  in nuclear cap, 27
Rootlets, striated, 26
Rozella, 458
Rozella polyphagi, 458
Rumposome, 27
Russula, 204, 500
Russulaceae, Family, 204
Russula emetica, 477
Rust fungi, 177–84 (*see also* Uredinales, Order)

S

Saccharomyces, 294
Saccharomyces carlsbergensis, 294, 507
Saccharomyces cerevisiae, 14, 114–17, 250, 294, 312–13, 334, 341–42, 350, 354, 385–86, 463, 465, 505–7, 511–12, 523
Saccharomyces rouxii, 509
Saccharomycodes ludwigii, 116–17
Saccobolus, 383
Saccobolus platensis, 351–52
Saprolegnia, 59, 60–61, 64
Saprolegnia ferax, 15, 60
Saprolegniales:
  classification, 32
  evolution, 265–66
  morphology, 60–66
  sexual reproduction, 56, 64
Saprolegnia litoralis, 57
Saprolegnia mixta, 340
Saprolegnian galaxy, in evolution, 266
Sapromyces androgynus, 62
Saprophytes:
  deterioration of foods and materials, 409–18
  fungi as, 393–418
  lichen fungi as, 473
  in nature, 398–408
  parasitic fungi as, 433–34, 441, 460
  in soil, 400–408 (*see also* Soil fungi)
Sarcoscypha, 150

Sarcoscypha coccinea, 141, 150
Schizophyllum commune, 173, 185, 255–56, 343, 345, 351, 433
Scleroderma, 224
Sclerodermatales, Order:
  classification, 177
  morphology, 224
Sclerotinia, 353
Sclerotinia fructicola, 354
Sclerotinia sclerotiorum, 148
Sclerotinia trifoliorum, 348
Sclerotinia whetzelii, 148
Sclerotium:
  in Ascomycotina, 143–44
  in Basidiomycotina, 190
  growth, 276–77
  luminescence, 212
  in soil, 402
Septobasidiales, Order:
  classification, 177
  evolution, 270
  morphology, 193–94
Septobasidium, 194, 484–87
Septonema, 232
Septoria, 242
Septum:
  Ascomycotina-type, 17–18
  Basidiomycotina-type, 18–19
  lysosome breakdown, 9
  in Trichomycetes, 93, 95
  in Zygomycetes, 79
Serpula lacrymans, 281–82, 411
Shikimic acid pathway, 321–22, 327
Shoyu production, 510
Sibling species, 260–61
Side body, 28
Simblum, 219
Sirenin:
  effects, 334
  as lipid, 330
Slime molds, 2–3
Smittium, 93, 96–97
Smittium culisetae, 93–94
Smut fungi, 184–86
Snow molds, 297
Soil fungi:
  antagonism, 404–6
  antibiosis and antibiotics, 404–6
  cellulose decomposition, 396–97
  characteristics, 402
  competition, 405–6
  dormancy, 402–3
  exploitation, 404
  growth, 403–4
  isolation, 401
  lignin decomposition, 397–98
  methods for study, 400–1
  nitrogen utilization, 395
  parasites as, 425, 433–34, 441
  and soil fertility, 398–99
  succession, 406–8
  sugar fungi, 408
Sordaria, 129, 137, 383
Sordaria brevicollis, 340
Sordariaceae, Family, 129
Sordaria curvula, 383
Sordaria fimicola, 294, 299, 344, 367–69
Sordaria-type development, 136
Sorus, 177
Spathulospora, 269
Speciation:
  allopatric, 258–59
  characteristics, 257–58
  ecological adaptation and, 261

  by hybridization, 263–64
  isolating mechanisms, 257–58
  sympatric, 259–63
Spermatia, characteristics, 103 (*see also* Reproduction, sexual)
Sphaeriaceae, Family, 129
Sphaeriales, Order:
  classification, 113, 129
  evolution, 268
  morphology, 127–39
Sphaerobolus, 224, 345, 374–76, 383
Sphaerobolus stellatus, 376–77
Sphaeropsidales, Form-Order, 240–42
Sphaerotheca castagnei, 125
Spindle pole bodies, 14
Spongospora subterranea, 53
Sporangiole, 77, 83, 85 (*see also* Sporangium)
Sporangiophore, 77, 86
Sporangiospores (*see* Sporangium)
Sporangium:
  characteristics, 77–78
  sporangiospore formation, 82–86
Spores (*see also* names of specific spore types):
  activation, 385–86
  dispersal:
    by air, 378–80
    by insects, 382
    by large animals, 381, 383, 480, 482
    role in speciation, 259
    by water, 380–81
  dormancy:
    constitutive, 383–87
    exogenous, 384, 387–88
    inhibitors, 384
    metabolic control, 385–87
    soil fungistasis, 402–3
  germination:
    morphological changes, 2, 388
    physiological changes, 388–89
  release:
    active, 362, 364–77
    air currents, effects, 362
    Ascomycotina, 367–71
    Basidiomycete forcible, 371–74
    cleistothecia, 367
    eversion mechanism, 374–77
    explosive mechanisms, 364–71
    ostiolate ascocarps, 367–68
    passive, 362–63
    raindrops, effects, 362–63
    rhythms, 368–69, 378
    rounding-off mechanism, 371
    tensile water mechanism, 377
  in soil, 400–2
  structure, 2, 3, 5
Sporocarps (*see also* Sporulation):
  acervuli, 241
  ascocarps, 106–7, 110–12
  basidiocarps, 186–90, 213
  development in nature, 355
  growth, 276–77
  pycnidia, 240–41
Sporodochium, 240–41
Sporomiella, 343
Sporomiella australis, 161
Sporophagy, 461
Sporotrichum schenckii, 433
Sporulation (*see also* Reproduction):
  biochemical changes, during, 350–52
  genetic control, 256–57
  hormonal control, 333–39
  incompatibility mechanisms, 351–57